COMPSTAT 1982

5th Symposium held at Toulouse 1982

Part I: Proceedings in Computational Statistics

Edited by
H. Caussinus
P. Ettinger
R. Tomassone

Physica-Verlag · Wien 1982

CIP-Kurztitelaufnahme der Deutschen Bibliothek

COMPSTAT:
COMPSTAT : proceedings in computational statistics ;
. . . symposium. – Wien : Physica-Verlag

5. Held at Toulouse 1982.
Pt. 1. Proceedings. – 1982
ISBN 978-3-7051-0002-2 ISBN 978-3-642-51461-6 (eBook)
DOI 10.1007/978-3-642-51461-6

ISBN 978-3-7051-0002-2

Preface

Pour le cinquième congrès de la série, COMPSTAT 82 réunit environ 500 participants d'origines scientifiques et géographiques très variées, prouvant à l'évidence l'intérêt persistant de la communauté scientifique pour tous les problèmes de calculs statistiques.

Le Comité de Programme chargé de l'organisation scientifique du Congrès était composé de:
S. Apelt (République démocratique d'Allemagne) – Å. Björck (Suède) – H. Caussinus (France), Président – Y. Escoufier (France) – A. de Falguerolles (France), Secrétaire – J.W. Frane (U.S.A.) – J. Gordesch (République Fédérale d'Allemagne) – Th. Havránek (Tchéchoslovaquie) – N. Lauro (Italie) – C. Millier (France) – R.J. Mokken (Pays-Bas) – R. Tomassone (France) – D. Wishart (Royaume Uni)

Ce Comité a décidé d'augmenter le nombre des conférenciers invités, cherchant de la sorte une représentation des diverses écoles ainsi que l'introduction de nouveaux thèmes. La tâche la plus difficile a ensuite été de sélectionner une soixantaine de contributions parmi 250 soumissions. Là encore le Comité de Programme s'est efforcé de favoriser des voies qui semblaient les plus nouvelles et a essayé de maintenir une bonne répartition scientifique et géographique. Cependant, comme dans les précédents congrès COMPSTAT, il a donné la préférence aux propositions clairement marquées simultanément du double aspect Statistique et Calcul. Dans bien des cas, ces deux aspects sont très liés rendant en particulier difficile et peu pertinente toute classification fine des contributions. Pour cette raison, nous n'avons pas cherché à séparer cet ouvrage en Chapitres si ce n'est la distinction habituelle en communications invitées et libres.

Les contributions de bonne valeur qui n'avaient pu être retenues faute de place ont pu s'ajouter aux affiches (posters) proposées. Elles font l'objet d'affiches et communications courtes. Une nouveauté de COMPSTAT 82 est l'édition (par Physica Verlag) d'un volume complémentaire contenant les résumés de ces travaux.

Comme précédemment, des démonstrations de logiciels ont été organisées tout le long du Congrès afin de compléter le programme scientifique. Des résumés descriptifs de ces démonstrations ont été publiés par le Comité d'Organisation.

Fifth in the series, COMPSTAT 82 gathered about 500 participants from very diversified scientific and geographical origins, thus proving the continuing surge of interest in statistical computing.

The programme for COMPSTAT 82 was selected by an international committee, the members being:
S. Apelt (German Democratic Republic) – Å. Björck (Sweden) – H. Caussinus (France), President – Y. Escoufier (France) – A. de Falguerolles (France), Secretary – J.W. Frane (U.S.A.) – J. Gordesch (Federal Republic of Germany) – Th. Havrânek (Czechoslovakia) – N. Lauro (Italy) – C. Millier (France) – R.J. Mokken (The Netherlands) – R. Tomassone (France) – D. Wishart (United Kingdom)

This committee decided to increase the number of invited speakers in order to achieve the representation of different schools and the introduction of new themes. The most difficult task was then to select about sixty papers out of 250 projects. Here also the committee tried to support novel tracks and to maintain a satisfactory scientific and geographical representation. Like in other COMPSTAT meetings, the committee gave its preference to papers simultaneously containing both statistical and computational aspects. In many cases, these two aspects are very linked, thus making difficult and somewhat irrelevant any refined classification of the contributions. Therefore this book is not divided into chapters apart from the usual distinction between invited papers and contributed papers.

Good contributions which could not be retained by lack of space were added to submitted posters. They are presented in the format of short communications and posters. A COMPSTAT 82 novelty is the editing (by Physica Verlag) of a supplementary volume containing summaries of these contributions.

Like in former COMPSTAT, demonstrations of statistical software and computing facilities have been organized. Summaries of these demonstrations have been published by the organizing committee.

Toulouse, juin 1982

H. Caussinus
P. Ettinger
R. Tomassone

Contents

1. Invited Papers

2. Contributed Papers

1. Invited Papers

On Some Problems in Analysing Non-orthogonal Designs

T. Caliński, Academy of Agriculture, Poznań, Poland

SUMMARY: Discussion on some controversial problems in analysing and interpreting data from non-orthogonal designs is reviewed and certain suggestions are made on how to overcome the difficulties. Geometric approach is adopted to show possible reconcilations.

KEYWORDS: non-orthogonal designs, unbalanced data, non-orthogonal analysis of variance, linear models.

1. Introduction

Association of a linear model with the experimental data is fundamental for the statistical analysis in many fields of research. Though the origin of the methodology dates back to Gauss (1809) and the literature on the subject is vast, not many statistical techniques have caused so much controversy as the least-squares fit of a linear model. In fact, as far as the analysis of variance is applied to a properly balanced experimental design there is a common agreement on how the analysis should be performed and how its results can be interpreted (although there may be disagreements when distinguishing between the fixed effects and the random effects models - the matter that will not be discussed here). The troubles start when the data are unbalanced and the usual orthogonal analysis of variance can not be applied uniquely. As Bock and Brandt (1980) put it, "to move from balanced to unbalanced designs in analysis of variance is not only to lose the ease of computation of the orthogonal solution, but also the intuitively appealing equivalence of the observed marginal means to the least-squares estimates of effects and, perhaps more important, also the uniqueness of the additive partition of the total sum of squares".

As the unity of method and clarity of interpretation disappear, various diverging approaches are adopted and controversies emerge.

In this paper attention is drawn to certain controversial issues that may influence preparations of analysis of variance programs. An attempt is made to clarify some of the problems by using the geometric approach to linear models.

2. Some of the Controversies

As an illustration to one of the most controversial problems, let us look at the results obtained when applying five different analysis of variance programs to the same set of data coming from an unbalanced 2x5 (Sex x Religion) classification originally analysed by Francis (1973) and then subsequently discussed by several authors. The results are reproduced in the table, given here in the form due to Aitkin (see his contribution to the discussion on Nelder's, 1977, paper), who added the last column. Aitkin explains the various results as follows: "Column (a) arises from a computing method which ignored the non-orthogonality, substracting the unadjusted main

ANOVA RESULTS FROM FRANCIS AND AITKIN

Source	d.f.	Sums of squares obtained by different programs				
		(a)	(b)	(c)	(d)	(e)
Mean	1	Not given	7305.78	12982.73	12982.73	12982.73
Sex	1	43.58	11.17	43.58	28.71	28.71
Religion	4	69.36	57.74	54.49	69.36	54.49
SxR	4	- 4.61 (!)	10.25	10.25	10.25	10.25
Error	1300	2988.95	2988.95	2988.95	2989.00	2988.95

effects SS from the among cell SS. Columns (c) and (d) are hierarchical analyses, giving respectively sex, religion adjusted for sex and interaction adjusted for sex and religion, and religion, sex adjusted for religion, and interaction adjusted for religion and sex. Column (b) arises from a "regression" analysis in which each effect is adjusted for all others in the model: thus sex is adjusted for religion and the interaction, and religion is adjusted for sex and the interaction (and the mean is adjusted for all effects!)". As to the column (e); "Here the main effects are adjusted for each other as in column (b), but are not adjusted for the interaction".

Nobody would support method (a), though the existence of such a program was in a way "fortunate", by initiating examination of some mysteries of the various programs (see Francis, 1973). According to their attitude towards the remaining four methods, statisticians can be clustered into three schools: (i) of the "hierarchical" partition (c) and (d), (ii) of the "full-regression" approach (b), and (iii) of the "experimental design" solution (e). Some representatives of the schools are: Bock (1963), Searle (1971), Overall and Klett (1972), O'Brien (1976), Aitkin (see Nelder, 1977) of school (i); Afifi and Azen (1972), Francis (1973), Kutner (1974) and Speed and Hocking (1976) of school (ii); Rao (1973), Nelder (1974, 1976, 1977) and Gianola (1975) of school (iii). (Also see Bock and Brandt, 1980.)

Another important problem that splits statisticians is the choice between the non-full rank overparameterized model (termed Model A by Hocking and Speed, 1975) and the full rank cell means model (termed by them Model B). Model A represents the present conventional approach to linear models. But Urquhart, Weeks and Henderson (1973), giving a historical account of the models, trace back Model B to the pioneers (particularly to the early works of Yates, 1933, 1934). They argue that the Model B approach "clarifies what common techniques are really estimating without the necessity of introducing estimability or imposing restrictions", and in a more recent paper Urquhart and Weeks (1978) conclude that "the cell means model gives the researcher the flexibility of specifying precisely the functions of interest". Even more outspoken critics of the Model A approach are Bryce, Scott and Carter (1980), who express their objections as follows: "In the early 1950's the use of the new common overparameterized model came into vogue. With it came concepts of estimability, testability, generalized inverse, non-unique solutions, etc. While these served to broaden the understanding of mathematical statisticians and supplied them with endless topics for papers in technical journals, the practicioner of statistical science either threw away data until balance was achieved or applied regression or other methods without knowing the hypotheses being tested or even realizing that different approaches implied different hypotheses".

This brings us to one of the main issues in computations of the non-orthogonal analysis of variance, to the "know-what-you-are-testing" problem.

Several Statisticians have examined what kind of hypotheses are really tested when applying the various analysis of variance procedures implemented in widely available computer programs (see, e.g., Francis, 1973, Kutner, 1974, Speed and Hocking, 1976, Speed, Hocking and Hackney, 1978, Hocking, Hackney and Speed, 1978, Frane, 1980, Goodnight, 1980, Searle, 1980, and Bock and Brandt, 1980). Their opinions show that we are far from a common agreement on how the analysis should proceed and how results can be interpreted. In particular, the gap between the overparameterized modelers and the cell means modelers has not been bridged, though most seem to agree that the only logical procedure is first to set out a hypothesis of interest and then to test it. This view is strongly advocated by Searle(1980), from the position of the Model A camp, and by Hocking, Speed and Coleman (1980), from the Model B camp. This contradicts with the, at present, prevailing practice of "producing" estimable functions and testable hypotheses by the programmes themselves, often without much control and understanding by the user.

It seems that a common agreement would have a chance to be reached only if we could look at the two models (A and B) as two sides of the same think, one rather supporting than contradicting the other. Moreover, it might be helpful to return to the basic principles and have a fresh look at the linear model, without the inherent prejudices of the two camps. Since both, A and B, models are algebraic in nature, the reconcilation might be sought in the geometric approach. This was already tried in the past (see Corsten, 1958). In fact, the use of geometric approach in the theory of linear models has a long and interesting history, as revealed recently by Herr (1980). Unfortunately, this approach has not yet become very popular.

3. The Principles of Geometric Approach

In this section we would like to draw attention to some features of the geometric approach to linear models that can help in profound understanding of the least-squares estimation and hypotheses testing. We do not want to go into details, as they have already been exposed in a number of papers (see, e.g., Monlezun and Speed, 1980.). We will only indicate certain principle ideas, leaving their explorations to those who become interested.

In the classical, algebraic, approach to linear models the attention is focused on a parameter space, and the parameter vector is estimated (not necessarily uniquely) by minimizing a quadratic form. The estimate of the mean (expectation) for the observation vector can then be obtained by transforming an estimate of the parameter vector with the use of the design matrix. In the geometric approach, on the contrary, attention is focused on observation space, where the mean vector of observations is estimated directly by orthogonally projecting the vector of observations onto an appropriate subspace. From this estimate the parameter vector, or its linear function, can be estimated by an appropriate linear transformation, provided such a transformation exists, i.e. the parameter vector, or its function, is identifiable.

To use some notation, suppose y is an nx1 vector of observations, in the n-dimentional Euclidean space E^n, with a mean vector $E(y) = \theta$. Taking $e = y - \theta$, the model can be written $y = \theta + e$. But the model is not set completely until we do two thinks more: define a p-dimensional subspace of E^n, Ω, in which θ is supposed to lie, and assume some distribution for e. As for the latter, the usual assumption is that it is multivariate normal with a covariance matrix $\sigma^2 I$ (though further generalizations are possible). As to Ω, its choice is usually not a straightforward matter. It has to be defined in such a way to reflect the basic structure of the experimental data and the interest of the researcher.

Technically, the definition of Ω can be accomplished by imposing any of the two types of relations, $A\theta = 0$ or $\theta = X\beta$, with appropriately chosen matrices A or X. In the first case, Ω is termed the null space of A, written $\Omega = N[A]$; in the second, it is called the range cpace of X, written $\Omega = R[X]$. (For details see Seber, 1980.) Depending on circumstances, one or the other way may be more convenient. For example, in a simple 1-way classification of data, with 2 classes, we have initially the model $y_{ij} = \theta_{ij} + e_{ij}$ $(i = 1,2,;\ j = 1,2, \dots ,n_i)$. But usually it will be natural to assume that $\theta_{i1} = \theta_{i2} = \dots = \theta_{in_i} = \mu_i$ (say), for 1,2. This can be imposed on the model either by the $n_1 + n_2 - 2$ relations $\theta_{ij} - \theta_{in_i} = 0$ ($i = 1,2$; $j = 1,2, \ldots, n_i - 1$) or by the $n = n_1 + n_2$ relations $\theta_{ij} = \mu_i$ ($i = 1,2$; $j = 1,2, \ldots, n_i$). The form of the $(n - 2)\times n$ matrix A and that of the $n \times 2$ matrix X are obvious. Also it is evident that the vector β is here $(\mu_1,\mu_2)'$. Both definitions of Ω are in this case simple, and evidently equivalent.

In a more complex case of a 3×4 crossed classification (as discussed by Searle, 1971, Chapter 7) with n_{ij} observations in the (i,j)-cell, we can first ignore the fact that the classes cross and define the model as for a 1-way classification with 12 classes (cells). If the vector θ is written

$$\theta = (\theta_{11}',\theta_{12}',\theta_{13}',\theta_{14}',\theta_{21}',\theta_{22}',\theta_{23}',\theta_{24}',\theta_{31}',\theta_{32}',\theta_{33}',\theta_{34}')'$$

where $\theta_{ij}' = (\theta_{ij1}, \theta_{ij2}, \ldots, \theta_{ijn_{ij}})$, then the $(n-12) \times n$ matrix A is of the form $A = \mathrm{diag}\,\{A_{11},A_{12},A_{13},A_{14},A_{21},A_{22},A_{23},A_{24},A_{31},A_{32},A_{33},A_{34}\}$ where the $(n_{ij} - 1) \times n_{ij}$ submatrix A_{ij} represents $n_{ij}-1$ contrasts within the (i,j)-cell ($i = 1,2,3$; $j = 1,2,3,4$). Using the other method of defining Ω, we could use as matrix X the matrix

$$W = \begin{bmatrix} 1_{11} & 0 & \cdots & 0 \\ 0 & 1_{12} & \cdots & 0 \\ \vdots & \vdots & \ddots & \vdots \\ 0 & 0 & \cdots & 1_{34} \end{bmatrix},$$

where 1_{ij} is the $n_{ij} \times 1$ vector of ones ($i = 1,2,3$; $j = 1,2,3,4$). As the corresponding vector β we would then have the 12×1 vector $\mu = (\mu_{11},\mu_{12}, \ldots, \mu_{34})'$, where $\mu_{ij} = \theta_{ij1} = \theta_{ij2} = \ldots = \theta_{ijn_{ij}}$ ($i = 1,2,3$; $j = 1,2,3,4$). (We have changed here from X to W and from β to μ to keep with the usual notation of that approach.) In either of the two ways of defining Ω we have arrived at the so called full rank cell means model (Model B). Usually the more convenient definition $\Omega = R[W]$, instead of $\Omega = N[A]$, is used in this case. To get the non-full rank overparameterized model (Model A) we would use for this example the common $n \times (1+3+4+3\times4)$ matrix X that corresponds to the vector $\beta = (\mu,\alpha_1,\alpha_2,\alpha_3,\beta_1,\beta_2,\beta_3,\beta_4,\gamma_{11},\gamma_{12}, \ldots ,\gamma_{34})'$. Now the subspace for θ is defined as $\Omega' = R[X]$. But since the matrix W is equal to the last 12 columns of X and since $p = \mathrm{rank}\ X = \mathrm{rank}\ W\ (= 12)$, $\Omega = \Omega'$. Hence the two models are equivalent (assuming the same distribution for the random part e) and the overparameterization of Model A is evident. The only apparent advantage of the $R[X]$definition seem to be that X reflects the structure of the data more explicitly than W. But this has no effect, whatever, on the estimation of θ. Under Model B the orthogonal projector of E^n onto Ω, P_Ω, is of the form $P_W = W(W'W)^{-1}W'$, while under Model A it is of the form $P_X = X(X'X)^{-}X'$, where $(X'X)^{-}$ is any generalized inverse of $X'X$ (see, e.g., Seber, 1977, § 3.8, or Seber, 1980, § 3.4). Since P_Ω is necessarily unique, $P_W = P_X$ (= P_Ω). As to the use of the definition $\Omega = N[A]$, $P_{A'} = A'(AA')^{-1}A$ is the orthogonal projector of E^n onto $\Omega^\perp$, the orthogonal complement of Ω. Hence, as another form, we have $P_\Omega = I - P_{A'}(= P_W)$ (see again Seber, 1980, § 3.4). Neither P_W nor $P_{A'}$ presents any computational difficulty. (Note that $W'W$ is a diagonal matrix.)

Now, if we are interested in estimating any linear function of the elements of β, this can uniquely be done if and only if such a function is identifiable (is a linear function of the components of θ), i.e. is estimable in the terminology of Model A. Whether an estimable function $c'\beta$ (c being such a vector that $a'\theta = c'\beta$ for some vector a) has more meaning for an experimental problem than the equivalent function $a'\theta$ may be a matter of controversy. Nevertheless, it seems obvious that if we are unable to explain something in terms of θ, or here perhaps better in terms of the cell means (expectations) $\mu = (W'W)^{-1}W'$, neither we will be able to explain it in terms of the, usually, undefined parameters β. Attemps to avoid the difficulty by imposing some explanatory (non-estimable) conditions on β usually bring us only back to functions of the elements of θ or μ (see Scheffé, 1959, Chapter 4, and Urquhart, Weeks and Henderson, 1973, p. 322).

So far, our examples have not shown any evident advantage of Model A over B. But let us continue further the discussion on the 2-way crossed classification example. Suppose now that because of some apriori knowledge coming from the experimental context we want to incorporate into the model some additional known restrictions on the expectation vector θ. In the 3x4 example such restrictions will be imposed when we rightly assume that the cell means are all additive with respect to the two crossed classifications.

The assumption will reduce Ω to some ω_1, defined according to the restrictions imposed. In the Model A approach the definition will be $\omega_1 = R[X_1]$, where X_1 comes from the partition $\beta = (\beta_1', \beta_2')'$ and $X = [X_1, X_2]$, accordingly, where $\beta_1 = (\mu, \alpha_1, \alpha_2, \alpha_3, \beta_1, \beta_2, \beta_3, \beta_4)'$. In the Model B approach the definition may be written $\omega_1 = N[C_1]$, with $C_1' = (A', A_1')$, where $A_1 = G_1(W'W)^{-1}W'$, if $G_1\mu = 0$ represents the additivity restrictions (usually confined to linearly independent only). The orthogonal projector of E^n onto ω_1, P_{ω_1}, is of the form $P_{X_1} = X_1(X_1'X_1)^-X_1'$ according to Model A, and of the form $I - P_{C_1'}$, where $P_{C_1'} = C_1'(C_1C_1')^-C_1 = I - P_W + P_{A_1'}$, according to Model B. The equivalence of the two models depends on the choice of G_1. If all cells are filled ($n_{ij} > 0$ for all i,j) the restrictions are simply $\mu_{ij} - \mu_{i'j} - \mu_{ij'} + \mu_{i'j'} = 0$, for all i,j,i',j', but,if there are some empty cells (as in Searle's, 1971, Table 7.6), they will not be so obvious. In fact, if $\theta \in \omega_1 \subset \Omega$, then θ is orthogonal to $\varpi_1 = \omega_1^\perp \cap \Omega$, the orthogonal complement of ω_1 in Ω. Thus, since the orthogonal projector onto ϖ_1 is $P_\Omega - P_{\omega_1}$ (see Seber, 1980, p. 27), we have $[I - X_1(X_1'X_1)^-X_1']\theta = 0$ from Model A, and $P_{A_1'}\theta = 0$ from Model B. Hence, the two models are equivalent if and only if $R[A_1'] = \{R[X_1]\}^\perp$, i.e. $G_1\overline{X}_1 = 0$, where $\overline{X}_1 = (W'W)^{-1}W'X_1$ is obtained from X_1 by taking one row from each set of identical rows. It may be noticed that matrix G_1 satisfying the condition determines interaction contrasts of the type defined by Searle (1971, p. 311). So it is now evident that with properly chosen G_1 the two models will be equivalent. Is there, however, any definite advantage of one of the model over the other? For estimating θ, there may be not much difference between computing the projector P_{X_1} and the equivalent projector $P_W - P_{A_1'}$, though with the latter, use can be made of the form $P_{A_1'} = W(W'W)^{-1}G_1'[G_1(W'W)^{-1}G_1']^-G_1(W'W)^{-1}W'$. But finding G_1 can be an onerous task if there are many empty cells. Therefore the Model A approach may appear more convenient. The traditional definition of additivity used in Model A seems also to be convincing, and one can always take the projector $I - P_{X_1}$ to see the additivity restrictions imposed on θ. Actually, an equivalent for G_1 can also be obtained by taking linearly independent rows of $(I - P_{X_1})X_2$.

Now suppose that the restrictions in our example are not known to be true and have to be confronted with data as a statistical hypothesis. In the geometric approach the test is performed with the use of two projectors. The numerator sum of squares for testing the hypothesis of no interactions is $\|P_{\varpi_1} y\|^2 = y'P_{\varpi_1} y$, where P_{ϖ_1} may be taken from any of the two approaches. The error sum of squares is $\|P_{\Omega^\perp} y\|^2 = y'P_{\Omega^\perp} y$, where $P_{\Omega^\perp} = I - P_\Omega$. The corresponding degrees of freedom are: $q = \dim[\varpi_1]$ and $n - p = \dim[\Omega^\perp]$, and the test statistic F has a non-central distribution $F_{q,n-p}$, with the non-centrality parameter $\delta = \|P_{\varpi_1} \theta\|^2/\sigma^2$. The distribution is central when $\delta = 0$. From the previous discussion we already know what we are testing here.

If the hypothesis is not rejected, the additivity restrictions will become a part of the model. We can then test another hypothesis, for example that there are no effects of the second factor in the classification. Geometrically this would mean, that the subspace for θ can be restricted even further, to $\omega_2 \subset \omega_1 \subset \Omega$. In the Model A approach $\omega_2 = R[X_{11}]$, where X_{11} is the submatrix of X_1 corresponding to $\beta_{11} = (\mu, \alpha_1, \alpha_2, \alpha_3)'$. In the Model B approach $\omega_2 = N[C_2]$, with $C_2' = (C_1', A_2')$, where $A_2 = G_2(W'W)^{-1}W'$, if $G_2\mu = 0$ are the new restrictions. The numerator and denominator sums of squares for testing the present hypothesis are given by the orthogonal projectors, P_{ϖ_2} and $P_{\omega_1^\perp}$, respectively, where $\varpi_2 = \omega_2^\perp \cap \omega_1$ and $\omega_1^\perp$ is the orthogonal complement of ω_1 in E^n. (For details see Monlezun and Speed.) The hypothesis tested is $P_{\varpi_2}\theta = 0$, which in the Model A approach takes the form $[I - X_{11}(X_{11}'X_{11})^- X_{11}']X_{11}\beta_{11} = 0$, while in the Model B approach it can be written $G_2\{I - (W'W)^{-1}G_1'[G_1(W'W)^{-1}G_1']^- G_1'\}\mu = 0$ (cf. Rao and Mitra, 1971, § 7.3).

4. Concluding Remarks

The geometric approach to estimation and hypothesis testing in linear models may be helpful in concise and clear presentation of the basic assumptions and methods involved in the analysis. Differences between approaches can be seen more acutely, and advantages or disadvantages of particular methods can be brought into light without confusing terminology.

Only some points of the vast area of disagreement has been indicated and viewed here from the geometrical angle. Though no final conclusions can be drawn yet, it seems that neither of the two types of Models, A and B, has a uniform superiority over the other. Perhaps Model A is more suitable for testing hypotheses stated in terms of certain auxiliary parameters representing some structure of the data, and concerned with fitting the model. The Model B approach seems to be more suited for testing particular hypotheses expressed in terms of well defined linear functions of cell means.

Though the problem of testing several hypotheses has not been examined here, the geometric approach seem to favour the hierarchical partition, as most natural in fitting the model. This however does not exclude testing certain parametric functions when a model is fitted.

References

Afifi A.A. and Azen S.P. (1972), Statistical Analysis: A Computer Oriented Approach, Academic Press, New York.

Bock R.D. (1963), Programming univariate and multivariate analysis of variance, Technometrics, 5, 95-117.

Bock R.D. and Brandt D. (1980), Comparison of some computer programs for univariate and multivariate analysis of variance. Handbook of Statistics, Vol. 1, Krishnaiah P.R. (ed.), North-Holland, Amsterdam, 703-744.

Bryce G.R., Scott D.T. and Carter M.W. (1980), Estimation and hypothesis testing in linear models - a reparameterization approach to the cell means model, Commun. Statist.-Theor.Meth., A 9, 131-150.

Corsten L.C.A. (1958), Vectors, a tool in statistical regression theory, Mededelingen van de Landbouwhogeschool te Wageningen, Nederland, 58, 1-92.

Francis I. (1973), A comparison of several analysis of variance programs, J. Am. Statist. Ass., 68, 344-865.

Frane J.W. (1980), Some computing methods for unbalanced analysis of variance and covariance. Commun. Statist.-Theor. Meth., A 9, 151-166.

Gauss K.F. (1809), Werke 4, 1-93, Göttingen.

Gianola D. (1975), Letter to editor, Amer. Statist., 29, 133.

Goodnight J.H. (1980), Tests of hypotheses in fixed effects linear models, Commun. Statist.-Theor. Meth., A 9, 167-180.

Herr D.G. (1980), On the history of the use of geometry in the general linear model, Amer. Statist., 34, 43-47.

Hocking R.R. and Speed F.M. (1975), A full rank analysis of some linear model problems, J. Am. Statist. Ass., 70, 706-712.

Hocking R.R., Hackney O.P. and Speed F.M. (1978), The analysis of linear models with unbalanced data, Contributions to Survey Sampling and Applied Statistics, David H.A. (ed.), Academic Press, New York, 133-151.

Hocking R.R., Speed F.M. and Coleman A.T. (1980), Hypotheses to be tested with unbalanced data, Commun. Statist.-Theor. Meth., A 9, 117-129.

Kutner M.H. (1974), Hypothesis testing in linear models (Eisenhart Model I), Amer. Statistician, 29, 98-100.

Monlezun C.J. and Speed F.M. (1980), The geometry of estimation and hypothesis testing in the constrained linear model - the full rank case, Commun. Statist.-Theor. Meth., A 9, 213-230.

Nelder J.A. (1974), Letter to editor, J.R.Statist.Soc., C, 23, 232.

Nelder J.A. (1976), Letter to editor, Amer. Statist., 30, 103.

Nelder J.A. (1977), A reformulation of linear models (with Discussion), J.R.

Statist. Soc., A, 140, 48-76.
O'Brien R.G. (1976), Comment on "Some problems in the nonorthogonal analysis of variance", Psychol. Bull., 83, 72-74.
Overall J.E. and Klett J.C. (1972), Applied Multivariate Analysis, Mc Graw-Hill, New York.
Rao C.R. (1973), Linear Statistical Inference and Its Applications, 2nd Ed., Wiley, New York.
Rao C.R. and Mitra S.K. (1971), Generalized Inverse of Matrices and its Applications, Wiley, New York.
Scheffé H. (1959), The Analysis of Variance, Wiley, New York.
Searle S.R. (1971), Linear Models, Wiley, New York.
Searle S.R. (1980), Arbitrary hypotheses in linear models with unbalanced data, Commun. Statist.-Theor. Meth., A 9, 181-200.
Seber G.A.F. (1977), Linear Regression Analysis, Wiley, New York.
Seber G.A.F. (1980), The Linear Hypothesis: A General Theory, Griffin, London.
Speed F.M. and Hocking R.R. (1976), The use of R( ) notation with unbalanced data, Amer. Statist., 30, 30-33.
Speed F.M., Hocking R.R. and Hackney O.P. (1978), Methods of analysis of linear models with unbalanced data, J.Am. Statist. Ass., 73, 105-112.
Urquhart N.S., Weeks D.L. and Henderson C.R. (1973), Commun. Statist., 1, 303-330.
Urquhart N.S. and Weeks D.L. (1978), Linear models in massy data: Some problems and alternatives, Biometrics, 34, 696-705.
Yates F. (1933), The principle of orthogonality and confounding in replicated experiments, J. Agric. Sci., 23, 108-145.
Yates F. (1934), The analysis of multiple classifications with unequal numbers in the different classes, J. Am. Statist. Ass., 29, 51-66.

Analytical Computing: Its Nature and Needs

J.M. Chambers, Bell Laboratories, Murray Hill, U.S.A.

ABSTRACT

The term *analysis* is commonly used in business, economics and elsewhere to mean the process of understanding a system quantitatively. A diverse collection of people (managers, economists, financial consultants,social scientists, etc.) and equally diverse systems (business projects, corporate or national economic activity, accounting and financial planning, political or social entities) have remarkably similar needs; namely, for a repeated cycle of data acquisition and management, data analysis, and presentation of results.

The purposes of this talk are:

1. to indicate the need for analytical computing, and argue that this is an important application area for computing;
2. to consider currently available computer systems and application software to meet analytical needs; and
3. to outline some promising directions for future work on analytical computing.

1. Analysis.

In business, finance, economics and other fields, the term analysis is commonly used to denote, roughly, the application of quantitative methods, mathematical models and statistical summaries to study the behavior of a system or a process. Analysis is an important activity, both as a practical part of how complex systems are understood and managed and for the technical challenges required to support it well. This paper considers the challenges in providing the important computing suppoprt to analysis. While the applications of analysis are diverse, there are common characteristics that have relevance for analytical computing.

- Usually, the analysis is not donė from idle curiosity or even from scientific curiosity; rather, it represents a part of an organization's need to plan, make decisions and understand important issues. The stake in good results is high.
- The ability to obtain results quickly is often critical. Much of the analysis supports decisions made in the presence of deadlines.
- The analysis must be communicated to an audience who are typically intelligent, busy and unevenly sophisticated in the underlying technical questions.

Important applications are typically non-routine. The need for analysis often arises because unusual events have occurred.

These characteristics imply the need for high-quality computing support, In particular, work on any of the following areas can greatly benefit analysis.

1. User interfaces.

2. Knowledge-based software.

3. Workstations and computing power.

4. Data acquisition and management.

5. Graphics and reports.

We will examine each of these areas and consider the needs and prospects for analytical computing.

2. User Interfaces and Knowledge-based Software.

The characteristics of analytical needs put a premium on getting the appropriate questions asked, quickly. While there are useful languages and software packages for analysis, current interfaces tend to introduce either delays or compromises into this process. If the available application languages and user interfaces are not sufficiently general or accessible to non-computer-specialists, there may be a time-consuming, imprecise and costly activity of program specification and implementation before answers are available. To reduce or eliminate such programming activities requires making the analytical software accessible to those directly involved in the analysis; that is, to a diverse group of analytical (but not computer) specialists and managers. Improvements in a number of areas in the list above will increase availability, particularly by putting more of the information and detailed knowledge needed to carry out the analysis into the supporting software.

Making analytical computing available by improving the human interface can be viewed in three stages: analytical languages, dialogue interfaces and knowledge-based software. These are overlapping and mutually reinforcing, but in general represent successively more advanced stages in designing the user interface.

2.1 Analytical Languages.

When analysts attempt to "program" the computations needed to obtain some results, they should be able to express the exact analysis they want, in a simple and natural way. The better current systems for statistical analysis, graphics, report generation and data acquisition have made progress in providing helpful languages, although barriers to the analyst remain. Some characteristics that should be provided in the language and the supporting system to provide a good high-level language include:

1. Simple analysis should be simple to express; i.e., the language should not introduce complexity where the analyst expects simplicity.

2. At the same time, the range of possible analysis should not be restricted. If an analysis makes sense, it should be feasible to express it in the language or extend the language to include the analysis.

3. The language should be uniform and consistent, so that the analyst will not have to learn a different way to express the same notion, depending on the analytical context.

4. The analyst should only need to describe the objects to be manipulated (data sets, reports, graphs) in terms of their analytical content, rather than their computational representation. In particular, the analyst should not be concerned about how, where or when storage management is done, or about how data is moved from one computer system to another (e.g , from a database system to a statistical system)[Robson 1981].

5. The terminology used and the description of the analysis to be performed should be natural to the particular class of users (financial terms for financial analysis, social science terms for the social scientist, etc.). It should not be the user's responsibility to learn the preferred jargon of the implementers of the language.

Most of these requirements can largely be met, for the component parts of the analysis by current languages for data analysis, database management, graphics and report generation. The criteria (with the exception of point 5) were among the design goals of the S language [Becker 1982] and have been achieved to a reasonable degree. The better data extraction and report generation systems have likewise achieved a reasonable quality of user interface by these criteria. None of the systems are near perfection, but experience shows that a wide range of analysts have been able to use them effectively.

The currently available analytical languages have at least three related, fundamental weaknesses. They are essentially passive participants in the analysis, waiting for the user to think of something to do and to formulate the request for the system. They tend to be technology oriented in terminology (statistical languages speaking in statistical terms, database systems in database terms). the analyst needs a unified interface that speaks in analytical terms. They are not knowledge-based in the sense that the knowledge guiding the choice and interpretation of the analysis resides with the users, not with the software. Two steps that will address these weaknesses are outlined in the next subsections.

2.2 Dialogue Interfaces.

A relatively straightforward improvement can be made in the user interface by augmenting the current passive interface with one which includes an interactive dialogue. Users would be provided with information about plausible next steps to take in the analysis, and could specify the desired action in a few, simple keystrokes or through the use of special input hardware. Such "menu" systems have been available for many years in one form or another. In the past their popularity has been limited, largely because of delays involved in presenting the menus, inflexibility in specifying the desired analysis, lack of sufficient information in the description accompanying the menus and the enforced repetitiveness of many tasks.

Workstations that provide local processing power and fast terminal interaction invite much improved menu systems, however. Much more information can be presented, the system can respond rapidly to the user's requests and the flexibility, for example, of a multiple-window display can be used to reduce the "straightjacket" feeling of early menu systems. The dialogue systems designed for this environment must also try to adapt to the user's level of sophistication so that repetitive, tedious sequences of question-response are replaced by simple commands whenever possible. Some of the processing power available should be used to anticipate the likely next steps and to incorporate information about what the user has done in the past. The menu system must be highly dynamic in its construction and use of the underlying menu data. Menu systems can also be the focus to start to use analytical terminology more appropriate to the local users.

A good, dynamic structure for menu interfaces should be able to make "thesaurus" changes of terminology more straightforward for the local analytical computing people. Since the data used in communicating with the menu user is part of a database, tools can be developed to aid in replacing unfamiliar terms with locally more common ones. At present, this is still a labor-intensive process, but the combination of good workstations and good menu data structures should help.

Dialogue interfaces are partners with the high-level application languages, not replacements for them. Whenever a user finds it more natural to express a request by extended input to the language, as opposed to a sequence of menu picks, the dialogue interface should be able to accept this input. (Even those who believe that a well-designed menu system could largely dominate a conventional language in convenience should realize that most users will not immediately transfer completely to such a system.)

2.3 **Knowledge-Based Software.**

Analytical languages and dialogue interfaces are steps in simplifying the specification of analysis. By the syntax and actual procedures in the language, and by the structure and contents of the dialogue menus, some implicit guidance is provided to the analysis. Things that are relatively easy to express are those that are likely to be done. In this way, the designers of the user interface are exercising judgement about the substance of the analysis.

This implied guidance, however, is at most only partly intentional on the implementers' part, and not specialized to the particular analysis. A different kind of guidance in many areas of analysis is given by written handbooks specialized to the kind of analysis (financial, marketing, etc.) and often to the particular organization involved. Such handbooks take analysts through a rather rigid sequence of steps, aimed at resolving a particular analytical goal.

A major need in analysis for the future is to supplement both these kinds of guidance by the use of knowledge-based software that engages a user in a dialogue aimed at carrying out the analysis according to the user's needs and understanding, supplemented by a structured subset of heuristic knowledge about analysis, both general and specialized to the current range of interest. The need for such software is emphasized by the success of analytical computing in reaching a wider audience and involving more and more users of varying sophistication. For more discussion of these issues, see [Chambers, 1981].

3. **Computing Environment for Analysis.**

The workstation (computer terminal, display and keyboard/input facilities) and the computing power to support it need to be sufficient to take advantage of the interface features.

a. The analysis should take place in a responsive, friendly interactive environment. This implies useful immediate feedback from analysis and helpful on-line documentation.

b. The analyst should have enough guaranteed computing power that needed analyses are not precluded by the extensive delays required (subject to the realization that users' computational ambitions expand to fill the computing power available).

c. The information coming back to the analyst should be presented in clear, attractive terms. This means high-quality plots and reports. It is not cost-effective for analysts to struggle with crude plots or illegible reports, still less to labor to extract information from crude numerical printouts.

Most of the features outlined are obtainable currently, at reasonable cost. Adequate computer power and terminal quality can be provided at a cost on the order of $25,000 to $50,000 per analyst, for example, by giving a group of about 4 analysts a 32-bit minicomputer with each user having a good raster terminal. This is a reasonable investment in providing the kind of analysts we are discussing with good computer support. An alternative, at about the same cost, is a network of personal computers. The technology supporting this approach is less mature, but may offer some advantages: guaranteed responsiveness to the individual user, increased redundancy and reliability, and possibly better display through higher bandwidth to the individual terminal.

4. **Data Acquisition and Management.**

In most analytical problems, obtaining the correct data and managing it so analysis can be done absorbs a large fraction of time and human skill, primarily because the data needed is transactional and is generated in the everyday operation, rather than being observations collected for the purpose of analysis. The computing environment and/or the people managing the computing environment for this data may be unfriendly to analysts trying to extract data. The data resides usually in a variety of non-analytical systems (assuming it is available in computerized form at all), and these systems are usually hard to get at and too occupied with everyday business to be taken over for analysis in any event. The various systems are typically heterogeneous and incompatible. Also, the full data is often much more than can be analyzed in a reasonable time.

A first data acquisition requirement therefore is to extract subsets of the underlying data and bring these back to an analytical environment. This is basically the task of generating data extraction requests for a variety of transaction-oriented database systems. At present some user-oriented interfaces do exist for a number of the more popular database systems (IMS, RAMIS, etc.), and analytical users may construct data-extraction scripts in these interfaces. An improved approach will be to generate such data extraction more generally through a single user-oriented interface capable of adapting to a wide variety of database systems.

Once retrieved, it is important that the data be stored in an analytical database system that emphasizes ease and efficiency in further extraction of subsets for analysis. Because of the difficulties mentioned in extracting from the original transaction databases, the best strategy will be to bring over as much data as practical to the analytical environment, with further extraction done there. The analytical database must be able, therefore, to extract subsets in a reasonably efficient manner. Also, the non-routine nature of analysis implies a need for great flexibility.

On the other hand, transaction issues are of little importance in the analysis, so that complex strategies for concurrency control and updating are not needed.

5. Graphics and Reports.

A major need for analytical graphs and reports is to produce high-quality output directly from the analysis. Current software increasingly includes graphical facilities, but these often are poor in quality or limited to unsatisfactory output devices. Similarly, report generation is usually limited to simple tables insufficiently clear in presentation and not designed for high-quality output devices such as photo-typesetters. High quality specialized graphics and report-generating systems exist, but these are usually separate from analytical systems and difficult to use. Improving the aesthetic quality of output without generating a large extra programming load will require integration into analytical systems of better quality graphical hardware and more attention to the visual quality of the graphs and reports. Since a purely algorithmic approach to such quality is unlikely to work in all examples, this is another area in which knowledge-based software may find application. Interactive user dialogue with a system having considerable knowledge of aesthetic preferences will allow improved quality to be obtained when needed with less human effort.

References.

[Becker 1982]
R. A. Becker and J. M. Chambers, "S: A Language and System for Data Analysis" Bell Laboratories Technical Information Service, Murray Hill, New Jersey.

[Chambers 1981]
J. M. Chambers, "Some Thoughts on Expert Software" Computer Science and Statistics, 13th Interface Symposium

[Robson 1981]
David Robson, "Object-oriented Software Systems" Byte, vol. 6 (August 1981), 74-89.

Updating Formulae and a Pairwise Algorithm for Computing Sample Variances

T.F. Chan, Yale University, New Haven, U.S.A.
G.H. Golub and **R.J. LeVeque**, Stanford University, U.S.A.

Abstract. A general formula is presented for computing the sample variance for a sample of size m+n given the means and variances for two subsamples of sizes m and n. This formula is used in the construction of a pairwise algorithm for computing the variance. Other applications are discussed as well, including the use of updating formulae in a parallel computing environment. We present numerical results and rounding error analyses for several numerical schemes.

1. Introduction.

In computing the variance of a sample of N data points $\{x_i\}$, the fundamental calculation consists of computing the sum of squares of deviations from the mean. This quantity, which for brevity will be referred to as "the sum of squares" or simply as "S", is defined as

$$S = \sum_{i=1}^{N} (x_i - \bar{x})^2, \tag{1.1a}$$

where

$$\bar{x} = \frac{1}{N} \sum_{i=1}^{N} x_i. \tag{1.1b}$$

This computation can be easily performed directly by the *two-pass algorithm* (1.1) provided that (a) N is small compared to the amount of core memory available and (b) the variance is not too small relative to the norm of the data, $\|x\|_2 = (\sum_{i=1}^{N} x_i^2)^{1/2}$. If either of these conditions is violated, however, the situation changes. If N is large, this approach to computing S may be unsatisfactory since it requires passing through the data twice: once to compute $\bar{x}$ and then again to compute S. This problem is sometimes avoided by use of the following *textbook algorithm*, so called because, unfortunately, it is often suggested in statistical textbooks:

*Department of Computer Science, Yale University, New Haven, CT 06520
**Department of Computer Science, Stanford University, Stanford, CA 94305.

This work was supported in part by Army contract No. DAAG29-78-G-0179 and by a National Science Foundation graduate fellowship. The paper was produced using TEX, a computer typesetting system created by Donald Knuth at Stanford.

$$S = \sum_{i=1}^{N} x_i^2 - \frac{1}{N}\left(\sum_{i=1}^{N} x_i\right)^2. \qquad (2)$$

This rearrangement allows S to be computed with only one pass through the data, but the computation may be numerically unstable and should almost never be used in practice. This instability is particularly troublesome when S is very small compared to $\|x\|_2$, in which case even the two-pass algorithm can be unstable.

In discussing the stability of a numerical scheme for computing S, a useful concept is that of the *condition number* κ of the data. This quantity was first introduced by Chan and Lewis[2] who give a thorough discussion. Briefly, κ is a measure of the sensitivity of S to changes in the data. The quantity κu is an upper bound for the relative perturbation which would occur in the exactly computed sum of squares if the input data contained relative errors of size u. If the true sum of squares is S, then κ is given by

$$\kappa = \frac{\|x\|_2}{\sqrt{S}}. \qquad (1.3)$$

It is easy to see that $\kappa \geq 1$ and that in general κ grows as the variance decreases.

An error analysis of the textbook algorithm[2] shows that the relative error in S can be bounded by something on the order of

$$3N\kappa^2 u,$$

where u is the machine roundoff unit (see section 6). This algorithm is therefore seldom useful, as confirmed by the experimental results of Table 1.

The error analysis of the two-pass algorithm found in section 6 shows that the relative error in the sum of squares computed using that algorithm can be bounded by

$$Nu + N^2\kappa^2u^2.$$

The second term in this bound has traditionally been ignored in error analyses of the two-pass algorithm as being of second order. But in the case we are interested in here, when N and κ are both large, this term can easily dominate. Table 2 shows this happening in practice.

During the preparation of this manuscript, a simple modification of the two-pass algorithm was found by Professor Åke Björck which reduces this bound. Based on the error analysis of section 6 for the standard two-pass algorithm, he suggested computing S by

$$S = \sum_{i=1}^{N}(x_i - \bar{x})^2 - \frac{1}{N}\left(\sum_{i=1}^{N}(x_i - \bar{x})\right)^2. \qquad (1.4)$$

In exact arithmetic the second term is zero, but computationally it is a good approximation to the error in the two-pass algorithm. Note that (1.4) can also be viewed as the textbook algorithm applied to the data $\{(x_i - \bar{x})\}$. The error analysis of section 6 shows that the relative error in S computed by (1.4) can be bounded by

$$Nu + 4N^2\kappa u^2.$$

This modification adds only N additions and 2 multiplications to the cost of the two-pass algorithm (already $3N-2$ additions and $N+1$ multiplications) and can be very useful when the data is poorly conditioned. See table 3 for some numerical results.

Of course formula (1.4) is still a two-pass algorithm. For large N it may be desirable to compute S with only one pass through the data. A number of papers have appeared recently on "updating" algorithms for computing S. These are algorithms which are based on formulae for adding one new data point to a sample and computing the value of S for the combined sample by updating the (presumably known) value of S for the original sample. By starting with a sample of size 1 and applying this formula repeatedly, we get a one-pass algorithm for computing S for a sample of arbitrary size. Youngs and Cramer[6] have investigated several such algorithms and have found the following algorithm to be the best:

$$\begin{aligned}
&S := 0\\
&T := x_1\\
&\text{for } j := 2,3,\ldots,N \text{ do}\\
&\quad T := T + x_j\\
&\quad S := S + \frac{1}{j(j-1)}(jx_j - T)^2.
\end{aligned} \tag{1.5}$$

This is based on the updating formula

$$S_{1,j} = S_{1,j-1} + \frac{1}{j(j-1)}(jx_j - T_{1,j})^2 \tag{1.6}$$

where $S_{i,j}$ stands for the sum of squares for the data points x_i through x_j and $T_{i,j}$ is the sum of x_i through x_j. This notation will be used throughout.

One imporant characteristic of this updating formula is that $S_{1,j}$ is formed from $S_{1,j-1}$ by adding to it a nonnegative quantitiy. In the textbook algorithm (1.2), on the other hand, S is formed by a subtraction which can lead to gross cancellation and even to negative values of S being computed.

In practice the method (1.5) generally performs on a level comparable to the two-pass algorithm (1.1). Chan and Lewis[2] present detailed error analyses of some similar updating methods.

In the next section we present a generalization of the updating formula (1.6) for combining two samples of arbitrary size. Then in section 3 we describe a pairwise algorithm for computing S which is essentially still a one-pass algorithm but which numerically is often more stable than the standard two-pass algorithm.

2. A General Updating Formula.

The method of Youngs and Cramer depends on an updating formula which allows one to compute S for $j+1$ points when given the value of S for j points and one new point. In other words, we can combine a sample of size j with a sample of size 1 and determine the value of S for the combined sample.

This formula can be easily generalized to allow us to combine two samples of arbitrary size. Suppose we have two samples $\{x_i\}_{i=1}^{m}$, $\{x_i\}_{i=m+1}^{m+n}$ and we know

$$T_{1,m} = \sum_{i=1}^{m} x_i, \qquad T_{m+1,m+n} = \sum_{i=m+1}^{m+n} x_i,$$

$$S_{1,m} = \sum_{i=1}^{m} (x_i - \frac{1}{m} T_{1,m})^2, \qquad S_{m+1,m+n} = \sum_{i=m+1}^{m+n} (x_i - \frac{1}{n} T_{m+1,m+n})^2.$$

Then, if we combine all of the data into a sample of size $m+n$, it can be shown that

$$T_{1,m+n} = T_{1,m} + T_{1,m+n} \tag{2.1a}$$

$$S_{1,m+n} = S_{1,m} + S_{m+1,m+n} + \frac{m}{n(m+n)}\left(\frac{n}{m}T_{1,m} - T_{m+1,m+n}\right)^2. \tag{2.1b}$$

If we rewrite the latter formula as

$$S_{1,m+n} = S_{1,m} + S_{m+1,m+n} + \frac{m}{n(m+n)}\left(\frac{m+n}{m}T_{1,m} - T_{1,n+m}\right)^2,$$

then we see that for $m = 1$, $n = j - 1$, this reduces to the formula of Youngs and Cramer, since $S = 0$ for any single data point. The form (2.1b) is more stable numerically, however.

Regardless of what method is used to compute S, the formulae (2.1) may be useful in their own right whenever two samples must be combined. One possible application is to parallel processing. If one has two or more processors available, the sample can be split up into smaller subsamples, and the sum of squares computed for each subsample independently using any algorithm desired. The sum of squares for the original sample can then be calculated using the updating formulae.

However, even in the case of a single processor, it is very desirable to compute S using (2.1). The method (1.5) may be generalized to compute S by processing the data in groups of m elements: compute the sum of squares for each group using the two-pass algorithm and then update the global S accordingly. Traditional updating algorithms such as that of Youngs and Cramer have used $m = 1$.

We have found, however, that the stability of the algorithm is increased by taking $m > 1$. One can easily see that the total number of arithmetic operations performed on the data is minimized by taking $m = \sqrt{N}$. We might expect that this choice of m will minimize the resulting error. Although we do not have a satisfactory error analysis of this algorithm, the experimental results of table 4 do tend to confirm this prediction. Strictly speaking, with $m > 1$ this is no longer a one-pass algorithm, but we see that only m data values at a time need to be kept in core, and m can be as small as necessary.

3. The Pairwise Algorithm.

Table 4 shows that choosing $m > 1$ not only gives more accuracy than using $m = 1$, but can actually give significantly more accuracy than the two-pass algorithm. This suggests that when computing the sum of squares for the subsamples of size m, we should not use the two-pass algorithm when S is small. Rather, we should split the subsample into yet smaller groups. Taking this idea to the limit yields a pairwise algorithm analogous to the well-known pairwise algorithm for computing the sum of N numbers. Let $S_{i,j}$ stand for the sum of squares of elements x_i through x_j, and let $m = \lfloor N/2 \rfloor$, the largest integer not exceeding $N/2$. Then the method consists of computing $S_{1,N}$ by first computing $S_{1,m}$ and $S_{m+1,N}$ and then combining these by means of (2.1). Each of these latter quantities has been computed by a similar combination of still smaller subsamples.

The algorithm can be implemented as just described, but for reasons which we will explain shortly it is actually best to perform the pairwise algorithm in a somewhat modified manner. Consider the following example with $N = 13$. Schematically, we compute from left to right in the tableau (3.1). The intermediate $T_{i,j}$ are also computed in a similar tableau for use in updating the $S_{i,j}$. The final value $T_{1,N}$ will be the sum of all the data points as computed by the pairwise summation algorithm. In practice we can compute from top to bottom in these tableaux requiring only one pass through the data and using only $O(\log_2 N)$ storage locations for intermediate results. We require one such location for each column in each tableau. The computation for the tableau (3.1) would proceed as follows:

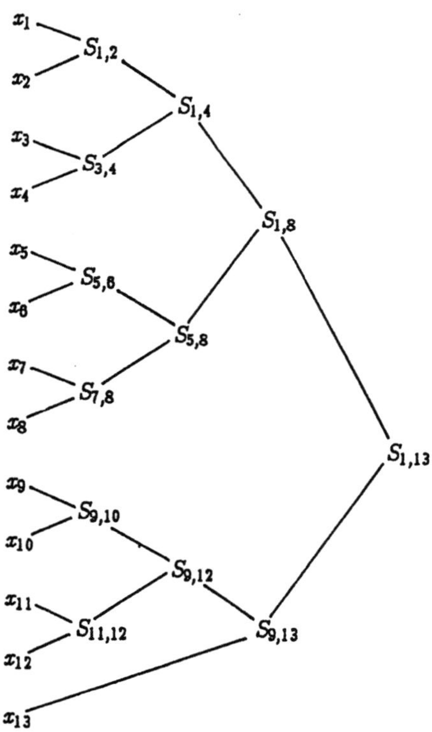

(3.1)

(a) Compute $S_{1,2}$ and store in S[1].
(b) Compute $S_{3,4}$, combine with S[1] to get $S_{1,4}$, and store this in S[2].
(c) Compute $S_{5,6}$ and store in S[1].
(d) Compute $S_{7,8}$, combine with S[1] to get $S_{5,8}$ and then combine this with S[2] to get $S_{1,8}$, which is then stored in S[3].
(e) Compute $S_{9,10}$ and store in S[1].
(f) Compute $S_{11,12}$, combine with S[1] to get $S_{9,12}$, which is then stored in S[2].
(g) Clean-up (necessary when N is not a power of 2):
Combine x_{13} with S[2] to get $S_{9,13}$. Combine this with S[3] to get $S_{1,13}$.

The final step (g) of our algorithm requires the combination of samples of quite disparate sizes. Such a calculation would be avoided if we adopted the algorithm as described in the first paragraph of this section. For the pairwise summation algorithm, Tsao[4] points out that the corresponding method gives a decreased "average error complexity" and presents an implementation based on the binary expansion of N. That strategy could be adopted in the present context as well, but its usefulness here is questionable. We feel that the small increase in accuracy which might result would be more than offset by the increased work which we would thus incur. For the updating formula (2.1b), it is desirable to have $n = m$ whenever possible, since that formula then becomes simply

$$S_{1,2m} = S_{1,m} + S_{m+1,2m} + \frac{1}{2m}(T_{1,m} - T_{m+1,2m})^2.$$

In this respect, the tableau (3.1) gives the preferable computational scheme. In fact, the amount of work required to perform the pairwise algorithm as described here is not significantly more than that required for the two-pass algorithm. An operation count shows that roughly $2N$ additions and $5N/2$ multiplications are required, as opposed to $3N$ additions and N multiplications for the two-pass algorithm. In addition, some bookkeeping operations are required to manage the pairwise algorithm.

Although we are not able to provide any error bounds proving the superiority of the pairwise algorithm, our experimental results have been quite satisfactory. Some of these results are shown in table 5 of section 8.

5. Extensions.

Often one wants to compute a weighted sum of squares of deviations from the mean,

$$S_{1,N}^{(W)} = \sum_{i=1}^{N} w_i(x_i - \bar{x})^2. \tag{5.1}$$

The updating formulae (2.1) still hold with only a few minor modifications. Let $W_{j,k} = \sum_{i=j}^{k} w_i$. Then (2.1) is replaced by

$$
\begin{aligned}
T_{1,m+n} &= T_{1,m} + T_{m+1,m+n} \\
W_{1,m+n} &= W_{1,m} + W_{m+1,m+n} \\
S^{(W)}_{1,m+n} &= S^{(W)}_{1,m} + \frac{W_{1,m}}{W_{m+1,m+n}(W_{1,m} + W_{m+1,m+n})} \\
&\quad \times \left(\frac{W_{m+1,m+n}}{W_{1,m}} T_{1,m} - T_{m+1,m+n} \right)^2 .
\end{aligned}
\tag{5.2}
$$

Another quantity which is often of interest is the covariance of two samples $\{x_i\}$ and $\{y_i\}$. For this it is necessary to compute

$$
C_{1,N} = \sum_{i=1}^{N} (x_i - \bar{x})(y_i - \bar{y}).
$$

If we let $T^{(x)}_{j,k} = \sum_{i=j}^{k} x_i$, $T^{(y)}_{j,k} = \sum_{i=j}^{k} y_i$, then the updating formula for C is

$$
\begin{aligned}
C_{1,m+n} &= C_{1,m} + C_{m+1,m+n} + \frac{m}{n(m+n)} \\
&\quad \times \left(\frac{n}{m} T^{(x)}_{1,m} - T^{(x)}_{m+1,m+n} \right)\left(\frac{n}{m} T^{(y)}_{1,m} - T^{(y)}_{m+1,m+n} \right).
\end{aligned}
\tag{5.3}
$$

6. Error analysis of the two-pass algorithms.

We assume throughout our error analyses that we are dealing with a machine with a guard digit and relative precision u. On a base β machine with a t digit mantissa and proper unbiased rounding, $u = \frac{1}{2}\beta^{1-t}$.

Roman letters with tildas over them will be used to denote quantitics actually computed numerically. The same letter without a tilda will indicate the corresponding exact quantity.

In this section we present error analyses for both the standard two-pass algorithm (1.1) and Björck's modification (1.4). Let

$$
S_1 = \sum_{i=1}^{N} (x_i - \bar{x})^2,
$$

$$
S_2 = \frac{1}{N}\left(\sum_{i=1}^{N} (x_i - \bar{x}) \right)^2,
$$

$$
S = S_1 - S_2.
$$

The standard two-pass algorithm is S_1. We first compute a value $\tilde{\bar{x}}$ for the mean of $\{x_i\}$. If this is computed in the standard manner we have

$$
\begin{aligned}
\tilde{\bar{x}} &= \frac{1}{N}\sum_{i=1}^{N} x_i(1+\xi_i), \qquad \text{with } |\xi_i| < Nu + O(u^2), \\
&= \bar{x} + \frac{1}{N}\sum x_i \xi_i .
\end{aligned}
\tag{6.1}
$$

The computed value $\tilde{S}_1$ is then given by

$$
\begin{aligned}
\tilde{S}_1 &= \sum (x_i - \tilde{\bar{x}})^2 (1+\eta_i), \qquad |\eta_i| < (N+2)u + O(u^2) \\
&= \sum \Big((x_i - \bar{x}) + (\bar{x} - \tilde{\bar{x}}) \Big)^2 (1+\eta_i) \\
&= \sum \Big((x_i - \bar{x})^2 + 2(x_i - \bar{x})(\bar{x} - \tilde{\bar{x}}) + (\bar{x} - \tilde{\bar{x}})^2 \Big)(1+\eta_i) \\
&= S + \sum (x_i - \bar{x})^2 \eta_i - \frac{2}{N}\Big(\sum x_i \xi_i\Big) \sum (x_i - \bar{x})(1+\eta_i) \\
&\quad + \Big(\frac{1}{N}\sum x_i \xi_i\Big)^2 \Big(N + \sum \eta_i\Big).
\end{aligned}
\tag{6.2}
$$

The $O(u^2)$ terms in the bounds for $|\eta_i|$ and $|\xi_i|$ turn out to be unimportant in the present error analysis and will be dropped below. Note that $\sum (x_i - \bar{x}) = 0$ and that the following inequalities hold:

$$
\begin{aligned}
&\Big|\sum (x_i - \bar{x})^2 \eta_i\Big| \le S\|\eta\|_\infty < S(N+2)u, \\
&\Big|\sum x_i \xi_i\Big| \le \|x\|_2 \|\xi\|_2 \le N^{1/2}\|x\|_2\|\xi\|_\infty < N^{3/2}\|x\|_2 u, \\
&\Big|\sum (x_i - \bar{x})\eta_i\Big| \le S^{1/2}\|\eta\|_2 \le S^{1/2}N^{1/2}\|\eta\|_\infty < S^{1/2}N^{1/2}(N+2)u.
\end{aligned}
$$

So from (6.2) we obtain the bound

$$
\|\tilde{S}_1 - S\| \le S(N+2)u + 2N(N+2)S^{1/2}\|x\|_2 u^2 + N^2\|x\|_2^2 u^2 (1 + (N+2)u).
$$

Recalling the definition (1.3) of κ, we see that

$$
\begin{aligned}
\left|\frac{\tilde{S}_1 - S}{S}\right| &< (N+2)u + 2N(N+2)\kappa u^2 + N^2\kappa^2 u^2 (1 + (N+2)u) \\
&\sim Nu + N^2\kappa^2 u^2 + 2N^2 \kappa u^2.
\end{aligned}
\tag{6.3}
$$

When $N >> 1$, $\kappa >> 1$, the term $N^2\kappa^2u^2$ may cause problems as was seen in table 2. Note that this term results from the term $N(\frac{1}{N}\sum x_i\xi_i)^2$ in (6.2). We will now show that the computed value $\tilde{S}_2$ is a good approximation to this error. We have that

$$
\begin{aligned}
\tilde{S}_2 &= \frac{1}{N}\Big(\sum (x_i - \tilde{\bar{x}})(1+\gamma_i)\Big)^2 \qquad \text{with } |\gamma_i| < (N+2)u + O(u^2). \\
&= \frac{1}{N}\Big(\sum ((x_i - \bar{x}) + (\bar{x} - \tilde{\bar{x}}))(1+\gamma_i)\Big)^2 \\
&= \frac{1}{N}\Big(\sum (x_i - \bar{x})\gamma_i - \frac{1}{N}\Big(\sum x_i \xi_i\Big)\Big(N + \sum \gamma_i\Big)\Big)^2 \\
&= \frac{1}{N}\Big(\sum (x_i - \bar{x})\gamma_i\Big)^2 - \frac{2}{N^2}\Big(\sum (x_i - \bar{x})\gamma_i\Big)\Big(\sum x_i \xi_i\Big)\Big(N + \sum \gamma_i\Big) \\
&\quad + \frac{1}{N^3}\Big(\sum x_i \xi_i\Big)^2 \Big(N^2 + 2N\sum \gamma_i + \Big(\sum \gamma_i\Big)^2\Big)
\end{aligned}
$$

Note that $\tilde{S}_2$ contains a term $\frac{1}{N}(\sum x_i\xi_i)^2$, so, using Björck's modification of the two-pass algorithm we compute $\tilde{S}$ as

$$\begin{aligned}\tilde{S} &= (\tilde{S}_1 - \tilde{S}_2)(1+\delta), \qquad \text{with } |\delta| < u \\ &= \Big(S + \sum(x_i-\bar{x})^2\eta_i - \frac{2}{N}\Big(\sum x_i\xi_i\Big)\sum(x_i-\bar{x})(1+\eta_i) \\ &\quad + \frac{1}{N^2}\Big(\sum x_i\xi_i\Big)^2\sum\eta_i - \frac{1}{N}\Big(\sum(x_i-\bar{x})\gamma_i\Big)^2 \\ &\quad - \frac{2}{N^2}\Big(\sum(x_i-\bar{x})\gamma_i\Big)\Big(\sum x_i\xi_i\Big)\Big(N+\sum\gamma_i\Big) \\ &\quad - \frac{1}{N^3}\Big(\sum x_i\xi_i\Big)^2\Big(2N\sum\gamma_i + \Big(\sum\gamma_i\Big)^2\Big)\Big)(1+\delta).\end{aligned}$$

Bounding these quantities as before gives the following bound for the relative error:

$$\begin{aligned}\left|\frac{\tilde{S}-S}{S}\right| &< (N+2)u + (N+2)(4N\kappa + 2(N+2))u^2 \\ &\quad + (N+2)(3N^2\kappa^2 + (6N+4)\kappa + (N+2))u^3 + O(u^4) \\ &\sim Nu + 4N^2\kappa u^2 + 3N^3\kappa^2 u^3.\end{aligned}$$

The modification has thus reduced the "second order term" by roughly a factor of κ.

7. Calculation of the mean in double precision.

A greater accuracy can be achieved from any algorithm for computing the sum of squares by simply using higher precision arithmetic. It is important to note, however, that a large increase in accuracy can often be achieved by shifting only some of the calculations to double precision. From the error analysis of the two-pass algorithm, we see that computing the sample mean in double precision would replace the bound $|\xi_i| < Nu + O(u^2)$ in (6.1) by $|\xi_i| < Nu^2 + O(u^3)$. If the remainder of the calculations are still computed in single precision, the error bound (6.3) will nonetheless be replaced by the improved bound

$$\left|\frac{\tilde{S}_1 - S}{S}\right| < Nu + O(u^3).$$

The difference which this can make in practice is evident from table 6 in section 8, which gives the results of some numerical experiments.

The generalized updating algorithm and the pairwise algorithm are also improved by calculating the corresponding running sums in double precision. Numerical results for these modified algorithms are given in tables 7 and 8 respectively.

8. Experimental results

All of the results presented in this section were computed on an IBM 370/168 computer at the Stanford Linear Accelerator Center. The data used was provided by a random number generator with mean 1 and a variety of different variances σ^2. For this choice of the mean, $\kappa \approx 1/\sigma$. In each case the results have been averaged over 20 runs. Single precision was used in most of the tests except in the cases where the mean was computed in double precision (tables 6-8). In single precision, $u \approx 5 \times 10^{-7}$. The "correct" answer for use in computing the error was computed in quad precision. We report the number of correct digits in the calculation, defined as $-\log_{10}(E)$ where E is the relative error.

Table 1: Number of correct digits for the textbook algorithm on N data points chosen randomly from N(1.0, σ^2).

σ^2 \ N	64	256	1024	2048
1.0	5.4	4.3	4.1	4.1
10^{-1}	4.2	4.7	3.0	3.0
10^{-2}	3.2	3.2	2.0	2.0
10^{-3}	2.2	2.2	1.0	1.0
10^{-4}	1.2	1.1	0.0	0.0
10^{-5}	0.2	0.2	−1.0	−1.0
10^{-6}	−0.8	−0.8	−2.0	−2.0
10^{-7}	−1.8	−1.9	−3.0	−3.0
10^{-8}	−2.8	−2.8	−4.0	−4.0

Table 2: Number of correct digits for the two-pass algorithm on N data points chosen randomly from N(1.0, σ^2).

σ^2 \ N	64	256	1024	2048
1.0	5.2	5.1	4.0	4.0
10^{-1}	5.4	4.5	4.2	4.2
10^{-2}	5.6	4.5	4.4	3.7
10^{-3}	5.6	4.6	4.5	3.6
10^{-4}	5.2	4.8	4.4	4.0
10^{-5}	5.5	5.3	3.1	3.0
10^{-6}	4.5	4.4	2.1	1.9
10^{-7}	3.5	3.3	1.1	0.9
10^{-8}	2.5	2.3	0.1	−0.1

Table 3: Number of correct digits for Björck's two-pass algorithm on N data points chosen randomly from $N(1.0, \sigma^2)$.

σ^2 \ N	64	256	1024	2048
1.0	5.2	5.1	4.0	4.0
10^{-1}	5.4	4.5	4.2	4.2
10^{-2}	5.6	4.5	4.4	3.7
10^{-3}	5.6	4.6	4.4	3.6
10^{-4}	5.2	4.8	3.9	3.7
10^{-5}	5.2	5.0	3.9	3.8
10^{-6}	5.4	5.0	4.1	4.0
10^{-7}	5.7	4.6	4.2	4.2
10^{-8}	6.2	4.6	3.8	3.3

Table 4: Number of correct digits for the generalized updating algorithm on 1024 data points chosen randomly from $N(1.0, \sigma^2)$ with various values of m. (Note that $m = 1$ corresponds to algorithm (1.5) while $m = 1024$ is just the two-pass algorithm).

σ^2 \ m	1	2	4	8	16	32	64	128	256	512
1.0	4.0	4.0	4.3	4.6	4.9	5.0	5.0	5.1	5.0	4.2
10^{-1}	4.2	4.2	4.5	4.8	5.0	5.1	5.2	5.3	4.5	4.3
10^{-2}	4.5	4.4	4.7	5.0	5.1	5.3	5.4	4.9	4.5	4.4
10^{-3}	4.1	4.2	4.5	4.8	5.0	5.2	5.2	4.8	4.6	4.6
10^{-4}	3.6	3.7	4.0	4.3	4.5	4.8	5.0	4.8	4.8	4.6
10^{-5}	3.2	3.4	3.8	4.1	4.3	4.6	4.8	5.1	5.1	3.4
10^{-6}	2.5	3.0	3.3	3.7	4.0	4.3	4.4	4.6	4.6	2.4
10^{-7}	1.4	2.1	2.5	2.9	3.6	3.6	3.5	3.4	3.3	1.4
10^{-8}	0.4	1.0	1.7	2.4	3.0	2.8	2.5	2.4	2.3	0.4

Table 5: Number of correct digits for the pairwise algorithm on N data points chosen randomly from $N(1.0, \sigma^2)$.

σ^2 \ N	64	256	1024	2048
1.0	5.8	5.8	5.6	5.6
10^{-1}	6.0	5.7	5.7	5.7
10^{-2}	6.2	5.8	5.7	5.6
10^{-3}	5.9	6.0	5.6	5.6
10^{-4}	5.5	5.8	5.9	5.8
10^{-5}	4.7	5.2	5.4	5.4
10^{-6}	4.5	4.7	4.8	4.9
10^{-7}	3.9	4.2	4.3	4.4
10^{-8}	3.2	3.7	3.8	3.9

Table 7: Number of correct digits for the generalized updating algorithm on 1024 data points chosen randomly from N(1.0, σ^2) with various values of m. In this test the running sums were computed in double precision. (Note that $m = 1$ corresponds to algorithm (1.5) while $m = 1024$ is just the two-pass algorithm).

σ^2 \ m	1	2	4	8	16	32	64	128	256	512
1.0	4.0	4.0	4.3	4.6	4.9	5.0	5.1	5.1	5.0	4.2
10^{-1}	4.2	4.2	4.5	4.8	5.0	5.1	5.2	5.3	4.5	4.3
10^{-2}	4.4	4.4	4.7	4.9	5.2	5.3	5.4	4.9	4.5	4.4
10^{-3}	4.4	4.4	4.7	4.9	5.2	5.3	5.3	4.8	4.6	4.6
10^{-4}	3.9	3.9	4.2	4.5	4.8	5.0	4.9	4.8	4.8	4.8
10^{-5}	3.9	3.9	4.2	4.5	4.8	5.0	5.0	4.9	4.9	4.3
10^{-6}	4.1	4.1	4.3	4.6	4.9	5.0	5.1	5.1	4.8	4.2
10^{-7}	4.2	4.2	4.5	4.8	5.0	5.2	5.2	5.3	4.5	4.3
10^{-8}	4.4	4.4	4.7	5.0	5.2	5.3	5.4	4.9	4.5	4.4

Table 8: Number of correct digits for the pairwise algorithm on N data points chosen randomly from N(1.0, σ^2). In this test the running sums were computed in double precision.

σ^2 \ N	64	256	1024	2048
1.0	5.9	5.8	5.7	5.7
10^{-1}	5.9	5.7	5.7	5.7
10^{-2}	6.0	5.7	5.7	5.6
10^{-3}	6.0	5.7	5.6	5.5
10^{-4}	5.9	5.8	5.7	5.6
10^{-5}	5.9	5.8	5.6	5.6
10^{-6}	5.9	5.8	5.7	5.6
10^{-7}	6.0	5.8	5.7	5.7
10^{-8}	6.1	5.8	5.8	5.6

Acknowledgements.

We are particularly indebted to Å. Björck for his suggestions concerning the two-pass algorithm.

References.

[1] Chan, T.F.C., and Lewis, J.G. Computing standard deviations: accuracy. CACM 22,9(Sept. 1979), 526-531.

[2] Chan, T.F.C., and Lewis, J.G. Rounding error analysis of algorithms for computing means and standard deviations. Tech. Rep. No. 284, Dept. of Mathematical Sciences, The Johns Hopkins University, Baltimore, Md., April 1978.

[3] Hanson, R.J. Stably updating mean and standard deviations of data. CACM 18,8(Aug. 1975), 458.

[4] Tsao, N. On "accurate" evaluation of extended sums, manuscript, Dept. of Computer Science, Wayne State University.

[5] West, D.H.D. Updating mean and variance estimates: an improved method. CACM 22,9(Sept. 1979), 532-535.

[6] Youngs, E.A., and Cramer, E.M. Some results relevant to choice of sum and sum-of-product algorithms. Technometrics 13(Aug. 1975), 458.

Convex Hull Peeling

W.F. Eddy, Carnegie-Mellon University, Pittsburgh, U.S.A.

Given a set of n points in the plane, a method is described for constructing a nested sequence of $m < n/2$ convex polygons based on the points. If the points are a random sample, it is shown that the convex sets share some of the distributional properties of one-dimensional order statistics. An algorithm which requires $O(n^3)$ time and $O(n^2)$ space is described for constructing the sequence of convex sets.

Key words: convex polygon, order statistic, intersection.

1. Introduction.

The basic idea of convex hull peeling is to construct, from a set of data points, a (possibly nested) sequence of convex sets which may contain useful information about the structre of the data points. Since this idea was introduced by Tukey (see Huber, 1972), a variety of specific methods have been proposed; for example, see Barnett (1976), Eddy (1981) and Tukey (1975). A thorough review of the peeling literature and more extensive list of references is given in Green (1980). While all of the proposed methods have a certain appeal to exploratory data analysts, a major drawback for the less adventurous has been the lack of an associated distribution theory.

In this paper, I will define a method for determining a nested sequence of convex sets; I will present some distribution theory for these sets which closely parallels the distribution theory for ordinary one-dimensional order statistics; and I will give an algorithm for constructing this sequence of sets from n points in the plane which requires $O(n^3)$ time and $O(n^2)$ space. The method has been presented by Tukey (1975, p. 530) and Green (1980, p. 10) but these authors give neither a distribution theory nor an algorithm for the sets. I will confine my attention throughout this paper to the case where the data points are in the plane; the methods and results obviously generalize to higher dimensions.

2. The Method.

Let $\mathcal{H}$ be the collection of half-planes defined, for $0 \leq \gamma < 2\pi$ and $-\infty < p < \infty$, by $H_{\gamma,p} = \{(r,\theta) | r \cos(\gamma - \theta) \leq p,\ 0 \leq \theta < 2\pi,\ 0 \leq r < \infty\}$. Let $(R_1,\theta_1), \ldots ,(R_n,\theta_n)$ be n independent observations with common distribution $\mathcal{P}$. Suppose also that under $\mathcal{P}$

$$\Pr\{(R,\theta) \in H_{\gamma,p}\} = G(\gamma,p);$$

I will call $G(\gamma,p)$ the "half-plane" function of $\mathcal{P}$. In the usual manner define the empirical half-plane function of the data by

$$G_n(\gamma,p) = \frac{1}{n}\sum_{i=1}^{n} I\{(R_i,\theta_i)\in H_{\gamma,p}\}.$$

where $I\{.\}$ is the indicator function of its argument. For each integer k greater than $(n - 1)/2$ define

$$\mathcal{H}_k = \{H_{\gamma,p} | G_n(\gamma,p) \geq \frac{k}{n}\}$$

and finally define

$$Q_k = \bigcap_{H\in\mathcal{H}_k} H.$$

The condition that k be greater than $(n - 1)/2$ eliminates these cases where Q_k is always empty.

I will call Q_k the kth layer for a reason that will become clear. This is approximately the reverse of Tukey's convention; he calls Q_k the $n + 1 + \frac{1}{2} - k$ polygon. Notice that Q_k is convex by construction. Notice also that

$$Q_n \supseteq Q_{n-1} \supseteq \cdots$$

so the $\{Q_k\}$ are ordered by inclusion. Furthermore, Q_n is the convex hull of the data; that is, Q_n is the smallest convex polygon containing the data.

Now for any fixed (γ,p), $Q_k \subseteq H_{\gamma,p}$ if and only if $n\, G_n(\gamma,p) \geq k$. Since $n\, G_n(\gamma,p)$ is binomially distributed with parameters n and $G(\gamma,p)$ it is immediate that

$$\Pr\{Q_k \subseteq H_{\gamma,p}\} = \Pr\{n\, G_n(\gamma,p) \geq k\} \tag{2.1}$$
$$= \sum_{i=k}^{n} \binom{n}{i} [G(\gamma,p)]^i [1 - G(\gamma,p)]^{n-i}.$$

This is immediately familiar as the distribution of the kth order statistic of n (see, e.g., David, 1970, p. 7). For fixed γ, the half-plane function $G(\gamma,p)$, when regarded as a function of p, is exactly the distribution function of $R\cos(\gamma - \theta)$. If (R,θ) is circularly symmetric so that θ is uniformly distributed on $[0,2\pi)$ and independent of R, then $G(\gamma,p)$ is independent of γ and (2.1) becomes

$$\sum_{i=k}^{n} \binom{n}{i} [G(p)]^i [1 - G(p)]^{n-i}.$$

3. Some Distribution Theory.

The support function of a convex set K is

$$b_K(\gamma) = \sup_{\underline{k}\in K}\{k_1 \cos\gamma + k_2 \sin\gamma\},\ 0 \leq \gamma < 2\pi,$$

where $\underline{k} = (k_1,k_2)$. If K is compact, then b_K is continuous and if K is not compact, then b_K takes the value $+\infty$ on a set U which is a union of half-open intervals of length π. In particular, if $K = H_{\alpha,p}$, then

$$b_K(\gamma) = \begin{cases} p & \text{if } \gamma = \alpha \\ +\infty & \text{otherwise.} \end{cases}$$

A single point $\underline{k} = (k_1,k_2) = (r,\theta)$ is a convex set and has support function

$$b_{\underline{k}}(\gamma) = k_1\cos\gamma + k_2\sin\gamma = r\cos(\gamma - \theta).$$

The importance of support functions here derives from the following fact. If K_1 and K_2 are convex sets with support functions b_{K_1} and b_{K_2}, the support function of the convex set $K_1 \cap K_2$ is (the largest support function less than or equal to) $\min(b_{K_1}, b_{K_2})$. The parenthetical condition applies whenever $K_1 \cup K_2$ is not convex. See Matheron (1975, p. 91).

The n observations $\{(R_i,\theta_i)\}$ have support functions $\{B_i\}_{i=1}^n$ where

$$B_i(\gamma) = R_i\cos(\gamma - \theta_i).$$

For each fixed γ the $\{B_i(\gamma)\}$ can be ordered

$$D_n(\gamma) \geq D_{n-1}(\gamma) \geq \cdots$$

so that $\{D_i(\gamma)\}$ are exactly the order statistics of $\{R_i\cos(\gamma - \theta_i)\}$ and have distributions given by (2.1) above. Moreover, if the $\{D_i\}$ are thought of as functions of γ, then

$$D_n \geq D_{n-1} \geq \cdots$$

for all γ, simultaneously. Except for D_n the D_i are not, in general, support functions.

At this time, little is known about the small sample structure of the D_i beyond the marginal property given by (2.1). Somewhat more is known about the asymptotic behavior.

The asymptotic behavior of the process D_n has been studied in Eddy (1980) and Eddy and Gale (1981). Under the assumption that (R,θ) has a circularly symmetric distribution it is known that

$$a_n[D_n(\gamma_0 + c_n t) - b_n] \xrightarrow{w} M_j(t)$$

where M_j is one of three possible extreme-value processes. This closely parallels the well-known theory of extreme-values in one-dimension. The sequences $\{a_n\}$ and $\{b_n\}$ are exactly those that apply to the random variable $\max_i R_i\cos(\gamma_0 - \theta_i)$. The sequence $\{c_n\}$ is proportional to the number of vertices of the convex set with support function D_n. The properties of three processes M_j, $j = 1,2,3$ are given by Eddy and Gale (1981).

Since the collection of support functions is closed with respect to maximum, there is a unique greatest support function not greater than a given function D_i. Without loss of generality and for simplicity also call this support function D_i. For fixed k the support function D_{n-k} has an asymptotic distribution related to that of D_n. By using the convergence of an underlying point process to a Poisson point process (described more fully in Eddy, 1980 and Eddy and Gale, 1981), it can be shown that

$$a_n[D_{n-k}(\gamma_0 + c_n t) - b_n] \overset{w}{\to} M_{jk}(t).$$

The processes M_{j0} are the M_j described above; they are the largest of the Poisson collection of functions. The processes M_{jk} for $k \neq 0$ are related to the M_{j0}; M_{jk} is the $(k + 1)^{st}$ largest of the same Poisson collection of functions.

The asymptotic behavior of D_i for intermediate values of the form $i = n\lambda$ $(0 < \lambda < 1$, fixed) is not known. Generalizing from one-dimensional results I suspect that

$$n^{\frac{1}{2}}[D_{n\lambda}(\gamma_0 + c_n t) - r_\lambda] \overset{w}{\to} X_\lambda(t).$$

The constant r_λ is the solution of

$$G(\gamma_0, r_\lambda) = \lambda$$

and X_λ is a Gaussian process. I suspect that $c_n = 1$ but much work remains.

4. The Algorithm.

The convex polygon Q_k can be thought of in several ways:

(1) The intersection of the $\binom{n}{k}$ convex hulls of subsamples of size $n - k$;

(2) The convex set corresponding to (the largest support function not greater than) $(k + 1)^{st}$ largest of the support functions of the n data points;

(3) The intersection of all half-planes with $n - k$ points on one side.

Repeated use of a standard convex hull algorithm such as that described in Eddy (1977) requires $O(2^n)$ time to construct the sequence $\{Q_k\}$ using the first approach. At this time I do not know how to use the second approach to construct the $\{Q_k\}$. The third approach yields the following algorithm.

For each of the $n(n - 1)/2$ pairs of points, count the number of points on each side of the line joining the pair. Let i be the minimum number for a particular line. For each integer $k = 0,1, \ldots, \lfloor\frac{n-2}{2}\rfloor$ form the intersection of all halfspaces determined by lines having $i = k$.

The intersection can be easily performed in a manner analogous to the Graham (1972) scan. Counting the number of points on each side of all lines requires $O(n^3)$ time and a list of all lines requires $O(n^2)$ space. In p dimensions these increase to $O(n^{p+1})$ time and $O(n^p)$ space for the analogous algorithm.

If only the 2nd layer is required, then one can use a convex hull algorithm requiring $O(n \log n)$ time and $O(n \log n)$ space on each of the c subsets obtained by deleting each of the c vertices of the convex hull in turn. Then the intersection algorithm yields Q_2. If $c = o(n)$, then this requires $o(n^2 \log n)$ time and $O(n \log n)$ space. If other specific layers are required, then it may be possible to generalize the $O(n)$ order statistic algorithm of Blum et al (1972).

5. An Example.

Figure 1 below depicts the result of convex hull peeling applied to a specific set of seven data points. The data points are indicated by large dots. The three layers: Q_7, Q_6, Q_5 are indicated by solid lines. The extension of each of the solid lines to the two data points which determine it are indicated by dashed lines.

Figure 1
Convex Hull Peeling
An Example for n = 7

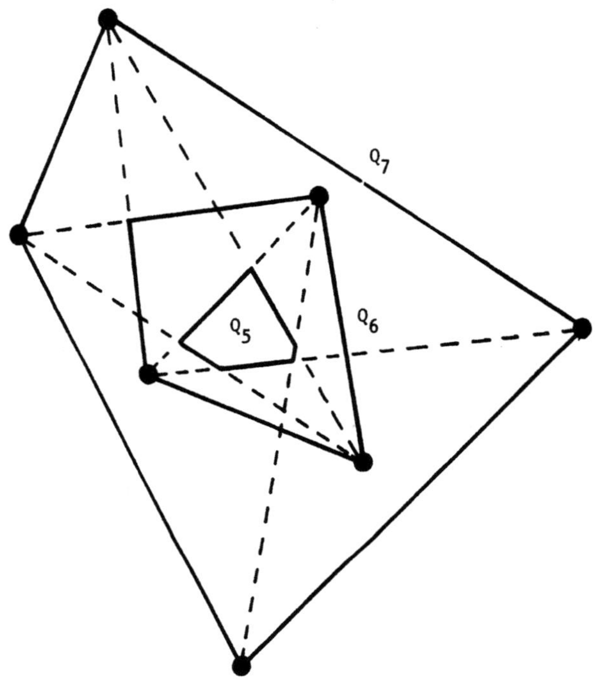

References

Barnett, V. (1976). The ordering of multivariate data. J. R. Statis. Soc. A, 139, 318-354.

Bebbington, A. C. (1978). A method of bivariate trimming for robust estimation of the correlation coefficient. Appl. Statist. 27, 221-226.

Blum, M., R. W. Floyd, V. R. Pratt, R. L. Rivest and R. E. Tarjan (1972). Time bounds for selection. J. Computer and System Sciences, 7, 448-461.

David, H. A. (1970). Order Statistics, Wiley, New York.

Eddy, W. F. (1977). A new convex hull algorithm for planar sets. ACM Trans. Math. Software, 3, 398-403.

Eddy, W. F. (1980). The distribution of the convex hull of a Gaussian sample. J. Appl. Prob., 17, 686-695.

Eddy, W. F. (1981). Discussion of "Graphics for the Multivariate Two Sample Problem" by J. H. Friedman and L. C. Rafsky, J. Amer. Statist. Assoc., 76, 287-289.

Eddy, W. F. and J. D. Gale (1981). The convex hull of a spherically symmetric sample. Adv. Appl. Prob., 13, 751-763.

Graham, R. L. (1972). An efficient algorithm for determining the convex hull of a planar set. Inform. Proc. Letters, 1, 132-133.

Green, P. J. (1980). Peeling bivariate data. Proc. Sheffield Conf. Multivariate Data Analysis, Chapter 1, 3-19.

Huber, P. J. (1972). Robust statistics: A review. Ann. Math. Statist., 43, 1041-1067.

Matheron, G. (1975). Random Sets and Integral Geometry, Wiley, New York.

Tukey, J. W. (1975). Mathematics and the picturing of data. Proc. Int. Cong. Math., Vancouver, 1974, Vol. 2, 523-531.

Statistical Computing at INSEE

J. Gerard, and J.-M. Grosbras, INSEE, Paris, France

SUMMARY :

For statistical computing at INSEE, there are several sets of software systems. According to our needs, they have been either produced by ourselves, either obtained from other organizations, with adjustement to our works. Basic statistical methods (descriptive statistics, econometry, macro economic models) are provided by some purchased packages (CENSUS, TSP, SAS, IMSL, SIMSYS, MODULECO...). On the other hand, processing census and sample surveys data, managing aggregated data (National accounts, time series...) have required specifing programs written at INSEE. The most important of them are LEDA and ARGOS-C. Because ARGOS-C is easy to handle aggregated data and because it has been connected to the main analysis packages, it is a suitable tool in an official organization such as INSEE.

Keywords : statistical packages.

I - INTRODUCTION

The matter of this paper is related to the choices of the Institut National de la Statistique et des Etudes Economiques (INSEE) about equipment in statistical programs. The variety of functions to assume led INSEE to orientate its policy in different ways, according to the nature of tasks. Schematically, we can classify them in three parts.

a) collecting raw materials : census, sample, surveys, other data sources

b) analyzing data by various methods : simple descriptions, or advanced econometrics, or adjusting macro-economic models

c) in addition to these two classical services, an official organization such as INSEE has to construct, manage, store, analyze elaborated data and aggregates, such as tables of National Accounts, time series, and so on.

A main point is to decide if INSEE has to make its own software systems, of if it must use other products, fitting them to its needs. We also must observe that statisticians need to be self sufficient in their use of informatic resources.

II - COLLECTING DATA (mentioned as a reminder)

Mainly the data are provided by :

- census, of populations, firms,...
- great sample surveys (employment, behaviour of households and firms, industriel structures, opinious about overall economic situation,..)
- other governmental source (fiscal data, vital statistics,..).

The work to be done is to collect, to check, to edit data and organize them in a suitable form. Files to be processed are generally large and studies to be made particular to INSEE. Therefore the Institute has normalized its file and their processing and has conceived, produces and developped a software called LEDA, which special use is to process census and sample surveys data. LEDA is able to :

- check validity and coherence of data
- edit, recode, regroupe
- create new variables
- build normalized files with a hierarchical structure
- publish a great deal of result tables (sums, means, standard deviations, medians, deciles, cross tables,...).

With that last function, simple description methods are available.

III - STATISTICAL ANALYSIS

In that field, there is a great variety of purposes, methods and data we can be satisfied with basic descriptive methods, but also need multivariate analysis, or econometric technics. Moreover data must be processed differently according to their type : time series, individual data or aggregates. Finally the theories, methods and technics are constantly improved, because of the progresses in the computer world.

For all these reasons, and after having written some analysis programs, INSEE thought it would be less expansive and more reliable to look for external programms, even if it would be necessary to fit them to our needs. At this respect, widely distributed packages give serious guarantees for reliability, maintenance and improvement, with larger means that INSEE. Could have at its disposal. One of the tasks of the Software Division is to be sure that statisticians at INSEE have the most suitable tools for their works. Without going into all the details, we can describe the main features.

a) Times series

- CENSUS and X11ARIMA for seasonal adjustment

- TSP for econometrics. It is a widely used package, connected with the main time series databases and it helps for the teaching of econometry at the school of statistics of INSEE.

b) Macro economic models

DATABANK, MASSAGER and SIMSYS packages have allowed the carrying out of the important models : DMS, METRIC, PROPAGE,...

At the present time, a powerful software system called MODULECO is being developped in France. In its operational release, MODULECO will offer, with a simple language, the most advanced methods in the building, estimating and simulating models, and the possibility to insert now programs and new commands. INSEE thinks that it is a very interesting project fo its own use as well for the use of others. Therefore it has contributed at its design and is now testing an experimental release.

c) Individual data

The main purchase that has been made is SAS. In spite of the fact it is only suitable for IBM machines, we have selected SAS because of its powerful language, the easiness of handling data files, its wide catalog of statistical methods and the possibility of including new items.

In order to add French methods in factor and cluster analysis for which we already had written programs such as BIBLOS, INSEE cooperate on the design and the realization of SAS/PAD project, including, among others, correspondence, principal components and cluster analysis.

Furthermore, we are interested by MODULAD, with other public organizations, for collecting in a normalized form, the programs of French research workers.

d) Particular programs

Howewer large they are, general packages cannot solve all the problems. Moreover it is important not to be restricted by their sole possibilities and to be able to compare their results with other methods. Therefore, in order to help the users who need specific applications, we have collected an important library of scientific routines and subroutines, including, for instance

- IMSL and HARWELL libraries
- ARIMA programs
- LISREL
- varied algorithms such as those written in ACM publications
- a selection of universities supported programs.

IV - THE AGGREGATED DATA

Lastly, for its specific needs in managing, processing and printing aggregated data, INSEE has been led to develop its own software. It is a table manipulation oriented sotfware, used either in time sharing or in batch processing. It includes some basic statistical functions, but its main purpose is to provide a tool, available to statisticians and economists of the institute for their own use. It has been tailored as an interface between the other analysis softwares in use at the institute.

The software ARGOS-C is now used in that way in the following fields :

- national accounting (yearly and quarterly)
- time series
- regional economic indicators
- work on tables resulting of various surveys, census or administrative forms.

The most important role of this software is

- to provide an easy access to the available data and to the related documentation.
- to give the users the opportunity to create and to manage their own data
- to provide a large range of common handling and computation functions.
- to offer a standard of file, easy to interface with other statistical softwares like TSP, CENSUS-X11 and data analysis packages.

Observing the present use of this software, it can be said that the original aim has been completely reached. It has become an actual communication tool within the institute. It can be observed either by the simple exchange of various machine reports easily available through ARGOS-C, or by the possibility to provide the user with the data and at the same time with the fitted query language.

In fact, the access to an ARGOS-C file is standardized.Structure of data is the same from one object file to the other, and the user does not need to learn a new system when he wants to use a new data set. In addition, this standardization allows the user to access simultaneously to more than one file in the same request step, independently of the origin of the data. So, general purpose data can be complemented with data mobilized for a particular study.

To give further details on the role played by that software we must give a short description of it. We are now going to describe the query system, consisting of the access to the data and of the calculation functions.

1) The main characteristics of ARGOS-C

To take an image, one can say that the design of ARGOS-C, on the point of view of data management, is rather near a software like SAS. The user, in a working session accedes to data sets which all have the same logical structure. These data sets called "bases" include two parts, one receiving the so called data, the other called "dictionary" receiving the data description. The results of calculation are stored in a work space called "internal base" which only exists during a working step. To keep data for a longer time one can store it in an "output base" which can be created in any step.

2) The standardization of data

An ARGOS-C "base" is a set of tables which are described in the dictionary. This description includes a part of text (titles, comments, title-lists corresponding to the values of codes) and a part of structural information closely associated to the query language. Each table has a unique name within a given "base". Each entry of a table is identified with the well known concepts of criterion and variable. A table can have up to 10 criteria and 100 variables. The examples below show how the concepts could be used.

Example 1 : time series are tables with only one criterion (which name is for example DATE) and one variable corresponding to the values of the series. On can also put in the same table the raw data and several types of corrected data (corrected from seasonal variations, working days, etc) In this case the table will contain several variables.

Example 2 : In the national accounting field, the table giving the intermediate consumption has got 4 criteria (type of price, product, branch of industry, year) and one variable.

Example 3 : In the demographic field, a table can take for example 3 criteria, geographic area, age, sex and several variables like the whole population, the number of household chiefs, the number of employed persons, etc...

To each criterion corresponds a list of values which represents the related classification. Different classifications or in general more aggregated classifications can be associated to a criterion. They are assumed by ARGOS-C through the notion of "sub-criterion". These are codes which depend on the criterion and which take a definite value for each value of the criterion. That value is obtained either by simple redefining on the criterion, or by associating a list of values to the criterion. We show examples below.

Example 4 : the criterion DATE used above in example 1 can take several "sub-criteria".

- YEAR redefined on DATE on the first two digits
- MONTH redefined on DATE on the last two digits.

Example 5 : More aggregated classifications can be associated to the product classification in example 2 mentioned above. The most detailed classification is of course assumed by the criterion. In the same way, different geographic area classifications can be defined in example 3 mentioned above. The number of entries of a table is always equal to the cardinal number of the cartesian product of the value-list of the criteria. The missing values are noted with a particular code. They are recognized during calculation and printing.

3 - The language

Because of the dictionary, the logical structure of the data is well known by ARGOS-C. This allows a very important simplication of the programming language. That is particularly true for basic statements, like table displaying, sub-table extraction, common calculations (evolution rates, sums, indexes, etc).

These different characteristics are shown in the following example, taken from the national accounting field.

Let CEB-T be the production account of branches.

This table has got 4 criteria :

- PRICE the type of price which takes 3 values respectively : current price, constant price, 1970 price.
- OPE the type of operation which takes the following values : intermediate consumptions of branches, actual production of branches, gross added value, etc
- BRANCH the activity branches
- YEAR the year.

If we want to know the actual production of branches which corresponds to the value 'PE10' of the criterion OPE we have only to write :

```
PEB = CEB-T (# 'PE10');
```

This statement will create a table named PEB in the work area. PEB is a subset of the entries of CEB-T. Its logical structure is automatically derived from the previous table. Only three criteria are kept : PRICE, BRANCH, YEAR.

To display the data on a report or on a screen we just write :

```
EDITION PEB ;
```

The title-list attached to the prices, branches and years will be displayed if the corresponding option has been coded before.

To show an example of calculation let us take the one of the input-output ratio. Given CI-T, the table of intermediate consumptions, having the following criteria :

PRICE	type of price
PRD	product
BRANCH	branch of activity
YEAR	year

The input-output ratio table is obtained by the following statement.

```
CT = CI-T * 100/PEB;
```

The mapping between the entries of the two operand tables is automatically realized by ARGOS-C. It is based on the values of the criteria.

CONCLUSION

To show how the different softwares are related to each other, we would briefly describe the organization chosen for model DMS.

The data used for this model are provided by the national accounting datasets. They are stored in a slightly different accounting framework. Because the national accounting data were stored in ARGOS-C "bases", it was very convenient to store also the DMS data into ARGOS-C "bases" :

Collecting data to update the model is then simplified : it can be made with ARGOS-C language.

Keeping the table structure is useful to show the main economic functions (consumption, production, investment, etc).

Conversely, the table structure is not suitable to the management of the model and to its forecast functions which are assumed by the softwares TSP and SIMSYS. But it was easy to build an interface acceeding to the time series which name and location had been noted in an index file.

So the econometric estimates and the analysis of different macroeconomic projections, for example, comparison of two models like DMS and PROPAGE are made easier.

Artificial Intelligence and Data Analysis

P. Hájek, ČSAV, Prague, Czechoslovakia
J. Ivánek, School of Economics, Prague, Czechoslovakia

Abstract. Possibilities of application of Artificial Intelligence (AI) to Exploratory Data Analysis (EDA) are discussed. Expert systems as understood in AI appear very suitable for automated consultations for users of statistical software packages as well as for simulation of EDA as an intelligent activity. This is documented on two small (and implemented) systems and one big project (GUHA-80) in development.

Key words: Exploratory data analysis, artificial intelligence, expert systems, software packages, GUHA.

Introduction. Exploratory Data Analysis (EDA) can be safely said to be both science and art; it requires both theoretical knowledge and clear intuition, good guesses and various rules of thumb. In addition, it is a cognitive activity:the data analyst wants to understand his data, reveal some structure, patterns and dependences. All this makes EDA to an activity suitable for investigation by means of Artificial Intelligence (AI). In seventees, a considerable part of activity in AI was concentrated to the development of computer programs for solving problems that require big amount of expert knowledge (as medical diagnosis, chemical analysis etc.etc.). Such programs are called *expert systems*. At present, more and more particular expert systems are being developed; they document the shift of AI activity from "toy problems" to practical applications.

In this paper, we are going to discuss the question, how AI can be applied to EDA (and also conversely, how EDA can be applied to AI). Our discussion is rather general, but we document our point of view by examples concerning the GUHA software package and its relation to AI.

Part 1. Consulting systems.

We start by an example. Let us briefly describe MYCIN, perhaps the best known consulting system (Shortliffe (1976)). MYCIN is a program for medical diagnosis of infectious diseases. It "knows" some tens of microorganisms and is able to conduct a consultation with the physician. During the consultation the program asks various questions concerning the investigated patient;it uses physician's answers to select one or more microorganisms that are most likely to be present in the patient and also recommends some antibiotic therapies.

The program heavily uses medical knowledge, represented by some hundreds of simple rules as, for example,

```
IF: 1) THE STAIN OF THE ORGANISM IS GRAMPOS, AND
    2) THE MORPHOLOGY OF THE ORGANISM IS COCCUS, AND
    3) THE GROWTH CONFORMATION OF THE ORGANISM IS CHAINS
THEN:THERE IS SUGGESTIVE EVIDENCE (.7) THAT
     THE IDENTITY OF THE ORGANISM IS STREPTOCOCCUS.
```

This knowledge base is the *domain-specific* part of MYCIN; on the other hand, MYCIN uses a general control mechanism (inference machine) to control stepwise selection and application of rules for (1) asking questions to the user, (2) propagating uncertain evidence through the net of rules, (3) deriving certainty degrees of some *goal statements* like "the identity of the organism is streptococcus" and, finally, explaining reasons for its decisions (if desired).

MYCIN's inference machine has been separated from the medical context (EMYCIN, van Melle (1980)) and combined with other knowledge-bases, e.g. PUFF (Osborn et al.(1979), pulmonary diseases); other consultation systems with various theoretical and technical modifications and improvements were constructed, e.g. PROSPECTOR (Hart et al. (1978) geology). In our context, SACON (Bennet et al. (1978)) deserves a particular attention;it is a direct application of the empty inference machine EMYCIN but to an interesting domain: it is a consulting system for users of a powerful but complex system of computer programs, namely the system MARC of programs for structural analysis, used e.g. in designing airplain wings. SACON's rules have the same structure as those of MYCIN, e.g.:

```
IF: 1) THE ANALYSIS ERROR (IN PERCENT) THAT IS TOLERABLE
       IS BETWEEN 5 AND 30, AND
    2) THE NON-DIMENSIONAL STRESS OF THE SUBSTRUCTURE IS
       GREATER THAN .7
THEN: IT IS DEFINITE (1.0) THAT MATERIAL IS ONE OF THE
       TYPES OF NONLINEARITY IN THE SUBSTRUCTURE.
```

An example of a question posed by SACON: "Is heating or cooling of interest in determining structure responses?". The result of a consultation with SACON is a selection of analysis strategies including recommendations for the use of MARC in the user's specific situation.

Thus the following suggestion is immediate: *construct a consulting system for users of various statistical software packages*. Clearly, software packages are both powerful and complex and they are user-oriented;their authors try to make the use of the package simple and enjoyable. But the packages themselves can rarely advise the user what to select from the immense variety of possibilities offered (which prodecures, with what parameters) in user's specific situation. This advice can hardly be put into the user's manual;the variety of particular possibilities is too wide and the practical rules are often only poorly theoretically founded. But the knowledge on how to use a software package in many different situations can be made accessible to the user through a consulting system.

To obtain practical experience, a miniature MYCIN-like consulting system for choice of a certain parameter of the GUHA-procedure ASSOC (searching for associations in the data) has been developed in Prague (GQUANT, Hájek and Hájková, see Hájek(1982)). The present reader is not supposed to know GUHA; roughly, to run ASSOC the user must choose (among other things) a statistic to be used, like a test of independence of two dichotomous variables (chi-square, Fisher's test). There are few strict rules for such a choice; but one can formulate various rules of thumb, like:

```
IF: 1) THE KIND OF STATISTIC IS SYMMETRICAL AND
    2) THE SIZE OF DATA IS SMALL
THEN: IT IS DEFINITE THAT THE (IDENTITY OF THE)
      STATISTIC IS FISHER.
```

ASSOC offers 6 possibilities for the choice of a statistic. Three ones concern independence of two dichotomous variables (and are called symmetrical) and three concern high conditional probability (implicational). At present, GQUANT has 34 rules expressing expert knowledge about the 6 statistics used in ASSOC. GQUANT asks the user up to 9 questions and produces recommendations for the choice of a statistic. In the example of a rule above, the question to the user is not "are your data small?" but "what is the size of your data?". The user tells the size of his data (the number of cases) andthen the certainty degree of the proposition "the size of data is small" is computed using an ad hoc formula.

The miniature system has been tested by questioning 11 users of ASSOC; each of them answered questions asked by GQUANT and also told his choice of a statistic. Then the 11 questionnaires were processed by GQUANT. The results were promising, even if not exciting: in all cases, the choice really made by the user obtained maximal recommendation by GQUANT.

GQUANT is being further developed, both as an empty inference machine and in connection with the specific knowledge concerning use of ASSOC. It serves as a source of experience for designing a normal-size consulting system for users of the GUHA package, planned for near future.

The GUHA package seems well-suited for the construction of such a consulting system because it is non-standard in some sense (for a short description of GUHA, the interested reader may consult Hájek, Havránek(1978), Havránek(1981)):GUHA is oriented mainly to dichotomous (or nominal) data and GUHA-procedures *generate* plausible hypotheses rather than testing some particular hypotheses asked by the user. A novice in GUHA needs urgently some advice - a consulting system is a promising possibility.

On the other hand, these specific features of GUHA make possible an *application of EDA to AI*: having large empirical data, one could process them by ASSOC (with an implicational statistic) and obtain automatically rules of the form IF (condition) THEN (conclusion) with some weight (certainty degree). They can be used as (a part of) the knowledge base of an expert system.

Part 2. Expert systems in general.

There are many important expert systems that are not MYCIN-like: let us name DENDRAL (Buchanan,Feigenbaum(1978)) - an AI system for determining the molecular structure of a chemical sample on the base of its mass spectrum (and of the expert knowledge of mass spectrometry theory) and AM (Lenat(1976)) - an AI system that creates and develops mathematical concepts. One difference of these systems in comparison with those mentioned in Part 1 is that both DENDRAL and AM have to *produce* (generate) best candidates rather than to select them from a given list. For this purpose, other AI techniques than those of MYCIN are used. Nevertheless, one can isolate features and problems common to all (or many) expert systems (see Feigenbaum(1980)). An important feature is that knowledge base is separated from the control (inference machine). Very often, the knowledge base consists of many *heuristic rules* of the form IF (condition) THEN (action). (Observe that claiming a conclusion is a particular action.) An inference machine selects in each situation a rule (or a group of rules) to be actually applied.

The rules may bear a rich structure. AM is a good example:it has an agenda of jobs, jobs work with concepts, and for each job and each rule one can decide whether or not the rule is relevant for the job or not. All this is used for the control of the work of the rules.

Observe that the aim of an AI system is often *ill-defined* (cf. Simon(1973)): It consists in maximizing *total interestingness*, but interestingness is not given explicitly. Rather, it is represented by the set of heuristic rules and by the way how they are used. One can test the system from outside by experimenting with the program, whether it behaves intelligently or not.

It should be now more clear that EDA fits well into the domain of intelligent activities that can be modelled by AI. Thus a second suggestion is immediate: *build an expert system modelling the pro-*

cess of EDA as a process, not only a consulting system for a subsequent use of a given software package.

There exists a rather detailed project of a fully fledged AI system of data analysis, called GUHA 80 (see Hájek, Havránek(1982)); we give a very sketchy desctiption of it.

GUHA 80 is largely inspired by AM;the paradigm of GUHA is direct rephrasing of the paradigm of AM:

i) The activity of open-ended data analysis is viewed as a search, an exploration in a space of partially developed concepts. The goal of this search is ill-defined: it is to maximize the interestingness value of what's being worked on in the moment.

ii) GUHA 80 is guided in this process by a collection of some hundreds of heuristic rules. They are relatively general rules of thumb which guide it to define and study the most plausible thing next.

iii) In each situation, GUHA 80 accesses potentially relevant heuristic rules, finds which of them are truly relevant and then follows them.

In contradistinction to AM, GUHA 80 will have to join the AI structure (AI mechanism) with (1) input data to be analyzed and (2) a system of effective modules able to process the data (e.g. various GUHA procedures, clustering procedures etc.).

The AI structure makes possible

- to imitate important steps of data analysis as debugging, reduction of data, main processing, interpretation (and iteration of these);

- to imitate stepwise precization of analyst's goals and methods;

- to separate a suggestion (made by the computer) to apply a certain procedure to the data from actual application of it;

- to make suggestions of new procedures dependent on the results of previously applied ones etc.

An example of a heuristic rule: "If you have to search for associations and the variables are somehow decomposed into groups then look for associations of variables from one group to those from another group."

The system produces a trace of its activity (description of what has been successively done) and results of the work of some effective modules (proper results).

All this requires many design choices and not all of them have yet been made. But the interested reader may consult Hájek,Havránek(1982) where he finds also a fictious run of the system, clarifying how the system should work.

The system GQUANT described above is a very small particular subsystem of the future syst m GUHA 80: automatic specification of parameters for a run of the GUHA-procedure ASSOC (or of another data-analytic procedure) is one of the tasks GUHA 80 must be able to fulfil. There is one more implemented miniature system called GSYNTR (Hájek,Hlavešová) that suggests two sets of variables to be jointly processed by ASSOC. Some information on GSYNTR can be found in Hájek (1982); GSYNTR is not MYCIN-like and generates possible candidates for the processing.

The two miniature systems mentioned serve not only as particular subsystems of future and larger AI systems but also may be directly used with the present GUHA package. They form a bridge between the two main aims of the GUHA circle:development of the GUHA software package of EDA and development of an expert system GUHA 80 of EDA.

Part 3. Conclusion.

We have tried to show that there are various promising possibilities of applying AI to EDA (and also of EDA to AI). First large expert systems of EDA may not be directly user-oriented; rather, they present and inquiry in AI. But if understood well, they can form a basis for building new user-oriented versions. Moreover, as usual in expert systems, knowledge acquisition for an expert system of EDA will contribute to explicit formulation of various rules and techniques of EDA.

References.

Bennet J., Croary L., Engelmore R., Melosh R. (1978), SACON - a knowledge based consultant for structural analysis, report No. STAN-CS-78-699, Computer Science Dept., Stanford Univ.

Buchanan B. G., Feigenbaum E. A. (1978), DENDRAL and META-DENDRAL: their application dimensions, Artificial Intelligence, 11, 5-24

Feigenbaum E. A. (1980), Knowledge engineering: the applied side of Artificial Intelligence, Report No. STAN-CS-80-812. Computer Science Dept., Stanford Univ.

Hájek P. (1982), Applying Artificial Intelligence to Data Analysis (GUHA in context of expert systems), Proc. ECAI-82, Orsay, France.

Hájek P., Havránek T. (1978), The GUHA method, its aims and techniques, Int. J. Man-Machine Studies, 10, 3-22.

Hájek P., Havránek T. (1982), GUHA 80 - an application of artificial intelligence to data analysis, Počítače a umelá inteligencia (Computers and artificial intelligence, 1, No. 2.

Hart P. E., Duda R. O., Einaudi M. T. (1978), PROSPECTOR - a computer based consulting system for mineral exploration, Math. Geology, 10, 589-610

Havránek T. (1981), The present state of the GUHA software, Int. J. Man-Machine Studies, 15, 253-264.

Lenat D. (1976), AM:an artificial intelligence approach to discpvery in mathematics as heuristic search, STAN-CS-76-570, Stanford Univ.

Osborn J., Fagan L., Fallat R., McClung D., Mitchell R. (1979), Managing the data from respiratory measurements, Medical Instrumentation, 13.

Shortliffe E. (1976), Computer-based medical consultation? MYCIN, American Elsevier, New York

Simon H. A. (1973), The structure of ill-structured problems, Artificial Intelligence, 4, 181-201.

VanMelle W. (1980), A domain-independent system that aids in constructing knowledge based consulting programs, STAN-CS-80-820, Stanford Univ.

Classification of Multivariate Autoregressive Processes

M. Krzyśko, Adam Mickiewicz University, Poznań, Poland

SUMMARY: This paper describes discrimination between multivariate autoregressive processes by the Bayes method. The asymptotic distribution of the discriminant function and estimation of the probability of misclassification are investigated.

KEYWORS: discriminant analysis; multivariate autoregressive process; asymptotic distribution.

1. INTRODUCTION

The classical problem of discrimination is associated with the identification of objects and phenomena described by a set of features. However, in many practical situations, problems occur in which it is necessary to identify objects and phenomena characterised by functions of time, that is continuous or discrete processes.

This paper is concerned with discrimination between multivariate autoregressive processes by the Bayes method. The asymptotic normality of the discriminant function is proved and the probability of misclassification is estimated depending on the values of the parameters of the autoregressive processes.

In the univariate case these problems were investigated by Kligienė (1977).

2. CLASSIFICATION METHOD

Let π_i denote a class of p-dimensional stochastic processes $\{\underline{X}_i(t),\ t = 0, \pm 1, \pm 2, \ldots\}$ with expectation $E\{\underline{X}_i(t)\} = \underline{\mu}_i$ and covariance matrix

$$E\left\{[\underline{X}_i(m) - \underline{u}_i][\underline{X}_i(n) - \underline{u}_i]'\right\} = \underline{G}(n-m) = \underline{G}'(m-n)$$

described by an autoregressive equation of order q_i

$$\sum_{u=0}^{q_i} \underline{B}_i(u)\left[\underline{X}_i(t-u) - \underline{\mu}_i\right] = \underline{e}_i(t), \tag{1}$$

where

$\underline{B}_i(0) = \underline{I}_p$, $i = 1,2,\ldots,k.$

The following conditions are imposed on this model:

(i) The $p \times p$ matrices $\underline{B}_i(u)$, $u = 1,2,\ldots,q_i$, $i = 1,2,\ldots,k$ are constants.

(ii) $\{\underline{e}_i(t),\ t = 0,\pm1,\pm2,\ldots\}$ is a sequence of independent $p \times 1$ random variables with common normal distribution, $N(\underline{0},\underline{V})$, $\underline{V} > 0$, $i = 1,2,\ldots,k.$

(iii) The parameters $\underline{\theta}_i = (\underline{B}_i(1),\ldots,\underline{B}_i(q_i),\ \underline{\mu}_i,\ \underline{V}\ )$, $i=1,2,\ldots,k$ are such that the solution $\underline{X}_i(t)$ to (1) is strictly stationary and ergodic.

Let us further assume that the parameters $\underline{\theta}_1,\underline{\theta}_2,\ldots,\underline{\theta}_k$ are known, and that we have a realization $\underline{X}(1),\underline{X}(2),\ldots,\underline{X}(T)$ of a process belonging to one of the classes $\pi_1, \pi_2,\ldots, \pi_k$. In relation to this realization, we construct k alternative hypotheses $H_1,H_2,\ldots,H_k$, forming a complete system of disjoint events, that is if P_i is the a priori probability of the acceptability of hypothesis H_i, then $P_1+P_2+\ldots+P_k = 1$, $(P_i > 0)$, $i = 1,2,\ldots,k$. Hypothesis H_i states that the realization $\underline{X}(1),\underline{X}(2),\ldots,\underline{X}(T)$ is a realization of a process belonging to class π_i, $i = 1,2,\ldots,k$. Our intention is to verify the acceptability of one of these k alternative hypotheses. In the statistical literature this problem is known as classification. We will make use of the Bayesian method of classification dividing the space R^{pT} into non intersecting classification regions $R_1,R_2,\ldots,R_k$ defined so as to minimize the Bayesian risk:

$$(2) \qquad r = 1 - \sum_{i=1}^{k} P_i \int_{R_i} f\left[\underline{x}(1),\dots,\underline{x}(T) \mid \underline{\theta}_i\right] d\underline{x}(1)\dots d\underline{x}(T),$$

where

$$(3) \qquad f\left[\underline{x}(1),\dots,\underline{x}(T) \mid \underline{\theta}_i\right] = (2\pi)^{-\frac{pT}{2}} \left|\underline{G}_{q_i}\right|^{-\frac{1}{2}} \left|\underline{V}\right|^{-\frac{T-q_i}{2}} \times$$

$$\exp\left\{-\frac{1}{2}\left[(\mathrm{vec}\,\underline{x}_{q_i} - \mathrm{vec}\,\underline{\mu}_{q_i})' \underline{G}_{q_i}^{-1} (\mathrm{vec}\,\underline{x}_{q_i} - \mathrm{vec}\,\underline{\mu}_{q_i}) + \right.\right.$$

$$\left.\left. + \sum_{t=q_i+1}^{T} \left(\sum_{u=0}^{q_i} \underline{B}_i(u)(\underline{x}(t-u) - \underline{\mu}_i)\right)' \underline{V}^{-1} \left(\sum_{u=0}^{q_i} \underline{B}_i(u)(\underline{x}(t-u) - \underline{\mu}_i)\right)\right]\right\}$$

is the density function of the joint distribution of variables $\underline{X}(1),\dots,X(T)$ in class π_i, $i = 1,2,\dots,k$, where

$$\underline{x}_{q_i} = (\underline{x}(1), \underline{x}(2), \dots, \underline{x}(q_i)), \quad \underline{\mu}_{q_i} = \underline{\mu}_i \, \underline{1}', \quad \underline{1}'_{1 \times q_i} = (1,1,\dots,1),$$

$i = 1,2,\dots,k$, $\mathrm{vec}\,\underline{A}$ is a vector formed by stacking the columns of $\underline{A}$, one on top of the other, in order from left to right. Matrix $\underline{G}_{q_i}$ is a block matrix of q_i^2 blocks, where block (u,v) contains matrix $\underline{G}(v-u)$, $i = 1,2,\dots,k$.

The classification region R_i minimizing the Bayesian risk (2) is

$$(4) \qquad R_i = \left\{ \underline{x}(1),\dots,\underline{x}(T) : v_{ij}\left[\underline{x}(1),\dots,\underline{x}(T)\right] \geqslant \ln(P_j/P_i), \ \forall j \neq i \right\}$$

where v_{ij} is a discriminant function of the following form:

$$v_{ij}\left[\underline{x}(1),\dots,\underline{x}(T)\right] = \ln \frac{f\left[\underline{x}(1),\dots,\underline{x}(T) \mid \underline{\theta}_i\right]}{f\left[\underline{x}(1),\dots,\underline{x}(T) \mid \underline{\theta}_j\right]}, \quad i,j = 1,2,\dots,k; j \neq i.$$

According to Bayesian method, hypothesis H_i is accepted when $(\underline{x}(1),\dots,\underline{x}(T)) \in R_i$, $i = 1,2,\dots,k$.

The probability α_i of the acceptance of hypothesis H_i when H_i is false is equal to

$$(5) \quad \alpha_i = 1 - \Pr(v_{ij} \geqslant \ln(P_j/P_i), \ \forall_{j \neq i} \mid \underline{\theta} = \underline{\theta}_i), \quad i = 1,2,\dots,k.$$

3. ASYMPTOTIC DISTRIBUTION OF THE DISCRIMINANT FUNCTION

Let us introduce the following notation:

$$A_{ij} = \frac{1}{2} \ln (|\underline{G}_{q_j}| \,/\, |\underline{G}_{q_i}|) + \frac{1}{2}(q_i - q_j) \ln |\underline{V}| ,$$

$$y_i^2(t) = \left\{ \sum_{u=0}^{q_i} \underline{B}_i(u) [\underline{X}(t-u) - \underline{\mu}_i] \right\}' \underline{V}^{-1} \left\{ \sum_{u=0}^{q_i} \underline{B}_i(u) [\underline{X}(t-u) - \underline{\mu}_i] \right\},$$

$$S_{ij}(n) = \sum_{t=1}^{n} Z_{ij}(t)$$

$$Z_{ij}(t) = y_j^2(t) - y_i^2(t), \quad n = T - q, \quad q = \max_{1 \le i \le k} q_i, \quad i,j = 1,2,\dots,k;\ j \ne i.$$

We first observe that the asymptotic distribution of v_{ij} as $T \to \infty$ is the same as the asymptotic distribution of

$$A_{ij} + \frac{1}{2} S_{ij}(n) \ \text{as} \ n \to \infty , \quad i,j = 1,2,\dots,k; \quad j \ne i .$$

The asymptotic normality of $S_{ij}(n)$ may be obtained by using a central limit theorem for dependet random variables due to Ibragimov (1975).

We must prove that

$\sigma_{ij}^2(n) = nh(n), \quad \lim_{n\to\infty} \frac{h(an)}{h(n)} = 1$ for every $a > 0$ (that is $h(n)$ is a slowly oscillating function), $\lim_{n\to\infty} \sigma_{ij}^2 = \infty$,

where

$$\sigma_{ij}^2(n) = E\{[S_{ij}(n) - nm_{ij}]^2 \mid \underline{\theta} = \underline{\theta}_i\} ,$$

$$m_{ij} = E\{ Z_{ij}(t) \mid \underline{\theta} = \underline{\theta}_i \}$$

and for every $\delta > 0$

(6) $E|[Z_{ij}(t) - m_{ij}] \mid \underline{\theta} = \underline{\theta}_i|^{2+\delta} < \infty$ holds for $i,j = 1,2,\dots,k;\ j \ne i$.

It is easy to see that

$$E\,|Z_{ij}(t) \mid \underline{\theta} = \underline{\theta}_i|^r < \infty \quad \text{for every} \quad r \geq 1.$$

But this shows that (6) holds.

We have

$$m_{ij} = \mathrm{tr}\Big(\underline{V}^{-1}\sum_{r=0}^{\infty} \underline{D}_{ij}(r)\ \underline{V}\ \underline{D}'_{ij}(r)\Big) + d_{ij}^2 - p,$$

$$\sigma_{ij}^2(n) = n\Big[R_{ij}(0) + 2\sum_{v=1}^{n-1}\Big(1 - \frac{v}{n}\Big)R_{ij}(v)\Big] = nh(n),$$

$$R_{ij}(v) = 2\mathrm{tr}\big(\underline{V}^{-1}\underline{K}_j(v)\underline{V}^{-1}\underline{K}'_j(v)\big) - 2\mathrm{tr}\big(\underline{D}'_{ij}(v)\ \underline{V}^{-1}\ \underline{D}_{ij}(v)\ \underline{V}\ \big) +$$

$$+ 4\underline{c}'_{ij}\ \underline{V}^{-1}\ \underline{K}_j(v)\ \underline{V}^{-1}\ \underline{c}_{ij},\ v \neq 0,$$

$$R_{ij}(0) = 2\mathrm{tr}\big(\underline{V}^{-1}\underline{K}_j(0)\underline{V}^{-1}\underline{K}'_j(0)\big) + 4\underline{c}'_{ij}\ \underline{V}^{-1}\underline{K}_j(0)\ \underline{V}^{-1}\ \underline{c}_{ij} - 2p,$$

$$\underline{K}_j(v) = \sum_{m=0}^{q_j}\sum_{n=0}^{q_j} \underline{B}_j(m)\ \underline{G}(v+m-n)\underline{B}'_j(n),$$

$$\underline{c}_{ij} = \sum_{u=0}^{q_j} \underline{B}_j(u)(\underline{\mu}_i - \underline{\mu}_j), \quad d_{ij}^2 = \underline{c}'_{ij}\ \underline{V}^{-1}\underline{c}_{ij},$$

$$\underline{D}_{ij}(v) = \sum_{r=0}^{v} \underline{B}_j(v-r)\ \underline{\Lambda}_i(r), \quad \underline{B}_j(l) \equiv 0 \quad \text{for} \quad l > q_j,$$

$\underline{\Lambda}_i(r)$ is the matrix of coefficients of $\underline{e}_i(t-r)$ in the Wold decomposition of the $\underline{X}_i(t)$ (see e.q. Hannan (1970)):

$$\underline{X}_i(t) = \underline{\mu}_i + \sum_{r=0}^{\infty} \underline{\Lambda}_i(r)\underline{e}_i(t-r), \quad i,j = 1,2,\dots,k; \quad j \neq i.$$

Since the process $\underline{X}_i(t)$ is ergodic, then $\sum_{v=1}^{\infty} R_{ij}(v) < \infty$. Hence

$$\lim_{n\to\infty}\frac{h(an)}{h(n)} = 1 \quad \text{for every } a > 0 \quad \text{and} \quad \lim_{n\to\infty} \sigma_{ij}^2(n) = \infty .$$

Hence by Ibragimov (1975) theorem we have

$$\lim_{n\to\infty} \Pr\left\{ \frac{S_{ij}(n) - nm_{ij}}{\sigma_{ij}(n)} < x \,\Big|\, \underline{\theta} = \underline{\theta}_i \right\} =$$

$$= \lim_{T\to\infty} \Pr \left\{ \frac{v_{ij} - A_{ij} - \frac{T-q}{2} m_{ij}}{\frac{1}{2}\sigma_{ij}(T-q)} < x \,\middle|\, \underline{\theta} = \underline{\theta}_i \right\} = \Phi(x),$$

for $i,j = 1,2,\ldots,k;\ \ j\neq i$, where $\Phi(x)$ is the standard normal distribution function.

Let us introduce the following notation:

$$L_{ij} = 2\ln(P_j/P_i) - \ln(|\underline{G}_{q_j}| / |\underline{G}_{q_i}|) - (q_i - q_j)\ln|\underline{V}|,$$

$i,j = 1,2,\ldots,k;\ \ j\neq i$.

Thus for T sufficiently large

$$\alpha_i \approx \Phi\left(\frac{L_{ij} - (T-q)m_{ij}}{\sigma_{ij}(T-q)} \right), \quad i,j = 1,2,\ldots,k;\ \ j\neq i.$$

REFERENCES

Hannan E.J. (1970), Multiple time series, Wiley, New York.
Ibragimov I.A. (1975), A note on the central limit theorem for dependent random variables, Teor. Veroyatn. Primen., XX, 134-139 (Russian).
Kligienė N. (1977), The probability of the error in the Bayes classifying procedure of the random sequences, in: Random sequences recognition problems (L. Telksnys, Ed.), Institute of Mathematics and Cybernetics of the Academy of Sciences of the Lithuanian SSR, Vilnius, 59-79.

Exploratory Analysis of Large Sparse Matrices with Application to Textual Data

L. Lebart, C.R.E.D.O.C., Paris, France

Abstract : This paper deals in a first part with the description of large sparse matrices through correspondence analysis. Whereas the usual algorithms may destroy the sparseness and require in-core diagonalization, the procedure presented here works only on a reduced coding of the basic array (put on an external file). The second part is devoted to the processing of textual data which leads frequently to such sparse arrays.

Keywords : Sparse matrices, correspondence analysis, textual data.

Numerous fields of investigation involve the acquisition or the construction of sparse matrices. These matrices are generally large, and their description through eigen-analyses or classification implies the use of specific procedures. The purpose of this paper is to present one of these procedures, permitting the treatment of textual data.

The information provided by the set of responses to open ended questions in socio-economic surveys leads to such sparse arrays :

The general cell entry of the matrix Z, z_{ij}, takes the value 1 if the individual i has used the word j in his response, and the value 0 otherwise.
The data set Z we are dealing with contains a low percentage of non-zero elements, since the number of utilized words in such surveys may easily reach the value 3000 (the number of words used at least three time in the sample being, for instance, 600) whereas a specific response contains a relatively small number of words (up to 60 in practice).

The information in a sparse vector z_i' such as :

0000100011000001100100000110001000000000000010

can be coded by the sequence of indices :

5, 9, 10, 16, 17, 20, 26, 27, 31, 43

which describes the position of the "1" in the binary vector.
The computational procedure does not alter the structure of the data (avoiding the storage of a moment product matrix in the central core of the computer).
It uses the indices instead of the real elements of the matrix Z, hence a smaller amount of calculations.
In case of textual data, a preliminary program performs directly the transformation of the words (sequences of characters between blanks, commas, points,...)

into indices. Further information about this step will be given later on in Part II.

Part I : Correspondence analysis (C.A.) of sparse matrices :

We shall focus here on one method, correspondence analysis (also named reciprocal averaging, dual scaling) : the principles and properties of this technique are supposed to be known by the reader.

Z denotes the (n,p) matrix of data (n individuals or cases, p items)

R denotes the (n,q) matrix of indices : r_{ij} is the index of the jth "1" in the row i, q is the maximum number of items chosen by an individual.

The matrix Z is taken into account in the theoretical formulas, whereas R is used by the algorithms.

Let us denote by D (resp M) the (n,n) diagonal matrix (resp the (p,p) diagonal matrix) the diagonal elements of which are the sums of the rows (resp. the columns) of the matrix Z.

We have : $d_{ii} = \sum_j z_{ij}$

$m_{jj} = \sum_i z_{ij}$

The factors issued from C.A. are the eigenvectors of the matrix A

$$A = M^{-1}Z'D^{-1}Z \qquad (1)$$

If z_i' denotes the ith row of the matrix Z,

A can be written as a weighted sum of matrices A_i of rank 1 :

$$A = M^{-1}\sum_i A_i \qquad (2)$$

$$\text{with } A_i = \frac{1}{d_{ii}} z_i z_i'$$

The post multiplication of A_i by a vector u consists of a scalar product $z_i'u = \sum_{k=1}^{q} u_{r_k}$ (followed by the multiplication of the non-zero component of z_i by this quantity) and therefore implies a very small number of operations.

Determination of a quasi-invariant subspace :

The iterated power algorithm leads to a low convergence rate and a poor accuracy in the case of close eigenvalues, which occurs frequently when performing C.A. of sparse matrices. The rate is even worse in the case of reciprocal averaging algorithm, which utilizes the decomposition of the matrix A into two probabilistic transitions :

$$(D^{-1}Z) \text{ and } (M^{-1}Z')$$

We shall confine ourselves to the search of a t-dimensional subspace containing the s first eigenvectors of the matrix A ($s \le t$), instead of directly trying to estimate these eigenvectors.

We shall then project the set of n profile-points onto this subspace, and diagonalize the corresponding (t,t) symmetric moment-product matrix to produce at least s well defined eigenvalues and eigenvectors.

Let us denote by H a (t,t) matrix with orthonormalized columns ($H'H = I_t$) which span a subspace v(H) containing the s first eigenvectors of A.
From the equation $A u_\alpha = \lambda_\alpha u_\alpha \qquad (\alpha = 1, \ldots, t)$
we can deduce the set of relationships :

$$u_\alpha = H v_\alpha$$
$$H'AHv_\alpha = \lambda_\alpha v_\alpha \qquad (\alpha=1, \ldots, t)$$

The (t,t) matrix B = H'AH has yet the same largest eigenvalue as A (a projection being a shrinking transformation, the (t - s) remaining eigenvalues can only be smaller...).
The determination of v(H) is conducted through simultaneous iteration with two acceleration devices : stochastic approximation and shift of origin. The different step of the procedure can be summarized as follows :
(The content of each step will be developed below)

1) First estimate H_1 of the (p,t) matrix H through stochastic approximation
2) Improvement of H_1 through k simultaneous iteration with shift of origin (producing H_2)
3) First estimate of the s dominant eigenvectors and eigenvalues after projection onto $v(H_2)$ (diagonalization using a classical in-core algorithm)
4) A supplementary simultaneous iteration is performed
5) Second estimate of the s eigenvectors, and computations of the cosines of their angles with the previous estimates.

Two parameters are used to control the process : t, dimension of the subspace v(H), and k, the number of iterations during the step 2.
Usual values giving satisfactory results are : t = s + 5 and k = 10.
We shall begin with the description of the step 2. It will be easier to deal with the principles of the step 1 afterwards (as a matter of fact, the step 1 is optional).

Step 2 : This step consist of an adaptation of classical simultaneous iteration (HOTELLING (1933), RUTISHAUSER (1969), WILKINSON (1965), CLINT et al. (1970)) to a row by row procedure.
An iteration spans the n rows of the reduced matrix R. The set of trial vectors will be the columns of a (p,t) matrix U (U is derived from H_1 by a GRAM-SCHMIDT M^{-1}-Orthonormalization).
The decomposition (2) of the matrix A leads to the statement, where the index i take the value 1 to n, for the iteration k :

$$U_k^{(i)} \leftarrow U_k^{(i)} + A_i U_{k-1} \qquad (3)$$

At the end of the iteration, U_k is premultiplied by the matrix M^{-1}.
The statement (3) is the translation of the formula :

$$U_k = M^{-1}(\sum_i A_i)U_{k-1}$$

The rate of convergence will be in fact governed by $|\lambda_{t+1}/\lambda_s|$. This ratio can be much smaller if we introduce a suitable shift of origin.
This ratio become $|\lambda_{t+1}-\beta/\lambda_s-\beta|$.
(The transformation is obtained by replacing U_k by $U_k - \beta U_{k-1}$ at the end of each iteration). Without a priori knowledge, a commonly adopted value is β = trace A/2p (Half the average eigenvalue).
Since we do not intend to directly estimate the eigenvalues, it is not necessary to orthonormalize U_k after each iteration. This operation which has a low cost is however recommended to reduce the effects of rounding errors.

Step 1 : It would be possible to start the whole process by performing a simultaneous iteration analogous of (3) :
For i = 1 to n : $U_1^{(i)} \leftarrow U_1^{(i)} + A_i U_0$ (4)

We are starting with an arbitrary matrix U_0, and one feels it is a waste of time to keep U_0 all along the first reading of the data file, without taking into account the contribution of each case i to the determination of the subspace H.
We can change the statement (4) into a statement (5) which reads as follows :

$$U_1^{(i)} \leftarrow U_1^{(i-1)} + f(i)\ A_i U_1^{(i-1)}$$

(with $U_1^{(0)} = U_0$) (5)

The operator $\sum_i A_i$ in (4) is now replaced by $\prod_i (I+f(i)A_i)$.
It has be shown (BENZECRI (1969), LEBART (1974)) that such procedure converges and leads to an estimate of the first eigenvectors in the case of an infinite series of values of the index i (which can be obtained by repeating the reading of the array Z, or equivalently R) under the following conditions :

The series f(i) must diverge, whereas the series $f^2(i)$ must converge.
The asymptotic efficiency of this algorithm being weak, it will be used here as a mere device of acceleration, giving a quick and coarse approximation of the subspace H. The series f(i) most commonly used is the harmonic series.

Step 3 : The (t,t) covariance matrix V of the projections onto v(H) is also computed in one reading of the matrix R row by row, using the formula :
(H being the result of a M^{-1} -Orthonormalization performed on H_2)

$$V = \sum_i 1/d_{ii}\ .\ H'\ z_i z_i'\ H$$

It is not necessary to center the projections, because we take advantage of the orthonormalization to eliminate the trivial factor corresponding to the eigenvalue 1. The projector H therefore operates in a direction perpendicular to the first bissector of the axes.

The diagonalization of V is done by a classical procedure (Householder, Jacobi). If v_α denotes the eigenvector of V corresponding to the eigenvalue λ_α , we deduce the original eigenvector u_α from the relation : $u_\alpha = H\, v_\alpha$.

Step 4 and 5 : These steps allow the user to control the quality of the estimates, and, if necessary, to carry out a new set of iterations.

Numerical example :

The following numerical results relate to the example of application developed in Part II :

The "theoretical" array of data Z has n = 500 rows and p = 144 columns. The reduced array of index has only q = 13 columns (the maximum number of non-zero elements for a row is less than 14).
The size of Z makes possible a comparison with a direct correspondence analysis of Z (cf. column : "exact eigenvalues").
The control parameters are :

s = 5 (5 factors)
t = s+4 = 9 (H will have 9 dimensions)
k = 10 (10 iterations)

Axis	"Exact eigenvalues"	"Appr.eigenvalues" (k = 11)	"Cosines between iterations (10 and 11)
1	0.5226	0.5225	0.99995
2	0.5058	0.5054	0.99989
3	0.4957	0.4934	0.9983
4	0.4883	0.4825	0.9974
5	0.4736	0.4727	0.9994

The space occupied in central core is 21.000 words for the exact analysis of Z, and 2980 for the iterated algorithm using the reduced array R.
The computer times are equivalent (18 sec. on I.B.M 370-168), but the approximate algorithm is faster when the number of column p increases, the number of operation being approximately a linear function of p.

The graphical displays produced by the printer are identical with both algorithms for the first two axes (see figure 1 below), and contain only very slight differences for the remaining axes.
Of course, a higher level of accuracy may be attained with larger values of t and k.

Part II : Analysis of textual data

The method described above will provide a valuable help for the processing of certain files of textual data, especially when these files contain numerous short texts. A concrete example is given by the set of responses to an open ended question in a socio-economic survey. The aim of the works presented here was precisely to settle an automatic treatment of such data. Since only textual data relating to french surveys were available, we have used instead of real responses a rather artificial example : 500 titles of papers drawn from nine statistical reviews :

1.journal of stat. Comp. and Simul. ; 2.Applied statistics ; 3.J.A.S.A. ; 4.J. Royal. Stat. Soc. (B) ; 5.J.Royal. Stat. Soc. (A) ; 6. The statistician ; 7.Biometrika ; 8.Psychometrika ; 9.Biometrics.

The analysis can be divided into four main phases :

Phase 1 : Conversion of the words into indices

Table 1 : First rows of the input data set (the dotted line indicates the end of a title)

```
SMALL SAMPLE COMPARISON OF ESTIMATION METHODS FOR THE BETA DISTRIBUTION
----
ESTIMATORS FOR THE CORRELATION COEFFICIENT IN A BIVARIATE EXPONENTIAL
DISTRIBUTION
----
THE GENERATION OF PIECEWISE LINEAR APPROXIMATIONS OF PROBABILITY
DISTRIBUTION FUNCTIONS
----
CONCENTRATION BANDS FOR UNIFORMITY PLOTS
----
THE GENERATION OF PSEUDO RANDOM CORRELATION MATRICES
----
RANDOM SEQUENTIAL PACKING IN R
----
ESTIMATING SWITCHING REGRESSIONS : A COMPUTATIONAL NOTE
----
WEIGHTED ESTIMATORS IN REGRESSION WITH MULTICOLLINEARITY
----
ROBUST ESTIMATION OF STANDARD DEVIATION
----
ESTIMATION OF SEEMINGLY UNRELATED REGRESSION WITH LAGGED
DEPENDENT VARIABLES AND AUTOCORRELATED ERRORS
----
FORWARD AND BACKWARD STEPPING IN VARIABLE SELECTION
----
A METHOD FOR CONSTRUCTING EXACT TESTS FROM TEST STATISTICS THAT HAVE UNKNOWN
NULL DISTRIBUTIONS
----
ON GROUPING SETS OF EXPONENTIAL POPULATIONS
----
```

From the input data as shown on table 1, a machine independent program (see for instance LEBART (1981)) can derive a sorting of all the utilized words (ranked in decreasing order of frequencies), select the w most frequent words (w is a parameter), ascribe an index value to each word, and eventually construct the reduced matrix R. We have here 144 words appearing at least five times. Among the selected words, only 13 are used simultaneously in a title. Hence, R will be a (500,13) matrix.

Figure 1 : Visualization of the associations between words within the titles. (Plane of the first two axes). The 500 titles are replaced by the centroids of the 9 corresponding reviews (supplementary nominal elements).

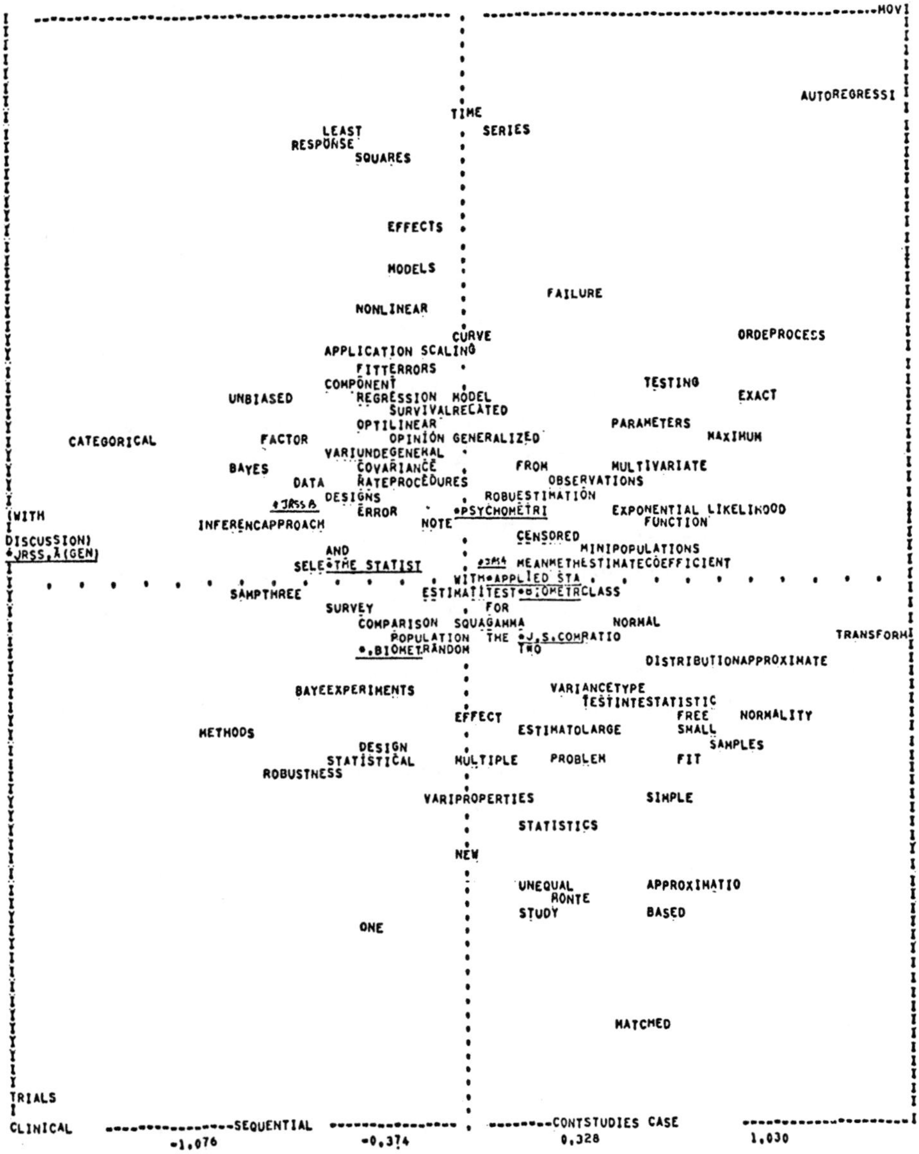

Phase 2 : Visualization of the sparse matrix (500,144) Z, using R
(cf. Numerical example of Part 1)

This phase leads to several diagrams analogous of figure 1. The listing of the coordinates of words and papers on the first axes is also obtained.
As it is always the case when analysing binary data, the percentages of variance explained by the first axes are very low (resp. : 1.83 %, 1,77 %, 1,74 %,...). They are nevertheless significant. It would be pessimistic to interpret them as "percentage of information". Well known counter-examples (matrices associated with graphs) show that the whole initial structure can be restored by axes associated with low percentages of variance.

Phase 3 : Projection of supplementary nominal data

In a sample survey, the individuals are described by numerous nominal variables such as sex, occupational groups, type of housing, etc. These variables can be represented on the same diagram as the words.
If φ_{α}^{i} denotes the coordinate of the individual i (here : the paper) on the axis α corresponding to the eigenvalue λ_{α} , a supplementary modality is a centroid of the concerned individuals, dilated by the coefficient $1/\lambda_{\alpha}$.
The only nominal data available in this example is the review. The nine corresponding modalities are plotted on figure 1.

Phase 4 : Pointing out modal responses (modal titles)

A classical weakness of lexical statistics is their inability to take into account the contexts of the words, which determine their real meaning. In such analysis, the complete responses, being the rows of Z, could be plotted on the same displays as the words (simultaneous representation). It is easier to print out a selection of responses which characterizes each modality of a nominal variable.
The aggregation of the rows of Z corresponding to a particular modality leads to the definition of an average lexical profile of the modality. The "modal responses" will be the nearest responses of this average profile (obtained from a ranking of the chi-square distances between each row and the average profile in the space R^{p} (here : R^{144}).
We get for example the following modal titles :

Journal of Stat. Comp. and Simulation :
"Monte-Carlo estimates of the distribution of the randoms polygones of the VORONOI tessellation with respect to a poisson process".

Applied Statistics :
"Fitting two-parameters discrete distribution to many data sets with one common parameter".

Psychometrika :
"A model and algorithm for multidimensionnal scaling with external constraints on the distances".

Conclusions :
With a less artificial example, the reader could probably better appreciate the interest of the different phases. From the responses to an open ended question in a sample survey, we can automatically produce, without preliminary coding or manual treatment :

- a sorting of the utilized words, with their frequencies,
- an overview of the association between words, between responses, between these elements and all the available nominal characteristics of the interviewed persons,
- a selection of responses which characterize each group or categories, enabling us to replace the words in their real context, without being submerged by redundant information.

  The computational procedures permit to carry out the different phases on medium - sized computer, without severe limitation upon the size of the input data file.

REFERENCES

Part I :

BENZECRI J.P. (1969). "Approximation stochastique dans une algèbre normée non commutative". Bulletin de la société mathématique de France. N° 97, pp. 225-241.

CLINT M., JENNINGS A. (1970). "The evaluation of eigenvalues and eigenvectors of real symmetric matrices by simultaneous iteration". Computer journal 13. pp. 76-80.

HOTELLING H. (1933). "Analysis of a complex of statistical variables into principal components. J. Educ. Psychol. 24. pp. 417-441 and pp. 498-520.

LEBART L. (1974). "On the BENZECRI's method for finding eigenvectors by stochastic approximation". Proc. in computational statistics. Physica Verlag. WIEN (AUSTRIA).

RUTISHAUSER H. (1969). "Computational aspects of F.L. BAUER's simultaneous iteration method". Numerische Mathematik N°13. pp. 4-13.

SCHWARZ M.R., RUTISHAUSER H., STIEFEL (1973). "Numerical analysis of symmetric matrices". Prentice Hall.

WILKINSON J.H. (1965). "The algebraic eigenvalue problem". Oxford University Press.

Part II :

BENZECRI J.P. et Coll. (1980). "Linguistique et Lexicologie". Pratique de l'analyse des données : Tome 3. Dunod.

LEBART L. (1981). "Une procédure d'analyse lexicale écrite en langage FORTRAN". Les cahiers de l'Analyse des Données. Vol. VI. N° 2. pp. 229-243. Dunod.

LEBART L. (1982). "L'analyse statistique des réponses libres dans les enquêtes socio-économiques". Consommation - Revue de Socio-Economie. N°1. pp. 39-62. Dunod.

Nonlinear Principal Component Analysis

J.d. Leeuw, University of Leiden, The Netherlands

SUMMARY

Two quite different forms of nonlinear principal component analysis have been proposed in the literature. The first one is associated with the names of Guttman, Burt, Hayashi, Benzécri, McDonald, De Leeuw, Hill, Nishisato. We call it multiple correspondence analysis. The second form has been discussed by Kruskal, Shepard, Roskam, Takane, Young, De Leeuw, Winsberg, Ramsay. We call it nonmetric principal component analysis. The two forms have been related and combined, both geometrically and computationally, by Albert Gifi. In this paper we discuss the relationships in more detail, and propose an alternative algorithm for nonlinear principal component analysis which combines features of both previous approaches.

KEYWORDS

correspondence analysis, nonmetric scaling, multidimensional scaling, eigenvalues, principal component analysis

1: Introduction

We suppose that the reader is familiar with multiple correspondence analysis and with nonmetric principal component analysis. For multiple correspondence analysis we refer to Cazes a.o. (1977) and to Lebart a.o. (1977) or to Hill (1974) and Nishisato (1980). We only discuss a comparatively simple case, the many interesting generalizations developed especially in Toulouse could in principle also be fitted into our framework. For nonmetric principal component analysis we refer to Kruskal and Shepard (1976) or Young a.o. (1978). Again we study a simple special case of this technique. For previous attempts to integrate the two approaches we refer to De Leeuw and Van Rijckevorsel (1980), De Leeuw a.o. (1981), and Gifi (1981).

2: Nonmetric principal component analysis

Suppose $L_1,\dots,L_m$ are closed subspaces of a separable Hilbert space H, with inner product $<.,.>$, norm $|.|$, and unit sphere S. For each choice of elements $y_j \in L_jS$, the intersection of L_j and S, we can compute the matrix $R(y_1,\dots,y_m)$ with elements $r_{j\ell}(y_1,\dots,y_m) = <y_j,y_\ell>$. This matrix is a *correlation matrix*, in the sense that it is positive semi-definite and has diagonal elements equal to unity. The problem of *nonmetric principal component analysis* (NCA) is to find $y_j \in L_jS$ in such a way that the sum of the p largest *eigenvalues* of the matrix $R(y_1,\dots,y_m)$ is maximized (or, equivalently, the sum of the m - p smallest eigenvalues is minimized). Observe that for different choices of p this defines a different problem. In some cases we shall not only be interested in solutions that maximize our criterion, but we shall be interested in all solutions of the stationary equations corresponding with the maximization problem.

It is clear that NCA generalizes ordinary principal component analysis, in which the subspaces L_j are one-dimensional. Our formulas deal with a somewhat restricted form of NCA, because most of the literature we have mentioned treats the more general case in which L_j is a closed convex cone. Tenenhaus (1982) even discusses the case of a possibly infinite number of convex cones. There is another serious restriction of generality in most of our paper: we suppose that all L_j are finite-dimensional. For notational convenience we shall even suppose that $\dim(L_j)$ is the same for all j, but this last assumption is in no way essential. We use orthonormal bases for each of the L_j, and collect them in the $n \times q$ matrices G_j. Here q is the common dimensionality of the L_j, and n is the dimensionality of H. We do not necessarily assume that n is finite. If n is not finite, then G_j is simply an ordered q-tuple of elements of H, and all 'matrix operations' we use have their obvious interpretations.

Thus $y_j = G_j\alpha_j$ is in L_jS if and only if the q-vector α_j satisfies $\alpha_j'\alpha_j = 1$. Then $r_{j\ell}(y_1,\dots,y_m) = \alpha_j'C_{j\ell}\alpha_\ell$, where $C_{j\ell}$ is the q-matrix defined by $C_{j\ell} = G_j'G_\ell$. Observe that for all j it is true that $C_{jj} = I$, the identity of order q. It is also convenient to define the $n \times mq$ supermatrix $G = (G_1|\dots|G_m)$ and the $mq \times mq$ supermatrix $C = G'G$. The supermatrix C is sometimes called the *Burt table* of the NCA problem, observe that it depends on the choice of the bases.

A useful operation for our purposes is the *direct sum* of a number of matrices (cf MacDuffee, 1946, p. 81). If X and Y are matrices of dimension $a \times b$ and $c \times d$, then the direct sum $X \dotplus Y$ is the $(a + c) \times (b + d)$ matrix $\begin{vmatrix} X & 0 \\ 0 & Y \end{vmatrix}$. The extension to direct sums of more than two matrices is obvious. We now define A as the $mq \times m$ matrix $\alpha_1 \dotplus \dots \dotplus \alpha_m$. In this notation $R(y_1,\dots,y_m) = A'CA$, provided that $A'A = I$.

It follows that $\sigma_p(y_1,\dots,y_m)$, the sum of the p largest eigenvalues of the correlation matrix $R(y_1,\dots,y_m)$, has the representation

$$\sigma_p(y_1,\dots,y_m) = \max\ \{\mathrm{tr}\ T'A'CAT\}, \tag{1}$$

with T varying over the $m \times p$ matrices satisfying $T'T = I$. Thus maximization of $\sigma_p(y_1,\dots,y_m)$ over $y_j \in L_jS$ amounts to the same thing as maximization of $\mathrm{tr}\ T'A'CAT$ over all $m \times p$ matrices T satisfying $T'T = I$ *and* all $mq \times m$ matrices A of the form $A = \alpha_1 \dotplus \dots \dotplus \alpha_m$ also satisfying $A'A = I$. If A and T satisfy these restrictions, and U is the $mq \times p$ matrix AT, then $U'U = I$, and U consists of m submatrices U_j, of dimension $q \times p$, and of the form $U_j = \alpha_j t_j'$, with t_j' row j of T. Thus U is *blockwise of rank one*, each subspace L_j defines a block.

We can now define

$$\sigma_p(L_1,\dots,L_m) = \max\ \{\mathrm{tr}\ U'CU\}, \tag{2}$$

with U varying over all orthonormal mq × p matrices which are blockwise of rank one. The NCA problem is to compute $\sigma_p(L_1,\ldots,L_m)$, and to find the maximizer U of the required form.

We now derive the stationary equations of the NCA problem. From (1) we must have

$$RT = T\Omega, \tag{3}$$

with Ω a symmetric p-matrix. For identification purposes we can actually assume that Ω is diagonal. Also remember that $R = A'CA$. Differentiation of (1) with respect to the α_j gives the equations

$$\sum_{\ell=1}^{m} \gamma_{j\ell}C_{j\ell}\alpha_\ell = \theta_j\alpha_j, \tag{4}$$

where $\Gamma = \{\gamma_{j\ell}\}$ is defined by $\Gamma = TT'$, where the θ_j are undetermined multipliers, and where the α_j satisfy $\alpha_j'\alpha_j = 1$. It follows from (3) and (4) that at a solution $\Theta = \text{diag}(T\Omega T')$, and thus $\text{tr}(\Theta) = \text{tr}(\Omega)$. Most algorithms for NCA are based on alternating solution of (3) for T, given A, and solution of (4) for A, given T.

Although (3) and (4) can be used to construct convergent algorithms for finding stationary points (cf Gifi, 1981) they give little insight into the mathematical structure of the NCA problem. It is not at all clear, for example, if (3) and (4) have more than one solution, and if so, how these different solutions are related. There is one fortunate exception. If p = 1, i.e. if we choose $y_j \in L_jS$ in such a way that the largest eigenvalue of $R(y_1,\ldots,y_m)$ is maximized, then the requirement that U is blockwise of rank one is no restriction at all, and thus the problem is maximizing $u'Cu$ over $u'u = 1$, where we have written u in stead of U, because U is of dimensions mq × 1. The solutions to the NCA problem with p = 1 are consequently the eigenvectors of C. A little reflection shows that p = m - 1, i.e. the solution for $y_j \in L_jS$ minimizing the smallest eigevalue of $R(y_1,\ldots,y_m)$, gives the identical result. If v is an eigenvector of C, $v'v = 1$, eigenvalue μ, then we can partition it into blocks v_j, each with q elements. If a block is nonzero we set $\alpha_j = v_j/(v_j'v_j)^{\frac{1}{2}}$, if a block is zero α_j is an arbitrary unit length vector. Moreover $t_j = (v_j'v_j)^{\frac{1}{2}}$, i.e. $u_j = v_j$. Remember that in this case T is m × 1, its elements are written simply as t_j. Also $\theta_j = \mu v_j'v_j$ and $\omega = \mu$. It is of some interest to observe that μ is not always the largest eigenvalue of the corresponding matrix R with elements $r_{j\ell} = \alpha_jC_{j\ell}\alpha_\ell$. More precisely: if μ is the largest eigenvalue of C then $\omega = \mu$ is also the largest eigenvalue of the corresponding R, and the same thing is true for the smallest eigenvalue of C. But no such thing is true for the intermediate eigenvalues.

3: Multiple correspondence analysis

The last paragraph brings us naturally to *multiple correspondence analysis* (MCA). This is defined simply as the computation of some or all of the eigenvalues and eigenvectors of C. Accordingly each solution of

$$Cv = \mu v \tag{5}$$

will be called an MCA solution. In a more restricted sense we can also call the problem of computing

$$\rho_p(L_1,\ldots,L_m) = \max\ \{\mathrm{tr}\ V'CV\}, \tag{6}$$

with V varying over all orthonormal mq × p matrices the MCA problem. This makes it more similar to the NCA problem. The main difference between the two, from this point of view, is that the MCA problem is *nested*, i.e. the solution for p - 1 are the first p - 1 dimensions of the solution for p. Although each choice of p defines a different MCA problem the various solutions are closely related. In fact they can all be constructed from the mq solutions of (5). Another obvious difference between NCA and MCA is that computing (6) gives p different elements of each L_jS, and thus p different correlation matrices. Each of these correlation matrices has at least one eigenvalue equal to the corresponding μ of (5), and usually only one. Computing (2) on the other hand gives only one solution for the y_j, thus only one correlation matrix, of which the p largest eigenvalues are equal to $\Omega = U'CU = T'RT$. In Gifi's terminology there is *multiple selection* from L_jS in MCA, there is *single selection* from L_jS in NCA (for *selection* Gifi also uses *transformation* or *quantification*, depending on the context, cf infra).

Before we study the relations between NCA and MCA in more detail, we must make one important comment. Interpreting NCA and MCA as generalizations of ordinary principal component analysis is only one possibility. We then use the fact that these techniques can be interpreted as optimizing some property of the correlation matrix before performing the actual principal component analysis. This particular interpretation is natural in some contexts, for example if L_j is defined as a subspace of possible transformations or quantifications of an element of H. Sometimes, however, the notion of quantification or transformation does not make much sense, because the variable is purely nominal and a one-dimensional arrangement of its possible values is somewhat far-fetched. In these cases it is usually more sensible to interpret NCA and MCA as multidimensional scaling techniques which map objects or individuals into low-dimensional space in such a way that some criterion defined in terms of distance and/or separation is optimized. Such interpretation are possible and fruitful, they are discussed in detail by Gifi (1981) and Heiser (1981).

In the previous section we have already one relationship between NCA and MCA. If p = 1 then NCA and MCA are the same, basically because mq × 1 matrices are always blockwise of rank one. If p > 1 then comparing (2) and (6) shows directly that $\rho_p(L_1,\ldots,L_m) \geqslant \sigma_p(L_1,\ldots,L_m)$ with equality if and only if the eigenvectors corresponding with the p largest eigenvalues of C are blockwise of rank one. In the next section we shall try to find out if this condition for equality is ever likely to be met in practice. We shall also discuss another problem connected with MCA. The mq solutions to (5) give mq correlation matrices, and thus mq principal component analyses. Gifi calls this *data production* in order to distinguish it from *data reduction*. A pertinent question is if the mq correlation matrices are related in some simple way, so that it is easy to see where the redundancy in this representation is.

4: Relationships in an important special case

Suppose that there exist α_j and $r_{j\ell}$ such that $\alpha_j'\alpha_j = 1$ for all j and

$$C_{j\ell}\alpha_\ell = r_{j\ell}\alpha_j \tag{7}$$

for all j,ℓ. Suppose moreover that T is of order m, satisfies $TT' = T'T = I$, and $RT = T\Omega$, with Ω of order m and diagonal. Thus (7) is the critical assumption, it defines α_j and $r_{j\ell}$, T is then defined simply as the matrix of eigenvectors of R. Now define $U = AT$, with $A = \alpha_1 \dotplus \ldots \dotplus \alpha_m$. By using (7) it is now simple to verify that $CU = U\Omega$. Thus U constructed in this way defines m solutions to (5), i.e. to MCA or to NCA with p = 1. The m solutions to (5) moreover induce the same correlation matrix R. If we select p columns from T, then these p columns obviously satisfy (3). We can also verify that they satisfy (4), if taken together with the α_j, and that consequently we can construct $\binom{m}{p}$ solutions to the NCA equations (3) and (4) if (7) can be satisfied. All these NCA solutions have the same R.

Now suppose v is another solution of (5), not one of the m solutions constructed by using the α_j of (7). By orthogonality $\sum v_j'U_j = \sum v_j'\alpha_j t_j' = 0$, which is possible only if $v_j'\alpha_j = 0$ for all j. Thus, in the terminology suggested by Dauxois and Pousse (1976), v is not only *weakly* but actually *strongly orthogonal* to the m columns of U. In particular this implies that if there is a second solution of (7), i.e. if $C_{j\ell}\beta_\ell = s_{j\ell}\beta_j$ for all j,ℓ, then $\alpha_j'\beta_j = 0$ for all j. Such a second solution of (7) again defines m solutions to (5) and $\binom{m}{p}$ solutions to (3) and (4). In total, of course, (7) can only have q solutions, all strongly orthogonal, each of them corresponding with m solutions of (5), and each of them having a single induced correlation matrix R. If we find q solutions to (7) then we have found mq, and consequently all, solutions to (5). We have also found $q\binom{m}{p}$ solutions to (3) and (4), but this need not be all possible solutions.

We now investigate if (7) is a realistic condition, with interesting interpretations. In the first place we repeat (7) in words: it says that if two of the matrices $C_{j\ell}$ have a subscript in common, then they have a singular vector in common. Another interpretation of (7) is also quite useful. Suppose α_j and $r_{j\ell}$ satisfy (7) and define $y_j = G_j\alpha_j$. Then (7) says that the projection of y_ℓ on L_j is equal to the projection of y_ℓ on the one-dimensional subspace through y_j. In the next section this interpretation will be related to linearity of the regression between two random variables.

A condition stronger than (7) is that there exist m matrices K_j, orthonormal, of order q, and diagonal matrices $D_{j\ell}$, such that
$C_{j\ell}K_\ell = K_jD_{j\ell}$. (8)
This is equivalent to the condition that (7) has q different solutions, it says that if two matrices $C_{j\ell}$ have a subscript in common then they must have all their singular vectors in common. By using familiar results this can be reduced to the condition that several matrices must commute (cf Bellman, 1960, p. 56), but in itself this commutation result does not give much insight. We know already that if (8) is true then we can construct all mq solution to (5). We now show how this can be done systematically.

Suppose (8) is satisfied. Let $K = K_1 \dotplus \ldots \dotplus K_m$. Then K'CK has submatrices $D_{j\ell} = K_j'C_{j\ell}K_\ell$, and as we know these submatrices are all diagonal. Moreover they are all of order q, there are m^2 of these $D_{j\ell}$, they all have q nonzero elements at most. It is now possible to construct a permutation matrix P such that P'DP = P'K'CKP = E is of the form $E = E_1 \dotplus \ldots \dotplus E_q$, where each of the E_t is a symmetric matrix of order m. Thus D consists of m^2 matrices, each with at most q nonzero elements, and E consists of q matrices, each with at most m^2 nonzero elements. Element j,ℓ of matrix E_t is equal to element t,t of matrix $D_{j\ell}$. As the next step in our solving of (5) we construct $L = L_1 \dotplus \ldots \dotplus L_q$, with each L_t orthonormal of order m, such that L_t diagonalizes E_t. Thus F = L'EL = L'P'DPL = L'P'K'CKPL is of the form $F = F_1 \dotplus \ldots \dotplus F_q$, with all of the F_t diagonal. It follows that KPL diagonalizes C, thus KPL contains the eigenvectors of C, and the matrices F_t contain eigenvalues *in some order*. If we look somewhat closer to KPL, we find that it consists of mq submatrices, all of order q × m, and all of rank one. Conversely this particular block structure of the eigenvectors of C, together with the rank one property of the blocks, implies that (8) is satisfied. It also implies that there is no data production, there are only q different induced correlation matrices, not mq. We also know that this block structure makes it possible to find $q\binom{m}{p}$ solutions to (3) and (4). It is a useful exercise to find out what happens to the results of this section if the L_j have different dimensionalities, making some $C_{j\ell}$ rectangular in stead of square.

5: Even more special cases

We now apply the three-step or KPL-diagonalization in some cases in which we can easily show that it works. If m = 2 we use the singular value decomposition $C_{12} = K_1D_{12}K_2'$. Here D_{12} is diagonal of order q, with singular values on the diagonal. Thus the E_t are correlation matrices of order two, each E_t has one singular value as off-diagonal element. The F_t are diagonal of order two, the two elements are one plus the singular value and one minus the singular value. This is the familiar result linking multiple correspondence analysis with two variables with ordinary correspondence analysis of the cross table of the two variables. We emphasize that in ordinary correspondence analysis we often use a very special basis of indicator functions or step functions. This basis has some attractive properties, we mention for example the centroid principle of Benzécri, and we mention the fact that (7) can be interpreted as the assumption that all bivariate regressions can be linearized by suitable choice of the $y_j \in L_jS$. The stronger assumption (8) means that all bivariate distributions can be diagonalized simultaneously.

Now consider the case q = 2 with a basis of normalized indicators for each of the L_j. Thus $C_{j\ell}$ is a 2 × 2 table with bivariate frequencies, divided by square root of the product of the univariate marginals. These $C_{j\ell}$ satisfy (8), one column of K_j is proportional to the square roots of the marginals, the other columns must be orthogonal to it and is consequently also determined completely by the marginals. It follows that $E = E_1 + E_2$, where E_1 is of order m and has all elements equal to one and where E_2 is likewise of order m and has as its elements the matrix of phi-coefficients or point correlations of the binary variables. Thus F_1 has one element on the diagonal equal to m and m - 1 diagonal elements equal to zero. The elements of F_2 are the eigenvalues of the matrix of phi-coefficients. This is the familiar result that multiple correspondence analysis or nonmetric principal component analysis of binary variables is the same thing as ordinary component analysis of phi-coefficients. A matrix E_1, of order m, with all elements +1, occurs whenever we use normalized indicators as a basis. Thus there is always a trivial eigenvalue equal to m and m - 1 trivial eigenvalues equal to zero in this case.

As a final special case suppose we have m standard normal variables, and suppose L_j is the linear space of all polynomial transformations of degree not exceeding q - 1. As a basis of each of the L_j we take the Hermite-Chebyshev polynomials of degree 0,...,q - 1. It is well known that transform s of variable j and transform t of variable ℓ are uncorrelated (orthogonal) if $s \neq t$. If s = t then their correlation is $\rho_{j\ell}^s$, where $\rho_{j\ell}$ is the correlation in the original multinormal distribution. Thus (8) is satisfied, E_1 has all elements equal to one, E_2 has elements $\rho_{j\ell}$, E_3 has elements $\rho_{j\ell}^2$, and so on. The eigenvalues of the MCA problem are those of E_1 and those of E_2 and so on. The largest nontrivial one is the largest eigenvalue of E_2, the smallest nontrivial one is the smallest of E_2 (Styan, 1973), but the order of the others is undecided. In fact the second largest MCA eigenvalue can be the largest of E_3, in which case all transformations on the second dimension are quadratic functions of the first (horse-shoe or Guttman effect). But the second largest MCA eigenvalue can also be the second largest eigenvalue of E_2, the second set of transformations is then linear with the first. For the $q\binom{m}{2}$ NCA solutions we can compute from this representation the transformations in two dimensions are both linear, or both quadratic, and so on. NCA rank one restrictions make sure that both or all p transformations come from the same E_t (cf Gifi, 1981, chapter 11). The multinormal example has been generalized in many directions by Lancaster and his pupils (Lancaster, 1969).

It is remarkable that in all three special cases condition (8) is satisfied, and KPL-diagonalization consequently works. Our results indicate that in general NCA solutions can easily be interpreted in a principal component analysis framework, the same thing is consequently true for MCA solutions with p = 1. But multidimensional MCA solutions are difficult to interpret in this framework if (8) is not approximately true. The rest of our paper studies if (8) is approximately true in a wide variety of examples.

6: An algorithm

In order to investigate in how far (8) is satisfied in practical examples, i.e. in how far KPL-diagonalization is possible we have written an APL-algorithm PREHOM. A FORTRAN-version is currently being developed. The technique is fairly simple: if $F = L'P'K'CKPL$, then we want to maximize the sum of squares of the diagonal elements of F on the condition that $L = L_1 \dot{+} \ldots \dot{+} L_q$ and $K = K_1 \dot{+} \ldots \dot{+} K_m$, where all matrices in the direct sums are square orthonormal. Observe that P is a fixed permutation matrix. For fixed K we maximize the criterion by choosing L_t as the eigenvectors of E_t, the maximum value is the sum of squares of the elements of E_t, which means that we maximize our criterion by maximizing the sum of squares of all diagonal elements of all $D_{j\ell} = K_j'C_{j\ell}K_\ell$. This is done by a straightforward generalization of the Jacobi-procedure which cycles through all the relevant plane rotations. As an initial estimate of K_j we use the eigenvectors of $\sum C_{j\ell}C_{\ell j}$, summed over $\ell \neq j$. This initial estimate is usually very good, but for precise convergence the program usually needs another five to ten cycles of plane rotations.

We have analyzed a large number of examples with PREHOM, although APL-restrictions imply that our examples cannot be too large (C must have order less than 35). The conclusion of the examples is partly as we expect it to be: if the number of subspaces (variables) is small (three or four) then KPL-diagonalization works very well, and if the average number of categories (dimensionalities) is close to two the KPL-diagonalization also works well. Generally KPL-diagonalization seems to work best for ordinal or rating scale type variables, and much less well for purely nominal variables. It came as a surprise to us that even in the least favourable circumstances KPL-diagonalization still works quite well, and it can predict the first three MCA-solutions (largest eigenvalues) and the last three MCA-solutions (smallest eigenvalues) very well.

PREHOM can be used to predict MCA-results very well. Our experience (cf Gifi, 1981, ch 13) suggest that NCA is often very similar to MCA with p = 1, and thus PREHOM can be supposed also to approximate NCA very well. It follows from these two conclusions that in ordinal and rating scale examples NCA and MCA actually compute the same solutions, but in a different order. The NCA eigenvalues can be found in the list of MCA eigenvalues, but they are only very rarely the p largest ones. If we want to relate MCA and NCA in the same dataset we can most efficiently do this by using PREHOM or a similar program.

There are four actual examples in the tables on the next page. Example 1 are three variables with 3 + 4 + 3 categories, 100 observations, from Burt's classical MCA-paper. The variables are somewhat between nominal and ordinal. Example 2 are 6 variables with 2 + 5 + 3 + 2 + 5 + 2 categories, 25 observations, describing screws, nails, and tacks, taken from John Hartigan's book on cluster analysis. Most of the variables are clearly nominal. Example 3 has 5 ordinal variables with 2 + 4 + 5 + 4 + 6 categories, about 30000 observations, data from a school career survey by the Dutch Central Bureau of Statistics. Example 4 are 8 rating scales with three categories each, 110 observations, taken from the Dutch Parliament Survey. All four tables have four columns: in the first column the eigenvalues of the E_t are given (without the trivial ones), the second column has the same numbers but ordered, the third column has the actual MCA eigenvalues, and the fourth column has diagonal elements of V'KPL, cosines between actual and predicted MCA eigenvectors.

7: References

Bellman R. (1960), Introduction to matrix analysis, McGrawHill, New York.

Cazes P., Baumerder A., Bonnefous S., Pagès J.P. (1977), Codage et analyse des tableaux logiques, Cahiers du B.U.R.O., no 27.

Dauxois J., Pousse A. (1976), Les analyses factorielles en calcul des probabilites et en statistique, Thèse, Universite Paul-Sabatier, Toulouse.

De Leeuw J., Van Rijckevorsel J. (1980), HOMALS and PRINCALS. In Diday E. et al (eds), Data analysis and informatics, North Holland Publishing Company, Amsterdam.

De Leeuw J., Van Rijckevorsel J., Van der Wouden, H. (1981), Nonlinear principal component analysis using B-splines, Methods of Operations Research,23,211-234.

Gifi A. (1981), Nonlinear multivariate analysis, Department of Data Theory FSW, University of Leiden.

Heiser W. (1981), Unfolding analysis of proximity data, Department of Data Theory FSW, University of Leiden.

Hill M.O. (1974), Correspondence analysis: a neglected multivariate method, Applied Statistics, 23, 340-354.

Kruskal J.B., Shepard R.N. (1974), A nonmetric variety of linear factor analysis. Psychometrika, 39, 123-157.

Lancaster H.O. (1969), The chi-squared distribution. Wiley, New York.

Lebart L., Morineau A., Tabard N. (1977), Techniques de la description statistique Dunod, Paris.

McDuffee C.C. (1946), The theory of matrices, Chelsea, New York.

Nishisato S. (1980), Analysis of categorical data: dual scaling and its applications, University of Toronto Press, Toronto.

Styan G.P. (1973), Hadamard products and multivariate statistical analysis, Linear algebra and its applications,6,217-240.

Young F.W., Takane Y., De Leeuw J. (1978), The principal components of mixed measurement level multivariate data, Psychometrika, 43, 279-281.

2.61	2.61	2.61	1.00
0.39	1.90	1.90	1.00
0.00	1.42	1.43	1.00
1.90	0.78	0.78	0.99
0.78	0.58	0.58	0.99
0.33	0.39	0.49	0.78
1.42	0.33	0.22	0.77
0.58	0.00	0.00	0.99

table 1: Burt data

3.68	3.68	3.73	0.99
1.40	2.04	2.21	0.84
0.68	1.83	1.97	0.59
0.23	1.49	1.67	0.69
0.01	1.40	1.18	0.59
0.01	0.68	0.77	0.57
2.04	0.67	0.51	0.75
0.67	0.51	0.50	0.63
0.29	0.29	0.33	0.75
1.49	0.23	0.11	0.89
0.51	0.17	0.00	1.00
1.83	0.01	0.00	1.00
0.17	0.01	0.00	1.00

table 2: Hartigan data

2.61	2.61	2.62	1.00
1.01	1.99	1.99	1.00
0.84	1.84	1.84	1.00
0.41	1.01	1.03	0.71
0.13	1.01	1.02	0.52
1.99	1.00	1.00	0.61
1.00	1.00	1.00	0.91
0.75	1.00	1.00	0.54
0.26	0.99	0.98	0.66
1.84	0.98	0.97	0.68
1.00	0.84	0.83	0.96
0.98	0.75	0.75	0.99
0.18	0.41	0.41	0.97
1.01	0.26	0.27	0.90
0.99	0.18	0.19	0.92
1.00	0.13	0.10	0.87

table 3: CBS data

3.87	3.87	3.88	1.0
1.37	2.09	2.42	0.9
0.82	1.37	1.40	0.7
0.55	1.31	1.28	0.7
0.43	1.14	1.16	0.8
0.39	1.03	1.05	0.6
0.37	0.84	0.90	0.8
0.19	0.82	0.72	0.7
2.09	0.67	0.66	0.8
1.31	0.57	0.61	0.8
1.14	0.55	0.51	0.8
1.03	0.43	0.39	0.5
0.84	0.39	0.33	0.7
0.67	0.37	0.27	0.6
0.57	0.34	0.22	0.7
0.34	0.19	0.19	0.9

table 4: rating scales

A Programming Language for Multivariate Data Analysis

K. Momirović and **J. Štalec**, University of Zagreb, Yugoslavia
E. Zakrajšek, University of Ljubljana, Yugoslavia

A command language for multivariate data analysis, named SS, is designed and implemented on UNIVAC and CDC systems.

SS is organized round a unique kernel which enables simple performance and control of input and output operations, creating and maintenance of data bases, control of syntactical and logical correctness of commands, and control of performing the operations. SS performs this automatically, thus freeing the programmer of the greatest part of activities connected with the creation of programs for the problems which can be solved by linear transformations and by application of standard distribution functions.

SS contains these types of commands:
(1) INPUT-OUTPUT COMMANDS
(2) COMMANDS FOR MANIPULATION WITH DATA BASES
(3) CONTROL COMMANDS
(4) ALGEBRAIC COMMANDS
(5) COORDINATE SYSTEMS FORMING AND TRANSFORMATION COMMANDS
(6) STATISTICAL COMMANDS
(7) MACRO COMMANDS

All commands have identical syntactical form.
Programs written in SS are written in simple sequence of commands. These programs can, virtually, be of arbitrary length and complexity and may be organized in macro programs.

Key words: MULTIVARIATE DATA ANALYSIS, COMPUTER PROGRAMMING, GENERAL LINEAR MODEL, STATISTICAL SOFTWARE

1. INTRODUCTION

Due to the constantly increasing price of designing and maintenance of software and due to the need to, as fast as possible, implement new methods, models and algorithms, standard procedural programming lanugages are becoming less efficient for the design of programs for multivariate data analysis. This is particularly true because the semantic distance between these languages and symbolic systems in which algorithms for data analysis are written is too large for those computer scientists who are predominantly oriented toward models and methods and much less toward programming problems. For these reasons great effort, particularly in the past ten years, has been made in the development of specialized meta lanugages suitable for multivariate data analysis and an, often unjustified, effort in the production of packages of statistical programs which, because of their rigidity, sometimes slow down rather than help the advancement of mathematical and statistical methods in this field of informatics.

SS programming language was designed in 1968, as a system for the design of programs, particularly for factor and regression analysis. Until 1974 SS was an internal language of the institutions which have worked on it. After 1974 SS, into which basic functions present even now were built in, became a language for the construction of complex algorithms for data analysis. In it most of the programs for multivariate data analysis has been written in the institutions where it has been implemented.

SS has been designed as a non-procedural command language whose syntax, although resembling formally the syntax of some macro assemblers, is very close to the syntax of matrix algebra.

The set of SS commands enables a simple control of input/output operations, the scope of data treatment essential for multivariate analysis, all operations over scalars, vectors and matrices, relatively wide scope of built in statistical functions, sufficient possibility of control over the program flow and simple forming and exploitation of macro programs.

2. SYSTEM ORGANIZATION

SS is organized round an unique kernel which enables simple performance and control of input and output operations, creating and maintenance of data bases, control of syntactical and logical correctness of commands, and control of performing the operations. SS performs this automatically, thus freeing the programmer of the greatest part of activities connected with the creation of programs for the problems which can be solved by linear transformations and by application of standard distribution functions.

All commands in SS have the same or the generic syntactical form. Each one of them starts with the command name, after which, in parentheses, in an arbitrary order potential arguments are associated with arbitrary names and logical parameters are activated. Most commands have as potential arguments built in assumed names and assumed values for logical parameters. Due to this it is often possible to end the command with a period, without explicitly stating the argument names and values of logical parameters. Each SS command communicates with the surroundings created by the SS system in such a way so as to take the arguments, as a rule matrices, from the internal SS data base, while it allocates products of operations to the internal SS data base. SS has its own virtual memory system built in and its own system for dynamic use of main memory.

SS communicates with the surroundings in only two different ways. The first one contains two input modalities, the first of which takes over standard system files, while the second one takes over system files under virtual control of the SS system.

The other way contains two output modalities. The first one is result output, over standard system outputs, to standard system output units which can be addressed. The second way is result output in system files which are still under virtual control of the SS system.

SS communications with its internal control are entirely transparent. However, two types of control record give to the SS programmer all information about all processes under way during program performance, and which are as a rule, of small or no interest for the standard user.

SS has built in systems for lexic, semantic and logic control, which disturb or interupt program execution if the program contains lexic, semantic or logic errors.

Diagnostic messages and warnings are written in understandable language and do not require any special decoding. The system has a built in optimizer of the internal data base which eliminates from the base all elements created during the program and will not be used further.

3. SS STATEMENTS

All SS statements can be divided into following logical blocks:

- input/output commands
- commands for manipulation with data bases
- algebraic commands
- coordinate systems forming and transformation commands
- statistical commands
- control commands
- commands for formina and exploitation of MACRO programs

3.1. Input/output commands

SS has built in and for user transparent input-output functions which are automatically activated during program performance. A special set of commands enables, however, controlled input and output. These commands are:

1. INPUT, which enables reading of a data matrix under the chosen format and with the control of the data. INPUT can or must be associated with commands for organization of records (SEQUENCE); for the determination of variable names, format and optionally, control of records (VARIABLE), as well as commands for the forming of the submatrices (SELECT).

2. PRINT enables the printing of the obtained results in the previously defined form with the optional table headings. Implicit transposing of the matrix which is printed is allowed, as well as explicit deletion of that matrix from the SS system.

3. OUTPUT addresses the output device and controls the number of reprints and copies.

4. DUMP writes the required matrix into the defined external file, which is under SS control. It is possible to record an arbitrary number of matrices in each one of such files by the DUMP commands, formated or non-formated.

5. STORE reads from the associated external medium matrices with the defined CODE which are formed by the DUMP commands.

6. HEADING writes the text, in three rows of 60 characters each, which can be arbitrarily changed depending on the need to designate the heading of the pages follow-

ing the HEADING command. SS automatically writes on every page of results the name of installation, version of the implemented SS system, as well as the date and the time of generation of the page.

7. TEXT writes, in capital letters, in three rows of 10 characters each, the given text. It is used to designate the program name and the names of particular program blocks.

3.2. Commands for Manipulation with Data Bases

SS considers the data base as a set of matrices with the associated alphanumeric records which describe the names of rows and columns, by numeric records which determine the content matrices and by alphanumeric records which define the names of all these elements. The manipulation of thus defined data bases is possible by the following commands.

1. DELETE explicitly delets the particular matrix or some part of the data base associated with it. Implicit DELETE is possible in the PRINT command.

2. COPY replicates, under another name any matrix created by any SS command as a result of that command or which is brought into SS from some external medium.

3. SORT sorts out rows or columns of a matrix by values of arbitrary variable contained in that matrix. As a standard SORT orders the matrix by alphanumeric sorting of records, describing row names.

4. MERGE forms a supermatrix of two or more matrices after it has previously sorted records of row names and eliminated records which are not common for all matrices forming the supermatrix.

5. CONFORM coordinates two matrices after having sorted records of row (or column) names, so that the matrices maintain rows with identical names.

6. SELECT chooses, from an external matrix, rows (or columns) whose record corresponds to the specified conditions, and forms a new matrix from them.

7. REWIND positions an external associated file at the beginning of the record in that file.

3.3. Algebraic Commands

SS contains a set of commands which enable algebraic operations over matrices, vectors and scalars. These commands have been formed so as to reproduce standard operations of linear algebra in SS language syntax.

1. COMPUTE enables all operations of scalar algebra written, after this command, in standard FORTRAN notation, including built in FORTRAN functions as well. This offers a possibility to write FORTRAN subprograms in SS, which are still further under the control of SS language.

2. TRANSPOSE explicitly transposes a matrix. Implicit TRANSPOSE is possible in the PRINT command as well as in commands for elementary operations over matrices.

3. LINEAR sums up two matrices. Any matrix can be previously multiplied with any real scalar. Due to this LINEAR enables subtraction of matrices as well as reflection of vector of a matrix.

4. HADMULT performs Hadamard (elementwise) multiplication of two matrices. Any of these matrices can be, by elements, raised to power to an arbitrary real constant. In this command only one matrix can be given, and this enables raising matrix elements to any power, including zero. Due to that this command makes possible elementwise division of two matrices.

5. DIAG forms diagonal matrix from the main diagonal of a given square matrix or a vector. The matrix formed in this manner can be raised to power ba an arbitrary real constant.

6. DIAGMULT multiplies a matrix which can be implicitly transposed from the left or/and right by the diagonal formed from the main diagonal of a square matrix or a vector. Elements of the diagonal matrix formed in such manner can be raised to power by a real constant.

7. SCALE norms vectors of a square matrix on the basis of the main diagonal of that matrix.

8. MULT multiplies two matrices each of which can be previously implicitly transposed.

9. INVERSION inverts a square non-singular matrix and computes the determinant of that matrix.

10. SYMINVERSION inverts a symmetric non-singular matrix by the Cholesky decomposition method. The lower triangle obtained by the decomposition can if required be the explicit result of that operation. Apart from that, this operation forms the diagonal matrix recorded as a vector from inverse diagonal elements of the inverted matrix, and forms the diagonal matrix of differences between diagonal elements of the original matrix and inverted diagonal elements of the inverted matrix.

11. EIGEN determines eigenvalues and eigenvectors of a symmetric matrix. Only eigenvalues greater than zero and the associated eigenvectors can be maintained by a special option.

12. GENEIGEN solves the general problem of eigenvalues of the type $(\mathrm{A} - \lambda \mathrm{B})\mathrm{x} = 0$ under the condition that matrix B is regular. By a special option it is possible to maintain only eigenvalues greater than zero with the associated eigenvectors.

13. SUM forms a vector (or scalar) by summation of columns (or rows) of a matrix (or a vector). The elements can be previously multiplied by a real constant and raised to an arbitrary real constant power.

14. TRACE forms the trace of a square matrix. The elements of this matrix can be previously multiplied and raised to power by a real constant.

15. CONGRUENCE computes cosines of angles between vectors of any two matrices.

Pseudoinverses of any type can be computed by a combination of MULT, EIGEN and SYMINVERSION commands. Spectral decomposition of any matrix is performed by a combination of MULT and EIGEN commands.

3.4. Coordinate Systems Forming and Transformation Commands

All data analysis methods can be reduced to compression and linear transformation of data matrices which extremizes a criterion function. From the geometric standpoint these methods are essentially formation of optimal coordinate systems for a vector configuration. Due to this SS contains a special class of commands for the compression of matrices and linear transformations which extremize come typical criterion functions.

1. HOTELLING norms eigenvectors of a matrix to the value of their eigenvalues. The selection of vectors which are to undergo this transformation is possible on the basis of a very large number of criteria, including the criteria based on the image transformation of original variables. Due to that, this command makes possible the acquirement of the basic solution within the framework of nearly all procedures belonging to component or factor analysis.

2. ORTHOSCORES orthogonalizes a set of variables under the least squares criterion, forming thus, in fact, a Mahalanobis variable transformation.

3. VARIMAX orthonormally transforms a matrix in such a way as to maximize the Kaiser varimax function.

4. QUARTIMAX orthonormally transforms a matrix in such a way as to maximize the Carrol quartimax function.

5. OBLIMIN nonorthogonally transforms a matrix in such a way so as to minimize the oblimin function by the direct Jenrich-Sampson procedure.

6. ORTHOBLIQUE orthonormally transforms eigenvectors of a matrix in such a way so as to extremize the varimax or quartimax function for any type of the Harris-Kaiser transformation procedure.

7. TAXON transforms, by orthogonal and nonorthogonal transformation, a data matrix by extremization of varimax and quartimax function of by minimizing the oblimin function.

8. PROMAX transforms a matrix in such a way so as to, under the least squares criterion, maximizes the function of likelihood of that matrix and a target matrix allowing thus procedures belonging to the confirmative component of factor analysis.

All mentioned procedures form matrices of pattern, structure, correlations or covariances of latent dimensions.

3.5. Statistical Commands

Although all statistical methods can be written in SS by input/output commands,

commands for manipulation of data bases, algebraic commands and commands for compression and linear transformation, SS contains built in statistical commands which make possible standard statistical analyses. This is due to the fact that the great majority of these methods require built in statistical tests and standard result outputs.

1. STATISTICS transforms data into the standardized form, forms variable distributions and computes these distributional parameters, test the hypothesis on normality of distributions and upon special request performs the single factor analysis of variance according to the arbitrary selector variable. This command upon special request computes norms in standardized values, normalized standardized values, standardized groups and percentiles.

2. NORMALIZATION determines parameters of discrete ordered variables and transforms values of these variables into a normalized form on the basis of the inverse function of normal distribution.

3. CORRELATION computes matrix of correlation and/or covariance matrix. This command allows the computation of all types of product-moment correlation coefficients.

4. CROSSCORRELATION computes the matrix of crosscorrelations and/or covariances of two data matrices. This command allows the computing of all types of product-moment correlation coefficients.

5. PARTIAL computes the matrix of partial correlations for any pair of variables if the other variables are constant, or for any pair of variables if a set of variables is constant.

6. IMSCORES transforms a set of variables into image form and explicitly forms the transformation matrix which turns the data matrix into image form.

7. IMAGE computes the matrix of covariances of variables transformed into image values.

8. REGRESSION carries out regression analysis of up to 250 criterion variables to up to 250 predictor variables. The program tests the significance of partial regression coefficients and the associated partial correlation coefficients, coefficients of correlation of the zero order between predictor variables and criterion variable and, naturally, of coefficients of multiple correlation. A moduel for computation of regression factors is associated with the program. Upon request the program forms the matrix of partial regression coefficients for all criterion variables and the matrix of regression factors. Depending on the type of predictor variables, this command can perform any analysis under the general linear model, including analysis of variance under the different experimental designs.

9. BETAR computes the matrix of regression coefficients for the estimation of factor values under the least square criterion.

10. BETAF produces the matrix of regression coefficients for the computation of factor values under the component model for, in whatever way transformed, set of

principal components.

11. SCORES computes factor values on the basis of the given data matrix and the given matrix of regression coefficients. This command can compute the estimation of criterion variables as well if the matrix of regression coefficients has been generated by the REGRESSION command.

12. MATRIX REDUCTION modifies a correlation matrix or any other square symmetric matrix by exchanging the diagonal values with the values of a given diagonal matrix or with values of the maximal coefficient in the row of the matrix itself. As a standard, this command produces the matrix of crosscovariances of standardized variables and of these variables transformed into image form.

13. ITERCOMMUNALITIES computes eigenvalues and eigenvectors of a square symmetric matrix with iterative adjustment of diagonal values of that matrix for a fixed number of factors.

14. OBLICOMP determines matrices of pattern, structure and intercorrelation of latent dimensions by iterative application of a procedure based on the general Guttman theorem on factorization of a square symmetric matrix.

15. RAO LAWLEY carries out canonical factor analysis by Rao algorithm if the initial structure matrix has been defined on variables scaled to the anti-image metric. If the initial structure matrix has been defined by standardized variables this operation performs factor analysis by the method of maximal likelihood of Lawley and Maxwell. The command automatically tests the hypothesis on the number of factors by the Bartlett test.

16. RESIDUAL computes the residual covariance matrix for a given covariance matrix and a given factor matrix.

17. CANCORR carries out the canonical correlation analysis of two sets of variables and tests canonical correlations by the Bartlett test. On the basis of results of that test the command reduces the number of significant canonical dimensions and determines the matrices of canonical weights and factors for the thus determined canonical dimensions. Depending on the type of variables in one of the sets this command can perform canonical discriminative analysis or canonical decomposition of contingency tables as well.

3.6. Control Commands

SS contains, although designed as a command, non-procedural language, a limited set of commands for the control of program flow.

1. IF...THEN...ELSE...END. These control commands make possible the branching of the program flow on the basis of the outcome of logical operations which can be given on the matrices as well.

2. REPEAT...UNTIL. This structure of commands activates the loop which is being carried out until the logical condition defined over matrices or scalar has been

satisfied.

Standard control commands are built into individual standard commands of SS. The structure of SS makes the GOTO or JUMP commands entirely unnecessary.

3.7. Forming of Macro Programs

SS allows a set of commands to be organized as a procedure. Such macro programs can be activated by any SS program and behave as standard SS commands.

1. MACRO declares that all commands form the macro which has been associated with the declared MACRO command. In this command it is necessary to give the names of input matrices and scalars. All results computed by a macro program which are not explicitly deleted from the program can be used in the program using that macro program.

2. CALL inserts a set of commands organized into the macro program in the place of the call. The command must contain the name of the called macro program and the names of matrices and scalars which are the result of macro programs and are necessary in further computing. All undeclared matrices or scalars are considered as local results of macro programs called upon ba the CALL command.

The set of commands in SS is not final. In the course of further development other commands will be built in as well, particularly in the group of statistical commands. Certain supplements in the set of commands for manipulation of data bases are possible as well.

4. PROGRAMMING IN SS LANGUAGE

Programs in SS are written in simple sequence of commands. These programs can, virtually, be of arbitrary length and complexity and may be organized in macro programs. The manner of programming is equal to that in all command languages.

REFERENCES

Zakrajšek E., Štalec J. and Momirović K. (1974), SS - Programming system for multivariate data analysis, Proceedings of the 1st International Symposium "Computer at the University", Zagreb, C8, 1-18.

Štalec J., Momirović K. and Zakrajšek E. (1981), Statistical system, programmers reference, University Computing Centre and Faculty of Physical Culture, Zagreb

Procedures for Optimal Collapsing of Two-way Contingency Table

D. Pokorný, ČSAV, Prague, Czechoslovakia

Summary : The paper presents basic ideas of the GUHA style processing of two-way contingency table. Methods and algorithms of collapsing two-way tables are discussed. Concrete proposals are made and experience with implemented procedures is shortly described.

Key words : contingency tables, two-way contingency tables, collapsing, GUHA .

1. INTRODUCTION .

After rejecting the independence hypothesis by an appropriate test, the analyst is usually asked to explain a dependence structure in the table. Various methods of the exploratory data analysis are available, ranging from simple displaying tables of residuals to more or very sophisticated ones, like graphical display of residuals, correspondence analysis, approaches using Gabriel's results or Brown's method of identifying sources of dependence. Cluster analysis using chi-square or similar statistics as a distance measure can be mentioned in this connection, too.

On the other hand, a collapsing of contingency tables - i.e. joining some rows and/or coloumns - is sometimes performed during an analysis for various reasons : (i) expected frequencies are very small, (ii) large size tables are expected to be non-significant, (iii) large tables can be hardly comprehended by a human intellect. Moreover, categories of cross-tabulated variables are sometimes defined by the researcher ad hoc and in unnecessary detail. To a limited extent, some programs (cf. the BMDP package) can do the collapsing automatically.

We shall introduce a concept of an optimal collapsing of a two-way contingency table, show some its properties and attempt to demonstrate that it is possible to base on this concept procedures for explaining dependence structure.

2. PROBLEM OF AN OPTIMAL COLLAPSING - A BASIC PROCEDURE .

Let us consider a R*C contingency table T and parameters r, c, N, CRIT, QUOT, MEASURE, where $2 \le r \le R$, $2 \le c \le C$, $N \ge 1$ are natural numbers, $CRIT \ge 0$, $0 \le QUOT \le 1$ are real numbers and MEASURE is a function associating with each r*c table a real number.

Suppose that all possible r*c tables, collapsed from the original R*C table, are ordered in the sequence with respect to decreasing value of MEASURE. Consider now just the initial segment of the mentioned sequence, obtained by the following restrictions :

(i) the length of the segment is less than or equal to N,

(ii) the value of MEASURE is positive and greater than or equal to CRIT,

(iii) if QUOT is positive, then the value of MEASURE is greater than or equal to

QUOT * (value of MEASURE for the first table) ,

(iv) it is the longest initial segment satisfying (i), (ii), (iii). This initial segment is supposed to be a solution of the task of optimal collapsing.

The first table in this segment represents an optimal collapsing of the given table, the subsequent tables represent suboptimal (alternative) collapsings. Intuitively said, optimal collapsed table is a simplification of a two-way table with minimal loss of information.

3. APPROPRIATE MEASURES.

3.1. MEASURES.

We shall consider the following measures for R*C tables :

The Pearson's chi-square statistic: $CHISQ=\Sigma\Sigma(n_{ij}-e_{ij})^2/e_{ij}$, where n_{ij} are observed and e_{ij} expected frequencies.
The maximal likelihood chi-square statistic: $MLCHISQ=2\Sigma\Sigma n_{ij}\ln(n_{ij}/e_{ij})$, and the Goodman-Anděl maximal interaction statistic:

$$WSQ = \max\ ((\Sigma\Sigma\ a_{ij}\ln n_{ij})^2/(\Sigma\Sigma\ a_{ij}^2\ /n_{ij}))\ ,$$

where a maximum is taken over all possible interaction matrices (a_{ij}).
(An interaction matrix is a non zero matrix of real numbers with zero marginal sums. For an efective procedure for computing WSQ see Andel (1974)).
Moreover, for 2*2 tables we shall consider the chi-square statistic with the continuity correction:

$$\begin{aligned} YATESSQ &= (\ |n_{22}n_{11}-n_{12}n_{21}|\ - n../2)^2\ n../(n._1 n._2 n_1. n_2.) \\ &\quad ...\text{if}\ \ |n_{22}n_{11}-n_{12}n_{21}|-n../2 > 0\ , \\ &= 0\quad \text{else.} \end{aligned}$$

3.2. PROPERTIES.

(a) Monotonicity of CHISQ w.r.t. collapsing :
Let C_1 be a value of CHISQ statistic in a R*C table T and C_2 a value of CHISQ in a table collapsed from T. Then $C_1 \geq C_2$ and equality holds if and only if rows collapsed together have identical vectors of row percentages in the table T. Similarly for columns.
(b) The analogous assertion is true for the MLCHISQ statistic.
(c) YATESSQ ≤ CHISQ for any 2*2 table.
(d) WSQ ≤ CHISQ is true for any R*C table with non-zero frequencies n_{ij}. (WSQ is in fact undefined for zero frequencies, it is recomended to replace zero frequence by 0.5 or to add e.g. 0.5 to each cell. If we do the same for CHISQ, inequality is preserved).

4. PARTICULARIZATIONS AND MODIFICATIONS OF A BASIC PROCEDURE .

4.1. A GENERALIZED PROCEDURE.

Its aim is to simplify the two-way table. Since sizes of suitable collapsed table are a priori not known, more attempts can be desirable. Here is a possibility for a slight generalization of the proposed procedure: the procedure can simultaneously perform optimal collapsing for a set of sizes $r_1{*}c_1\ ,...,\ r_s{*}c_s$ of collapsed tables. After reporting the best collapsed tables for each size $r_i{*}c_i$, tables best at all can be reported, compared significancies attained.

4.2. A PROCEDURE FOR SYMETRICAL TABLES.

Sometimes we have to deal with R*R tables, where categories 1,2,...,R of both variables are corresponding in the natural way. Such tables are e.g. social mobility tables, summarizations of discretely valued time series, comparisons of eyes, married couples etc.

Thus it can be useful to have a procedure searching optimal r*r collapsed tables where both rows and columns are collapsed in an exactly same way. Note that associations on the diagonal of the collapsed table need not be positive.

4.3. NON-HIERARCHICAL CLUSTER ANALYSIS.

Let us consider a particular case of the basic procedure: c=C. In this case rows are collapsed only, in fact we perform a non--hierarchical cluster analysis of categories of the row variable.

4.4. HIERARCHICAL CLUSTER ANALYSIS.

Let us take a futher particularization: r=2, c=C. Now the procedure divides rows 1,...,R into two sets of similar rows. Thus the initial R*C table is divided into two subtables R_1*C and R_2*C, the procedure can be used for both of them recursively again.

So, the divisible hierarchical cluster analysis of rows is performed. Stopping rules for this hierarchical procedure will not be discussed here.

4.5. 2*2 COLLAPSED TABLES.

It is recomendable to arrange the order of rows and columns in a collapsed 2*2 table in such a way that $a_{11}a_{22}-a_{12}a_{21} > 0$ holds. Then we can simply speak about association between some set of row categories and some set of column categories.

4.6. HIERARCHICAL DEBLOCKING PROCEDURE.

Having an optimal 2*2 collapsed table, we shall consider two subtables R_1*C_1 and R_2*C_2 of an initial R*C table, corresponding to two cells of the main diagonal of collapsed 2*2 table. These subtables are collapsed to 2*2 tables etc. As a result we obtain a tree, leaves of which are sets of corresponding categories. Results of this procedure are in some way similar to those of the correspondence analysis.

This procedure corresponds to the COLLAPS procedure, proposed earlier (Pokorný (1978), Havránek and Pokorný (1978)).

Stopping rules for this procedure are discussed by T. Havránek elsewhere (Havránek (1979) and Havránek (1980)). Cf. also Ecob's approach for the AID procedure here (Ecob (1978)).

5. ALGORITHMS.

5.1. AN ALGORITHM FOR THE BASIC COLLAPSING PROCEDURE.

An algorithm for the basic procedure collapsing the R*C table to the r*c tables is based exclusively on the monotonicity of the considered measure with respect to the collapsing operation. The construction of an algorithm follows.

(a) Coding of partitions.

A partition of a set (1,...,R) can be coded by the vector p(1),...,p(R) , an element i then belongs to the class p(i) . If we shall consider only vectors having following properties : (i) $1 \le p(i) \le i$ and (ii) $p(p(i)) = p(i)$, then we shall have an one-one mapping between vectors and partitions. For example, the partition ((1 2 5) (3 4) (6) (7)) will be represented by the vector (1 1 3 3 1 6 7) . It must be p(1) = 1 , anytime.

We shall divide the set of indices having the property p(i) = i (set of class representants) into two subclasses :
Let $k = \max \{j ; p(j) \neq j\}$ if the set mentioned is non-empty, k=1 otherwise. Then the non-empty set $\{ j ; 1 \le j \le k , p(j) = j \}$ will be called the head and the (maybe empty) set $\{ j ; k < j \le R \}$ will be called the tail.

For example, the considered vector (1 1 3 3 1 6 7) has the head (1 3) and the tail (6 7) . Each class of the partition belongs either to the head or to the tail.

(b) The lattice of all partitions.

We shall consider a relation < on the set of all partitions of the set (1,...,R) inverse to the refinement relation. So we have for example ((1 2 3 4) (5 6)) < ((1 2) (3 4) (5 6)). We can speak on the lattice of all partitions. Similarly for the set (1,...,C). The cartesian product of these two lattices is also a lattice ;this lattice corresponds to all possible collapsings of the R*C table. As we have noted, both the CHISQ and MLCHISQ statistics are monotone w.r.t. the ordering in this lattice.

(c) The tree of all partitions.

A tree is as a rule a structure more suitable for a searching algorithm than a lattice. We shall construct a tree in following steps :
- Transpose the R*C table, if R > C.
- Reorder rows of the R*C table with respect to their decreasing "importance". The criterion can be e.g. the CHISQ or another measure in 2*C collapsed table, where the given row is compared with all other rows collapsed together.
- Do the same for columns.
- Construct a tree of row partitions by forgetting some <-comparisons in the lattice of row partitions.

A node in the tree is represented by the vector p(1),...,p(R) having the head H = (x(1),...,x(h)) and the tail T = (y(1),...,y(t)) . Succesors of this vector will represent some k-1 class partitions, when the vector p represent k-class partition. Succesors will be generated in the following order:

```
          for  j = y(1), ... , y(t)
          for  i = x(1), ... , x(h), y(1), ... , y(j-1)

               construct from p the new vector w, putting
               w(j) := i
               w(k) := p(k)  for k=1,..,j-1,j+1,..,R

          end
          end
```

(See the example of the tree of partitions in Fig. 1.)
- Consider the subtree of just defined tree, consisting from r-class partitions and partitions located in the tree over some r-class partition, i.e. partitions between some r-class partition and the bottom of the tree. Note: A subtree under the right brother of a given node is not deeper than its own subtree. Hence, a mentioned subtree

will be easily constructed by cutting off certain right and certain lower part of the tree of all partitions.

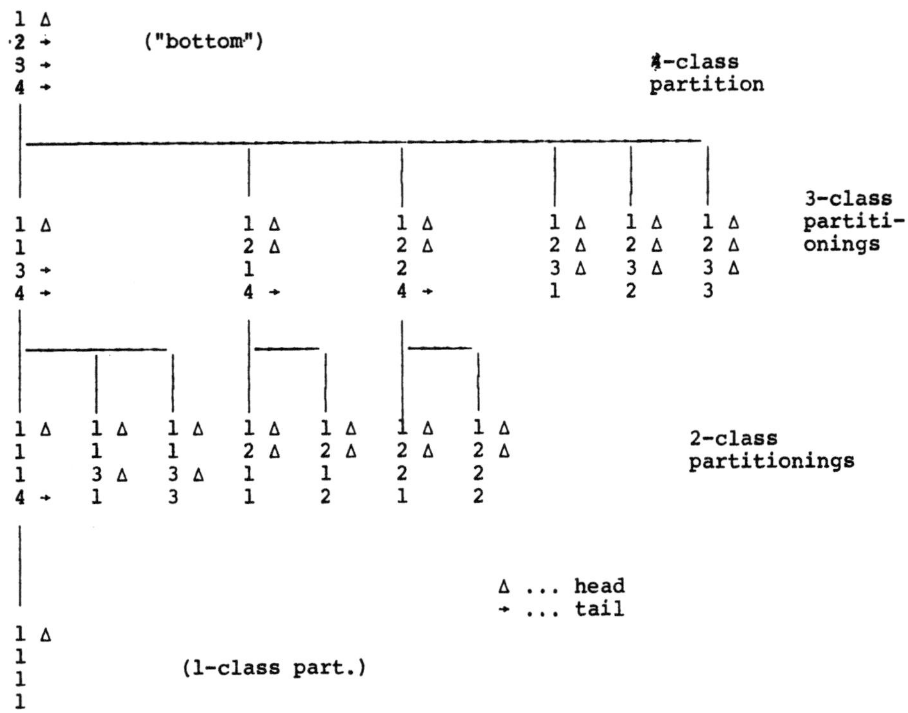

Fig. 1 :The tree of all partitionings of the set (1 2 3 4).

- Construct the tree of column partitions analogously.
- Append a column tree to each leaf of the row tree, identifying the leaf of the row tree with the bottom of the column tree.
Our tree is constructed.

(d) <u>A searching algorithm.</u>
Let the R*C table T and parameters r, c, N, CRIT, QUOT and MEASURE ∈ { CHISQ, MLCHISQ [,...]} be given.

Consider the above described tree and prepare a stack for N hypotheses (=pairs of row and column partitions), set CRITICAL := CRIT .

Search in the tree using the depth-first technique.

A successor differs from its ancestor only by collapsing two rows (resp. columns). Hence a quick recurent recomputation of the collapsed table is possible. (By the way, the CHISQ is advanteously computed from two matrices $(n_{ij}n_{..}-n_{i.}n_{.j})$ and $(n_{i.}n_{.j})$ here).

In a node, ask if the value of MEASURE is less than the current CRITICAL value, in this case leave the node and the whole subtree determined by it. If the value of MEASURE is not less than CRITICAL, and the node is not a leaf, then investigate in the same way all successors of the node. Leaves of the tree represent the hypotheses, i.e. r*c collapsed tables. If a value of MEASURE greater than or equal to the current CRITICAL value is found here, update the stack and set a new CRITICAL value.

Clearly, the tree is rather a control structure then a data structure occupying the computer memory.

We shal call the ratio

(# of existing r*c tables) / (# of visited nodes)

the speed-up factor. From the point of view of maximal computational complexity this algorithm is very bad. However, substantial time savings can be reached for real contingency tables. The results of the proposed procedure are - unlike stepwise procedures - exhaustive, i.e. really optimal.

5.2. NOTES ON ALGORITHMS FOR MODIFIED PROCEDURES.

(a) A generalized procedure (4.1).

It is possible to generalize the described algorithm for a set of sizes r_1*c_1 ,..., r_s*c_s of collapsed tables. Some difficulties arise here, e.g. comparing the values of chi-square statistics with different degrees of freedom - since significances are often "computer zeros".

(b) Non-monotone measures.

We can weaken the condition on the used measure. A measure need not be monotone with respect to collapsing, but must have a monotone majorant. This is a case of WSQ and CHISQ (cf. 3.2). An algorithm is then modified in the following way: In the node we ask firstly: "Is the majorant large enough ?" (if not, skip the subtree immediately) and then "Is the measure large enough ?". The second question is asked in leaves only.

(c) A procedure for symetrical tables (4.2).

The algorithm is in fact a simplification of the presented one, we have to search in the tree of partitionings of the set (1,...,R).

(d) A procedure for 2*2 tables (4.5, 4.6).

Here it is possible to apply more hints when the measure is CHISQ (and consequently YATES and WSQ). A tree of all subsets rather than a tree of all partitions is then used ; a tree of this type is discussed in Furnival and Wilson (1974). An algorithm was described in Pokorný (1978) and in Pokorný and Havránek (1978).

(e) A procedure for 2*C tables.

A proposal of a quick algorithm for CHISQ measure is intended.

6. IMPLEMENTATION .

The hierarchical procedure for collapsing to 2*2 tables is implemeneted. It is written in the PL/1 language (level F) and occurs in the GUHA package. Great care is given to readability and interpretability of printed output. Both the great efficiency (tables 12*12 in cca 20 sec. on IBM 370/135) and usefulnes from the EDA point of view were proven in the practical use. The procedure will be improved by introducing reasonable statistically motivated stopping rules, suggested by T. Havránek.

The "basic" procedure for collapsing R*C contingency tables to r*c tables is written in PL/1 too and its efficiency is being tested now.

REFERENCES

Anděl J. (1974), The most significant interaction in a contingency table, Aplikace matematiky, 19, 264-252.
Brown M.B. (1974), The identification of sources of significance in two-way contingency table, Appl.Stat., 23, 405-413.
Dixon W.J. and Brown M.B. (EDs.) (1979), BMDP-79 Biomedical Computer Programs, Univ. of Calif. Press, Los Angeles.
Ecob R. (1978), The distribution and power properties of the AID criterion, in:COMPSTAT 78 Proceedings, Physica Verlag, 253-258.
Gabriel K.R. (1966), Simultaneus test procedures for multiple comparisons on categorial data, JASA, 61, 1081-1096.
Hájek P. and Havranek T. (1978), Mechanizing hypothesis formation, Springer Verlag, Berlin Heidelberg New York.
Havránek T. (1979), Approximate distribution of the maximum of 2*2 chi-square statistics derived from an R*C contingency table, in: Proceedings of Second Prague Symposium on Asymptotic Statistics, North-Holland, Amsterdam, 212-219.
Havranek T. (1980), Some comments on GUHA procedures, in:Victor N., Lehrmacher W., van Eimeren W. (EDs.), Explorative Datenanlyse, Medizinische Informatik und Statistik, vol. 26, Springer Verlag, Berlin Heidelberg, New York, 156-177.
Havránek T. and Pokorny D. (1978), GUHA style of processing of mixed data, Int.J.Man-Machine Studies, 10, 47-58.
Pokorný D. (1978), The GUHA method and desk calculators, Int.J. Man-Machine Studies, 10, 75-86.
Pokorný D. (1980), Knowledge acquisition by the GUHA method, Int.J. Policy Analysis and Information Systems, 4, 379-399.
Pokorný D. (1981), The COLLAPS procedure, in:Havránek T. (ED.), GUHA reference manual, Center of Biomathematics, Prague.
Pokorný D. and Havránek T. (1978), On some procedures for identifying sources of dependence in contingency tables, in:COMPSTAT 78 Proceedings Physica Velag, Wien, 221-[illegible].

Computer Science as a Tool Improving Data Analysis Researches and Uses

Y. Schektman, A. Ibrahim, J. Jockin, J. Pastor and D. Vielle,
Paul Sabatier University, Toulouse, France

Summary : *Analyser des données est une pratique pluridisciplinaire. Les chercheurs en Analyse de Données doivent (i) acquérir des compétences dans plusieurs disciplines scientifiques, et (ii) apprécier la pertinence de leurs idées, de leurs découvertes, des techniques proposées, en traitant effectivement sur ordinateur le type de données et de problèmes qui ont provoqué et sous-tendu leurs recherches. Ils ne doivent jamais négliger le dialogue avec les utilisateurs de leurs méthodes et logiciels. Notons que ceux-ci deviennent de plus en plus exigeants : ils veulent que les logiciels puissent être utilisés en libre service et demandent souvent des logiciels personnalisés, c'est-à-dire adaptés à leurs besoins voire à leurs habitudes de travail et à leur propre pratique de l'Analyse des Données. De son côté, le chercheur n'a pas toujours à sa disposition les outils informatiques adaptés à ses besoins.*

Dans ce contexte de travail difficile, le chercheur en Analyse des Données doit, pour pouvoir assumer correctement toutes ses responsabilités, être assisté par des outils d'aide à la conception et à l'utilisation des logiciels. Nous proposons des outils de ce type dans cet article. Pour illustrer les différents niveaux d'assistance que nous présentons, nous prenons des exemples dans les recherches que nous effectuons sur la construction de "mesures en facteurs principaux signifiants" c'est-à-dire sous contraintes et avec usage de métriques particulières.

Keywords : *Compiler-compiler, Interactive computing, Metrics, Principal components under constraints, Statistical programming language, Tool aids for producing softwares.*

1. Statistical computing, a priority for data analysis development

It is instructive for a French academic statistician to read the introduction to the first issue (may 1838) of the J.R.S.S. named, at that time "Journal of the Statistical Society of London". Besides idealistic points of view such as "...*the object of Statistical Science is to ..., with the view to determine those principles upon which the well-being of society depends*", or social and economical objectives such as "...*Journal (J.R.S.S.) devoted to the collection and comparison of facts which illustrate the condition of mankind, and tend to develop the principles by which the progress of society is determined*", one finds also "...*they* (statistics) *are, as it were, the link which connects them* (the other sciences) *with the practical purposes of life*." Indeed, this latter sentence is a consequence of the other two , but above all it reminds us that statistics constitute a multidisciplinary science.

During years and years one could only find economical, demographical, legal, etc... papers in J.R.S.S.. Then, applied mathematics, probability and more recently statistical computing papers became more and more numerous. Why this evolution ?

The difficulties encountered, to day, by an academic statistician researcher in facing up to all his tasks, is a possible explanation : because a statistician must acquire knowledge in various sciences, and above all appreciate whether his ideas, discoveries and designed software , fit the user's needs, by doing statistical by treating the kind of raw data and real problems which provoked and stimulated his own researches. So, for an academic statistician it is sometimes more pleasant to become a specialist in mathematics or probability. Today, too much time is wasted in managing magnetic files, in programming with unsuitable languages, or in fighting against inconvenient programs, etc.. Consequently, in this unfriendly environment, a statistician, if he is to do his job thoroughly, has to be helped by tools fitted to his own needs and also to the users' needs, in order to get as much free time as possible for his own researches. Now, one may say that Statistics master Probability, so in many fields important progress has been made. Nothing of the kind with computer science, for statisticians just begin to avail themselves of all the facilities offered. In the future, we think that most of the current tedious tasks and many of the mathematical problems which arise, will be solved with the progress of computer science and its ever growing applications in Statistical Computing. So, statisticians will be able to support all the above-mentioned responsibilities. As a conclusion, the development of Statistical Computing seems to us a priority for Statistics in years to come.

2. Cost of programming, reliability, adaptability, portability and maintenance of softwares are preoccupations not to be neglected.

While computer manufacturing is led by industrial production criteria, software is being produced in an amateur way. So, if the cost of hardware is decreasing, on the contrary the cost of programming is increasing a lot, whereas software supply and application area are also increasing. Studies done by the "Data and Analysis Center for Software" (USA) show that (i) half of the work put in to produce software is spent in checking, (ii) most of software cost comes from maintenance, (iii) error cost is multiplied by 10 when a software is in use, etc.. The above statistics show the importance of designing tool aids for producing cheap statistical softwares and of being able to fit easily old well-designed programs into a new one.

So, in spite of the increase in cost, no improvement in reliability and adaptability is achieved chiefly because of the lack of tools for generating automatically all or part of the software required. Now we must work in this area, for users (i) are harder and harder to please as far as the quality of the results and above all the speed and flexibility of the methods are concerned, and (ii) ask that application software be personalized, that is to say, be adapted to their specific needs and to their own work habits. Concerning portability, we hope that the French SOL project will be a success : this is a normalization project around Pascal and Unix with compilers and operating systems available on the most frequently used computers in France.

Finally, in order to simplify the maintenance, a statistical system must run in both interactive and batch modes without modifications and redundancy.

3. Some useful tool aids for producing statistical software

3.1. An algorithmic statistical language is a priority for doing statistical researches

3.1.1. A scientific context : let us present briefly the scientific context which led us to design a statistical programming language. Many sciences use data analysis and need more and more methods, models and adapted softwares. The evolution of Data Analysis, reveals a trend in getting, indeed reasonably complete, but also and above all most meaningful summaries (or factors) of whatever information reside in large arrays of data gathered in a given scientific environment : i.e. to produce what we could name "measures in factors". Spearman's G-Factor and afterwards Thurstone's Simple Structure, Common and Specific factors, Rotations, etc... are the well-known manifestations of researching meaningful factors, while Pearson's and Hotelling's results provide more complete summaries. In order to produce "measures in factors", we propose [12b] to gather these two approaches into one , which consists in finding principal components under constraints. So, anyone will be able to take into account hypotheses, like Simple Structure or order constraint for example [5,12g], in carrying out measures in factors. These researches, which are contributions to measure theory, mainly in the observational and experimental sciences, led us to introduce [12a] and study such a model [2,12b,13], and to research what information (from scientific environment of measures) metrics can carry. Note that all classical models in linear data analysis (Discriminant, Canonical, Correspondance analyses, Regression), Principal Components of instrumental variables or uncorrelated with an instrumental variable [11], and other models are particular cases of Principal Components under linear constraints with suitable choices of metrics, "intra" or "inter" unit subspaces associated to predictor and predictand sets of variables. In all predictand models, Mahalanobis metrics and metrics with relationship effects, which we introduce [12b] and study [2,12d], are of great interest. In this new research area many programs are to be written, and now we are doing this work [5,12g].

Designing a data analysis model is a dialectical process between analysing raw data of well-defined statistical studies, and improving the researcher's hypotheses, criteria, algorithms, etc.. that is to say improving his way of thinking. Given this, it is obvious that the development of data analysis methods depends very much on the ease with which the researchers will be able to write the programs they need. The programming languages available, even perhaps ADA, as general as they are, are never entirely fitted to statisticians' specific needs. Therefore designing algorithmic statistical languages seems to us very useful. A well-known and very useful example of such a language is GENSTAT (NAG-OXFORD). We mainly describe, in the next section, features of a statistical language through our project in this area [12a-b-c,f].

3.1.2. Some useful features of the LOM language for programming data analysis algorithms

LOM (**M**odules **O**rganization **L**anguage) is the experimental statistical language we are designing. We present briefly in this section only the features useful to researchers in programming algorithms. A statistical language, such as LOM, must dispose of :

. all the algebraic facilities available in Fortran,
. means interfacing with Fortran subroutines or Pascal procedures,
. some structured control instructions, as in Pascal,
. algebraic matrix operators, extended to matrix elements, and possibly executed mixing real numbers with matrix identifiers.

. operations of matrix handling (inversion, trace, transposition, concatenation by rows or columns, extraction or insertion of one element, one row, etc ...),
. mathematical operators (inner-product, distance, norm, projection, etc ...),
. statistical operators (mean, variance, covariance, correlation, etc ...).

Let us give, for instance, some LOM instructions : LOM statements, in figure 1, are extracted from a qualitative principal component analysis algorithm [5] useful in predictand clustering, while LOM statements, in figure 2, are part of an algorithm of principal components under a minimum correlation constraint.

FIGURE 1 : Some algebraic LOM instructions

```
$ INITIAL CRITERION VALUE $
CRITERION=0 ;
INVVX=INV(COV(X;DP)) ; X=CENT(X;DP) ;
FOR I=1 UNTIL NBUNITS DO
  INDI=ROW(I;X) ;
  CRITERION=CRITERION+EL(I;DP)*INDI*INVVX*TRS(X)*DP*CENT(CLASS;DP)*
                      INV(TRS(CLASS)*DP*CLASS)*TRS(CENT(CLASS;DP))*
                      DP*X*INVVX*TRS(INDI)
END
```

In figure 1 : *For easy reading LOM keywords have been translated into English and underlined. Elongated character string enclosed in matched "$" delimiters is comment;* INV, COV, TRS *are respectively inversion, covariance (with respect to* DP *inner-product matrix) and transposition operators ;* CENT(X;DP) *transforms all the vector columns of* X *into mean zero (with respect to* DP) *variables ;* ROW(I;X) *denotes the ith row of the matrix* X *and* EL(I;DP) *the (i,i) element of the diagonal matrix* DP. *The* CRITERION *value measures the contribution of the ith unit to the "correlation increase" between the given variables and the first principal qualitative component (a classification). The* CRITERION *formula is derived from the expression of the inter metric with relationship effect, when the intra metrics are Mahalanobis ones.*

In [12b] it is shown that there exists an infinity of functions (or criteria) of principal vectors, or components, which allow principal component analysis to be defined as an optimization problem. So, the p vectors $C_j = X\,M\,c_j$ j=1,2,...,p in $\mathbb{R}^n$, such that c_j j=1,2,...,p are orthonormalized with respect to the metric M, are principal components of the (n,p) data matrix X, iff $\sum_{j=1}^{p} (\mathrm{Var}\ C_j)^2 = \sum_{j=1}^{p} (c_j'\,M\,X'\,DP\,X\,M\,c_j)^2$ is maximum. This criterion is used to define one family of principal components under constraints : this latter is such that principal components may be correlated. The algorithm, we design [12a,2,5,12g], uses elementary rotations in planes as in VARIMAX rotations. So, LOM disposes also of special operators such as ROTation used, for instance, in figure 2.

FIGURE 2 : Some LOM statements

```
COSA=RAC2(1/(1+ALPHA*T)↑2) ; SINA=ALPHA*T*COSA ; ANG=ACOS(COSA) ;
$ DETERMINATION OF NEW A1 (A1T) $
(A1T,A2T)=ROT(A1,AJ;ANG) ;
I1T=IPROD(A1T;MVM) ;
$ A1T OUT OF CONSTRAINT CONE ? $
```

In figure 2 : A1T *and* A2T *are orthogonal vectors obtained by* ROTating, *with the angle value* ANG, *the orthogonal vectors* A1 *and* AJ. I1T *is equal to the square norm of vector* A1T *with respect to* MVM *inner product* (IPROD).

Notes : (i) *One can define Simple Structure by combining for instance minimum correlation and linear constraints.*

(ii) *The LOM language compiler is led by syntactic rules. From these rules (see, for instance, the rules of the GALI project in figure 7) and the semantic action associated, the LOM compiler is generated with the help of a compiler-compiler [1,8]. Now, the LOM language syntax is described with more than 400 syntactic rules.*

3.1.3. Some useful features of the LOM language for producing application programs : Data analysis uses are dialectical processings between choosing models and interpreting the statistical results of preceding ones : data coding and modeling parameters are often modified to get better results, moreover several data analysis programs must be linked together. Thus LOM allows the researcher to call an existing LOM program from a new one, and to transfer any data values from one LOM program into another, by using a set of flexible interface instructions. Note that any FORTRAN program can be transformed without difficulties into a LOM program by using the LOM interface instructions with FORTRAN subroutines. Furthermore, statistical software with command languages such as GLIM, SPSS or BMDP, can also be called in a LOM program ; using this facility, personalized program libraries [12e] can be set up, for input data structures supplied to these programs (via command statements) can be parametrized (see, for instance, figure 5).

FIGURE 3 : Some statements of a LOM linking program

```
1 IF RESULT PCADATA
2 THEN PCA : INPUT  MATRIX DATA ; WEIGHT
3             OUTPUT MATRIX INNERPROD ; PRINCIPALVECT
4          END
5          PRINCIPALCOMP=IPROD(DATA,PRINCIPALVECT;INNERPROD) ;
6          MREG : INPUT MATRIX DATA <- PRINCIPALCOMP ; WEIGHT  END
7 ELSE MREG : INPUT MATRIX DATA ; WEIGHT  END
8 END
```

In figure 3 : *we give, as a simple example, some statements of a LOM linking program. With this program one can evaluate a multiple regression* (MREG), *either on the raw predictor data (statement 7) or on the principal components* (PCA) *taken from the raw predictor data (statement 6). The linked programs* PCA *and* MREG *are invoked at statements 2,6 and 7. At statements 2,3,6 and 7 are data interface instructions. At statement 5,* PRINCIPALCOMP *is the matrix whose (j,k) element is the result of the inner-product* INNERPROD *on the columns j and k respectively of* DATA *and* PRINCIPALVECT *matrices.* MREG *is a LOM program which allows you to carry out different multiple regression programs, in particular the* BMDP1R *one with weighted units.*

Now, let us present the top down and nested programming which can be used with LOM : we start from the LOM program named MREG and attain in three steps the BMDP1R program (University of California). First, MREG calls the LOM program named BMDPREGWEIGHT (units are weighted), as in figure 4.

FIGURE 4 : Some interface statements with BMDPREGWEIGHT program

```
OBSERVATIONS = WEIGHT &COL VARPREDICTAND &COL DATA ;
BMDPREGWEIGHT :
  INPUT SCALAR NBUNITS ; NBVAR ; DATASAVED<-FALSE
        STRING COLWEIGHT<-"1" ; COLVARPREDICTAND<-"2"
        MATRIX OBSERVATIONS
END
```

In figure 4 : OBSERVATIONS *is the input data matrix built by concatenating in column* (&COL) *the* WEIGHT, VARPREDICTAND *and* DATA *matrices.* BMDPREGWEIGHT *is called at statement 2 ; 3 to 5 are interface data statements.*

Second, BMDPREGWEIGHT calls (figure 5) the LOM program BMDP1RLOM which is equivalent to the BMDP1R program. .

FIGURE 5 :

Some interface statements with BMDP1RLOM program

```
BMDP1RLOM  :
     "PROBLEM"
     "          TITLE IS "  TITLEPROBLEM   " ./"
     "INPUT"
IF DATASAVED
THEN "          UNIT IS "  FILEDATA  " ."
     "          CODE IS "  CODEDATA  " ."
     "          CONTENT IS "  CONTDATA  " ."
     "          LABEL IS "  LABDATA  " ."
     "          TYPE IS "  TYPDATA  " ./"
ELSE "          CASES ARE "  NBUNITS(INTEGER)  " ."
     "          VARIABLES ARE "  NBVAR(INTEGER)  " ."
     ...
     OBSERVATIONS (REEL 15.7)
     "FINISH/"
END
```

In figure 5 : IF, THEN, ELSE, INTEGER *are LOM keywords ; the character strings enclosed in matching quotes belong to BMDP command language ; identifiers in elongated letters are LOM objects. These latter play a similar role to that of arguments in a CALL Fortran subroutine statement.*

Finally, BMDP1RLOM chiefly consists in an interface (figure 6) with the Fortran BMDP1R program previously transformed into a Fortran subroutine.

FIGURE 6 :

Instructions of BMDP1RLOM program

```
BMDP1RLOM
  DECLARATION
    STANDARD INPUT  FILE INPUT
    STANDARD OUTPUT FILE OUTPUT
    FORTRAN EXTERNAL BMDP1R
  END
  CALCULATION
    EXECUTE BMDP1R(INPUT,OUTPUT) END
  END
END
```

Note : *the LOM system loads only one LOM program in core memory and thus allows it to dispose of the whole available memory space.*

3.2. Automatic generators of software

In programming statistical models, researchers wish to be aided by (i) automatic generators of program instructions or (ii) specialized language. We describe in the following subsections three different tools for doing this.

3.2.1. Automatic generator of interactive FORTRAN application programs [12c] : it so happens that a users' team wants a statistical system fitted to its needs. Specifications of such personalized data analysis systems cannot be defined entirely, for needs are often incompletely formulated ; moreover, as the project is getting on, the definition of the tasks becomes more and more precise. So, several releases of the system are to be written, thus increasing the chance of errors. Furthermore, the users want programming to be done quickly . The solution chosen consists in generating automatically, in a quick and reliable way, the required programs. This is what our GALI project (Automatic Generator of Interactive Programs) does, as soon as you give the program specifications. These programs (i) give the user the possibility of choosing statistical analysis by going down and up a treatment tree with the help of menus displayed on a terminal screen, and (ii) call selected users' subroutines previously gathered in a Fortran library. These programs are specified with the help of an easily used structured language, a simplified grammar of which is given in figure 7.

FIGURE 7 : A simplified grammar of GALI language

```
<PROGRAM>          = <PRINCIPAL MENU> : ( <DISH LIST> ) END
<PRINCIPAL MENU>   = IDENTIFIER               $ PROGRAM NAME $
<DISH LIST>        = <DISH LIST>  ;  <DISH>
<DISH LIST>        = <DISH>
<DISH>             = IDENTIFIER               $ SUBROUTINE NAME $
<DISH>             = <MENU> : ( <DISH LIST> )
<MENU>             = IDENTIFIER               $ MENU NAME $
```

In figure 7 : *The character strings enclosed by matched "$" delimiters are comments, while those enclosed by matched delimiters "<" and ">" are auxiliary symbols only used to define the suitable grammar : for instance, the two syntax rules, at statements 3 and 4, allow the user to put as many dishes as he wants into a menu.* "IDENTIFIER" *is a lexical unit, generally these latter are defined with lexical rules in another grammar.* " ( ) : END ; " *are keywords of the GALI language.*

Starting from the syntax rules, associated semantic actions (for instance, at statements 5, the semantic action consists in generating the user's subroutine CALL whose name is given by the lexical analyser when it encounters corresponding IDENTIFIER in the program specifications) and lexical rules, a compiler-compiler automatically generates as many programs as program specifications provided. Note that the aid brought to a software designer becomes more powerful by using together the facilities offered by both LOM and GALI.

3.2.2. Automatic generation of the declaration block of a LOM program : In a LOM program all objects used (reals, logicals, strings, full or symmetric or diagonal or scalar matrices temporary or permanent files, etc ...) must be declared ; for a researcher this task is often dull. So, we are designing with a compiler-compiler, an interactive generator of the LOM declaration block for a given program.

3.2.3. Automatic generation of a specialized statistical language compiler : In [12b] and as indicated in 3.1., it is shown that metrics can support part of the context of a problem, that is to say part of the specifications of a model. If data handle k sets of variables, a priori for defining a metric in the unit space, one has to choose k intra metrics (diagonal submatrices) and $\frac{k(k-1)}{2}$ inter metrics (extradiagonal submatrices). So, in order to make the input of the metric in a given model easy, it is useful to be helped with a specialized statistical language. Let us present (figure 8) a simple and well-known example which derives from our general Principal Component under linear constraints model [12b-c,13]. Note that the compiler of such a specialized language is also designed with the help of a compiler-compiler.

```
DECLARATION
   X  QUANTITATIVE  : WITHOUT VARIANCE AND CORRELATION EFFECTS
   Y  QUALITATIVE   : WITHOUT VARIANCE AND CORRELATION EFFECTS
END
STRONG PRINCIPAL COMPONENT OF  Y  UNDER  X  WITH RELATIONSHIP EFFECT END
```

FIGURE 8 : *Discriminant analysis specifications in our specialized language*

In figure 8 : X *and* Y *are respectively quantitative and qualitative binary variables taken into account for a discriminant analysis.* X *is the constraint subspace, the intra metrics respectively in X and Y subspaces are Mahalanobis (or statistically equivalent) ones, that is to say* $[VAR(X)]^{-1}$ *and chisquare* (χ^2_y) *; in this particular case the inter metric (X.Y) with relationship effect is equal to* $[VAR(X)]^{-1}\ COV(X,Y)\ \chi^2_y$.

4. Some useful tool aids mainly for using statistical software

For use and especially so for self service use of programs, an algorithmic statistical language must be enriched. We shall restrict ourselves to some of the more important tools which you can find in LOM . Note that they are also improvements in statistical programming. For commodity's sake, but also for the sake of reliability, it is a good policy that a program be self-using. With suitable instructions, LOM allows you to insert the program directions for use into a "bailiff" block. Whether you run your program interactively or in a batch mode, by leading the question answering protocol described in the bailiff, you can supply all useful data and parameter values to the program before running it.

```
BAILIFF
 "GIVE THE TITLE OF YOUR STUDY" : TITLE ;
 STANDARD BAILIFF "PREDICTOR DATA MATRIX" : DATA ;
 "IDENTICAL WEIGHT (YES-NO) ?"
  IFYES TYPMATRIXWEIGHT <- SCALAR ; WEIGHT <- 1/NBUNITS
  IFNO  TYPMATRIXWEIGHT <- DIAGONAL ;
        STANDARD BAILIFF "INDIVIDUAL WEIGHT" : WEIGHT
  END
END
```

FIGURE 9 : *A possible bailiff for the linking program of figure 3*

In figure 9 : *the elongated character strings are sentences to be printed on the terminal screen ;* STANDARD BAILIFF *consists with standard sentences relative to corresponding LOM objects.*

Of course, for a linked program, all data in the interface causes the bailiff to jump the corresponding enquiring instructions. This action is necessary when you link or nest programs as in 3.1.3..With the interactive approach, if you happen to get wrong answers, you can bypass the bailiff, correct the error and then restart the bailiff. Furthermore, the programmer must design the bailiff to jump over many secondary questions : for example, although the bailiff described above is short, the bailiffs of other programs may be long (more than 100 statements), perhaps with as many as four " IFYES , IFNO " nested levels. Finally note that, as is provided in LOM, a trained user must be able to avoid the bailiff by using a command language (like FORTRAN namelist), especially when he wants to rerun program with only few modifications. According to another point of view, when a linking program is running in interactive mode, it is useful to guide the evaluation with a previously defined question-answer protocol. In LOM, this is possible with the help of the LOM object named "result name". In the linking program of figure 3, PCADATA (statement 1) is a "result name". To each "result name" corresponds one or two character strings to be printed out on the terminal screen for each "if result" LOM statement. These character strings are gathered into the "guide block " of a LOM program. See figure 10, a possible "guide block" for the program of figure 3 : the second optional question is printed out on the terminal screen at the end statement corresponding to the "if result" statement : for instance at statement 8 on figure 3.

```
GUIDE
 PCADATA   : "DO YOU WISH MULTIPLE REGRESSION ON PRINCIPAL COMPONENTS (YES-NO) ?"
           / "DO YOU WANT TO EXECUTE AGAIN PRINCIPAL COMPONENT ANALYSIS ?"
END
```

FIGURE 10 : *A possible "guide block" for the linking program of figure 3*

It is obvious that other facilities can be proposed, it is only a question of need and imagination : for instance, in LOM, you can program several kinds of menus which are more powerful than result name. Note that with these possibilities the LOM statistical system runs in both interactive and batch modes in a reliable and very modular way. Finally, it is important to note that, in using a program a statistical data base management system is a very useful tool. So, the project designed by [7] is to constitute with the LOM system the framework of the MODULAD data analysis system [10].

5. Conclusion

While the use of a compiler-compiler is recent practice [6], statisticians are already familiar with these tools [7,9,12c]. There exists many compiler-compilers more or less powerful according to (i) the power of the grammars available LR [1,8,14] or LL [3,4] , or to (ii) the possibility of using semantic attributes [8]. These tools allow you to design very modular, adaptable and reliable, software : (i) if you want to introduce a new operator in LOM, for instance ACOS, it is sufficient to add in LOM grammar the following rule and semantic, as in figure 11.

```
<FCPRIM> = ACOS ( <EXP> ) ;
##S.DGCOD(<FCPRIM>)
       CALL AFONCF4G(S.DGCOD(<EXP>),'ACOS',1,S.NOTREXP(<EXP>),MIM,
      %STYPEXP(<EXP>),MIM,  DTAREF,ECTS,DCMAUX,  S.DGCOD(<FCPRIM>))
```

FIGURE 11 : *the syntax rule and the semantic of* ACOS *LOM operator*

In figure 11 : EXP *means algebraic* EXP*ression ;* S.DGCOD *is a semantic attribut ;* AFONCF4G *is a Fortran subroutine which generates the standard statements interface with the corresponding utilitary Fortran subroutine.*

(ii) you can delete or modify your syntax and semantic with the same facility, (iii) if you design a large scale project, as LOM for instance, you can provide easily a wide spectrum of more or less powerful or specialized languages, as for cars. Note that, since each grammar is extracted from the full LOM grammar, only one compiler needs to be maintained. In the present state of evolution of compiler-compiler techniques, the problems of rapidity have been satisfactorily solved. But storage memory space issues remain to be tackled, especially as far as projects on a small scale are concerned, such as in 3.2.3. One possible solution would seem to be the creation of a wide spectrum of very modular compiler-compilers . Finally, thanks to the programming tools presented in this paper, and to other tools, such as for instance graphic systems, the conceptual level in designing is increasing, therefore feelings of complexity and having to acquire particular knowledge are decreasing, while data analysis researches and programming productivity grows in a reasonably reliable way.

References

[1] BOULLIER P. (1980), Generation automatique d'analyseurs syntaxiques avec rattrapage d'erreurs. Genève : Act. Jour. Franc. Prod. Ass. logiciels, 136-192.

[2] CROQUETTE A. (1980), Quelques résultats synthétiques en Analyse des Données Multidimensionnelles : optimalité et métriques à effets relationnels. Thèse 3ème cycle, Univ. P. Sabatier, Toulouse.

[3] GRIFFITHS M. (1975), LL(1) Grammars and analysers - In : Compiler construction. ed. F.L. Bauer, J. Eickel - Berlin, Germany Springer-Verlag, 57-84.

[4] GUYOT J., ZAFFALON L. (1982), Génération assistée d'un traducteur pour EPGAL. Rapport interne CUI, Univ. de Genève.

[5] IBRAHIM A. (1982), Quelques applications des analyses en composantes principales sous contraintes et des métriques à effets relationnels en analyse des données. Thèse 3ème cycle, Univ. P. Sabatier, Toulouse (to be published).

[6] JOHNSON S.C. : Compiler-compilers (1976), where have we been and where are we going. Proc. North Interface Symp. on Computer Sci. and Statistics, 56-58.

[7] LEONARD M., LUONG B.T. (1981), Information system design approach integrating data and transactions. IEEE Very Large Data Bases, 7th Conf. Cannes.

[8] LORHO B. (1974), De la définition à la traduction des languages de programmation : méthode des attributs sémantiques. Thèse Univ. P. Sabatier, Toulouse.

[9] MALLET J.L. et al (1980), GEOL : a graphical and statistical retrieval language for GEODATA management. COMPSTAT, Physica Verlag, Wien.

[10] MODULAD (1980), Cahier des charges - INRIA - 78150 Le Chesnay - France.

[11] RAO C.R. (1964), The use and interpretation of principal component analysis in applied research. Sankhya, ser. A., 26, 329-358.

[12] SCHEKTMAN Y. and al.
a) (1977), A software project, some new ideas in statistical computing Structural hypothesis on factors in linear data analysis in "Recent developments in statistics" Barra and al. Ed. North Holland Pub. Co. 87-98.
b) (1978), Contribution à la mesure en facteurs dans les sciences expérimentales et à la mise en oeuvre automatique des calculs statistiques sur ordinateur. Thèse, Université Paul Sabatier, Toulouse, France.
c) (1980), Utilisation des meta traducteurs dans la conception des logiciels statistiques, présentation d'exemples dans trois domaines d'application. Data Analysis and Informatics. E. Diday et al., Ed. North Holland Pub. Co, 393-410.
d) (1980), Proposition d'une métrique à effet relationnel. Propriétés et applications au cas des analyses en composantes principales sous contraintes linéaires. Cahier du C.E.R.O., 22, 201-208, Bruxelles.
e) (1981), Setting up a program library using the GLIM and LAPUMS systems. GLIM Newsletters, issue n° 4, 33-38.
f) (1982), Aided Design and Use in Data Analysis Computing. Statistical Software Newsletters, 1, 3-14.
g) (1982), Optimality in principal component analysis. Applications to nonlinear constraints (Submitted to publication).

[13] TEBBOUNE F.O. (1981), Contribution aux analyses en composantes principales sous contraintes linéaires. Thèse 3ème cycle, Univ. P. Sabatier, Toulouse.

[14] YACC (1975), Yet another compiler-compiler. Bell Laboratories, CSTR 32.

Applied Graph Analysis in the Social Sciences: The Software Project GRADAP

C.J.A. Sprenger, University of Amsterdam, The Netherlands
F.N. Stokman, University of Groningen, The Netherlands

In Section 1 it will be shown that all main types of graphs have been used to represent social systems. Section 2 gives a survey of the possibilities of the package GRADAP (GRAph Definition and Analysis Package) for graph definition, management and analysis. GRADAP provides facilities to analyze all major types of graphs. A direct data interface with SPSS provides extra data manipulation facilities and ample opportunities for analyzing multigraphs and relating social network results to other characteristics of points and lines. In Section 4 the planned extension of GRADAP is discussed.

Keywords: graph analysis, social sciences, software project, social networks, computer package

1. Introduction

The seventies have shown an increasing number of applications of graph theory in the social sciences. A graph is an object that contains points and lines, each line being incident with two - not necessarily different - points. Lines that are incident with two identical points are called loops. Two points incident with the same line are adjacent. Lines incident with the same points are called parallel. In applied graph analysis in the social sciences the points of the graph represent social entities, the lines certain social relations between the entities. It depends on the concrete empirical application whether a direction is given to the lines (directed graphs), whether the lines have a sign (signed graphs) or a certain value (valued graphs), or whether different kinds of lines exist (multigraphs). By using graph-theoretical concepts and theories such graphs can be analyzed and new graphs may be generated. The results of these analyses are again empirically interpreted in terms of the social network itself.

All main types of graphs have been used to represent social systems. Moreover, applied graph analysis in the social sciences has been used not only to represent social systems between persons, but between organizations, roles and variables as well, as the following survey shows. The survey is not intended to be complete, but aims to illustrate the wide range of applications in the past.

In small group analysis and anthropology the points represent persons and the (directed) lines social relations. In case of one social relation, *e.g.* friendship, digraphs (directed graphs without loops and parallel lines) are used to represent the social system (Hallinan and Felmlee. 1975).

In case of several types of social relations and time series data multigraphs and valued graphs have to be used (see White *et.al.*, 1976). When positive and negative social relations are considered, signed graphs are used as representation (Cartwright and Harary, 1956; Hoede, 1981, a,b). For the analysis of communication relations simple undirected graphs without loops and parallel lines and undirected multigraphs can be used (Bavelas, 1968).

In elite studies again the points represent persons, whereas the lines represent communication or influence relations. Simple graphs, multigraphs and digraphs are used to represent these social relations between elites (Kadushin, 1974; Laumann and Pappi 1976). In studies of cognitive maps and knowledge graphs the points represent variables and the (directed) lines causal relations. Signed graphs are used to represent such cognitive maps (Axelrod, 1976). In legislative analysis multigraphs and valued graphs are used to represent common activities of delegates (Stokman, 1977). Simple (di)graphs, multigraphs and valued graphs have been used to analyze intra- and inter-organizational relations between organizations (Fennema and Schijf, 1979).

Recent applications of graph theory in the social sciences make increasing use of models from which certain restrictions for social structures can be derived to test the models empirically. This development can be accelerated by graph-theoretical elaboration of existing measurement and analysis models in the social sciences. Examples of such elaborations and applications are the graph theoretical elaboration of log-lineair models (Fienberg and Wasserman, 1979), multidimensional scaling techniques (Laumann, 1973; Laumann and Pappi, 1976), stochastic cumulative scaling models (Stokman, 1977, 1980) and logistic models (Scheiblechner, 1971; 1972; 1977; Holland and Leinhardt, 1981).

2. The GRAph Definition and Analysis Package GRADAP

The package GRADAP (GRAph Definition and Analysis Package) provides facilities to analyze the main types of graphs, up to 6000 points and 60000 lines (Stokman and Van Veen, 1980). GRADAP is a batch-oriented system that interprets a user program, written in the GRADAP (application) language, and performs the tasks specified in that program. The GRADAP language has been modelled after that of SPSS, mainly to simplify its learning and its use. A second reason for this close relationship with the SPSS language will become apparent later.

GRADAP contains a number of appropriate modes designed for the analysis of simple graphs, digraphs and valued graphs with special facilities and adjustments of coefficients for bipartite graphs. Burt (1978) summarizes the different modes of network analysis that can be distinguished in the enormous flow of recent literature on applied network analysis in a typology with six categories, based upon two fundamental dimensions. For the points he distinguished three levels of analysis: individual points, subsets of points and the whole set of points in the graph. For each of these levels two approaches can be distinguished: the positional or attributional approach on the one hand and the relational on the other. On a less abstract level Freeman (1978) offers another typology which is primarily concerned with point and graph centrality measures. In this typology three basic structural concepts are distinguished, which are the basis of the different modes of centrality measures. The concepts are: degree (*i.e.* neighbourhood relations), betweenness (*i.e.* characterization by the way a point links other points) and closeness (distance). GRADAP has analytic modes and

special facilities for all these kinds of analysis. Freeman's coefficients of graph centralization have certain undesirable properties, however. Therefore a new coefficient of graph centralization has been developed at the University of Groningen with considerable better properties than Freeman's coefficients (see Section 3).
The data that make up a social network and any further (documentary) information is specified to GRADAP in the graph definition phase of a user program. This information may be saved in a GRADAP system file and accessed in later user programs, a mechanism analoguous to that supported by SPSS and other packages.
Besides this system file GRADAP also allows users to request part of the social network data to be stored in an SPSS system file (for subsequent analyses in SPSS, STAP and other packages that can access such a file).
It is also possible to have GRADAP read additional network data from an SPSS system file. This interface allows users to combine social network data analysis techniques available in GRADAP with the more standard types of data analysis available in SPSS and STAP and to use the data management facilities of SPSS and SIR. This forms the second reason for keeping the GRADAP language "in line" with the languages of those packages.

2.1. Graph definition in GRADAP

The definition of a graph or network in GRADAP consists essentially of the definition of the elements of a graph: the points in the graph or network, the information associated to the points (pointinfos), the lines, and the information associated to the lines (lineinfos). A further extension to this basic scheme is the possibility to group the points and the lines (during their definition) into pointsets and linesets, respectively. These sets will in general be disjoint, but overlap is allowed. Sets play an important role in the GRADAP language and a number of statements exist to define new sets or redefine existing ones. These will be discussed in the next subsection.
Both points and lines may be defined, together with any associated info values, either free formatted as part of the user program or by fixed format input. In the latter case that data may be either embedded in the user program or reside on an alternate input medium.
It should be realized that the flexibility of GRADAP's data structure is particularly due to the fact that is is not based on the commonly used representation of a graph by its adjacency matrix between the points. GRADAP's data structure consists essentially of a point data matrix in which the points are the cases and the pointinfos the variables defined on the points, together with a line data matrix in which the lines are the cases and the lineinfos the variables defined on the lines. The possibility to define (overlapping) point- and linesets increases the flexibility even more, because it enables a user to submit any (partial) (sub)graph for analysis.

For identification in the user program the graph as a whole, point- and linesets, point- and lineinfos are identified by unique names of up to eight characters. Further documentation is possible by labels.

Points may be identified by such names and documented by labels, but in some situations GRADAP allows labels to be used for identification as well. In user programs lines can only by referred to by their sequence number, a possibility which also exists for points.

Once a graph is defined it may be saved and accessed in later programs, thus providing a "canned" definition. GRADAP also supports addition of new points and lines to an existing graph and addition of values of points and lines in new point and lineinfos. Both types of extension may be carried out via fixed format input or, in the case of new point- or lineinfos, through SPSS system files.

2.2. Data management

GRADAP has many facilities to generate new graphs from the original graph. These data management facilities may be divided in facilities that modify the basic structure of the graph or network, i.e. create new points or lines and/or delete existing points or lines, and facilities that modify the set structure by defining new pointsets or linesets, or redefining old ones. Modifications that affect the basic structure of the graph or network are:

- induction, where one group of points induces a network formed by the points in another group and the lines that are "carried" by points of the first group;
- combination, where parallel lines are replaced by single new lines;
- condensation, where points in user-defined equivalence classes are replaced by single new points;
- sampling of points and/or lines, where the points and/or lines that do not belong to the sample are deleted;
- reversion of the direction of lines.

New sets may be defined (and already existing sets may be redefined) by enumeration of their constituting elements, by giving their definition in the form of a set expression (using set operators as AND, OR, XOR, NOT and MINUS), or by specifying a selection clause. The latter form allows users to define pointsets on the basis of pointinfo values, and linesets on the basis of lineinfo values and on the basis of pointinfo values for the endpoints of the lines.

Direction is an attribute of linesets: each lineset may be declared DIRECTED or UNDIRECTED. In the latter case, GRADAP will treat lines in the lineset as having no direction. Although the endpoints of each line have a fixed meaning (one is the TAIL, the other is the HEAD and the line, if treated as directed, is directed from TAIL to HEAD), they are treated as equivalent when the line is part of an undirected lineset.

2.3. Analytic procedures

The analysis phase of a GRADAP user program consists of an unlimited number of analysis blocks, each of which may be optionally preceded by temporary data management statements. These temporary modifications remain in effect until the analysis block is completed. Each analysis block starts with an ANALYSE SETS statement, that specifies the (partial) (sub)graphs to be analyzed in the subsequent block. The block is terminated by an END ANALYSE statement. Within an analysis block one or more analytical procedures may be activated, of which the following are currently available:

- ADJACENCY, that computes the adjacency matrix and optionally lists the isolated points and for each non-isolated point its neighbours.
- CENTRALITY, that computes point and graph centrality measures based on adjacency, distance or all sequences in the graph.
- DISTANCE, that computes the distance matrix and corresponding statistics based on distance and neighbourhood.
- REDUCE, that reduces the graph in an iterative process to the null graph, according to criteria based on adjacency and incidence.
- RUSH, that computes the centrality measure rush for points and optionally for lines. This measure is based on the structural property betweenness.
- SUBGRAPHS that detects subgraphs like (strong) (weak) components, blocks and N-cliques.
- VARIANCE DEGREE that computes measures of graph heterogeneity based on the variance of the degrees of the points.

Moreover several informative procedures are available, such as list/write graphinfo (cf list/write file info of SPSS) and list/write points/lines (cf list/write cases of SPSS).

3. An example of applied graph analysis with GRADAP

In a study on intercorporate structure in the Netherlands two types of relations between the largest 250 corporations and financial institutions are considered: financial participations and personal interlocks between the highest decision making boards (executives and directors) in these organizations. In the literature it is assumed that the tightness of an interlock depends on the combination of positions of the multiple director in the two corporations. To test this assumption three types of positions are distinguished in the Dutch corporations on the basis of involvement in decision making within a corporation. Chief executives are assumed to be involved in all major decisions within a corporation on a day to day basis; these positions are therefore considered to be inside (IN) positions. Chairmen of boards of directors, delegated directors and members of governing boards of cooperatives are involved only in major decisions, but on a more regular basis than the remaining positions; these positions are therefore considered to be intermediate (MED) positions. All other positions are considered to be outside (OUT) positions. Interlocks can now be classified into six types on the basis of all possible combinations of positions of a multiple director in two corporations: ININ, INMED, INOUT, MEDMED, MEDOUT and OUTOUTinterlocks.

We assume that the tightness of a type of interlock is reflected in the number of times that it coincides with other types of interlocks. For each type of interlock we therefore determined how often it co-occurred with another type of interlock or with a financial participation between the two corporations. With the help of the interface between SPSS and GRADAP together with several management facilities within GRADAP such as the possibility to combine parallel lines, the lines in the GRADAP file were transformed in such a way that each pair of corporations for which at least one relation existed, was connected by an undirected line with seven lineinfos, containing respectively the percentage of financial participation and the number of interlocks in each of the six types that were defined above. The results of these analyses are given in Tabel 1. The more outside the combination of positions, the looser the interlock, but INOUT interlocks are clearly tighter than MEDMED interlocks. This is supported, if we consider with which types of interlocks financial participations co-occur. Only in 3 out of 47 times, financial participations coincide solely with interlocks of a looser type than INOUT interlocks, whereas in 39 out of 47 times INOUT interlocks are among the co-occurring interlocks. Together with the results presented in Table 1 this is a strong argument to consider ININ, INMED and INOUT interlocks as primary interlocks and the other types of interlocks as secondary interlocks.

Table 1 (Co-)occurrence of financial participations and types of interlocks in the Dutch network in percentages (absolute numbers in parantheses)

	Fin.Part.	ININ	INMED	INOUT	MEDMED	INOUT	OUTOUT
Together with other type of relation	55(47)	67(2)	49(23)	37 (81)	31(19)	21 (52)	14 (71)
Sole relation	45(38)	22(1)	51(24)	63(136)	69(43)	79(200)	86(445)
Total	100(85)	100(3)	100(47)	100(217)	100(62)	100(252)	100(516)

Which consequences has the elimination of secondary interlocks for the centralization of the graph? Freeman's coefficients of graph-centralization are all based on the sum of the differences between the point-centrality of the most central point and all other points in the graph. The highest graph-centrality is attained for the graph that is a star. For graphs with different numbers of lines his coefficients are difficult to compare and are inadequate therefore. Snijders' index H of graph heterogeneity is superior for two reasons. It is based on the degree variance, which implies that differences of centrality of all pairs of points are taken into account. Moreover, it is normalized on the basis of the expected degree variance under the null model of the uniform distribution of all graphs with that number of points and lines. This makes it possible to compare heterogeneity of graphs with different numbers of points and lines (Snijders, 1981). Elimination of secondary interlocks reduces the graph heterogeneity from .25 to .22 in the large component of 190 directly or indirectly connected corporations. It gives however far more heterogeneity among the central corporations within the large component. In the network of all interlocks centrality of the corporations was determined on the basis of their degree. The 38 corporations with the highest degree were considered as the center, being the maximal subgraph of central corporations with diameter 2 in the graph as a whole. With a density of .36 the graph heterogeneity among these 38 corporations was .13. After elimination of secondary interlocks the density decreased to .09, whereas the graph heterogeneity increased substantially to .26.

4. Extensions

At the end of 1982 the second release of GRADAP will be delivered with two additional analytic procedures:

-TRIAD COUNT, to perform the triad census, as developed by Holland and Leinhardt (1970, 1975), but with special adaptations in case bipartite graphs are analyzed;

-SPATIAL AUTOCORRELATION, to compute coefficients of spatial autocorrelation.

In 1983 and 1984 extensions are planned, specially focused on the analysis of signed graphs. Moreover, at that time, a FORTRAN 77 version of GRADAP may be expected, that is portable to all major computers.

We thank W. Molenaar and W.H. van Schuur for their comments on an earlier draft.

References

Axelrod, R. (ed.) (1976), Structure of decision. The cognitive Maps of political elites, Princeton University Press, Princeton N.J.

Bavelas, A. (1968), Communication patterns in task-orientated groups, Cartwright D. and Zander A. (eds.), Group dynamics. Research and theory, Third edition, Tavistock, London.

Burt, R.S. (1978), Applied network analysis: An overview, Sociological Methods and Research, 7, 123-31.

Cartwright, D. and Harary F. (1956), Structural balance: A generalization of Heider's theory, Psychological Review, 63, 277-93.

Fennema, M. and Schijf H. (1979), Analysing interlocking directorates: Theory and methods, Social Networks, 1, 297-332.

Fienberg, S.F. and Wasserman S.S. (1979), Categorical data analysis of directed graphs: models for single generator, A.S.A. Proceedings Social Statistics Section, 1979, 407-412.

Freeman, L.C. (1978), Centrality in social networks, Conceptual Clarification, Social Networks, 1, 215-239.

Hallinan, M. and Felmlee D. (1975), An analysis of intransitivity in sociometric data, Sociometry, 38, 195-212.

Hoede, C. (1981a), The integration of cognitive consistency theories. Memorandum nr. 353, THT, Enschede, The Netherlands.

Hoede, C. (1981b), Social atoms. Part I: Kinetics. Memorandum nr. 372, THT, Enschede, The Netherlands.

Holland, P.W. and Leinhardt S. (1970), A method for detecting structure in sociometric data, American Journal of Sociology, 76, 492-513.

Holland, P.W. and Leinhardt S. (1975), Local structure in social networks, Heisse, D.R. (ed.) Sociological Methodology, 1976, 1-45, Jossey-Bass, San Francisco etc.

Holland, P.W. and Leinhardt S. (1981), An Exponential Family of Probability Distribution for Directed Graphs, JASA, 76, 33-50.

Kadushin, C. (1974), The American intellectual elite, Little Brown, Boston.

Laumann, E.D. (1973), Bonds of pluralism: The form and substance of urban social networks, Wiley, New York etc.

Laumann, E.D. and Pappi F.U. (1976), Networks of collective actions. A perspective on community influence systems, Academic Press, New York etc.

Scheiblechner, H.H. (1972), Personality and system influences on behaviour in groups: frequency models, Acta Psychologica, 36, 322-36.

Scheiblechner, H.H. (1971), The separation of individual and system influences on behaviour in social contexts, Acta Psychologica, 35, 442-60.

Scheiblechner, H.H. (1977), The social structure of large groups, Kempf W.H. and Repp B.H. (eds.) Mathematical models for social psychology, Huber H., Bern etc.

Snijders, T.A.B. (1981), The degree variance: an index of graph heterogeneity, Social Networks, 3, 163-174.
Stokman, F.N. (1977), Roll calls and sponsorship. A methodological analysis of Third World group formation in the United Nations. Sijthoff, Leiden, The Netherlands.
Stokman, F.N. (1980), Graph theoretical elaboration of cumulative scaling techniques, Quality and Quantity, 14, 277-88.
Stokman, F.N. and Van Veen, F.J.A.M. (eds.) (1981), GRADAP, Graph Definition and Analysis Package USER'S MANUAL. Volume 1 and Volume 2. Interuniversity Project Group GRADAP: University of Amsterdam-Groningen-Nijmegen.
White, H.C., Boorman, S.A. and Breiger, R.L. (1976), Social Structure from Multiple Networks. I Blockmodels of Roles and Positions, A.J.S., 81, 730-780.

Microprocessors for Biomedical Research Database Management and Analysis

F.W. Stitt, Clinical Data Research Services Inc., San Francisco, U.S.A.

The software requirements of biomedical research data management are discussed, notably the complexity and diversity of biomedical data, with some possible approaches to the problem. Statistical packages, data management and database management systems can be interfaced together, and to report generators and tabulation programs.

A fully interfaced system for biomedical research studies using a newer approach, which includes a microprocessor-based system for data entry, data management, and database management has been implemented: the system uses a 16-bit microcomputer with a hard disk and tape backup.

The operating system is UNIX, developed at Bell Laboratories, which provides a productive environment for software development or for the support of software packages.

Keywords: Microprocessors; database management systems; UNIX; statistics; biomedical research

1. Introduction

The ever-increasing demands for more information on health-related problems have placed a great burden on those responsible for the collection, organization and reporting of the data required to answer these questions.

As well as government-sponsored cohort studies of cancer or diabetes treatment, rigorous tests are now required for pre-market approval of new drugs, devices and other new therapeutic entities. An increasing number of studies are also required for monitoring of safety in the environment and in the work-place.

Biomedical data is noted for its diversity and complexity.

Existing packaged software is oriented towards statistical analysis and hypothesis testing, whereas in biomedical studies the main effort is required for data entry, data validity checking, data reorganization, tabular reporting and hypothesis generation. While researchers have good and well-tested software available for statistical analysis there is a paucity of reliable, conversational, user-oriented data and database management software, required for the bulk of the research effort.

This paper will discuss the use of microcomputers to replace data-processing which could only previously be handled on large mainframe computers.

2. Previous experience with an interfaced system

The author (Stitt 1978, 1979, 1981) has previously described the assembly of a data management and analysis system which can supply easy-to-use, conversational tools for the management and analysis of a diversity of biomedical research studies, including surveys, registries, cohort studies and clinical trials.

The software modules chosen are characterized by English-like control languages and by interfaces between modules.

The system can be used by doctors and nurses as well as computer programmers and statisticians. It is currently installed on a DEC VAX 11/750 computer.

3. Microcomputer-based applications

For some time researchers have been waiting for the arrival of a microprocessor with the power of a mini-computer and the affordability of smaller machines.

The quite recent availability of 16-bit processors, with Winchester disks and good operating systems, has made possible the down-loading of many tasks of a more routine nature such as data management and text or document processing.

3.1. Hardware Considerations

Machines consist of: a central processing unit (CPU), solid state memory, online disk storage devices, other peripheral devices such as printers and visual display terminals, and communication facilities. The particular application sets bounds on the configuration. In addition to the CPU other processors are desirable to perform memory management and disk control functions. The design and features of this part of the architecture are important factors in selecting a system.

For any non-trivial use a hard rather than a floppy disk system should be obtained. If a hard disk based system is chosen then fast tape backup facilities are essential. For future expansion a system with extendable disk and memory as well as one providing an environment for multiple users should be selected.

A good operating system will support multiple users, provided that there is enough disk space and memory.

3.2. The UNIX Operating System

UNIX was designed at BELL Telephone Laboratories and has wide and respected use. Some features of UNIX which make it an attractive choice for 16- and 32-bit microcomputer operating systems are:

1) There is a large user perceived demand for UNIX systems.

2) UNIX is very well known within universities, and even though it is not supported by AT&T, the fact that it was developed by Bell Laboratories provides a cachet not present with other systems.

3) The price difference between other operating systems and UNIX is becoming less important.

3) Many software improvements have been made to UNIX, and the development of applications is particularly easy with it.

4) It is highly portable, allowing a smooth upward migration from existing computer systems to the latest processor development.

5) Lower hardware prices for 16- and 32-bit microprocessors make UNIX more justifiable.

6) UNIX-based systems offer users the insurance they need by providing flexible and interactive productivity tools.

7) It is simple enough to be comprehended, yet powerful enough to do most of the things its users want.

8) The user interface is clean and relatively surprise-free. It is also terse to the point of being cryptic.

The UNIX market increased 106% from 1980 to 1981, with a 65% increase from January to June 1981, when 2,538 systems had been installed. In January 1982, 3098 systems had been installed, of which only 11% were commercial licenses (the remainder being within the Bell system, the Federal Government, or universities).

UNIX has many useful features: among the more relevant are complete word processing and document preparation facilities; electronic mail and calendar scheduling; document retrieval by key or from free text; networking ability from one UNIX machine to another; and many excellent programming languages.

3.3. Communicating with other machines

For most activities the machine will function alone, however for those situations requiring access to a large mainframe communications interfaces must exist.

Three interface modes should be considered:

1) The simplest is when the machine is used as a terminal operating asynchronously. This conversational mode requires relatively low transmission rates. Commands can be sent to the main machine which will interactively communicate with the user; all data will be manipulated in the main machine with the microprocessor functioning as a terminal.

2) The microprocessor will function as a remote job entry station, transmitting blocks of data and commands to the main machine, bisynchronously. This latter mode is suitable for sending large quantities of data. Such data could be destined for a centrally maintained data base or for statistical analysis while that received would be the results of analysis or retrieved data.

3) For communications with other UNIX machines the UUCP (UNIX-to-UNIX copy) utility allows highly sophisticated networking of any number of microprocessors or UNIX mainframes, such as the DEC VAX 11/780 or 11/750. The author has participated in the creation of such a network in the San Francisco Bay Area.

The notion of using a central sophisticated data base system on the main machine which can be queried in such a way that the results are made available to the microprocessor for on site processing is the simplest form of distributed data processing. For complex multi-institutional studies this should be considered.

3.4. Experience

ONYX Systems Inc. of San Jose, California manufactures a microprocessor which fulfils the requirements that we have considered essential and additionally has features that are desirable.

It is an integration of six modules: main processor, memory mass storage controller, disk drive, cartridge magnetic tape drive, power supply. It uses a large scale integrated (LSI) design based on the 16 bit Z8002 chip (processor). Main memory is expandable up to 1024k bytes, in four 256k modules (each with four 64k byte banks). A Z80A chip is used as a mass storage processor with part of its read only memory (ROM) being used for device specific formatting and error recovery routines. The remaining memory of this processor is used as fast cache for disk address storage. This can reduce considerably apparent disk access time. The mass storage controller can, in addition to the disk drive contained within the C8002, support seven external disk drives. The hard disk drive (using "Winchester" type heads) has a formatted capacity of either 10, 18, or 40M bytes. An integral part of the system is the removable tape cartridge which can store 12M bytes.

Version 7 of Bell Labs' UNIX operating system has been adapted for use on this machine. This system (renamed the ONIX) differs from the Western Electric version only in the nucleus. This means that with only limitations of size practically all software written under the UNIX operating system will be compatible. An advanced editor, text handling, and typographic facilities are part of UNIX and most document manipulation can be performed using the built in UNIX software.

4. Application software for 16-bit microcomputers

4.1. The IMP data management system

IMP is a simple yet powerful data entry and data management system written in the "C" programming language for UNIX-based systems.

IMP allows a form to be described and formatted on the screen and can be used with "dumb" or intelligent terminals. Data is then entered from the keyboard and verified field-by-field; any errors are displayed at the terminal and can be immediately corrected.

Records are indexed using up to 10 keys; this means that the records (or cases) can be retrieved sorted or one at a time based upon any combination of the keys. Over forty utilities are available for searching, listing, data reorganization or report generation.

IMP is a powerful and flexible front end for clinical trial and occupational health data processing, where the data must be entered into the computer for statistical analysis and cross-tabulation.

4.2. The INFORMIX Relational Database Management System

Nearly all computer users store information on disks, modify it regularly, and wish to query it and have reports written from it. If the application is common then a packaged program may be available to perform these tasks. This means that customization is needed, with sometimes expensive programming.

Many consultants, system designers, and programmers are turning to UNIX to increase their custom programming activity. UNIX provides not only a pleasant environment for software development, but also provides a wide variety of programs that are already written, and designed to work together. These off-the-shelf components can be used to build custom systems without starting completely from scratch, resulting in lower costs and increased productivity.

As valuable as UNIX is it lacks the most important tool for building application software, a database management system (DBMS). A database management system is a collection of programs that work together to make the construction of an application system simply a matter of a flexible set of building blocks, using simple tools.

The resulting application is built faster and is more dependable because the building blocks are sturdier than newly-written code. The same tools used to build the software make it easy to modify and enhance the application as the user's needs mature and grow.

INFORMIX is a collection of such programs designed to aid the building of single or multi-user applications on UNIX operating systems.

INFORMIX's components include an interactive query language, and interactive data entry and maintenance program, a report writer, audit trail and recovery programs, and all the utilities needed to create, modify, and optimize databases.

This set of tools is flexible enough to allow many simple applications to be simply configured, and not programmed at all. For applications where detailed customization is needed, interfaces to high-level programming languages are provided.

INFORMIX uses the relational data model which is renowned for its elegance and simplicity. The relationships between data groups do not have to be rigidly defined in advance, and a large number of variables and files can easily be maintained within the database, based upon the creation of a data dictionary used by all the other programs within the system. Additional variables can be added or indexes created at any time, interactively.

The flexibility of a relational database system lends itself to many health-care oriented database applications, such as tumor registries and occupational health monitoring.

A disadvantage of relational databases is the requirement to create temporary files after a join of two files, based on a common key (unless the software permits virtual files). This can cause problems in a strongly update-oriented environment where all the files for a single case are to viewed and updated at once. A hierarchical or network database, with the ability to store implicit relationships with "pointers", may have some advantages for administrative databases.

INFORMIX is strongly user-oriented, with a simple structure, an excellent manual, and a simple but powerful report writer.

4.3. UNIFY

UNIFY is a powerful database development tool for UNIX programmers. It is srongly oriented towards high performance on microprocessors, and has several important features:

1) The system can bypass the normal UNIX file-handling mechanism, and write directly to disk.

2) It allows explicit relationships to be described with "pointers", and therefore uses a hybrid relational-network model.

3) It uses hashing algorithms rather than B-trees: while greatly improving update times, this has the disadvantage that partial matches on keys are not possible during retrievals.

4) It has some powerful development aids, including: a menu-handler, screen-formatter, and the most extensive C-language application library.

Some disadvantages are: a lack of a database load feature, and the inability to define secondary keys. The documentation is difficult to use, although a tutorial is now available. The report writer requires programming, and is not non-procedural as with INFORMIX.

4.4. LOGIX

This relational DBMS is closely integrated into UNIX: all commands appear to the user as UNIX utilities, and are invoked from the command level; it uses UNIX files rather than an internal format, making it easy to interface with other systems.

It appears that the system may be wasteful of disk space, because every operation creates a new relation, and the history of each file is stored with the data. There does not appear to be a recovery procedure from tape, but good privacy and security is possible. The system is strongly oriented to data management; there are no adequate sceen-oriented data entry features at all.

4.5. SEQUITUR

This DBMS allows processing of variable-length records of text. It has a unique text-editor built into the system, allowing large databases of text to be handled internally. It also has a Query-by-Example language allowing easy retrievals and updates by the user, and also permits virtual files resulting from relation joins (thus avoiding the creation of large temporary files.

5. Evaluation of DBMS's for UNIX-based microcomputers

An evaluation has been made of four major DBMS which will run on 16- and 32-bit microprocessors with UNIX. These evaluations will be found in the Appendix "DBMS Evaluations". The 73 criteria were selected from a variety of sources in the literature and from personal communications, and reflect the personal biases of the author, although guided by the work of Francis (1981) in statistical software evaluation.

The ratings are graded from + = Least desirable, to ++++ = Excellent; a "!" means that the system is outstanding on this criterion, probably because it was designed to optimize the feature. "?" means that the feature cannot be evaluated from the documentation, and "-" means that the feature is absent. For the section on vendor support, a "*" indicates that the author has used the system in a production environment. IMP is not graded in this group because it is not commercially available from UC Berkeley at present, although it has extensive use within the university (as has UNIX itself).

5.1. Statistical Analysis

Attempts have been made to convert statistical packages to 16-bit microcomputers (in particular the ONYX-C002), but the problems have been overwhelming in two areas: the poor quality of the FORTRAN compilers supplied by vendors, and the inherent limitations of the 64K address space on such machines.

The author is aware that MINITAB and SPSS-11 are being converted to the LSI 11/23, and that MINITAB and BMDP are being converted to the WICAT 32-bit machine, based upon the Motorola 68000 processor. A FORTRAN-77 compiler that allows some of the character-handling of FORTRAN IV seems to be essential for this effort, since very little software has been written in strict FORTRAN-77.

It seems that a statistical package written in C, which would be portable across all UNIX systems, would be desirable.

6. Conclusions

Although most large biomedical research databases are being managed with ad hoc study-by-study programs, or with statistical packages, increasing use is being made of database management systems.

Since these DBMS's have no statistical capability, interfaces must be constructed between the DBMS and the statistical system to be used for the analysis.

Our experience has shown that these interfaces can be constructed, and there is increasing experience of this approach now available. Some data management systems have been specifically designed with this goal in mind, and more can be expected.

Other computer aids for forms development and production, data dictionaries, text encoding with thesauri and table generation have also been found to be most useful, and can be recommended.

There is at present a lack of good available software for inter-variable consistency checking, and for the production of case-wise reports from complex hierarchical databases; it is to be hoped that progress will be made in these directions soon by workers active in this field.

Exciting new developments are taking place in microprocessor systems, the newer of which will support software development or packaged applications which would previously have required machines three-to-five times the cost.

A significant factor in the utility of these machines is the availability of the UNIX operating system, with its associated utilities and programming languages. Powerful database management systems are now available for a variety of biomedical research endeavors.

References

Francis I. (1981), Statistical Software: A Comparative Review. New York: Elsevier North-Holland, Inc.

Stitt F.W. (1978), Clinical Research Data Management and Analysis - a Physician's View". Drug Information Journal: April, 98-110.

Stitt F.W. (1979), Software for Data Management and Analysis of Large-scale Biomedical Studies. in "Techniques for the Processing or Analysis of Large Data Sets". Manila: Proceedings of the 42nd Session of the International Statistical Institute.

Stitt F.W. (1981), Clinical Database Management: a Clinician's View, in "Concepts Related to Database Management". Toronto: Canadian Organization for the Advancement of Computers in Health. Fifth Conference on the Health Professional and the Computer.

APPENDIX: DBMS EVALUATION

FEATURE	UNIFY	LOGIX	IMP	INFORMIX
MODEL: (R = Relational; I = ISAM; N = Network)	RN	R	I	R
1.0 DATA MANIPULATION:				
1.1 Data manipulation processes	N	N	N	N
1.1 Privacy/security techniques	+	++++	++++	-
1.3 Disaster recovery procedures	++++	++	++	+
1.4 Data integrity controls	?	++++	?	?
1.5 Format modification ability	+	++	+++	+
1.6 Redundancy/consolidation control	++++	++	++	?
1.7 File growth	++	++++	+++	+++
2.0 QUERY LANGUAGE:				
2.1 Availability of feature	Y	Y	N	Y
2.2 Ease of use	+++	+++	-	++
2.3 Capability	++	++	-	++
3.0 APPLICATION PROGRAMMING COMPLEXITIES:				
3.1 Program/data independence	+++	+++	+	+++
3.2 Methods used to define manipulation and retrieval	++++	++++	++	++++
3.3 Subsystem view development	-	-	-	?
3.4 Schema description process	+++	-	+	+++
3.5 Programmer skill requirement	+++	++++	+	++
3.6 C language interface, N of routines	63	?	40	10
3.7 Other language support	N	N	N	Y
4.0 PHYSICAL FILE DESIGN:				
4.1 File organization	++	+++	++	++
4.2 Access method: (H)ash; (B)trees	H	B	B+	B+
4.3 Indexing methods	+	++++	++	++++
N of secondary indexes	0	7	8	N of fields
4.4 Record types supported	++++	++++	+	+
4.5 Record change capability	++	++++	+	+
4.6 Ability to combine records	+++	++++	+	+++
4.7 Virtual fields	++++	+	-	-
4.8 File space management	++	-	+	-
4.9 Logical record definition	-	-	?	-
4.10 Logical structures	+++	++++	-	++++
5.0 SYSTEM INSTALLATION:				
5.1 Physical file distribution control	?	+++	-	-
5.2 Database loading facility	-	++++	++++	++
5.3 Hardware requirements (+ = Major; ++++ = Minimal)	++++	++	++++	++++
6.0 DBMS UTILITIES:				
6.1 Performance statistics gathering	-	++	-	-
6.2 Minimum reorganization for performance	+++	++	-	-
6.3 Data dictionary	+++	+	+++	++

7.0 SECONDARY FEATURES:				
7.1 System performance	++++!	?	++++	++
7.2 DBMS maintenance policy	O++	U+	-	U++
(O = OEM; U = End-user sales)				
7.3 Systems design and development time	++	+++	++	++++!
7.4 System designer training time	++	+++	++	++++
7.5 Ease of installation	++++	++++	+	++++
7.6 Documentation	++	+++	+	++++!
7.7 Vendor support	+++ (*)	?	-(*)	+++ (*)
7.8 Vendor responsiveness to change	?	?	-	++
7.9 Customer experience	++++	?	+	++++
8.0 COMPATIBILITY WITH UNIX:				
8.1 Integration into operating system	++++	++++!	++++	++++
8.2 Integration with file system	++++	++++!	++++!	++
8.3 Accessibility with C programs	+	++	+++	+
9.0 COMPATIBILITY WITH DISTRIBUTED PROCESSING:				
9.1 Modular back-end	++	?	-	-
9.2 Concurrent access	++	+++	+	++
9.3 Audit trails	++++	+	+++	++
10.0 SYSTEM DESIGN:				
10.1 Integration of features	++	++++!	+	++
10.2 Data dictionary support of UNIX files	-	++++	++++	-
10.3 Schema entry	++++	++	++	++
10.4 Ad-hoc creation of logical structures	-	+++	-	+
10.5 Variable length and nonexistent fields	++++	++++	-	-
10.6 System limits				
Files/database	100		1	30
Fields/file		128		
Records/file		1M	32K	
Fields/database	1300			300
bytes/record	25,600		16M	2048
Files/join	No limit	2	2	2
11.0 EFFICIENCY AND PERFORMANCE:				
11.1 Bias in favor of update vs retrieval	U++++	UR	U+++	R++
11.2 Performance implications of concurrency	++	?	++	-
11.3 No reorganization	+++	+++	++++	+
12.0 MAINTENANCE:				
12.1 Security of administrator functions	++	+++	++++	-
12.2 Backup to UNIX files and tape	++++	++	+++	++
12.3 Fine tuning requirements	++++	+++	?	?
12.4 Maintenance while system is running	++	++++	++	++
12.5 No skills required	+++	++++	+	++++
13.0 APPLICATIONS:				
13.1 Application software development(*)	++++!	?	+	+++
13.2 On-line update	++++	++++	++++!	++
13.3 Alias support	+++	-	-	-
13.4 "Canned" requests	+++	++++	-	+
13.5 Development aids				
Menu handler	++++	-	-	+
Screen formatter	++++	-	++++!	-
Internal controls	++++	?	?	?
13.6 Report writer	++	+	+	++++

14.0 DATA ENTRY & VALIDITY CHECKING:				
14.1 Data entry	++++	+	++++!	++
14.2 Implicit data types	+++	++++	++++!	+
14.3 Cross-field checking	++	?	++	-
14.4 External decode tables	-	-	++++	-

Parallel Processing in Statistics

J.D. Sylwestrowicz, University of London, G.B.

Summary

Many statistical calculations can be made more efficient when performed on a parallel processor. Since early 1980 the ICL Distributed Array Processor has been in operation at the Queen Mary College, London. In this paper a brief description of the ICL DAP and its programming language DAP-FORTRAN, is given. Elementary examples of statistical applications are also discussed.

Keywords: Array processors, Parallel processing.

1. Introduction

Although, theoretical work in the field of parallel processing is quite advanced, and many algorithms have been developed, only recently parallel processing computer systems become available for practical use. The ICL DAP belongs to the class of Single Instruction Multiple Data Streem (SIMD) processors. The SIMD processors have the following common characteristics:

a) A Master Control Unit (MCU), which brodcasts identical instructions to the processing elements (PE's).

b) A set of, say NP, processing elements, each of the PE's with its own memory.

c) System by which the processors are connected, and information is transfered between processors and their memories.

Different systems may have different numbers of processors, also interconnections of processors and memories can be different for different systems. Several existing SIMD processors are described and compared in Parkinson (1982).

In the following section we give a brief description of the ICL DAP and the DAP-FORTRAN. Section 3, discusses a parallel random number generator and its application in the Monte-Carlo computations. Finally, section 4 contains some concluding remarks.

2. The ICL DAP and the DAP-FORTRAN

The DAP consists of 4096 processing elements, each with the local store of 4096 bits, equivalent to about 120 32-bit floating-point numbers. All PE's obey simultaneously a single instruction stream broadcast by a Master Control Unit. The PE's are arranged in a 64x64 matrix form. Each PE has acces to the four neighbouring elements and to their stores, so the data can be efficiently transmitted along column and row highways. A very useful programmable feature of the processing element is that it can be switched off, so an instruction is executed only in a subset of PE's.

A high-level language has been provided by ICL; this is an extension of FORTRAN with additional features that enable parallel algorithms to be expressed naturally and efficiently. The implication of the hardware structure is that 64x64 matrices, 4096 element vectors and 64 element vectors can be processed as single entities. The DAP-FORTRAN declaration:

```
      DIMENSION A(,),B(,),C(,),D()
```

defines three 64x64 matrices A,B,C and a vector of 64 elements D. The equivalent declaration in the normal FORTRAN is:

```
      DIMENSION A(64,64),B(64,64),C(64,64),D(64)
```

The DAP-FORTRAN statements:

```
      C=A+B
      A(A.LT.0.0)=EXP(B)
```

are equivalent to a FORTRAN loop:

```
      DO 1 I=1,64
      DO 1 J=1,64
      C(I,J)=A(I,J)+B(I,J)
      IF(A(I,J).LT.0.0) A(I,J)=EXP(B(I,J))
    1 CONTINUE
```

The statement C=A+B is not only concise, it represents one operation performed simultaneously in all processing elements. In the DAP-FORTRAN a logical expression (or a logical matrix) can be used. In the above example, only negative elements of A will be replaced. Also, note that the standard function EXP when operating on a matrix returns a matrix.

A very useful facility of shift indexing has been introduced in the DAP-FORTRAN, for example: $S=\sum_i a_{i-1}*a_{i+1}$; can be represented by one statement:

```
S= SUM(A(-) * A(+))
```

Here, the matrix A is treated as a long vector. In all applications where two dimensional ordering of the PE's is not necessary a DAP matrix can be treated as a long vector. SUM is a DAP-FORTRAN standard function which calculates sum of elements of a vector (or a matrix). A number of new standard functions operating on matrices and vectors has been introduced in the DAP-FORTRAN. For more elaborate examples and discussion of the DAP software, see Gostick (1979). To summarise, the DAP-FORTRAN has facilities for concise and efficient description of parallel computations performed on large data samples and their subsets. Logical indexing is a very powerful feature which allows to avoid many constructs which use IF-statements in ordinary FORTRAN.

3. Parallel random number generator and its application in Monte-Carlo computations

Sequences of random numbers are required in many areas of statistical computations. Practice has shown that in most computer applications it is convenient to use deterministic sequences satisfying certain statistical requirements which guarantee "quasi-randomness" sufficient for a given class of problems. Such numbers are called pseudo-random numbers. Initial task is to generate uniformly distributed numbers, and they serve as basis for obtaining samples from other distributions.

One of the most popular among the currently available implementations, is the multiplicative congruential generator, defined by:

$$r_i = a\, r_{i-1} (\mathrm{mod}\ m); \quad i=0,1,\ldots \qquad \ldots (1)$$

where: r_0 - the seed of the generator, a - the multiplier, and m - the modulus, are given integer constants. Uniform numbers between 0 and 1, are obtained by setting $u_i=r_i/m$.

It is easy to show that the following formulae hold for generator (1):

$$r_k = a^k r_0 (\mathrm{mod}\ m) \qquad \ldots (2)$$

$$r_{N+k} = a^N r_k (\mathrm{mod}\ m) \qquad \ldots (3)$$

In order to define a parallel algorithm the numbers (r_i) are written as a sequence of N-vectors:

$$R_0 = (r_0, r_1, \ldots, r_{N-1})$$

$$R_1 = (r_N, r_{N+1}, \ldots, r_{2N-1})$$

$$\vdots$$

$$R_k = (r_{k*N}, r_{k*N+1}, \ldots, r_{(k+1)N-1})$$

It follows, from (2) and (3) that

$$R_k = a^N R_{k-1} (\text{mod } m) \qquad \ldots (4)$$

Provided that the seed vector R_0 is given, (4) defines a parallel algorithm which generates the same numbers as (1). The DAP implementation of (1) generates simultaneously 4096 numbers (ie. N=4096). The sequence is the same as that for the NAG-library generator, ie. the parameters are set as follows: r_0=123456789*(2**32+1); a=13**13; and m=2**59. The work is in progress on an implementation of the shift register generator, see Bright (1979), which is expected to be faster than the above procedure.

To illustrate applications of the parallel generator in Monte-Carlo computations, let us consider an elementary algorithm for Monte-Carlo approximation of an integral:

$$I = \int_0^1 f(x)dx$$

STEP 1: Generate N random numbers uniformly distributed in (0,1): $r_1, r_2, \ldots, r_N$

STEP 2: Calculate: $f(r_1), f(r_2), \ldots, f(r_N)$.

STEP 3: Compute:

$$\hat{I} = \frac{1}{N}\sum_{i=1}^{N} f(r_i)$$

$\hat{I}$ is the required approximation to I. Statistical properties of the algorithm are discussed in most of the handbooks on the Monte-Carlo methods, see for example Hammersley and Handscomb (1964).

Note that in the integration algorithm, N can be chosen as a multiple of NP (4096 in the case of the DAP), so that each set of NP random numbers in STEP 1 is generated simultaneously. Then in STEP 2 each NP functions are evaluated in parallel. This superposition of parallel random number generation and parallel evaluation of the function to be integrated, gives remarkable results. In numerical examples discussed in Sylwestrowicz (1981), the above algorithm implemented on the DAP was about 100 times faster than its sequential version on the ICL 2980 (ie. approximately 25 times faster than CDC 7600).

4. Concluding remarks

Monte-Carlo methods are particularly well suited for parallel processing, since computation of a "large number of independent trials" can be performed simultaneously. Also, the total number of experiments can be adjusted to the number of processors, so that full capacity of the parallel processor can be used. In many cases, however, full utilization of the parallel processor can critically depend on the size and dimensions of a particular problem. For example in case of non-linear econometric estimation the size and complexity of the model is determined by several parameters like: sample size (T), the number of equations (n), the number of unknown coefficients (p) etc.. Chong (1982) describes the DAP implementations of programs for the maximum likelihood estimation of such models. The results are quite encouraging. The parallel routines usually worked considerably faster than their serial implementations on CDC 7600 and ICL 2980. However, in order to make efficient use of the parallel processor, different algorithms were developed for large and small samples (T>2000, T<150 respectively). It was sensible to use different optimization algorithms in both cases. Magnitude of n and p also affects the efficiency of the parallel algorithms.

It is expected that the DAP can be very useful in the areas such as survey analysis. Assuming that each processing element handles one questionnaire (up to 4096 questionnaires processed in parallel), and that all the entries can be stored in PE's local memory (4096 bits), the DAP advantages can be easily exploited for both tabulations and statistical analysis of the data. For larger surveys, of say census size, a different approach must be taken, involving data transfers and analysis of separate subsamples.

Current experience on the DAP shows that for many problems it is inefficient to construct very general subroutines or packages. In such cases the final program is best constructed from more elementary utility routines. The subroutine library of statistical utility programs is now under developement in the DAP Support Unit at Queen Mary College.

Since 1980 the DAP was successfully used for many scientific and industrial computations. In most cases its performance was superior to that of other computers, for comparisons see Parkinson and Liddell (1982). Parallel processors are very well suited for many statistical problems. It is clear that they will play significant role in statistical computing, relaxing the current limits of computer time and space.

5. References

Bright H.S. (1979), Quasi-Random Number Sequences from a Long-Period TLP Generator with Remarks on Application to Cryptography, Computing Surveys, Vol.11, No.4.

Chong Y.Y. (1982) Comparative methods of computing maximum likelihood estimates for non-linear econometric systems. Ph.D thesis, London University.

Gostick R.W. (1979) Software and algorithms for the Distributed Array Processor, ICL Tech. J. 1(2), 116-135.

Hammersley J.M. and Handscomb D.C. (1964), Monte Carlo Methods, Menthuen.

Parkinson D. (1982), Practical parallel processors and their uses, in: Parallel processing systems, ed. D.J. Evans, Cambridge University Press.

Parkinson D. and Liddell H.M. (1982), The Measurement of Performance on a Highly Parallel System, to be published in IEEE Transactions on Computers, Special issue, Nov. 1982.

Sylwestrowicz J.D. (1981), Applications of the ICL Distributed Array Processor in econometric computations, ICL Tech. J., Vol.2, 280-286.

2. Contributed Papers

Software to Link Database Interrogation and Statistical Analysis

A.J.B. Anderson, University of Aberdeen, G.B.

Anderson, A.J.B., Aberdeen, GB.

Summary

This paper discusses the extent to which database handling facilities might usefully be incorporated in statistical analysis software. The rectangular data matrix so fundamental to almost all statistical analysis packages can certainly be output as a report file resulting from database interrogation. The circumstances in which this intermediate file could usefully be eliminated are discussed and arguments are presented on which of the three database models - network, hierarchical, relational - might be considered most appropriate for statistical work.

For an integrated system, retrieval is the vital database function and essentially involves row-by-row extraction of the data matrix. Data manipulation languages developed so far are generally hosted in some high-level language, often by means of a preprocessor. This paper considers such a mechanism in combination with Anderson's (1980) approach to preprocessor-driven software for statistical analysis and comments on the feasibility and usefulness of such a system.

KEYWORDS: Database retrieval; preprocessing; host language.

Introduction.

There is considerable current interest, particularly among social scientists, in the extent to which database maintenance systems might be integrated with statistical analysis software. This is in some ways surprising since most research in the social sciences is based on the one-off survey for which the effort of establishing a full-scale database is unlikely to be worthwhile. However, there are areas (e.g. education, health care, off-shore employment etc.) where studies are essentially longitudinal and may merit such treatment.

A number of advantages might be anticipated from use of a database approach:-

(i) Data validation is likely to be well-supported as a feature of the record modification aspect of the system. Such checks will be made once and for all and need not form part of any statistical analysis.

(ii) A new user of the data can remain innocent of all parts of the available information not accessible through his own subschema (unless the ongoing research stimulates widening of the investigation and hence redefinition of subschema).

(iii) Provided the documentation associated with a particular database is of sufficiently high standard, new users will benefit from design groundwork and development experience.

(iv) For flexibility in analysis, general access methods may be advantageous. For instance, it is generally true that analysis of the variate structure of data (e.g. regression, multivariate analysis) is best achieved by cycling through the units × variates data matrix row-by-row, whereas studies of the unit structure (e.g. analysis of variance, cluster analysis) benefit from column-by-column access (i.e. transposed file).

Of course, these advantages would be present in great degree even if the database system were used only to abstract a file of data in the classical rectangular format that provides the starting point for most statistical software. It would be an additional bonus if the system could be persuaded to hand over the usually extensive declarative material such as variable names, factor level codes (=domains) and other descriptors required by the analysis package. This paper addresses the question "Is there any benefit to be gained for the user by elimination of the need for an intermediate file?".

Data maintenance versus statistical analysis.

It seems useful to begin by reviewing the respective goals of the database designer and the statistical analyst. To some extent, the two viewpoints have been converging but it is not clear that they will meet and coalesce. We are in danger of having both database systems with (probably) crude statistical appendages and statistical packages with (probably) inadequate data management facilities.

(a) The database view.

In the past, the database system designer has placed greatest emphasis on identification of the individual record. The traditional application areas of company administration, on-line reservation systems and health care provision are dominated by the create/update/delete requirement and retrieval gives rise only to the simplest listing of units. Only comparatively recently have library functions and "GROUP BY" operations on tuples been proposed to permit very simple queries involving aggregation. (See, for example, SQL; Chamberlin et al, 1976).

(b) The statistical view.

In statistical analysis, on the other hand, individual units are seldom of interest except possibly in the investigation of outliers. Analysis of a subset of records (e.g. patients with a certain disease) is occasionally needed and this has traditionally been accomplished by scanning the whole of the data matrix, rejecting unwanted units; contrast the database approach where use of keys or transposed files can shortcut this. Much more common is the requirement to repeat an analysis between and/or within categories that exhaust all units; the relative efficiencies of the complete scan and the tuple-by-tuple analysis may be very different in the usual case where all possible tuples are present.

What should be provided?

Perhaps the comments of these last paragraphs suggest that the link between statistical analysis program and database system should continue to be the data matrix or "flat file", each software area being left to concentrate on what it does best. Yet there remains a feeling that the little concrete experimentation made so far with integrated systems should be extended so that "hands on" experience can be gained. Let us consider which database facilities might reasonably be incorporated in statistical software and, more importantly, which should not. We must bear in mind that the lack of confidence of the average user of statistical software is such that he regards the computer as a somewhat unpredictable tool and should, in any event, be concentrating on whether or not his proposed analysis is a correct one. We do not wish to add to his worries.

(i) Retrieval.

First, we are really interested only in retrieval, not the traditional forms of update that operate on only a small proportion of records. Such mechanisms demand random access operation with consequent problems associated with machine failure. In any case, the usual statistical analysis requirement is to modify all records and these can simply be written to a new sequential file if it is thought worthwhile to preserve the modification. This is a considerable simplification.

(ii) Data model.

Database protagonists would argue that the form of model to be adopted is the primary decision. The natural format of much survey data and its resulting analysis would seem to suggest the hierarchical model but that is to ignore the navigation burden that has usually been placed on users by both this and the network model where the intricacies of currency indicators are almost guaranteed to cause trouble. Relational models are not without their own problems but Lachovsky and Tsichritzis (1978) have demonstrated their superiority for naive users due possibly to their resemblance to familiar flat files and tables. Nevertheless, hierarchical ideas can be incorporated in the relational model and Date (1976) has argued that, at the external level, all three models can be handled in a consistent way.

(iii) Schema access.

By this is meant the ability of the statistical software to get at and use data item descriptors (e.g. domains of attributes, missing value flags) without re-specification. In a totally integrated system this would obviously be possible but we are not yet at that stage.

(iv) Simulation.

Users need to be able to check that the output from their proposed analysis is roughly what they expect without going through all the data; this is especially so when navigational complexities are present. Either random sampling of the data must be available or, perhaps better, a mechanism to generate statistically representative pseudo-samples. This requires that domains and even relations need to be associated with probability distributions (though excessive sophistication in this respect is unnecessary.)

(v) Interactive working.

This is not unrelated to the last paragraph, since it is clearly useful to build up at least some part of the analysis request conversationally. At the very least, interactive checking of the user's commands should be a separate preliminary to a batch run and Gray (1981) has suggested that code generation for database interrogation benefits particularly from this approach.

(vi) Language syntax.

Designers of database syntax have tended to regard statistical analysis in a slightly cavalier way as being a specific function hung on the end of, for instance, a "GROUP BY" operator. This may be appealing to the purist for simple counting or averaging but it is hard to conceive of a full analysis of variance being successfully glued into a data access statement in this way! The opposite view - traditional among statistical software designers and, therefore, users - regards data access as the trivial function, nebulously labelled "get". The difference between these is probably imagined rather than real (reflecting relational/non-procedural versus hierarchical/procedural approaches) and is therefore a worthy area for experimentation.

Preprocessor-driven software.

In this section, the case is argued that the requirements listed above are best served at present by means of preprocessor-driven software. The technique is well-known to database researchers and Anderson (1980) has previously discussed its usefulness for statistical work. The associated design flexibility is of great benefit during the current experimental phase of integration and development. If the analysis language is to be sufficiently powerful for use by professional statisticians it must be hosted in, say, Fortran, but this is easily disguised from the naive user.

We can assume that the statistical summary algorithms (but not necessarily the users) view the data as a single row of the data matrix. Even when the structure is hierarchical, all required ancestor data would be maintained as part of this vector so that work area specification is unnecessary. The formation of all requested summary statistics in a single data pass as recommended in Anderson's (1980) paper is thereby facilitated.

If we take the procedural approach, this data pass will be a "DO FOR EACH" loop with "GET" or "GET NEXT" calls at appropriate points. Obviously, in a retrieval-only situation, the operations are not difficult to implement and the preprocessing involved is relatively trivial.

For the relational model, currently-proposed syntax is a little clumsy, looking something like

```
<relation>:=<relational expression> group by <list of variables>

     creating <variable>:=<aggregate function>(variables expression)
```

e.g.

```
     AVGRANT:=STUDENT group by SEX,FACULTY creating AV:=avg(GRANT)
```

Many users would find this hard to cope with, especially as it does not produce a table in the usual statistical sense but rather a relation. Furthermore, it is not clear that all statistical summaries can reasonably be represented even as relations. Nevertheless, Gray's (1981) paper shows clearly that a preprocessor-driven system can utilise a relational formulation to interrogate databases with considerable flexibility and efficiency. The generated Fortran code eventuates in a call to a single routine that could obviously incorporate a further call to a preprocessed statistical analysis procedure.

Conclusion.

It may well be that the requirements of most users are such that efficiency is best achieved by use of the traditional data matrix abstracted from the database. This would partly depend on the extent to which users' lack of proficiency in database interrogation increases the number of computer runs needed to complete an analysis.

It therefore seems useful to conceive of databases, established and sustained by one or other of the relatively standard systems, but analysed by preprocessor-driven statistical software incorporating only the retrieval component of database interaction. By this means, experience would be gained in both computer efficiency and user reaction: the author's earlier work is being expanded in this direction.

Probably most model variants could be handled since the data accessing interface is usually relatively standard and modular and hence interchangeable. Unfortunately, software to interrogate the schema is often buried more deeply in the system and is therefore not at all standard. This makes "double description" of the data much harder to avoid.

References.

Anderson, A.J.B.(1980), The use of preprocessors in statistical software, COMPSTAT 1980 Physika-Verlag, Vienna.

Chamberlin, D.D., Astrahan, M.M., Eswaran, K.P., Griffiths, P.P., Lorie, R.A., Mehl, J.W., Reisner, P. and Wade, B.W. (1976), SEQUEL 2: A unified approach to data definition, manipulation and control, IBM J. Res. D., 20, 560-575

Date, C.J.(1976), An architecture for high-level language database extensions, Proc. 1976 ACM SIGMOD International Conference on the Management of Data.

Gray, P.M.D.(1981), Use of automatic programming and simulation to facilitate operations on Codasyl databases, Infotech State-of-the-Art report "Database" ed. M.P.Atkinson.

Development of a Statistical Language

R.J. Baker, M. Green, M.R.B. Clarke, M. Slater, R.P. White,
Rothamsted Experimental Station, Harpenden, G.B.

Summary: GLIM-3 is currently being superseded by Prism. This paper explains some of the new language features of Prism and describes the additional data structures and multidimensional array handling supported by the system.

Keywords: GLIM, Prism, multidimensional arrays, syntax, graphics.

1. Background

Since its first release ten years ago the statistical package GLIM has proved to be a most popular and useful tool for statisticians. It provides a simple, interactive framework for specifying and fitting generalised linear models (e.g. Normal regression, log-linear and logit models) and for displaying and manipulating the results of each fitting. The last release, GLIM-3 (see Baker and Nelder, 1978), was first distributed in 1978 and the number of copies now in use exceeds 500.

Since 1978, there has been a demand for an interactive analysis-of-variance package based on the same algorithm as the ANOVA directive in Genstat (see Alvey et al, 1977) but set into the syntax of GLIM. Similarly, interest in the use of computer graphics for statistical analysis led to the development of a graphics facility using a GLIM-like language (see Slater and Baker 1982) while work has continued on the fourth release of GLIM itself (see Clarke and Baker, 1982).

It soon became apparent that these developments would overlap unless the programming requirements common to all three programs could be centralized into a 'Kernel' of routines to which the three programs would have access. Such requirements include the creation and deletion of structures, the allocation and recovery of space, handling of input/output etc... .Furthermore, by performing syntax analysis in the Kernel it is possible to ensure a consistent language across the three programs, a bonus to users already familiar with the syntax of one of the programs.

The name Prism (**pr**ograms for **i**nteractive **s**tatistical **m**odelling) has provisionally been adopted for the whole system, and the individual programs (or 'modules') are known as AOV (the analysis-of-variance module) GRAPH (the graphics module) and GLIM (the module for generalised linear interactive modelling). Any combination of modules that includes the Kernel can be used to form an executable program, so that, for example, the Kernel plus the GLIM module is the replacement for GLIM-3, while the Kernel plus AOV plus GRAPH is an analysis-of-variance program with graphical facilities.

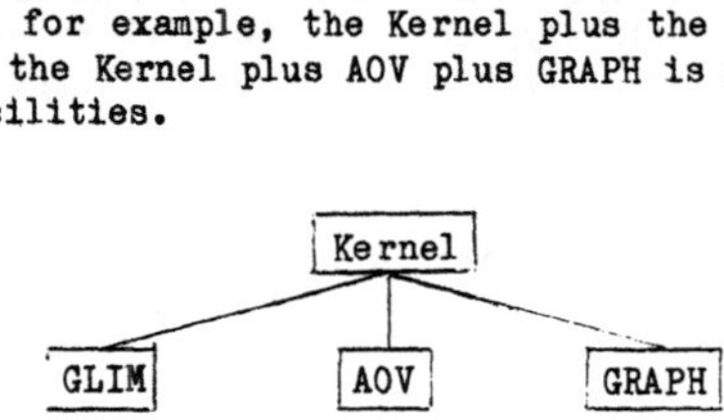

modular arrangement of Prism

There is full compatibility between modules loaded together in the same system, in that commands for one module may be freely interspersed with commands for other modules, and structures produced by one module are always accessible to other modules.

In the next two sections we discuss the type of structures that Prism now supports and some syntactic features of the language. Finally in section 4 we discuss the manipulation of data as multidimensional arrays.

2. **Prism structures**

The Prism language is based on that of GLIM-3 with certain changes and extensions to allow greater convenience in storing values and setting options, more flexibility in argument-passing for macros and a more English-language look in certain directives.

As in GLIM-3 the language is directive-based. That is, each command to the system takes the form of a directive-name (beginning with a dollar) to specify the action required, followed by certain 'items' (depending on the directive) to modify that action. Additionally, in Prism each directive-name may be followed by a 'temporary output-setting', to reset the output channel details and accuracy for the duration of the directive. And, instead of a directive, a 'direct-macro-invocation' may be used; this takes the form of a macro identifier (preceded by an ampersand) specifying the macro to be invoked, followed by the arguments for that macro; as macros can be defined during a job this effectively gives a dynamic extension to the set of commands available to the user. Directives and direct-macro-invocations are known collectively as 'statements', and the dollar, colon and ampersand are known as 'statement symbols'.

The items needed in GLIM-3 directives fell into 4 groups: identifiers (e.g. vector, scalar or macro identifiers), values (e.g. numbers, integers, strings), pathwords for naming actions (e.g. the items for the LINK and DISPLAY directives) and operators. In Prism there are more such groups, as well as a greater variety within each group.

The most noticeable difference between GLIM-3 and Prism is in the variety of structures now supported. A structure is a location for storing a value and is referenced by its identifier; each structure is of a specific 'type', which determines the sort of value it can store and the operations that can be performed on it. In Prism there are *scalar*, *vector*, *range*, *text*, *graphic* and *referential* structures. *Scalars* are: *numbers* for holding real numbers, *integers* for non-negative integer values and *labels* for denoting a member of a finite set. *Vectors* are: *variates*, which are n-tuples of real numbers, *counts*, n-tuples of non-negative integers, and *factors*, n-tuples of *label* values. The *vector* and *scalar* types reflect the kinds of measurements that occur in statistical analysis. *Number-ranges*, *integer-ranges* and *label-ranges*, the three types of *range* structure, are used to specify subsets of the real numbers, subsets of the non-negative integers, and finite sets, respectively; they may be used to set or restrict the values that a *vector* or *scalar* may take or to set the mappings for converting between the different types of *vector* and *scalar* values.

In general, conversion between the different types of *vector* and *scalar* value is done automatically by the system in an obvious way - thus, the ith distinct *label* value of a *factor* will be converted to the *integer* i, or to the *number* float(i). But the user has the option to specify his own conversion for each *vector* or *scalar*, or that no conversion is to be performed.

There are 3 *text* structures - *strings* (sequences of characters enclosed within quotation symbols), *macros* (sequences of statements) and *modules*; the latter are the 'compiled' form of *macros*, in the sense that the original statements have been converted into an internal code that executes more efficiently. *Strings* can be used for headings, captions etc or can be inserted in-line within a directive; they cannot contain a statement symbol. *Macros* begin and end with statement symbols and are used for program-control. *Modules*, outside of their defining directive, are interchangeable with *macros*.

Two *graphic* structures are supported: *figures* and *shapes*, the latter being an n-tuple of the former. A *figure* value is the specification of a two-dimensional graphical object (e.g. a circle, line, box etc) in a coordinate-free way. *Graphic* structures are usable only within the GRAPH module - see Slater and Baker (1982).

Finally, *referential* structures are either *pointers* or *records*, the latter being an n-tuple of the former. A *pointer* can store an identifier, a value (of almost any type) or a pathword. *Pointers* existed in GLIM-3 (e.g. %YV, %PW) in all but name; but could only store identifiers. *Records* are used to store lists -e.g. argument-lists for *macros* or data lists for positioning graphical *shapes* - see Slater and Baker (1982).

Additionally, whereas pathwords in GLIM-3 were restricted to one letter, Prism uses four letters but allows less if the pathword can still be recognised.

3. Prism syntax

As important as the structures themselves are the syntactic rules that determine how their identifiers, values, etc may appear as items in Prism directives. In GLIM-3 there were few general rules of syntax, and most directives set their own style. In the more diverse environment of Prism it would be undesirable to allow a similar proliferation of rules. At the same time, any attempt to impose some common syntactic framework on all directives could easily lead to the syntax for particular directives being unnatural and in the long term would hinder the development of new language features. Hence it was decided to retain each directive's individual style, but to allow variations only over a given selection of 'syntactic constructions', these being lists, option-lists, phrase-lists, formulae/expressions and clauses, as described below.

The needs of each directive would determine its syntax, but this must be expressed through the syntactic constructions provided. Putting it a different way, a small set of general rules determine the form of certain syntactic constructions, and the syntax rules for each directive determine which constructions will be used in that directive.

'Lists' are sequences of identifiers, values or pathwords, optionally interspersed with commas, as in the DATA directive. Identifiers, values and operators can also be combined into formulae/expressions as in the TABULATE and CALCULATE directives.

'Option-lists' are used to select from a linear sequence of settings. E.g.

```
$OUTPUT    channel=6,  height=24
```

An optionword specifies the setting required and is followed by an identifier, value or pathword as appropriate. If an optionword is omitted settings are taken in serial order.

When the choice of options is too complicated to be presented as a linear sequence then it may be structured as a list of 'phrases'. Each phrase begins with a phraseword (which follows the same construction rules as a pathword) and may be followed by further phrasewords or ultimately by an identifier, value or pathword, a formulae/expression, or a list or option-list. E.g. in

```
$TABULATE    the MEAN    of Y    by A.B.C
```

there are three phrases, begun by the phrasewords 'the', 'of' and 'by'. The GRAPH module provides an example of a nested phrase:

```
$DRAW    the    histogram    data=X,  intervals=*
```

where 'the' and 'histogram' are both phrasewords, which are then followed by an option-list.

Finally, in certain circumstances, a directive may be divided into a sequence of 'clauses', the clauses being separated by semi-colons, and each clause being composed of phrases. The principle purpose of the clausal construction is that options set in earlier clauses will remain in force through succeeding clauses until changed or until the end of the directive, whichever occurs first. E.g.

```
$TABULATE    the MEAN     of X    by A.B.C    ;
             of Y  ;  the TOTAL   of Z        $
```

will tabulate X, Y and Z by A, B and C producing the means of X and Y but the total of Z.

4. Multi-dimensional arrays

Although it is usually convenient to store statistical data as *vectors* (i.e. sequences of values), it is often convenient to manipulate them as if the values formed a multidimensional array. However, rather than create a new 'array' type of structure, it is simpler to append to each *vector* the attributes needed to manipulate the *vector* values as an array. Thus an array in Prism is a *vector* with extra attribute values to record the shape of the array, that is the number of dimensions and the number of levels for each dimension. These attribute values are either declared through the VARIATE, COUNT or FACTOR directive or may be set through the array calculate directive, ACALCULATE.

To comply with the rest of the Prism syntax no explicit looping over the elements is provided and hence there is a need for a powerful syntax within ACALCULATE to provide the common operations in a simple and concise form. This is achieved by the use of a syntax similar to that for the array operations in the language APL. Not all the facilities of APL are provided though some additional operations are included. Wherever possible the syntax has been made to comply with that of CALCULATE in GLIM, so that the usual arithmetic operations are element-by-element operations (direct dyadic). To retain portability Prism restricts itself to a small character set, so that to cover the multitude of operations needed, double operators, combinations of operators and functions are used. The conciseness of the resulting expressions far outweighs the initial difficulty the novice may experience in using ACALCULATE.

The direct dyadic operators are the usual arithmetic operators, a minimum and maximum operator and the relational operators which have a result of zero for false and one for true. For direct dyadic operations the left and right operands must have the same shape. The exception to this rule is assignment since both the array attribute values and data values are transferred on assignment and hence the left operand takes the shape of the right.

More complex operators are formed by combining these direct dyadic operators with a prime. These define general forms of outer product, inner product and margining operations in which the element by element operation can be any of the direct dyadic operations. Thus ordinary matrix multiplication becomes one example of an inner product and margining could, for instance, produce margins of the products of the elements.

Further generality is provided through the indexing of arrays which is taken to be a regular subset of the elements of the array forming another array. Functions are also provided for manipulating sets of elements that do not form a regular array. The shape of an array may be readily accessed and since the shape is itself defined as an array (in particular, a count) it may be manipulated within ACALCULATE. Operations are provided to facilitate the setting of data values without reading them in and to set or reset the shape of an array within an expression. Many other operations are provided through functions, for example %PERMUTE which is the multi-dimensional extension of transposition.

Another form of array is the table which in Prism is taken to be an array for which the dimensions have labels and the levels within a dimension also have labels. The attribute values contain pointers to the various label-ranges. An array is a table with the default labels corresponding to the integers and can always be treated as a table. On output through TPRINT the table will be labelled making such output easily understood. Tables can be manipulated through TCALCULATE exactly as with ACALCULATE except that dimensions and levels are referred to via their labels.

This labelling relieves the user of the task of keeping track of the dimension and level he/she should be considering. A further benefit is that wherever possible in dyadic operations the operands will be made conformable though permutation and expansion. Expansion is carried out by repeating the data values if a dimension is missing and inserting missing value if a level is missing. Missing values are allowed in all operations in TCALCULATE and ACALCULATE though clearly in certain cases a missing value will trigger a fault.

The TABULATE directive forms tables of summary statistics from raw data classified by factors and/or grouped variates. The combination of this with TCALCULATE or ACALCULATE provides powerful tools for data preparation and manipulation either as a precursor to analysis or as an aid to the interpretation of the results of a GLIM analysis.

References

ALVEY,N.G. *et al* (1977). *Genstat Manual* Rothamsted Experimental Station, Harpenden, Herts, U.K.,

BAKER,R.J. and NELDER,J.A. (1978). *The GLIM System, Release 3*. NAG: Oxford.

CLARKE,M.R.B. and BAKER,R.J. (1982) 'GLIM-4: a wider choice of methods and greater control of resources', COMPSTAT 82: Short Communications, Summaries of Posters, Physica-Verlag.

SLATER,M and BAKER,R.J. (1982) 'The GRAPH module of Prism: an interactive program based on GKS', Proceedings of Eurographics 82, Manchester: North-Holland.

MICROSTAT: A Microcomputer Conversational System for Statistical Data Analysis

S. Blumenthal, I.U.T. of Paris, France

Microstat is an entirely conversational software package designed for multidimensional data analysis. It is written in BASIC to be used on an APPLE II microcomputer. On the microcomputer level it offers the usual facilities of statistical packages such as BMDP, SPSS or SAS.

Multidimensional data analysis - conversational - microcomputer

1- At the Statistics Department of the Institut Universitaire de Technologie of Paris, the use of microcomputers for the teaching of statistics at the undergraduate level goes back to 1977.

It was the graphic facilities offered by the Apple II microcomputer which led us to write a software for the empirical study of convergence, tending to the Normal distribution, of some usual statistical laws. The software was to be used by students without the mathematical background normally required for the understanding of a theoretical proof.

The great interest shown by the students for this type of pedagogical approach encouraged us to go further in this direction and we are now able to offer the students, along with our traditional course of statistics, two types of microcomputer based activities, both of which aim at developping a feel for statistics, even at the undergraduate level.

In the first phase, students are asked to produce elementary statistics oriented programs to be used on the Apple II. In the second phase, they are introduced to Microstat, a statistical software package developed at the Statistics Department.

The purpose of Microstat is to teach students how to solve some of the usual statistical problems. Students having gone through this training programme should have acquired the methodology enabling them to cope with general use packages as BMDP, SPSS, SAS or GENSTAT.

2-Microstat was designed to be user friendly, mostly to undergraduate students, with little or no knowledge of computer science.
It enables one to go through the most important steps of a statistical survey analysis, beginning with the formatting of a questionnaire, the data handling facilities and going to various exploratory methods such as Principal Component Analysis or Factorial Analysis of Correspondance, as well as the sort of modelisation given by Multiple Regression, without forgetting the test facilities.

As it is, clearly, not possible to provide more than a limited number of standard procedures, our choice was not made on the basis of a strict professional efficiency, but rather by taking into account the undergraduate student's ability to cope with the theoretical background to such procedures. And indeed, in our course we emphasize the need for a thorough understanding not only of the practical but also the theoretical aspects of the statistical tools, which, ultimately, lead to more reliable results.
Of course it is not possible to describe a software system and demonstrate its abilities on just a few pages, so we will merely try to sketch the essential features of our system.

3-As already mentioned, the entire system was designed on a user-friendly basis,thus remaining conversational at every stage. At every stage, and whenever it is deemed necessary, it is entirely up to the user to supply relevant data and select whatever processing sequence is needed for the completion of his project, e.g. the values of optional parameters are to be supplied as you go along. For example, the computation of a couple of supplementary factors in a Principal Component Analysis can be done when, and only when, the user, having gone through a thorough inspection of the preceeding results, has decided to go on.
Due to constraints of microcomputers available at present we decided, whenever possible, to design programming structures that would allow us to confine the computations to the data entry process. And indeed, it is worth waiting a few extra seconds after each data batch has

been entered rather than waiting whatever number of minutes it will take to obtain the results after the last data input. Batch size was determined on the basis of screen size and with a view to a set of ergonomic criteria.

To facilitate the implementation of Microstat to various other microcomputers, all programs were designed according to the LCP-Warnier programming method (structured programming). Each one of them is supported by an LCP flow-chart, a program listing in Applesoft Basic and a typical application.

An optimal use of Microstat implies having two floppy disk drives and a line printer connected. It is nonetheless quite possible to use Microstat on a computer with only a single drive connected.

4-The present possibilities of Microstat are based on the conversational access to a set of inter-related frame of modules by mean of accessory selecting programs. Microstat therefore includes the following facilities:

-a module for the handling of statistical data files allowing one to create, update, merge, sort data files and select sub-files according to various criteria pertaining to variables and/or to sample units. The first module can also be used to generate those specific files which serve as entries of the actual data processing modules.

- various specific modules allowing the usual data processing of uni- or bi-dimensional descriptive statistics (table and graph production, correlation analysis etc.) or production of values of the most usual statistical distributions.

- a large module for multi-dimensional data analysis, such as Principal Component Analysis, Factor Analysis of Correspondance or Multiple Regression.

Finally to ensure its adaptability to a wide range of situations Microstat includes a module to update existing selection programs or even to create new ones according to the specific needs of any given user. Conversational possibilities of the system are then maintained throughout the process.

To illustrate the capabilities of MICROSTAT , we include below a sample of what a working session might look like on the screen when only one disk drive is available.
The problem to be dealt with was a factorial analysis including the re-configuration required by the use of a single disk drive. Total processing time took less than three minutes after input was completed. Of course it was not possible to reproduce here the purely graphical output.

```
                MICROSTAT
PROGRAMME CONVERSATIONNEL
D'ANALYSE DE DONNEES STATISTIQUES
  SERGE BLUMENTHAL  IUT DE PARIS
    1982 APPLE II 48K
VOULEZ-VOUS UN RAPPEL DES POSSIBILITES DE MICROSTAT? O/N?N
DATE:17/5/82


PORT/SLOT 6
LECTEUR DE PROGRAMMES 1
LECTEUR DE DONNEES 2
IMPRIMANTE 0

VOULEZ-VOUS RECONFIGURER? O/N?O
NUMERO DE PORT/SLOT ?6
NUMERO DE LECTEUR DE PROGRAMMES ?1
NUMERO DE LECTEUR DE DONNEES ?2
IMPRIMANTE EN ?0

DATE ?17/5/82


VOULEZ-VOUS :
1 TRAITER DES SERIES SIMPLES
2 TRAITER DES SERIES DOUBLES
3 TRAITER DES SERIES MULTIPLES
4 TRAVAILLER SUR UNE BASE DE DONNEES
5 CONSULTER DES TABLES
6 UTILISER DES PROGRAMMES SPECIAUX
7 QUITTER

DONNEZ LE NUMERO DE L'OPTION CHOISIE ?3


1 FAIRE DES ANALYSES FACTORIELLES
2 FAIRE DES REGRESSIONS MULTIPLES
3 FAIRE DES TRIS
4 REVENIR AU MENU PRINCIPAL
5 QUITTER

DONNEZ LE NUMERO DE L'OPTION CHOISIE ?1

VOULEZ-VOUS :

1 FAIRE UNE NOUVELLE ANALYSE
2 REPRENDRE UNE ANALYSE ANTERIEURE
```

```
VOTRE CHOIX :?1

PROGRAMME D'ANALYSES FACTORIELLES

1 AFC
2 ACP
3 ACPN
VOTRE CHOIX:?1

DONNEES DEJA SUR DISQUE? O/N ?O

NOM DU FICHIER DE DONNEES? VACANCES LEB306

ETAPE DE DIAGONALISATION DE LA MATRICE
CHARGEMENT
CHARGEMENT TERMINE
LES CALCULS COMMENCENT PRENEZ PATIENCE !

TRACE DES VARIABLES OU OBJETS
NUMERO DU FACTEUR X ? 1
NUMERO DU FACTEUR Y ? 2

APPUYEZ SUR RETURN


CONTRIBUTIONS ABSOLUES
VALEURS PROPRES       .0443     .0202
POURCENTAGES          53.00     24.22

HOTEL                 35.85     21.80
HABITANT              1.43      10.14
MAISON                26.88     6.21
PARENTS               3.77      41.67
AMIS                  1.7       .02
CAMPING               25.39     12.27
VIL-VAC               4.4       5.66
DIVERS                .54       2.19

RETOUR AU GRAPHIQUE? O/N?N

VOULEZ-VOUS DES FACTEURS SUJETS? O/N?N


VOULEZ-VOUS :
1 REVOIR DES GRAPHIQUES DES VARIABLES
2 REVOIR DES GRAPHIQUES DES SUJETS
3 CALCULER DES FACTEURS DES SUJETS
4 AJOUTER DES ELEMENTS SUPPLEMENTAIRES COLONNES
5 AJOUTER DES ELEMENTS SUPPLEMENTAIRES LIGNES
6 EXTRAIRE DE NOUVEAUX FACTEURS
7 REVOIR DES GRAPHIQUES SIMULTANES
8 REVENIR AU MENU PRINCIPAL
9 QUITTER

VOTRE CHOIX :?9
```

An Algorithm for Nonnegative Least Squares and Projection onto Cones

J.M. Bremner, University of Canterbury, G.B.

Keywords: Constrained least squares; quadratic programming; sweep operator; stepwise regression; convex regression; restricted maximum likelihood estimation.

Summary

An algorithm for fitting linear models in which some of the parameters are restricted in sign is described. The algorithm is based on standard computational routines for stepwise regression and may be implemented using a suitable interactive regression package.

1. Introduction

A problem which has been considered by, amongst others, Waterman (1974) and Armstrong and Frome (1976) is that of fitting a linear model when certain parameters are restricted to take nonnegative values. Another problem which has received some attention (see Barlow et. al., 1972, Section 4.1) is that of finding the maximum likelihood estimate of the mean of a multivariate normal distribution where some, or all, of the elements of the mean vector are constrained to be non-negative. Although this is a generalized least squares problem, it may be transformed into a least squares problem by premultiplying the observations by a square root of the inverse of the dispersion matrix. After this transformation the problem becomes that of finding the closest point to the transformed mean vector in a region defined by a number of homogeneous linear inequalities.

The two problems above may be solved by quadratic programming methods. There is, however, for the first problem in particular, a certain appeal to the statistician in a method of solution based on standard computational routines for regression. Although such methods have been described by a number of authors, we shall in this paper describe a new method which is more efficient than its predecessors. Due to the duality relationship between the two types of problem, the method can be used to solve problems of either type.

2. Projection onto Cones. Duality

The nonnegative least squares problem is that of finding, for a given vector of observations $\underline{y}$, the closest point $\underline{y}_1$ of the form $\underline{y}_1 = X\underline{\beta}$, for some $\underline{\beta} \geq \underline{0}$. It is thus the problem of finding the orthogonal projection of $\underline{y}$ onto a polyhedral cone (see Stoer and Witzgall, 1970, Chapter 2, for definitions and theory relevant to this section). To see that our second problem is also of this form, let $\underline{x} \sim N(\underline{\mu}, \sigma^2\Lambda)$, where $\underline{\mu} \geq \underline{0}$. Then $\underline{y} = \Lambda^{-\frac{1}{2}}\underline{x} \sim N(\Lambda^{-\frac{1}{2}}\underline{\mu}, \sigma^2 I)$, and the problem of maximum likelihood estimation of $\underline{\mu}$ is equivalent to that of finding the closest point $\underline{y}_2$ to $\underline{y}$ in the region defined by the inequalities $\Lambda^{\frac{1}{2}}\underline{y}_2 \geq \underline{0}$: again we have a problem of orthogonal projection onto a polyhedral cone. Note that, although we shall describe methods for problems with all parameters nonnegative, extensions to problems with only some parameters restricted, or with some taking nonpositive values, are straightforward.

It is well known that the problem of fitting the model $\underline{y} = X\underline{\beta} + \underline{e}$ by least squares may be regarded as one of decomposing a vector of observations $\underline{y}$ as $\hat{\underline{y}} + \underline{r}$, where $\underline{y}$ is the vector of fitted values lying in the subspace generated by the model, i.e. the range of the design matrix X, and $\underline{r}$ is the vector of residuals lying in the orthocomplement of that subspace, i.e. the null space of X'. The nonnegative least squares problem is equivalent to that of finding a similar decomposition of $\underline{y}$ into mutually orthogonal components lying in a dual pair of convex cones. To clarify this, we require a definition and a theorem.

Definition If C is a cone, its *dual cone* is

$$C^* = \{\underline{y} | \underline{y}'\underline{x} \le 0 \text{ for all } \underline{x} \in C\}.$$

Note that, if C is closed and convex, the property of duality is reflexive, i.e. $C^{**} = C$.

Theorem 1 If C_1 and C_2 are dual cones, then any vector $\underline{y}$ may be uniquely decomposed as the sum of orthogonal vectors $\underline{y}_1$, $\underline{y}_2$ where $\underline{y}_1 \in C_1$ and $\underline{y}_2 \in C_2$. The vectors $\underline{y}_1$ and $\underline{y}_2$ are the orthogonal projections of $\underline{y}$ onto C_1 and C_2 respectively.

The above theorem indicates that the problem of projecting onto a cone and that of projecting onto its dual are equivalent, in the sense that a solution to one provides a solution to the other. To see the relevance of this to our problems, we observe that the dual to the cone $C_1 = \{\underline{y}_1 \mid \underline{y}_1 = X\underline{\beta},\ \underline{\beta} \ge \underline{0}\}$ is the cone $C_2 = \{\underline{y}_2 \mid X'\underline{y}_2 \le \underline{0}\}$. Thus, although our notation in what follows will relate to the nonnegative least squares problem, for which the solution is, in terms of the decomposition of Theorem 1, $\hat{\underline{y}} = \underline{y}_1$, $\underline{r} = \underline{y}_2$, we may use the same methods for our second problem on substitution of $-\Lambda^{\frac{1}{2}}$ for X: the maximum likelihood estimate of $\underline{\mu}$ is then $\Lambda^{\frac{1}{2}}\underline{y}_2$.

3. Testing for Optimality. Computational Principles

Although a number of authors, including Theil and van de Panne (1960) and Fletcher (1971), have developed general quadratic programming algorithms in which the solution is obtained by projecting onto an appropriate subspace, such methods are relatively uncommon. They are, however, ideally suited to the nonnegative least squares problem, since efficient computational procedures for projecting onto the relevant subspaces (i.e. fitting unrestricted submodels) are well known. Waterman (1974) used such procedures as the basis of his method. He performed an 'all possible regressions' search, discarding all submodels having negative estimates for one or more parameters, and chose from the remaining submodels that with the smallest residual sum of squares. In order to improve on this approach we shall require an efficient strategy for searching submodels and a test to determine when an optimum has been reached.

We now require further notation and definitions. If our model has p parameters, we shall use subsets of $P = \{1,2,...,p\}$ to index submodels. For $I \subset P$, we shall use the term 'submodel I' when referring to the model $E(\underline{y}) = X_I\underline{\beta}_I$, where X_I and $\underline{\beta}_I$ are obtained by selecting columns and elements corresponding to elements of I from X and $\underline{\beta}$ respectively. Then the vectors of parameter estimates, fitted values and residuals which result from fitting submodel I are

$$\hat{\underline{\beta}}_I = (X_I'X_I)^{-1}X_I'\underline{y}$$

$$\underline{y}_I = X_I(X_I'X_I)^{-1}X_I'\underline{y}$$

and

$$\underline{r}_I = (I - X_I(X_I'X_I)^{-1}X_I')\underline{y}$$

respectively (we are assuming X to be of full rank).

We are now in a position to state a necessary and sufficient condition for submodel I to yield the solution to the nonnegative least squares problem. Since $\underline{y}$ is decomposed into the orthogonal vectors $\hat{\underline{y}}_I$ and $\underline{r}_I$, the conditions of Theorem 1 which remain to be checked are that these vectors lie in the appropriate cones. It is then easily seen that $\hat{\underline{y}} = \hat{\underline{y}}_I$ (and $\underline{r} = \underline{r}_I$) if, and only if

$$(X_I'X_I)^{-1}X_I'\underline{y} \geq \underline{0} \qquad (3.1)$$

and

$$X_{P-I}'(I - X_I(X_I'X_I)^{-1}X_I')\underline{y} \leq \underline{0} \qquad (3.2)$$

(with the obvious interpretations if $I = \phi$ or $I = P$).

The basis of our algorithm is the familiar sweep operator, as described by, amongst others, Seber (1977, p.354). We operate on a $(p+1) \times (p+1)$ array, initially set to

$$\begin{bmatrix} X'X & X'\underline{y} \\ \underline{y}'X & \underline{y}'\underline{y} \end{bmatrix} .$$

If we sweep this array on the diagonal positions corresponding to elements of I, the result is an array with the final column

$$\begin{bmatrix} -(X_I'X_I)^{-1}X_I'\underline{y} \\ X_{P-I}'(I-X_I(X_I'X_I)^{-1}X_I')\underline{y} \\ \underline{y}'(I-X_I(X_I'X_I)^{-1}X_I')\underline{y} \end{bmatrix} .$$

We see therefore that this column provides not only parameter estimates and the residual sum of squares for submodel I, but also all necessary information for checking the conditions (3.1) and (3.2).

While (3.1) has an obvious interpretation, it will be helpful to provide an interpretation for (3.2). It is straightforward to show that if, for some $i \in \bar{I}$, we fit submodel $I + \{i\}$, the resulting estimate of β_i is proportional to $\underline{x}_i'(I-X_I(X_I'X_I)^{-1}X_I')\underline{y}$ (where $\underline{x}_i$ denotes the ith column of X); we shall refer to the sign of this quantity as the *potential sign* of $\hat{\beta}_i$ when submodel I is fitted. Thus conditions (3.1) and (3.2) state that all parameter estimates in the model are nonnegative while no additional parameter estimates are potentially positive. Similar conditions appear in the termination rules for the quadratic programming algorithm of Theil and van de Panne

(1960) and for a cone projection algorithm due to Kudô and Choi (1975).

4. An Algorithm for Nonnegative Least Squares

Having obtained necessary and sufficient conditions for $\hat{\underline{y}} = \hat{\underline{y}}_I$ we now require a set of rules to direct our search for an appropriate I. We shall describe a search procedure which begins with the null model ($I = \phi$) and adds parameters to this, although it is possible to employ a dual version of this procedure which starts from the full unconstrained model ($I = P$) and deletes parameters; in the majority of circumstances the former procedure will prove more convenient and efficient.

It would be convenient to be able to proceed to the solution in a manner analogous to forward selection, adding one parameter with a potentially positive estimate at each stage. Though it can be proved, through a theorem similar to Theorem 2 below, that such a route does exist, there does not seem to be any simple rule for finding that route. We shall, instead, adopt a procedure similar to stepwise regression, adding parameters with potentially positive estimates and deleting parameters with negative estimates until a solution satisfying (3.4) and (3.5) is found. Our algorithm may be described as follows :

1. Set $I = \phi$.
2. Find an $i \notin I$ such that $\hat{\beta}_i$ is potentially positive. If no such i exists, the problem is solved; otherwise add i to I.
3. If $\hat{\underline{\beta}}_I \geq \underline{0}$, go to 2. Otherwise set $R = I$ and go to 4.
4. Use a suitable subsidiary algorithm to fit submodel R subject to the constraints $\underline{\beta}_R \geq \underline{0}$. Go to 4.

Note that step 4 will involve changes in I so that, on return to step 2, I will be a proper subset of R. The value of $\|\underline{r}_I\|^2$ will, however, have decreased since the last occasion on which step 2 was reached, ensuring the convergence of the algorithm.

The subsidiary algorithm we shall use is related to that of Theil and van de Panne (1960), and is based on the following theorem.

Theorem 2 If the least squares solution to the problem of fitting submodel R subject to the constraints $\underline{\beta}_R \geq \underline{0}$ is obtained by fitting the unconstrained model I then, for any $J \neq I$ such that $I \subset J \subset R$ and $\hat{\underline{y}}_J \neq \hat{\underline{y}}_I$, $\hat{\underline{\beta}}_J$ has at least one element $\hat{\beta}_j$, $j \in J-I$, which is negative.

Proof We note that $\hat{\underline{y}}_I$ is the closest vector to $\underline{y}$ of the form $X_I\underline{\beta}_I + X_{R-I}\underline{\beta}_{R-I}$, not only if nonnegativity restrictions are imposed on both vectors of parameters but also if only the restrictions $\underline{\beta}_{R-I} \geq \underline{0}$ apply. We also note that $\hat{\underline{y}}_J$ can also be written in this form with, unless the result of the theorem holds, $\underline{\beta}_{R-I} \geq \underline{0}$. The observation that $\hat{\underline{y}}_J$ is closer to $\underline{y}$ than $\hat{\underline{y}}_I$ concludes the proof.

We see from the above theorem that it is possible to proceed from an initial state in which the unrestricted submodel R is fitted towards a solution to the corresponding restricted problem by successively deleting parameters with negative estimates. However, when faced with a choice between a number of candidates for deletion, it is possible that we will choose one which is contained in the solution submodel; if this happens the error will eventually be detected and another choice may be made. The following subsidiary algorithm results.

4a. Set $I = R$ (this stage is not, strictly speaking, necessary: see step 3 of the main algorithm).

4b. Find an $i \in I$ such that $\hat{\beta}_i$ is negative. If no such i exists, go to 4c; otherwise delete i from I and repeat this step.

4c. If there is no element of R-I such that the corresponding parameter is potentially positive, go to 2; otherwise return to 4a and repeat the algorithm making a different sequence of deletions.

The relative inefficiency of this subsidiary algorithm is unimportant, since generally a single deletion will suffice. In both the main and subsidiary algorithms it is advisable (though not necessary) to choose, from a number of contenders for addition or deletion, the parameter producing the greatest change in the residual sum of squares; such changes may readily be calculated from the elements of the swept array.

5. Examples

In order to use our algorithm it is useful to have access to an interactive regression package which provides, amongst its output, information on the potential sign of parameter estimates not in the current model. Such information may be provided by, for example, partial correlation coefficients or, as in the case of the MULREG package used at the University of Kent, t-values for parameters not in the regression. The latter also provide information regarding the magnitude of the change in the residual sum of squares which will result from the addition or deletion of a parameter; we shall always choose, amongst candidates for addition or deletion, that with the largest (absolute) t-value.

We shall describe the operation of the algorithm by means of tables giving, at each stage, the t-values for all parameters; those for parameters in the current submodel will be indicated by an asterisk.

Example 1. Armstrong and Frome (1976) considered a problem with three regressor variables, and constrained the three regression coefficients (and also the intercept) to be nonnegative. With their data, but without constraining the intercept, we obtain the following.

Stage	t-values		
1	2.87	-4.21	-0.63
2	2.87*	-7.26	-0.39

The solution, involving only the first regressor variable, is obtained without use of the subsidiary algorithm.

Example 2. Draper and Smith (1981), amongst others, consider the relationship between the amount of heat evolved during the hardening of cement and the percentages of four constituents. Constraining the four regression coefficients to be nonnegative, we obtain the following.

Stage	t-values			
1	3.55	4.69	-2.10	-4.77
2	12.10	4.69*	-3.44	-0.66
3	12.10*	14.44*	1.35	-1.37
4	8.29*	14.85*	1.35*	-0.20

The solution, involving the first three regressor variables, is obtained without using the subsidiary algorithm.

Example 3. Given observations y = 43, 39, 18, 10, 11, 9, 19, 31, 38 corresponding to the x-values 0, ..., 8, we wish to fit a convex curve as the regression of y on x. The best-fitting curve will consist of straight-line segments, and may be parameterised as

$$E(y) = \alpha + \beta_o x + \sum_{j=1}^{x-1} \beta_j (x-j),$$

where $\beta_1, \ldots, \beta_7$ are restricted to take nonnegative values. We fit this model in the following stages (only the t-values corresponding to $\beta_1,\ldots,\beta_7$ are shown).

Stage	t-values						
1	1.53	3.63	6.72	7.37	5.76	3.12	1.63
2	-0.11	1.12	1.33	7.37*	0.84	0.54	0.10
3	-0.99	0.18	1.33*	1.73*	1.99	1.11	0.37
4	-1.38	0.22	2.33*	-0.36*	1.99*	-0.17	-0.49
5	-1.41	0.32	3.87*	-0.36	3.22*	-0.11	-0.49
6	-4.95	0.32*	1.57*	-0.26	2.87*	-0.11	-0.44

The fitted line has gradient discontinuities at x = 2, 3 and 5. The subsidiary algorithm is used in proceeding from stage 4 to stage 5.

References

Armstrong,R.D., and Frome, E.L.(1976), A branch-and-bound solution of a restricted least squares problem, Technometrics, 18, 447-450.

Barlow, R.E.,Bartholomew, D.J., Bremner,J.M., and Brunk, H.D.(1972), Statistical Inference under Order Restrictions, Wiley, London.

Draper, N.R., and Smith, H. (1981), Applied Regression Analysis (2nd ed.), Wiley, New York.

Fletcher, R. (1971), A general quadratic programming algorithm, J. Inst. Maths. Applics., 7, 76-91.

Kudô, A., and Choi,J.-R. (1975), A generalized multivariate analogue of the one sided test, Mem.Fac.Sci. Kyushu Univ., Ser. A (Mathematics), 29, 303-328.

Seber, G.A.F. (1977), Linear Regression Analysis, Wiley, New York.

Stoer,J., and Witzgall,C.(1970), Convexity and Optimization in Finite Dimensions I, Springer, Berlin.

Theil, H., and van de Panne, C. (1960), Quadratic programming as an extension of classical quadratic maximization, Management Science, 7, 1-20.

Waterman, M.S. (1974), A restricted least squares problem, Technometrics, 16, 135-136.

Non Parametric Decision Trees by Bayesian Approach

G. Celeux and Y. Lechevallier, I.N.R.I.A., France

Key-words: segmentation, discrimination, Kolmogorov-Smirnov distance.

1. INTRODUCTION

The problem of discriminant analysis arises when you make a number of measurements on a individual and you wish to classify the individual into one of several classes $P_1,\dots,P_k$ on the basis of these measurements.

Let $\{w_1,\dots,w_n\}$ be a sample of the population Ω where each individual w_i is considered as a random observation from the population Ω. Let $x_i = \{x_i^1,\dots,x_i^p\}$ be a p dimensional vector which represents the individual w_i of Ω in the space of variables X_j, j=1,p and Θ_i the class to which w_i belongs. The aim of discriminant analysis is to build in $\mathbb{R}^p$ (the space of measurements) k regions $R_1,\dots,R_k$ which minimize a criterion based on the percentage of misclassification.

Bayes Decision Rule of minimum Risk :

Let π_j be the prior probability of class P_j

f_j the probability density function of P_j

ℓ_{jh} the loss incured when an individual of P_j is assigned to P_h.

The discriminant method builds k regions $R_1,\dots,R_k$ which minimize the cost of misclassification $\sum_{j=1}^{k} \pi_j \sum_{h=1}^{k} \ell_{jh} \int_{R_h} f_j(x)\,dx.$

The density estimation problem is clearly important in implementing these decision rules. If we have multiple modes, the problem of estimating this density function can be difficult, but in the case of a Gaussian family of densities the mean vector and covariance matrix characterizing each mode can be estimated. An alternative approach is to measure the local density of each class, it is a k-nearest-neighbour procedure. However this technique assumes that the distance measure between two individuals of Ω is known. Furthermore the search of the nearest neighbours to a point in p-dimensional space is rather expensive computationally.

In this paper we present methods which estimate distribution functions instead of densities. So these methods are non parametric and fast.

2. NON PARAMETRIC DECISION PROCEDURE

In [Fri 77] Friedman has developped the following procedure :

Consider the case of two classes and let $F_1(x)$, $F_2(x)$ be their (corresponding) cumulative distributions. Suppose that $\ell_{12}\pi_2=\ell_{21}\pi_1$ and that $F_1(x)$ and $F_2(x)$ are known univariate distributions. Stoller [Sto 54] has shown that the point x^* which minimizes the Bayes' risk of misclassification is a point which maximizes the quantity

$D(x) = |F_1(x) - F_2(x)|$

that is $D(x^*) = \underset{x}{\text{Max}}\ D(x)$

The quantity $D(x^*)$ is the Kolmogorov-Smirnov distance between the two distributions. This distance is a measure of the separability of two distributions.

Usually the marginal cumulative distributions $F_1(x)$ and $F_2(x)$ are unknown. However they are easily estimated by

$$\hat{F}_i(x) = \begin{cases} 0 & x < x_1^{(i)} \\ k/n & x_k^{(i)} \leq x < x_{k+1}^{(i)} \\ 1 & x_n^{(i)} \leq x \end{cases}$$

where $x_k^{(i)}$ is the kth point of the ith class ordered by increasing values

$$x_1^{(i)} < \dots x_k^{(i)} < x_{k+1}^{(i)} \dots < x_n^{(i)}$$

<u>Algorithm used</u> :

A non parametric recursive partitionning algorithm [Fri 77] for two class discrimination can proceed as follows :

$$D(x_j^*) = \sup_x |\hat{F}_1^{\,j}(x) - \hat{F}_2^{\,j}(x)|$$

is evaluated for each variable X_j. At each step p allocation rules are considered, each of them based on one of the variable $X_j, j=1,\dots p$ and a variable X_{j*} which verifies $\quad D(x^*_{j*}) = \max\limits_{j=1,\dots p} D(x_j^*) \quad$ is kept.

The location of the cut is taken to be x^*_{j*}.

The population is cut into two subsets C_1 and C_2 which verify

$$C_1 = \{w \in \Omega \ / \ X_{j*}(w) \leq x^*_{j*}\}$$

and

$$C_2 = \{w \in \Omega \ / \ X_{j*}(w) > x^*_{j*}\}$$

In each subset C_1 and C_2 this algorithm is used until a stopping criterion (Number of elements into each class, number of steps..) is verified. A subset in which the stopping criterion is verified is called a terminal cell.
At the end it is possible to build the following binary tree.

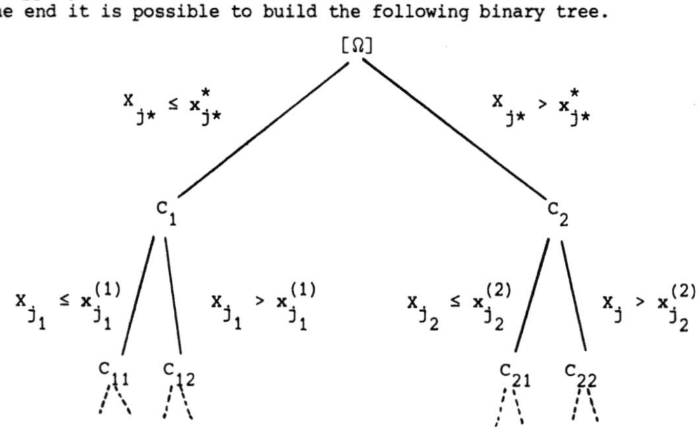

Variables containing little or no discrimating information are ignored.

Measurement of performance :

The decision rule for classifying a test vector is simply to go over the binary tree until a terminal cell is reached.

If the terminal cell represents a unique class the test vector is assigned to that class.

If the terminal cell represents a mixed class then the test vector is assigned to the most numerous class in this cell.

3. MULTICLASS DISCRIMINATION

An extension, used by Friedman, of this procedure to multiclass problem is to treat a k-class problem as a series of two class problems. For each two class problem, this algorithm is performed to separate one of the class P_i form all other classes. With this technique we obtain k binary trees and a test vector is classified into k terminal cells.In each terminal cell of each binary tree, the number M of individuals of the majority class and the number O of individuals of the other classes are stored and the test vector is assigned to the majority class where M-O is maximum over these k terminal cells. But it is possible to obtain two terminal cells which contain only individuals from a single class and that these classes are different [CeLe 80].

In this case it is difficult to apply the majority decision rule given by Friedman. So we choose another solution ; this solution is to build only one binary tree for k classes. This algorithm cutting the k classes into two classes processes as follows :

for each variable X_j we compute

$$D(x_j^*) = \sup_{A \in \mathscr{A}} \sup_x \left| \hat{F}_A^j (x) - \hat{F}_{\bar{A}}^j (x) \right|$$

where $\mathscr{A}$ is the set of the partitions into two classes of $\{P_1,..,P_k\}$ and $\hat{F}_A$ is the empirical cumulative distribution for a group A of a priori classes and $\hat{F}_{\bar{A}}$ is the empirical cumulative distribution for a group of another a priori classes then we search $D(x_{j*}^*) = \max_{j=1,p} D(x_j^*)$

the location of the cut is taken to be x_j^* .

Let N be the set of the indices of the a priori classes of A. The empirical cumulative distribution can be computed by

$$\hat{F}_A (x) = \frac{1}{\sum_{i \in N} \text{card}(P_i)} \sum_{i \in N} \text{card}(P_i) \cdot \hat{F}_i (x)$$

For a given x it is not necessary to compute the $2^{k-1}-1$ possibilities to divide k classes into two groups because the optimal solution is contained in (k-1) possibilities [CeLe 80]. Also we must add another stopping criterion which is the value M-O. At each step and under the hypothesis $\pi_A \ell_{A\bar{A}} = \pi_{\bar{A}} \ell_{\bar{A}A}$ the decision rule is asymptotically Bayes' risk efficient [CeLe 80].

Application in multiclass discrimination

The data are a set of 3000 individuals, on which is measured a set of 81 variables called characteristic Loci [SSSK 76]. Each individual is known to belong to one class among 10, and there are 300 samples in each class. The problem is to find a decision rule to recognize each of these classes. The result obtained is a decision tree based upon 9 variables L13, L14, L15, L22, L23, L28, L41, L47, L50 and the decision rule is described by following flowchart

```
if (L22>.22689076) then
  the individual is a 9(100%)
else
  if (L28>0) then
    the individual is a 1(100%)
  else
    if (L23>.05395683) then
      if (L15>.04491018) then
        the individual is a 3 (100%)
      else
        the individual is a 5 (100%)
      end
    else
      if (L41>.22712934) then
        the individual is a 0 (100%)
      else
        if (L50>.06760563) then
          the individual is a 8 (96.7%) or a 6(2%) or a 0(.7%) or a 2(.3%) or a 5(.3%)
        else
          if (L.14>.02409639) then
            if (L47>.01282051) then
              the individual is a 2 (100%)
            else
              the individual is a 7 (100%)
            end
          else
            if (L13>.07380074) then
              the individual is a 4 (99.7%) or a 7 (.3%)
            else
              the individual is a 6 (98.4%) or a 2 (1.3%) or a 0 (.3%)
            end
          end
        end
      end
    end
  end
end
```

We can see that the recognition rate (99,5%) is very high in all case.

4. TERNARY DECISION TREES

The method

When the a priori classes are somewhat mixed, the above method will not give good results.

In this section we present a method whose aim is to work out such cases. We consider here the two class discrimination problem.

The basic idea is the following : we build a decision tree by using two cutting points x^{*1}_{j*} and x^{*2}_{j*} instead of a single cut point x^{*}_{j*}.

So at each step of the algorithm we build three subsets C_1, C_2, C_3 which verify :

$$C_1 = \{w \in \Omega \ / \ X_{j*}(w) \leq x^{*1}_{j*}\}$$

$$C_2 = \{w \in \Omega \ / \ x^{*1}_{j*} < X_{j*}(w) \leq x^{*2}_{j*}\}$$

$$C_3 = \{w \in \Omega \ / \ X_{j*}(w) > x^{*2}_{j*}\}$$

The subset C_2 can be seen as a subset of indecision. So we define a cost of indecision q_k for each class P_k which verifies $aq_1 = \ell_{12}$ and $aq_2 = \ell_{21}$ with $a>1$

In this frame, we choose the two cutting points to minimize the Bayes' risk associated to the procedure. Under the hypothesis $\pi_1 \ell_{12} = \pi_2 \ell_{21}$ which implies $\pi_1 q_1 = \pi_2 q_2$, it amounts to select at each step the variable j^* and the two cutting point x^{*1}_{j*} and x^{*2}_{j*} so that :

$$D(x^{*1}_{j*}, x^{*2}_{j*}) \quad \sup_{j=1,p} \quad \sup_{x_1,x_2} \left[(a-1)\hat{F}^j_1(x_2) + \hat{F}^j_1(x_1) - \hat{F}^j_2(x_1)(a-1) - \hat{F}^j_2(x_2) \right]$$

Where $F_1(x)$ is the cumulative distribution function of the lower class determined previously as follows : P_1 is the lower class for the variable X_j if

$$\sup_x |\hat{F}^j_1(x) - \hat{F}^j_2(x)| = \hat{F}^j_1(x) - \hat{F}^j_2(x).$$

Remarks :

At each step and under the hypothesis $\pi_1 \ell_{12} = \pi_2 \ell_{21}$ the decision rule is asymptotically Bayes' risk efficient [CeLe 80]

When the classes are well separated the subset C_2 vanishes and the algorithm gives the same results as those described in §2.

To determine the two cutting points it is not necessary to examine all the couples of the sample and the complexity of the algorithm at each step is 2.n where n is the number of sample points in the cell [CeLe 80].

There is no difficulty to generalize the above method to the multiclass problem.

A medical application [CLL 82]

The data set is a sample of 211 patients with a bronchial disorder. There are two main diagnoses : bronchitis and emphisema. But there is an evolution between these two diagnoses. So the doctors have defined four classes : bronchitis (B), dominating bronchitis and emphisema (Be), dominating emphisema and bronchitis (Eb) and emphisema (E).

The patients are described by 11 clinical variables. To illustrate the above method we use it to recognize the two mixed classes : DB = B ∪ Be and DE = E∪Eb. We summarize here under the resulting decision tree :

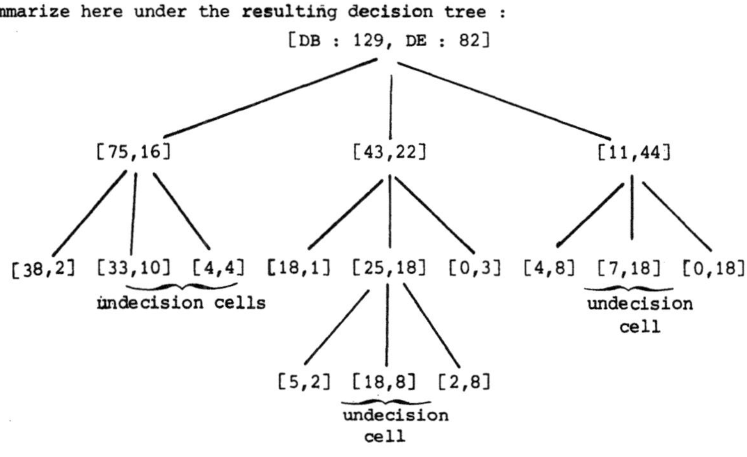

The detailed conclusions are the following :

9 misclassified points (B:0, Be:4, Eb:4,E:1)
102 unclassified points (B:11, Be:51, Eb:34, E:6)

Comments :

- Only one misclassified point belongs to a pure class (B and E).
- As expected, most misclassified points belong to mixed classes (Be and Eb).

5. PROGRAM

All experiments were performed on Multics/DPS8 computer with programs coded in Fortran IV. The first program builds the decision trees with the training sample and the second program assigns each individual of the test sample in one terminal cell. But it is possible to randomly extract a test sample from the data set.These programs are available from the authors.

BIBLIOGRAPHY

[CeLe 80] CELEUX G, LECHEVALLIER Y. - "Methodes de discrimination non parametriques assymptotiquement efficaces au sens de Bayes", Rapport de Recherche n° 52. INRIA 1980.

[CLL 82] CELEUX G, LAURO N, LECHEVALLIER Y. - "Contributi dell' analisi multidimensionale nello studio di gruppi clinici a priori mal définiti", Rivista di Statistica Applicata n°2 (1982).

[Fri 77] FRIEDMAN J.H. - "A recursive partitioning decision rule for non parametric classification" IEEE Trans. Computer, pp. 404-408 (1977)

[Fuk 72] FUKUNAGA K - "Introduction to statistical Pattern Recognition", Academic Press (1972).

[GoO 78] GORDON L, OHLSEN R.A. - "Asymptotically efficient solutions to the classification problem", Annals of Statisties Vol.6 n°3 (1978).

[Sto 54] STOLLER D.C. - "Univariate two-population distribution-free discrimination", J. Amer. Statist. Assoc. Vol. 49, pp. 770-775(1954).

[SSSK 76] SUEN C.Y, SHIAU C., SHINGHAL R. and KWAN C.C. - "Reliable Recognition of Handprint Data", Proceeding of the joint Workshop on Pattern Recognition and Artificial intelligence pp. 98-102, (1976).

A Comparative Assessment of Data Management Software

S. Clark, Timberlake Clark Ltd., Chelmsford, G.B.

This paper presents the results of a comparative evaluation of four modern software systems - SIR, P-STAT, SAS, and OSIRIS IV. The evaluation was designed to assess the data management facilities provided by each system, using a set of trial data drawn from the UK General Household Survey. The systems are compared across five overall categories: data structuring, data validation, data security, data independence and ease of use.

KEYWORDS : DATA MANAGEMENT, VALIDATION, SECURITY

INTRODUCTION

Facilities for data management are of central importance to effective data analysis, but relatively little attention has so far been paid to these facilities in studies of software evaluation. It is not surprising, then, to discover the wrong software being used for the wrong reason - most notably, the use of statistical packages for data management - and evidence of data analysis being carried out on data sets which are improperly managed. This paper sets out to redress the imbalance by offering the results of an evaluation of four software systems from a data management point of view. The systems are Scientific Information Retrieval (SIR), Princeton Statistical System (P-STAT), Statistical Analysis System (SAS), and the Organised Set of Integrated Routines for the Investigation of Social Science Data (OSIRIS).

METHODS

The approach taken was to concentrate on a small set of categories felt to be of central importance in a research and investigative environment, and to establish each system against a set of trial data drawn from the UK General Household Survey - one of the most complex, hierarchical data sets to face secondary analysts in the United Kingdom.

The overall categories for assessment were data structuring, data independence, data validation, data security, and ease of use. Each broad topic was broken down into a number of sub-categories, and these were further sub-divided into yet more detailed categories. At this most detailed level, the systems were ranked, and weights were attached to provide a measure of the importance of each category.

DATA STRUCTURING

Five sub-categories were identified within the overall category of data structuring. These were: data organisations, data types, data definition, data manipulation, and data display. Within each sub-category it was also possible to identify a number of further sub-categories. Thus, data organisations naturally sub-divided into networks hierarchies, and rectangular files, whilst data types broke down into characters, missing values, and other formats. The most extensive breakdown involved the area of data manipulation, where no less than ten sub-components were examined, ranging from normalising facilities to sorts and sampling. The overall ranking achieved by each system is given in Table One.

Table One : Rankings for Data Structuring

	Weight	SIR	OSIRIS	P-STAT	SAS
DATA STRUCTURING		1	4	3	2
Data Organisations	2	1	4	2"	2"
Data Types	1	2	4	3	1
Data Definition	1	2	1	4	3
Data Manipulation	2	1"	3"	3"	1"
Data Display	1	1"	4	1"	1"

From the figures in Table One, it can be seen that SIR emerges as the leading system for data structuring, followed by SAS, P-STAT, and OSIRIS. In the area of data types however, SAS is pre-eminent, whilst P-STAT scores well for data display, and OSIRIS for data definition. Attachment of weights has the effect of producing a clear final ordering, but use of a different set of weights would most likely change the result. If, for example, one was dealing with simple file structures, but wanting to support a wide variety of data types, then SAS would lead, whilst if data display was the ultimate aim, there would be little to choose between SIR, P-STAT and SAS.

What makes SIR best at supporting data organisations, and why is OSIRIS the least suitable ? To a degree, it all depends on what one is trying to use the software for. In terms of rectangular file structures, there is little to distinguish one system from another. But this is not so for hierarchies and networks. Only SIR will handle a hierarchical file directly; in the remaining systems, it is necessary to operate on sets of distributed flat files (and to reassemble these into a structured file in the case of OSIRIS). For network structures, it is not possible to use OSIRIS (unless the structure is reduced to a hierarchy).

In the handling of different data types, SAS offered the most extensive facilities, including Roman Numerals and Social Security Numbers as well as many other formats. SIR was particularly well-suited to applications requiring date and time formats, but had it not been for its inability (at the time of writing) to handle character variables, P-STAT would have mounted a serious challenge for overall leadership of the data types category thanks to its simple, yet highly flexible method of defining missing values.

Facilities for data definition were strongest in OSIRIS, where extensive codebook records are made available for storing information. The other systems, however, were generally superior in their variable naming facilities. For data manipulation, all four systems offered powerful facilities, but the wider availability of functions coupled with more elegant methods of hierarchical control, pushed SIR and SAS to the front. Finally, the data display component saw P-STAT scoring highly for tabulation and selective listing of output, with SAS and SIR scoring highly for report generation.

DATA VALIDATION

Three main categories were identified in the area of data validation: error recognition, error reporting, and error correction. The overall results are shown in Table Two.

Table Two : Rankings for Data Validation

	Weight	SIR	OSIRIS	P-STAT	SAS
DATA VALIDATION		1	4	2	3
Error Recognition	2	1	3	2	4
Error Reporting	2	1	4	3	2
Error Correction	1	1"	1"	1"	1"

All four systems offered powerful facilities for data validation. In terms of the set of validation criteria proposed by Roistacher (see Roistacher 1976), each system had the ability to recognise and recover from syntactic data errors automatically. SIR, SAS, and P-STAT performed this before any analysis of the data could take place. For structural checking, the SAS user is required to prepare program statements, whereas in the remaining systems, general purpose routines are available to make such a task easier and quicker to effect. Overall, SIR provided the most extensive set of automatic error recognition routines, and SAS provided the fewest. P-STAT edged above OSIRIS in allowing for more error recognition at the time of file creation.

In the reporting of errors, P-STAT provided the best error summary, SAS gave the user the most control over the reporting of errors, but SIR was ranked first thanks to its ability to store erroneous data in an error file (thereby making correction easier), and the high quality of the error messages. Both SAS and P-STAT had facilities for outputting bad data to a separate error file, but required more user intervention. All the systems provided automatic facilties for setting bad data to missing values. User-driven correction facilities were also well catered-for in each system, and no differences were noted.

DATA INDEPENDENCE

Data independence was assessed under four categories: multiple views, portability, physical independence, and scope for updating and adding to the data (variables and observations). Table Three summarises the results.

Table Three : Data Independence Rankings

	Weight	SIR	OSIRIS	P-STAT	SAS
DATA INDEPENDENCE		1"	4	1"	3
Multiple Views	2	1"	4	1"	1"
Portability	2	2	3"	1	3"
Physical	1	1"	4	3	1"
Updates	1	1	3"	3"	2

Each system permitted multiple views of the data to be taken by an end-user, but the underlying conceptual data model was different from one system to another. OSIRIS employed an entry definition language to traverse a tree structure in a specific order, SAS and P-STAT, however, set out to offer facilities akin to relational approaches. The strict hierarchical model employed by OSIRIS limited the variety of user-views that could be established. SIR, P-STAT and SAS were all capable of giving the user the ability to establish different views.

In terms of portability, the systems are very different. Both P-STAT and SIR have a design philosophy aimed at making the software available across a wide range of machines. In contrast, it is likely that formidable difficulties would beset attempts at providing conversions for the other systems. All share, however, limitations in providing for physical adjustments to the data. There are routines in SAS and SIR to allow users to inspect the physical characteristics of their data, but the scope for tuning is strictly limited.

Updates, amendments, and additions to the data can be handled by each system. SIR provides a full set of commands to enable these to be carried out during batch data input, or during an interactive retrieval. The other systems also provide for master data files to be updated by a transaction file - in SAS it is also possible to use a forms type facility to perform data updating (a similar tool is under development at SIR).

Assuming that we are willing to place heavier emphasis on portability and provision of multiple views (that is, making sure that the most people get the most use out of their data), SIR and P-STAT emerge as the leading systems for data independence. SAS would also score well, were it not for its total dependence on IBM and IBM-like mainframes.

DATA SECURITY

Amongst the four systems, only SIR set out to provide facilities for data security. The SIR data base administrator can protect the complete database against access by unauthorised persons, and can also establish read and write security down to the level of a named variable within a record type. The other three systems do not set out to provide data security facilities at all. However, Clive Payne of Oxford has rightly pointed out that the distributed file systems (P-STAT and SAS), can be set up in such a way that sensitive data files are placed on disk files protected by the operating system, and non-sensitive data files are placed on unprotected disk files. This, however, does have the effect of making design considerations dependent upon security considerations in that if one variable in a file is to be protected, all variables in that file must also be protected, unless the variable is moved to another file.

EASE OF USE

Three categories were examined to assess ease of use: documentation, interactive capabilities, and ease of general use. The latter was formed from an assessment of the ease of establishing a system file, and of manipulating a file once established. The results are shown in Table Four.

Table Four : Ease of Use Rankings

	Weight	SIR	OSIRIS	P-STAT	SAS
EASE OF USE		1	4	2"	2"
Documentation	1	1"	4	3	1"
Interactive Use	1	1"	3"	1"	3"
General Use - Creation	1	1"	4	1"	1"
- Manipulation	1	1	2"	2"	2"

The quality of documentation varied considerably from one system to another. P-STAT, SAS and SIR provided all the information required to use the system in one manual. Use of the OSIRIS structured file routines required access to two manuals - which appeared to have been written by people who had not spoken to one another for some time. The OSIRIS documentation was also felt to rather inaccessible to inexperienced users. P-STAT documentation took a decidedly tutorial approach, but had not totally kept pace with the many developments in the system since it was published (this is now being rectified).

Both P-STAT and SIR were easy to use interactively. Both have an internal text editor, extensive help facilities, and facilities for adjusting the software to suit the characteristics of the environment in which processing is being carried out. There were no major differences in general use, other than at the stage of creation of the data file, where OSIRIS was found to require time and patience.

CONCLUSIONS

The overall pattern revealed in the evaluation, suggests that there are marked differences between the systems in their suitability for particular data management tasks. As expected, the data base management system SIR achieved the highest scores across the broad categories, but this does not infer that the remaining systems are to be regarded as inferior - everything depends on what the software is being used for, and how important a particular feature is felt to be for a particular application. It can be shown that each system is capable of achieving the top ranking for some category, and that adjustment of the weights could change the final outcome.

It must also be recognised that all these systems are in a continuous state of development, with new features being added on a regular basis. SIR, for example, are in the process of adding a host language interface, a forms facility for easy updating, and TPL-like tabulation. P-STAT are introducing more data types, random data files and a report generator. In comparison with the data management facilities offered by traditional statistical packages, all four systems deserve greater attention as a means of ensuring that data analysis is carried out on properly managed data.

REFERENCES

Buhler R. and Buhler S. (1979), P-STAT User Manual. Available from P-STAT Inc. PO Box 285, Princeton, New Jersey, USA.

Institute for Social Research (1979), OHDS - An introduction to OSIRIS hierarchical data structures, Computer Support Group, University of Michigan, USA.

Institute for Social Research (1980), OSIRIS IV Statistical Analysis and Data Management System, University of Michigan, USA.

Robinson, B.N. et al. (1980), Scientific Information Retrieval User's Manual, available from SIR Inc. PO Box 1404, Evanston, IL, USA.

Roistacher, R. (1976), A general consistency check procedure for machine-readable data. Sociological Methods and Research, Vol 4, No 3, February 1976.

SAS Institute Inc. (1979), SAS User Guide, available from SAS Inc. PO Box 10066, Raleigh, N Carolina,USA.

From Data Analysis to Non Parametric Statistics: Recent Developments and a Computer Realization for Exploratory Techniques in Regression or Prediction

G. Collomb, Paul Sabatier University, Toulouse, France

In the recent development of statistics, exploratory techniques in regression or prediction has often been considered only from the data analysis point of view. At the same time mathematical results in non parametric regression or prediction problems has often been given without considerations of their use in a data analysis approach of a statistical problem. These results concern various methods and are more and more numerous and accurate.

Here we focus our attention on the use of these results for the construction or investigation of exploratory techniques. For instance, we see how results on the limit law of real random variables defined from these non parametric methods can be used in the model choice problem : specially the kernel method in non parametric regression estimation is investigated by a computer realization.

Key words : data analysis, non parametric, regression, density, prediction, bias, variance, rate of convergence, confidence interval.

1. INTRODUCTION

Let (X_i,Y_i), $i=1,\dots,n$ and Z_i, $i=1,\dots,n$ be two bodies of data which are described with the language of traditional statistics as follows : (X_i,Y_i), $i=1,\dots,n$ are independent observations of a $\mathbb{R}^p \times \mathbb{R}$ valued random pair (X,Y) and for all $i=1,\dots,n$, Z_i is the ith observation of a stationary process $\{Z_j,\ j=1,2,\dots,\}$ in $\mathbb{R}^q$. We consider the following *data analysis* problems :
(α) how does Y_i depend on X_i in average for all $i=1,\dots,n$. (α') for an observation (x,y) of (X,Y) predict y from x which only is known. (β) describe the distribution of the values X_i, $i=1,\dots,n$. (γ) for a fixed integer k, how Z_{j+1} depends on $Z_{j-k+1},\dots,Z_j$ in average for all $j=k,\dots,n-1$. (γ') predict Z_{n+1} from the sequence Z_i, $i=1,\dots,n$.

These problems are associated with the following *statistical problems* of *functional estimation* :
(a) estimation of the *regression* function $r(.)=E(Y/X=.)$.
(b) estimation of the *density* function of the distribution of X.
(c) estimation of the function $R(.)=E(Z_{n+1}/[Z_{n-k+1},\dots,Z_n]=.)$.
(c') "estimation" of the random vector $R(Z_{n-k+1},\dots,Z_n)$ when $\{Z_j,j=1,2,\dots,\}$ is a Markov process of order k or *prediction* of Z_{n+1}.

Our investigation of some *non parametric estimators* of the functions r, f and R is presented here as a contribution to a more general discussion "from data analysis to non parametric statistics ... and the converse". This discussion can be organized around the few following characteristics of data analysis :

(i) Contrary to traditional statistics, there is no *statistical model*. However parametrical data analysis involves implicitly the model of the existence of parameters. On the contrary non parametric statistics does not involve the existence of such a model and the probabilistic assumptions are only very few restrictive conditions of regularity : existence of EY^2, Doeblin's condition on the markovian process $\{Z_j, j=1,2,\ldots,\}$, continuity, Lipschitz condition or derivability for the functions r, f or R, ... These assumptions will not be given here but can be found in the referenced papers, for each result stated below.

(ii) Data *summarizations* and *graphical representations*, which play a prominent part in data analysis, are also important features of the non parametric statistical methods, which can also be only considered from a *descriptive statistics* point of view. We note here that these methods (except for the algorithm (A) at the end of the present paper) involve short computing time and little programming effort : this simpleness implies that a special package is useless.

(iii) To the aims of descriptive statistics data analysis adds what Tukey (1977) calls a *detective activity*, which can have the goal of *discovering structure*. Similar ideas are developped by Barra (1981). For this *exploratory analysis* (but also - see Tukey, 1977 - for *critical* and *confirmatory* data analysis) some mathematical results which are now available in non parametric function estimation can be useful tools : approximation of the *bias*, *variance* and *mean square error* (*m.s.e.*) of the estimators, evaluation of *rates of convergence*, definition of pointwise *confidence intervals* or of *confidence bands* (for the construction of tests).

Some preliminaries and conventions. For the problem (a) we shall consider the *conditional variance* v defined on the set $\mathbb{R}^p$ by $\mathbf{v(x) = E((Y-r(X))^2/X = x)}$, $\forall x \in \mathbb{R}^p$. For the problem (c) we denote $\mathbf{p = kq}$, so that the $\mathbb{R}^q$ valued function R and the real functions r and f are all defined on $\mathbb{R}^p$. In the definition of the estimators we use the convention $\mathbf{o/o = o}$ and the positive number $\mathbf{h = h_n}$ which is named *bandwidth* and satisfies $h \to 0$ and $nh^p \to \infty$ when n tends to infinity. The mathematical signification of the sign "$\approx$" is "equivalent as n tends to infinity".

The *convergence type results* (in probability, L_2 or with probability one pointwise or uniform convergence) are primary and most important mathematical justifications to the utilization of these methods. These results will not be considered for the now rather classical problems of density and regression estimation (see the review of Collomb, 1981a, §2 and 3). On the contrary they will be mentioned for non parametric prediction which is now a rather new area of non parametric statistics.

2. REGRESSOGRAM, HISTOGRAM AND PREDICTOGRAM

Let $\mathcal{S}_n$ be a partition of $\mathbb{R}^p$ into equal and disjoint cubes of side length h. The regressogram $\hat{r}_n$, the histogram $\hat{f}_n$ and the predictogram $\hat{R}_n$ are defined as

$$\hat{r}_n(x)=\frac{1}{|J|}\sum_{i=1}^{n} Y_i\, 1_{\{X_i\in J\}},\quad \hat{f}_n(x)=\frac{|J|}{nh_n^p}\quad\text{and}\quad \hat{R}_n(x)=\frac{1}{|J|}\sum_{i=k}^{n-1} Z_{i+1}\, 1_{\{(Z_{i-k+1},\ldots,Z_i)\in J\}}$$

where $J = J(x)$ is the cube of $\mathcal{S}_n$ such that $x \in J$ and $|J|$ denotes the number of X_i, $i=1,\ldots,n$ (and $(Z_{i-p+1},\ldots,Z_i)$, $i=k,\ldots,n-1$) falling in the cube J ($|J| = \sum_{i=1}^{n} 1_{\{X_i\in J\}}$)

The word "*regressogram*" was first introduced by Tukey (1961, p. 682) in the case p = 1 and from a data analysis point of view. An example of utilization of the regressogram is given by Pearson et al. (1903, diagram 1) for the pair (X,Y) = ("father's stature", "son's stature"). For n = 1078 observations of this pair this author obtains the regressogram which is associated with the partition $\mathfrak{I}_n=\{[57.5+j, 58.5+j[, j \in Z\}$. These summarization and display of data were used by Pearson for the calculus of the regression line which also appears on the above-mentioned figure. This study was at the origin of the statistical term "regression" by its utilization in the expression "*regression line*". We can remark that the regressogram brings out a sufficient information for the regression of Y on X, and that this information is doubtless more relevant than the information given by the linear regression when X is small or large. Therefore for the regression considered by Pearson et al. (1903) the direct use of the regressogram seems more reasonable than the introduction of a parametrical method.

About this historical work, we note that the regressogram was introduced in order to simplify the calculus (without computer at that time) of the coefficients of the linear regression but curiously the very recent investigation of the mathematical properties of the regressogram is due to the non parametric statistics development which itself comes from the development of computer calculus.

The *histogram* $\hat{f}_n$ is a very well known descriptive statistical technique, considered by Tukey (1977) from a modern data analysis point of view. Lastly the word "*predictogram*" was recently suggested by Collomb (1981b) for the extension $\hat{R}_n$ of the two previous non parametric estimators methods to prediction problems.

The statistical properties of the regressogram are reviewed by Collomb (1981a, §2). These properties are very similar to the properties of the kernel regression estimator considered below : here we note only that it can be shown that the convergence of the regressogram is less rapid than the one of the kernel method. For the predictogram $\hat{R}_n$, the convergence property $E|\hat{R}_n(Z_{n-k+1},\ldots,Z_n) - R(Z_{n-k+1},\ldots,Z_n)|^2 \underset{n\to\infty}{\to} 0$ is obtained by Collomb (1981b,§4) who also gives a majoration of the above m.s.e. .

3. KERNEL METHOD FOR REGRESSION AND DENSITY ESTIMATION AND PREDICTION

Let K be a symetrical *kernel* of $\mathbb{R}^p$, that is to say any usual bounded symetrical density function in $\mathbb{R}^p$. The regression estimator r_n, the density estimator f_n and the predictor R_n are defined with the kernel K by

$$r_n(x)=\frac{\sum_{i=1}^{n} Y_i K((x-X_i)/h)}{\sum_{i=1}^{n} K((x-X_i)/h)},\quad f_n(x)=\frac{\sum_{i=1}^{n} K((x-X_i)/h)}{nh^p} \text{ and } R_n(z)=\frac{\sum_{i=k}^{n-1} Z_{i+1}K((z-[Z_{i-k+1},\ldots,Z_i])/h)}{\sum_{i=k}^{n-1} K((z-[Z_{i-k+1},\ldots,Z_i])/h)}$$

The three estimators r_n, f_n and R_n are investigated by Watson (1964) from a data analysis point of view, besides this pionner work numerous authors consider the kernel *regression* estimator : for instance Bordet, Brenot, Cazes, Lacourly, Lebeaux, Butler, Duhamel or Sabaton who are referenced by Collomb (1981a,§10). Here we only mention the paper of Cazes (1976) who introduces a *data analysis* measure of the precision (of $r_n(x)$, when $K(z)=1_{\{z\in B(center=0,volume=1)\}}$) which is closelely related to the *statistical* result on the confidence interval (for r(x)) given below.

Always through a rather restrictive data analysis approach, the *density* estimator

f_n is used in *discriminant analysis* by Habbema et al. (1978) and in *clustering* by von Eye et al. (1980). Recently Saporta (1981) introduced methods similar to R_n for the study of *time series* from an exploratory data analysis point of view in *prediction*.

The following statistical properties of $r_n(x)$, x fixed in $\mathbb{R}^p$, are obtained by the author in various papers referenced by Collomb (1981a,§3). First we give the following approximate expressions of *bias* and *variance* :

$$(3.1)\quad Er_n(x)-r(x)\simeq h^2\sum_{j=1}^{p}\sum_{k=1}^{p}\frac{1}{2}\left[\frac{\partial^2 r(x)}{\partial x_j \partial x_k}+\frac{1}{f(x)}\left(\frac{\partial r(x)}{\partial x_j}\frac{\partial f(x)}{\partial x_k}\ \frac{\partial f(x)}{\partial x_j}\frac{\partial r(x)}{\partial x_k}\right)\right]\int u_j\, u_k\, K(u)\, du$$

where here x_j (resp. u_j), j=1,...,p denotes the jth coordinate of x (resp. u) and

$$(3.2)\quad E(r_n(x)-Er_n(x))^2 \simeq \frac{1}{nh^p}\frac{v(x)}{f(x)}\int K^2(u)\, du.$$

These results on $r_n(x)$ can be considered from an *intuitive perspective*. For instance the formula (3.1) in the case p=1 is coherent with the following statements : at x fixed with r'(x)=0, the bias is positive if the regression is convexe, negative if the regression is concave and the more important the convexity or concavity is, the more important the absolute value of the error will be.

The previous results and the formula m.s.e.=$(\text{bias})^2$+variance immediatly bring out the m.s.e. $E(r_n(x)-r(x))^2$. This expression of the m.s.e. leads to

$$(3.3)\quad \min_{h\in\mathbb{R}^+} E(r_n(x)-r(x))^2 \simeq An^{-4/(p+4)}$$

where A is a positive constant which does not depend on n. This last result shows that larger the dimension p is, smaller the *rate of convergence* will be, so that a relévant utilization of such methods involves a very large sample size when p is large.

Numerous other practical applications of the results (3.1) and (3.2) are given by Collomb (1978) : determination of the optimal bandwidth h_n^* from the values of n, r"(x),r'(x),f'(x),f(x) and v(x), estimation of that optimal bandwidth h_n from estimations of the above values, approximation of $h_{n'}^*$ from h_n^* when n'>n, determination of an optimal kernel, investigation of the integrated m.s.e., ...

Now we consider the definition of pointwise *confidence intervals* for the regression r. Let $\alpha\in[0,1]$, $\beta=\Phi^{-1}(1-\alpha/2)$ where Φ is the distribution function of the standard normal law (e.g. α=5% and β=1.96). Then

$$(3.4)\quad P(r(x)\in[r_n(x)-\beta\sqrt{V/N},r_n(x)+\beta\sqrt{V/N}]) \simeq 1-\alpha$$

where $N=\sum_{i=1}^{n} K((x-X_i)/h)$ and $V=\frac{1}{N}\sum_{i=1}^{n}(Y_i-r_n(x))^2 K((x-X_i)/h)\ (\int K^2)$ under the additional assumption $h^4=o(nh^p)^{-1}$. Recently Wandl (1980) obtained a result on the limit law of a r.r.v. similar to $\sup_x |r_n(x)-r(x)|$: this result is useful for the definition of a *confidence band* for the regression but presents a rather complex formulation. Lastly Collomb (1981b,§3) reviewed the few available results on the kernel predictor.

An example of use of the kernel regression estimator, in the case p=1 with K(.)=1 if $\cdot\in[-0.5,0.5]$, 0 otherwise. We consider an application of the result (3.4) to 962 data which are the observations of a pair of blood analysis variables : rate of potassium (denoted X) and rate of urea (denoted Y). The figure 1 visualizes the curve r_n asso-

Figure 1 : pointwise confidence intervals for the regression of urea rate on potassium rate for a security level equal to 95 %.

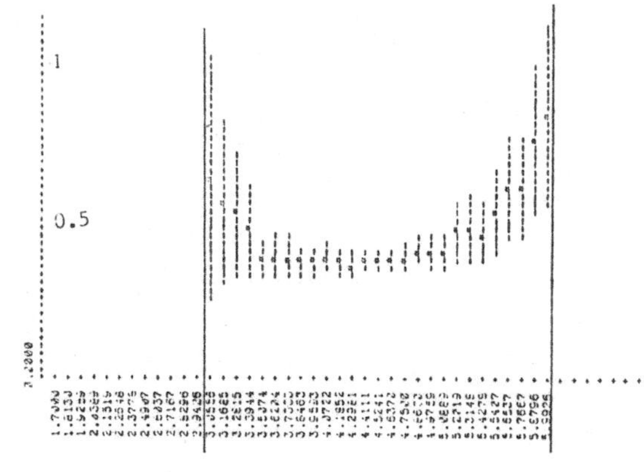

ciated with these data by signs "*" , the signs "|" display for each x the confidence interval of r(x) when α=5%. It is obvious that even from a descriptive statistics point of view this graphical procedure brings out an interesting information on the dependence of the urea rate on the potassium rate : the figure 1 suggests the quadratic model "$y=ax^2+bx+c$" for the investigated regression. Therefore this first exploratory data analysis gives a relevant solution to the problem of the *model choice*.

4. OTHER PROBLEMS AND OTHER METHODS

Among the numerous *estimation problems* which are related to the regression estimation and can be approached with similar methods we mention the problem of estimating the following functions

(F) derivatives of the regression (with applications in optimization), conditional quantiles and moments, conditional density or distribution function and conditional probabilities, with applications to discriminant analysis and empirical bayes procedures.

Parzen (1979) considers the density estimation problem from a statistical data analysis point of view. Problems which are similar to those of (F) can also be investigated in prediction : Collomb (1981b,§ 4 .4.b) considers various estimators defined from the predictogram.

Besides the methods considered in the paragraphs 2 and 3 various *other nonparametric regression estimators* are reviewed by Collomb (1981a). Here we only mention splines methods investigated by Schlee (1979) from a statistical point of view and by Militký et al. (1980) who propose a package.

Now we consider other regression estimators which are defined in a general form from the estimators $\hat{r}_n, r_n$ or any estimator associated with a positive number h.

An heuristic method for the choice of the bandwidth h. Let $r_n(h,.;(X_i,Y_i),\ i=1,\dots,n)$ be an estimator (for instance $\hat{r}_n$ or r_n) of the regression r, for every h in $\mathbb{R}^+_*$. We propose the following "optimal" choice of h (the following r.r.v. h^* exists in the usual cases of $\hat{r}_n$ and r_n) :

(A) let us define $S_n(h) = \sum_{j=1}^{n} (Y_j - r_{n-1}(h,X_j;(X_i,Y_i),\ i=1,\dots,n\ ,\ i\neq j))^2$ for every h in $\mathbb{R}^+_*$ and choose an h^* satisfying $S_n(h^*) = \min_{h\in\mathbb{R}^+_*} S_n(h)$.

It is clear that most of the practicians make approximatively this choice of h^* when they choose the number h so that on the same graphical visuzalization the curve $r_n(h,.;(X_i,Y_i)\ i=1,\dots,n)$ is "well centered" inside the set $\{(X_i,Y_i),i=1,\dots,n\}$. The algorithm (A) is only a formalization of this intuitive approach. This method is given by Collomb (1978) who also considers its extension to nonparametric discrimination. A similar procedure is developped by Collomb (1981b,Appendice) for the predictogram.

Lastly we note that the previous method is a data analysis algorithm for which no statistical properties are now available and we give some concluding remarks coming from the general considerations of Barra (1981) about the statistical methodology.

Exploratory data analysis uses various mathematical tools (formulas in algebra, geometry, optimization, statistics, ...) in order to discover an unknown structure of the data - and in general these data are very numerous. The object of this approach shows that data analysis ought to define and investigate mathematical tools which involve the least possible restrictive models (probabilistic or even descriptive) : this basic idea was the origin of many robustness investigations. The statistical models introduced here for various non parametric estimators are very unrestrictive. The results on these estimators - and even the result (3.3) concerning the necessity of a great number of observations - show how this kind of non parametric statistics search may be considered as a development of data analysis.

REFERENCES

BARRA,J.R.(1981). Phase inductive et phase décisionnelle dans une étude statistique. Statistique et Analyse des Données, 1981, 3, 1-4.

CAZES,P.(1976). Régression par boule et par l'analyse des correspondances. Revue de Statistique Appliquée, vol. XXIV n° 4, 5-22.

COLLOMB,G.(1978). Estimation non paramétrique de la régression : régressogramme et méthode du noyau. Publ. du Labo. de Stat. et Prob., Univ. P.Sabatier Toulouse,7,1-59.

COLLOMB,G.(1981a). Estimation non paramétrique de la régression : Revue Bibliographique. International Statistical Review, 49, 75-93.

COLLOMB,G.(1981b). From nonparametric regression to nonparametric prediction : survey of the mean square error and original results on the predictogram. Rencontre franco-belge de Statisticiens, Octobre 1981, Louvain, Belgique. To appear.

von EYE,A. and WIRSING,M.(1980). Cluster search by enveloping space density maxima. Compstat 80, Proceeding in Computational Statistics, Physica Verlag, Wien, 447-453.

HABBEMA,J.D.F., HERMANS,J. and REMME,J.(1978). Variable Kernel Density Estimation in Discriminant Analysis. Compstat 78, Proc. in Comp. Stat. Physica Verlag, Wien,178-185.

MILITKÝ,J., MILITKÁ,D.,CAP,J.(1980). User's package for non parametric data smoothing. Compstat 80, Proc. in Comp. Stat. Physica Verlag, Wien, 269-275.

PARZEN,E.(1979). Nonparametric Statistical Data Modelling. Journal of the American Statistical Association, 74, 365, 105-131.

PEARSON,K., LEE,A.(1903). On the laws of inheritance in a man II.Biometrika,II,4,357-462.

SAPORTA,G.(1981). Méthodes exploratoires d'analyse de données temporelles. Cahiers du Bureau Universitaire de Recherche Opérationnelle, n° 37-38.

SCHLEE,W.(1979). Estimation non paramétrique de la courbe de régression. Proceedings of the Sec. Intern. Symp. on Data Analysis and Informatics, North Holland,109-120.

TUKEY,J.W.(1961). Curves as parameters, and touch estimation. Proc. 4th Symp. Math. Stat. Prob. Berkeley, 681-694.

TUKEY,J.W.(1977). Exploratory Data Analysis. Addison-Wesley, New-York.

WANDL,H.(1980). On kernel estimation of regression functions. Akademie der Wissenschaften der D.D.R., Wissenschaftliche Sitzungen zur Stochastic WSS-03/80, Berlin.

WATSON,G.S.(1964). Smooth regression analysis. Sankhya, Ser. A, 26, 359-372.

Postgraduate Training for Statisticians – Database Methods

A.L. Dekker, Statistical Institute for Asia and the Pacific, Tokyo, Japan

Some problems relating to postgraduate training in computer aided management of statistical data resources are examined. Two statistical databases used for training purposes are explained, and the different approach required by macro- and microdata is discussed. The paper concludes with some remarks on the practical applicability of the methods taught.

KEYWORDS: *Database systems, Statistical Software, Time series, Microdata, Teaching*

1. Introduction

A person currently specialising in statistical computing is likely to be someone who was trained as either a statistician or a data processing specialist, and acquired enough knowledge about the other field to function on the interface. Since computer science is too often perceived as difficult to penetrate, statisticians are probably the rarer species in this rapidly developing interdiscipline. Clearly, there exists a need for professional statisticians to assimilate data processing know-how, and for them to start using computers as day-to-day tools.

Computer data processing in statistics dates back to an earlier time than the use of this technique in most other disciplines. However, while tabulation and analytical methods became much more powerful and editing applications were developed, over a number of years no other major fields of general application reached maturity. One limiting factor has been the relatively small direct-access storage capacity of computers, that did not allow other than sequential access to the large bodies of data typical for statistical data processing. This technological barrier is now gradually being removed, and fascinating new applications of statistical computing come within reach, not only through the construction of large random-access databases but also as a result of inquiry options provided by network architecture that reaches into the offices of planners and the homes of citizens.

2. Approach to training

The traditional approach to training in database methods begins with a discussion of the various basic design structures, such as the hierarchical, network and relation model. This is underpinned with references to relational algebra using the terminology of set theory and references to studies by E.F. Codd, C.G. Date and the CODASYL Committee. While this approach has merit in an academic environment, it is impractical for short postgraduage courses attended by statistical practioners. Two basic criteria have to be fulfilled for a subject to be well-received by such a group:

- it should have indisputable practical applicability;
- it should be understandable for intelligent people lacking recent formal training in mathematics or informatics.

This is not to say that the course should be superficial and simple, since when real interest is developing, the students will be willing to make considerable efforts to master the subject.

The simplest, and probably most effective, introduction of a statistician to database methods is to give him or her access to one or more existing statistical databases. In some experimenting with the new power at his hands, the student will experience pleasant surprises as well as limitations in terms of database flexibility and data resources. This will allow the explanation of intricacies of the present application and the software used. Logically, the question for specifications of other systems will arise, and comparisons can be made. The discussion may be concluded with a more fundamental exposé on design criteria and trade-offs. The participants are then ready to make a considered judgement about the usability of such methods in their own working environment.

At the Statistical Institute for Asia and the Pacific (SIAP) two model databases are maintained which both are normally accessed through display terminals. One is a macro-database of statistical indicators covering a number of countries. The other is a micro-database containing mostly individual demographic data.

3. Macro-databases

The variable time is of major inportance in nearly all statistical data collections, the primary use of which is often to supply information on which planning can be based. This may not be true for other types of databases: an airline company can not sell a chair that remained empty yesterday, and an automobile factory has little use for the bill of materials of last year's model. Therefore, statistical macro-databases tend to be organized in the form of time series - one series per record. This makes for very long record lengths (perhaps: very long and variable record lengths). In another design approach, a file may be used to store one (or several) time series with one record being used per term, but this poses the difficult question of which series to group together, and reduces the flexibility of the concept.

A time-series database requires associated software for statistical analysis to be really useful. Basic routines needed are seasonal adjustment, curve fitting (for forecasting purposes) and correlation/regression. There is a tendency to integrate these tools with the database in order to make a stand-alone system, user-friendly to analysts and planners.

Because of the specific characteristics of statistical macro-databases, a general DBMS is not very suitable as supporting software. The use of such a system for time-series management may pose more problems then it solves. A better approach could be to use the system files provided by packages such as SPSS, BMDP and SAS. It ensures the availability of ample analytical techniques, at present usually including seasonal adjustment procedures. We developed an application using SPSS system files, but had to cope with a file structure where a "case" (consisting of one or more records) is supposed to be a set of variable occurrences for a particular unit of information. Time series do not fit easily into this structure in which they have to be recorded vertically, adding a new "case" for each new term of the series to be included in the file. Nevertheless, it is possible and instructive to operate a time-series database with these simple means.

4. TIMS Time-series database

Packaged software specificly designed for statistical time-series management is rare (TSP, marketed by TSP International in Stanford, California, has sophisticated processing options but limited database facilities). Even while we do not favour this approach for a training institute, we are presently using manufacturer-specific software, FACOM's Time-series Information Management System. It uses a rather inflexible but adequate storage structure, accomodating a maximum of 14400 time series each consisting of up to 300 data terms. Four bytes are available for storage of each term, limiting the precision of data stored (in floating binary format) to about eight decimal figures. The data organization is indexed sequential, using two levels of index with 120 entries each. Since for processing purposes it is often convenient to have the data also in a simple sequential file and in external code, utilities are available that convert the database completely or partially to such a format, and vice versa. Two major types of such sequential files are defined. The time-series file contains the data one series per record. In the cross-section file one record consists of the terms of several series at one particular point of time (as in an SPSS System file). The cross-section file is suitable as input to the multiple-regression module of the package. If a user wants to apply lagging to one or more series, which is useful, say, to study delay between variation of interest rates and investment levels, this can be done easily by "back-dating" the series to be lagged in the cross-section file.

In Fig. 1 MAINT is the TIMS module that creates and loads a time-series database. DGN is the module used for various manipulations on the data such as adding two series, applying seasonal adjustment, deriving quarterly or yearly data from monthly series, applying exponentiation or taking logarithms, and so forth, a long list. The other modules are intended for analysis of the database contents. MAES for estimate and forecasting through moving averages and exponential smoothing methods (no explicit curve-fitting formula determined), LINE for determining a curve-fitting formula, CENSUS for seasonal adjustment using the CENSUS X-10 method of the U.S. Bureau of the Census, and EPA for seasonal adjustment using a slightly

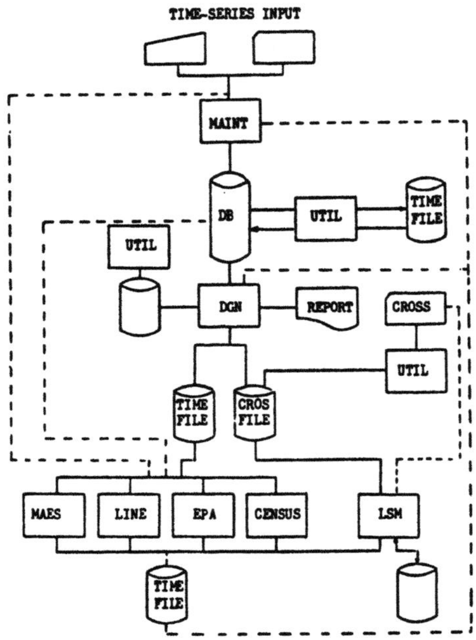

Fig. 1. Structure of TIMS package

different method developed by the Japanese Economic Planning Agency. LSM is the module used to apply regression using least-squares methods. There are some overlapping functions in the package, which suggests that it has probably been assembled by putting together a number of programs that existed already before. There is no reason why such a procedure could not be followed by other developers of time-series software, which would work to make the effort required less imposing. This combined with the relatively simple structure of the central database, makes in-house development appear to be a course that could be taken even by an organization with relatively limited programming resources.

Because of the simple structures and data formats, linking of the database with external software, like SPSS or SAS, is not complicated. The facilities of these packages would again duplicate a number of processing options available in TIMS, or - for those developing own software - such facilities can be taken advantage of minimalise the development effort.

TIMS is not an interactive package; each task, even the simplest updating operation, requires the submission of a batch job. This is today a bit of an anachronism. On the other hand, if such batch jobs can be run from a terminal supported by conversational time-sharing software and for small tasks the turn-around is virtually immediate, the practical difference is not very large.

The demonstration database contains a set of time series for a number of statistical indicators of Asian and Pacific countries. These data were mostly taken from the statistical Yearbook for Asia and the Pacific (United Nations 81). On preparing such a database, one encounters many irregularities in potential input data, like

- lacking terms;
- extreme values (printing errors?);
- changes of definition of the series;
- etc.

It is not simple to prepare a comprehensive set of comparable data for a number of countries, and it is nearly unavoidable to apply occasional editing and imputation based on less than perfect knowledge of the background of the series. This points to the fact that software is only part of the problem. Selecting and preparing the right time-series data and maintaining adequate quality control throughout the process of initial loading and periodic updating are of crucial interest to the usability of the database for its intended purposes of evaluation and planning. The larger and more ambitious the database is designed, the more the quality control problems increase.

5. ADABAS Demographic database

Micro-data management in statistics is not as specific as time-series management. Microdata can be stored in a general DBMS, and retrieved from there. However, some authors (Teitel 77, Cotton 80) have drawn attention to the fact that retrieval characteristics differ in that statistical retrievals tend to address large subsets of the database ("establishments in land transport", "women over 45"), while general database software is often designed for retrieval of individual records or small subsets ("flight ABC on date DEF", "Account of Mr. X"). This untypical use of database software for statistical purposes may lead to inefficiency in response times and otherwise.

It is probably beyond reach for most statistical organizations to develop their own variant of a general DBMS that would be specifically adapted to the pattern of usage in a statistical environment. On the other hand, computers are becoming faster and DBMS software is being improved in scope and variety. Therefore, there is hope that the problems now limiting the effectiveness of general software for statistical micro-databases will decrease in importance.

In our model ADABAS database six main files are defined. These are the usual system files, that is the checkpoint file, the security file and the data dictionary, and three data files. The latter are an Area file, a Household file and an Individual file. The database is of modest size - limited by availability of resources in terms of computer processing time and disk space. The data cover about 50000 individuals, 10% of the population of a province of a developing country, and were mostly obtained from a population census.

This simple database allows demonstration and practical excercises relating to database aspects of direct interest to statistical users, such as:

Querying and updating. Through command language processors interaction with the database can be easy and within the reach of casual users.

Coupling. Data are organized in the form of files, which entails the problem of file contents definition. However, files can be coupled through common fields. For example, the Area file and Individual file can be coupled through the common field area-code. In the Individual file there even is a choice from three area-codes: place-of-present-residence, place-of-former-residence and place-of-birth. Couplings can be dynamically effected and removed, with query possibilities depending on existing couplings. The resulting flexibility can be used to explain the notion of relational databases.

```
26 APR 82 11:57 AM   PAGE 1
CODE ROOF: 1   TYPE ROOF: CONCRETE               NUMBER CASES:       8
CODE ROOF: 2   TYPE ROOF: METALSHEET-OR-TILE     NUMBER CASES:   1,676
CODE ROOF: 3   TYPE ROOF: MUD                    NUMBER CASES:      73
CODE ROOF: 4   TYPE ROOF: THATCH-OR-GRASS        NUMBER CASES:   5,288
CODE ROOF: 5   TYPE ROOF: OTHER                  NUMBER CASES:      18
END OF REPORT    7063 RECORDS FOUND
```

Fig. 2 Query using code translation.

Data dictionary. Full use of the data dictionary helps in understanding the central role such a facility has to play in maintaining order in an organization's data resources. The ADABAS dictionary provides elementary support for editing-on-entry through range checking or the calling of user-written subroutines, a facility of major importance for most statistical applications. Useful also is the Code translation option, which displays or prints meaningful terms in stead of codes for the variables covered (Fig. 2).

Interfacing with application programs. It is possible to link the database with application programs, or even packages like COCENTS. This supports the possible role of a micro-database as a repository for all collected data in a particular statistical sector.

Userviews. It may be desirable to limit the access of certain user groups to parts of the database only, and to present such parts nevertheless as a logical universe. For example, a partial view of the database concentrating on education-related variables can be defined.

6. Practical applicability

In a training situation it is important to discuss database methods in statistics objectively, since it would be misleading to advocate techniques that may have technical beauty but do not address the urgent problems of today. This is even more important if students come from developing countries where resources are scarce. It may be better, for example, to use available resources to upgrade existing chains of data preparation and processing, than to start storing dubious data in an expensive database now.

On the other hand, the subject has become quite popular in recent years, and can no longer be neglected in statistical data processing. Our model databases serve as an introduction to the subject that can be easily understood and allows critical evaluation.

References

Cotton, R., M.J. Turner and R. Hammond (1980), RAPID: A Database Management System for Statistical Applications, I.S.I. 42th Session, Manila.

Teitel, R.F. (1977), Relational Database Models and Social Science Computing, 10th Annual Symposium on the Interface, NBS, Washington.

United Nations, ESCAP (1981), Statistical Yearbook for Asia and the Pacific, Bangkok.

Crossings, Orders and Ultrametrics: Application to Visualization of Consensus for Comparing Classifications

E. Diday, I.N.R.I.A., France

ABSTRACT

One of the most important and difficult problems encountered in automatic classification is that of comparison of classifications. The problem arises when we want to compare the same set of objects characterized by several data arrays. For instance, a time-series of data arrays or data arrays each depending on a different set of variables. This problem also arises when we wish to study the effect of different coding transformations, different choices of dissimilarity indices, the robustness of classification obtained, etc. The notion of crossing sheds new light in this framework. It allows us to relate the visual representation of a hierarchy and the notion of compatibility between an order and a dissimilarity index for which matrix characterizations are provided. The theoretical results provide simple and effective algorithms which facilitate the visual comparison of the classifications and the study of consensus between them.

KEYWORDS

Clustering, hierarchies, ultrametrics, orders, consensus.

INTRODUCTION

The visual representation of a hierarchy induces an order on the objects. If we want to represent the same hierarchy taking into account some external constraints (for example an order on the objects induced by another hierarchy , a partition, etc...) crossings may appear. There is a crossing in a hierarchy when the vertical branch associated with a class of the hierarchy is cut by a horizontal branch.

A complete report on that work may be found in [2]. Hubert (1974) gives a definition of the graph theory terminology which is used here. Rohlf (1981) gives a recent review on "consensus" trees for comparing classifications.

Let H be a hierarchy defined on a set $\Omega = \{w_1, \ldots, w_n\}$ and $\theta : w_{i_1} \ldots w_{i_n}$ a total order on Ω , the set of objects.

1. DEFINITION OF A CROSSING

The order θ produces a crossing in the hierarchy H if and only if there exist three elements w_{i_ℓ} , w_{i_j} , w_{i_k} , and an $h \in H$ such that w_{i_ℓ} and $w_{i_k} \in h$ while

$w_{i_j} \notin h$, for $\ell < j < k$.

2. ORDERS WITHOUT CROSSINGS

In a binary hierarchy there are 2^{n-1} orders without crossings (n = card Ω), which is much, but is still less than the number of possible hierarchies : $n!\,(n-1)!\,2^{1-n}$. The notion of crossing is related to the "compatibility" between an order θ and an index of dissimilarity d.

Definitions

By a dissimilarity index d we mean a map : $\Omega \times \Omega \to R^{+}$ which is symmetric (i.e., $d(w_1, w_2) = d(w_2, w_1)$) and such that $d(w,w) = 0$. Let d be such an index and $d_{ij} = d(w_{\ell_i}, w_{\ell_j})$. Let $M(d,\theta) = [d_{ij}]$ be the $n \times n$ matrix whose elements are the index values d_{ij} and where rows and columns are ordered according to θ. Intuitively speaking, the index d should be interpreted as a measure of dissimilarity of objects, in such a way that more similar objects would be assigned smaller index values. Let us now define three kinds of compatibility.

3. COMPATIBILITY BETWEEN AN ORDER AND A DISSIMILARITY INDEX

Definition 1

d and θ are weakly compatible if and only if for any ordered triple $w_{i_{\ell_1}}$ $w_{i_{\ell_2}}$ $w_{i_{\ell_3}}$ having two consecutive elements, $d_{\ell_1\ell_3}$ is larger than the dissimilarity between the consecutive elements.

Definition 2

We say that d and θ are semi-compatible if and only if for any ordered quadruple $(w_{i_{\ell_1}}, w_{i_{\ell_2}}, w_{i_{\ell_3}}, w_{i_{\ell_4}}) \in \Omega^4$ with $\ell_1 \le \ell_2 < \ell_3 \le \ell_4$ and $\ell_3 = \ell_2 + 1$ we have :

$$d_{\ell_2\ell_3} \le d_{\ell_1\ell_4}.$$

Definition 3 (Brossier (1980)).

d and θ are compatible if and only if for any ordered triple w_{ℓ_i} w_{ℓ_j} w_{ℓ_k} we have :

$$d_{\ell_i\ell_j} \le d_{\ell_i\ell_k} \quad \text{and} \quad d_{\ell_j\ell_k} \le d_{\ell_i\ell_k}.$$

4. THE MATRIX CHARACTERIZATION OF THE VARIOUS COMPATIBILITIES

In the present paper we say that $M(d, \theta)$ is a Robinson matrix if and only if in every row as we go towards the diagonal the elements do not increase (and of course the diagonal element is not greater than any other elements within that row). (We note that Hubert (1974) uses the matrix M in the sense of similarity). We say that a matrix $M(d, \theta)$ is SDR if the elements within any above-rectangle containing an element $d_{\ell_i\ell_{i+1}}$ of the above diagonal are not less than $d_{\ell_i\ell_{i+1}}$. We say that a matrix $M(d,\theta)$ is SDD if and only if in the above triangular matrix the elements of any row (resp. column) are not less than the element of the above diagonal which is lying on that row (resp column). (By the above diagonal we mean the above diagonal immediately above the main diagonal).

0	1	3	5	6
1	0	2	4	6
3	2	0	4	5
5	4	4	0	1
6	6	5	1	0

3a. Robinson

0	1	5	6	2
1	0	2	6	5
5	2	0	4	7
6	6	4	0	1
2	5	7	1	0

← above rectangle

3b. SDR

0	2	5	4
2	0	4	6
5	4	0	1
4	6	1	0

← above diagonal

3c. SDD

FIGURE 1 (The three kinds of matrices)

Table 1 summarizes the three definitions.

For $i \le j$

Robinson <==>	rows $d_{ij} \le d_{ij+1}$	SDR <==>	$\{d_{jj+1} \le d_{i\ell}\ \ j+1 \le \ell\}$
	columns $d_{ij} \le d_{i-1j}$	SDD <==>	rows $d_{ii+1} \le d_{ij+1}$
			columns $d_{j-1j} \le d_{i-1j}$

TABLE 1

It is easy to show that the set of Robinson matrices is included in the set of SDR matrices which turns to be included in the set of SDD matrices.

5. X-COMPATIBILITY IN THE CASE WHEN ONLY A SUBSET Ω' : $w_{i_1},\ldots,w_{i_\ell}$ (with $\ell \le n$) OF Ω IS ORDERED BY θ'

To obtain a general definition in this case we use the term "X-compatibility" which stands for any of the previous three compatibility types. We say that $w \in \Omega$ is left X-compatible if and only if the order $w\,\theta'$: $ww_{i_1}\ldots w_{i_\ell}$ is X-compatible on $\Omega' \cup w$; it is right compatible if and only if $\theta' w$ is X-compatible on $\Omega' \cup w$.

Definition

θ' and d are X-compatible on Ω if and only if the following two conditions are satisfied.

1 - $\forall w \in \Omega$, $w \notin \Omega'$, w is left X-compatible or right X-compatible.

2 - If w is left X-compatible and w' is right X-compatible then the order $w\,\theta\,w'$: $ww_{i_1}\ \ldots\ w_{i_\ell}\ w'$ is X-compatible on $\Omega' \cup \{w, w'\}$.

6. PROPERTIES

We denote by $C(d, \theta')$ the path which satisfies the order θ' on $\Omega' \subset \Omega$ weighted by the sum of weights assigned to its $\ell-1$ edges where the weight of an edge $w_{\ell_i}\, w_{\ell_{i+1}}$ is $d_{i,i+1}$.

The following three diagrams summarize the main properties which characterize the preceding definitions. In diagram 1 the equivalence (1) has been proved by Leclerc (1977) from another point of view. In diagram 2 the relations (1) & (2) are given in Hubert (1974) and Brossier (1981) ; the relation (3) has been given by Rosensthiel (1967) and Leclerc (1981) (the latter author uses the notion of T-minimax path).

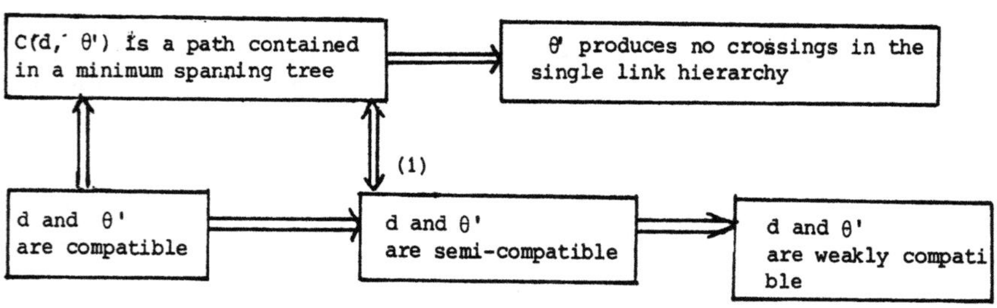

DIAGRAM 1 : θ' is an order on $\Omega' \subset \Omega$

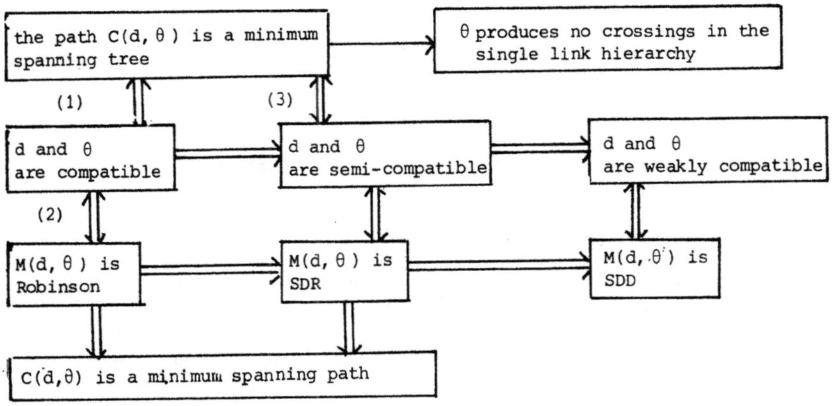

DIAGRAM 2 : θ is an order on Ω

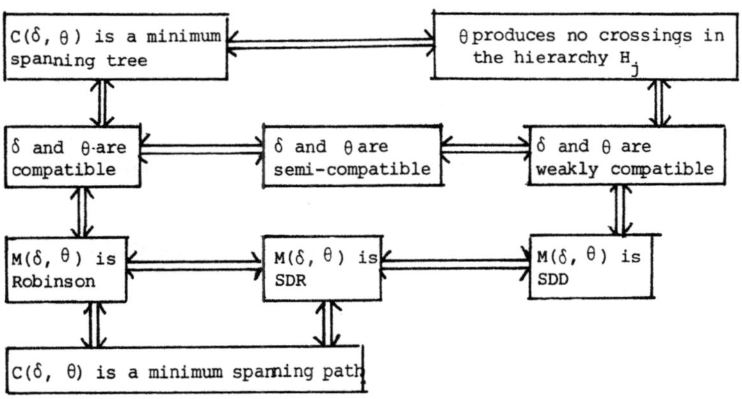

DIAGRAM 3 : θ is an order on Ω and δ is an ultrametric

7. A SIMPLE ALGORITHM WHICH PRODUCES A MINIMUM SPANNING PATH IN THE CASE OF AN ULTRAMETRIC

Given an ultrametric δ, the algorithm gives a path $w_{i_1} \dots w_{i_n}$ obtained as follows : w_{i_1} is chosen randomly, for $\ell = 1,\dots,n-1$, $w_{i_{\ell+1}}$ minimizes $\delta(w, w_{i_\ell})$ among the set of elements $w \in \Omega$ not belonging to $\{w_{i_1}, \dots, w_{i_\ell}\}$. It is easy to show that the matrix $M(\delta, \theta)$ with $\theta : w_{i_1} \dots w_{i_n}$ is SDR ; using the diagram 3 it follows that the algorithm gives a minimum spanning path.

8. AN ALGORITHM WHICH GIVES AN ORDER WITHOUT CROSSINGS FOR SEVERAL HIERARCHIES

Given several hierarchies $H_1, \dots, H_k$ and the corresponding ultrametrics $\delta_1, \dots, \delta_k$ the aim of the algorithm is to give an order θ on Ω without crossings in $H_1, \dots, H_k$. It consists of computing a minimum spanning path with respect to the ultrametrics $\delta_1, \dots, \delta_k$ simultaneously by using the algorithm of §7. We start from an element w_{i_1} choosen randomly, for δ_ℓ we note $A^\ell_{i_1}$ the elements of Ω which are at minimum distance of w_{i_1} ; we choose an element w_{i_2} in $A^1_{i_1} \cap \ldots \cap A^k_{i_1}$, which constitutes the second element of the path ; we iterate the process until one of the two following cases are reached.

i) $A^1_{i_\ell} \cap \ldots \cap A^k_{i_\ell}$ is empty : the preceding choices are reconsidered.

ii) the path contains all the elements of Ω (i.e. it is Hamiltonian) : the process stops.

The algorithm is described more precisely in (2).

It is shown that if there exists an order θ on Ω with no crossing in H the algorithm gives it. Any order θ' on $\Omega' \subseteq \Omega$ obtained by the algorithm produces no crossings in H, (we say that an order θ' : $w_{i_1} \ldots w_{i_\ell}$ (with $\ell \leq n$) produces no crossing in H if there exists $w_{i_j} \notin h \in H$ where h contains w_{i_k} and w_{i_m} such that $k < j < m \leq \ell$). If there is no common order without crossing in $H_1, \ldots, H_k$ it is possible to use the same algorithm to obtain several orders θ' on $\Omega' \subset \Omega$ which finally cover Ω (an example is given in [2]).

9. INTERSECTION OF HIERARCHIES

The intersection of several hierarchies is the set of all the common classes of those hierarchies. It is easy to show that this set is also a hierarchy. We denote $H = H_1 \cap \ldots \cap H_k$.

We denote H_i/H_j the intersection of H_i and H_j from the point of view of H_i ; it is defined as follows :

$$h \in H_i/H_j \iff \{\exists h' \in H_j : h' \subset h \text{ where } h \in H_i \text{ and has minimum cardinality}\}$$

It is easy to show that H_i/H_j is also a hierarchy.

10. CONSTRUCTIVE PROCEDURE

To describe the algorithm we use Table 2 where the notation is defined as follows :
The common order $\theta : w_{i_1} \ldots w_{i_n}$ is obtained as described in § 8; $h_{\ell\ell+1}$ is the smallest common class of $H_1, \ldots, H_k$ containing w_{i_ℓ} and $w_{i_{\ell+1}}$; $h^i_{\ell\ell+1}$ is the smallest class of the hierarchy H_i containing the smallest class of the hierarchy $H_1 \cap \ldots \cap H_k$ which contains w_{i_ℓ} and $w_{i_{\ell+1}}$. It is proved in [2] that :

i) the set of the classes $h_{\ell\ell+1}$ with $\ell = 1, \ldots, n-1$ produces the hierarchy $H_1 \cap \ldots \cap H_k$;

ii) The set of the classes $h^i_{\ell\ell+1}$ for $i = 1, \ldots, k$ and $\ell = 1, \ldots, n-1$ produces the hierarchy $H_i \;/\; H_1 \cap \ldots \cap H_k$.

Example

Suppose given the hierarchies H_1 and H_2 shown in figure 3, table 2 produces table 3. This table permits : i) the visualisation of the hierarchies H_1 and H_2 with the same order on the objects (figure 4) ; ii) the visualisation of $H_1 \cap H_2$, H_1 / H_2 and H_2 / H_1 (figure 5) where $f_H(h_i)$ is the height of h_i in H.

Conclusion

The introduction of the crossing notion has emphasized a clear link between ultrametrics, minimum spanning path, different types of compatibilities and their matrix characterization. These theoretical results provide practical and simple algorithms which facilitate the comparison of classifications for the user.

	$A^1_{i_1}$		$A^1_{i_\ell}$		$A^1_{i_{n-1}}$
	$A^k_{i_1}$		$A^k_{i_\ell}$		$A^k_{i_{n-1}}$
• w_{i_1}	w_{i_2}		$w_{i_{\ell+1}}$		w_{i_n}
$H_1 \cap \ldots \cap H_k$	h_{12}		$h_{\ell\ell+1}$		h_{n-1n}
$H_i/(H_1 \cap \ldots \cap H_k)$	h_{12}		$h^i_{\ell\ell+1}$		h^i_{n-1n}

TABLE 2

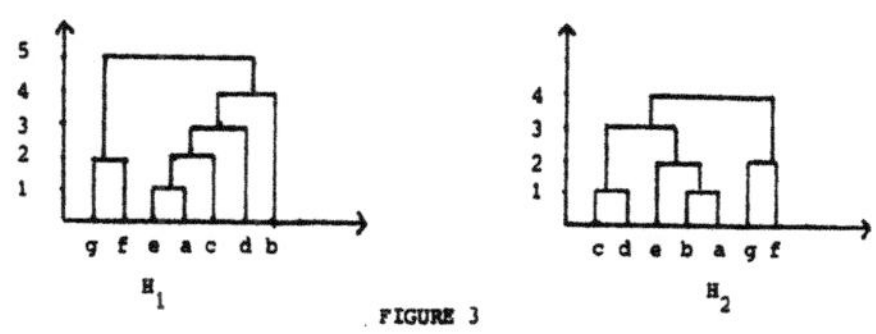

FIGURE 3

	eac	ea	a	b	gf	f
	c	eba	ba	b	gf	f
d	c	e	a	b	g	f
$H_1 \cap H_2$	(eacdb)	(eacdb)	(eacdb)	(eacdb)	Ω	(gf)
H_1/H_2	(eacd)	(eacdb)	(eacdb)	(eacdb)	Ω	(gf)
H_2/H_1	(eacdb)	(eacdb)	(eba)	eacdb	Ω	(gf)

TABLE 3

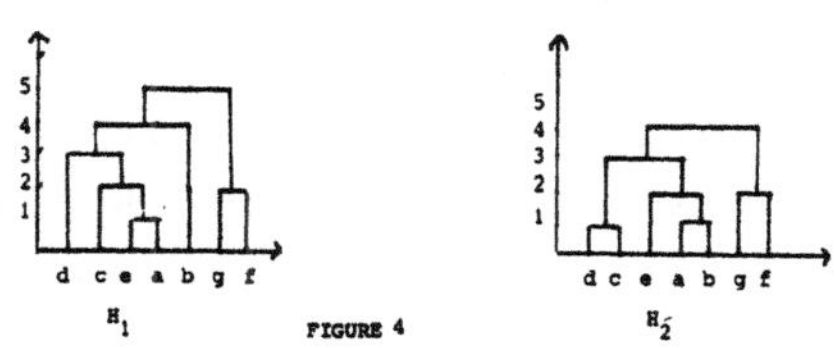

FIGURE 4

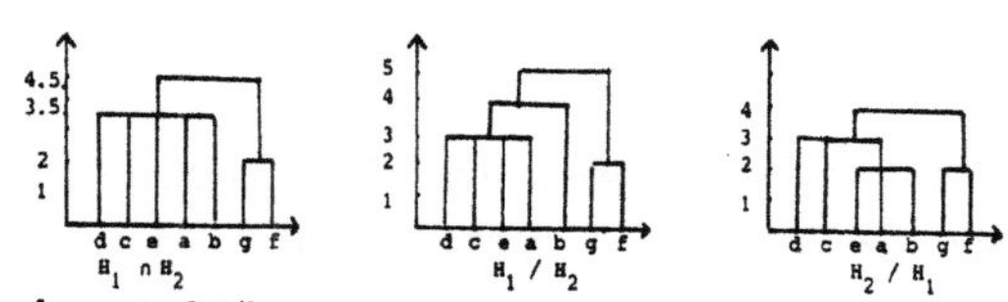

$f_{H_1 \cap H_2} = (f_{H_1} + f_{H_2})/2$ FIGURE 5 $f_{H_i/H_j} = f_{H_i}$

REFERENCIES

[1] BROSSIER (1980). "Représentation ordonnée des classifications hiérarchiques" Statistique et analyse des données. 2, 31-34.

[2] DIDAY E. (1982). "Croisements ordres et ultramétriques : application à la recherche de consensus en classification hiérarchique". INRIA, Report, Rocquencourt, 78150 LE CHESNAY (France).

Further references may be found in our poster summary published in COMPSTAT 82 - Short communications, summaries of posters, Physica-Verlag 1982.

Statistical Programs for Microcomputers: The Implementation of a Directory for Data Structures

P.G.N. Digby and **R.W. Payne**, Rothamsted Experimental Station, Harpenden, G.B.

Summary

A Pascal implementation is given of the data structures described by Lamacraft and Payne (1980).

Keywords: compound data structures, microcomputers, Pascal, statistical computing

1. Introduction

The variety and flexibility of data structures supported by statistical software greatly enhance the range of analyses that can be performed (Payne and Nelder, 1976). Payne and Nelder recognised eight different basic structures, including scalars, variates (or variables), factors (or grouping variables), data matrices, matrices (rectangular, triangular or diagonal), tables, etc. All these structures have their place in statistical work, yet to cater for such a wide range of structures may result in an unacceptable overhead of code, particularly if the program is intended to run on a microcomputer. One feasible solution (Lamacraft and Payne, 1980) is to work with a few basic types of data structure and represent the more complicated structures as compound structures. For example, a two-way table with margins would be a compound structure containing a two dimensional array for the body of the table, two one dimensional vectors for the row and column margins and a scalar for the overall margin (grand mean or total etc.).

In this paper we present an implementation of the Lamacraft and Payne data structures, appropriate for programs designed to run on microcomputers. On microcomputers it may not be feasible to construct large general-purpose packages and a better approach may be to write suites of compatible programs. If a single implementation of a directory for data structures, such as we present, is used for all these programs, they can then communicate via standard files.

Whilst a number of languages could be used, we have chosen Pascal because it offers the following advantages: it is widely available on micro (and other) computers; it is a structured language; it supports complex data structures, including pointers, which improve readability. In addition, it is likely that programs written in Pascal could easily be translated into another structured language, either by hand or automatically, should that become desirable (c.f. Albrecht et al., 1980).

2. Specification of data structures

Lamacraft and Payne (1980) indicate that the information about a data structure can be considered in four parts and they suggest that this information should be kept in four separate blocks of workspace. These blocks are: a header block, essentially an entry in a directory of structures; a reference block, giving information on the nature of the data structure; a data attributes block, containing information about the data values; and a data values block, holding the values of the structure.

The Pascal type declarations required to implement these blocks of information are given below. The header, reference, and data attributes blocks are Pascal records, given on lines 42 - 73.

The header block contains an identifier (lines 23 - 27), a pointer to a reference block (line 13) and two pointers to other header blocks (line 12). The identifier consists of an array of up to eight characters (lines 2, 22 and 25) and an integer suffix, as in Genstat (Alvey et al., 1977). This recognizes, for example, YIELD(1980) and YIELD(1981) as being distinct structures and permits the convenient shorthand of allowing YIELD(1971...1978) to represent a list of eight structures. The pointers to header blocks are used to build a directory of header blocks, in alphabetical order, as a binary tree; this generally gives shorter search times than for a linked list (Knuth, 1968).

The information stored in a reference block (lines 49 - 64) varies depending on the type of data structure (lines 51 and 18). If the structure is a scalar the information kept is its value (line 52) and the data attribute and data value blocks are not required.

The reference block for a vector contains five elements: vecaccess indicates the method of access (lines 53 and 21); novalues stores the number of values in the structure (line 54); levelnames may, if required, point to the header block of a structure giving names for the individual values (line 55); acclist may point to the header block of a structure containing a list of values indicating the accessing order (line 56); and vecattribs points to the data attribute block for the vector (line 57). The order in which the data values are accessed is controlled by two mechanisms: the element vecaccess indicates whether the values are to be taken in (normal) order or in reversed order; while the structure to which acclist points, if any, gives the order in which the values are to be accessed, possibly reversed if indicated by vecaccess .

The reference blocks for multiway structures (such as matrices) and compound structures are described in detail below.

In the attribute block integer novals (line 67) gives the number of values, mode (line 68) indicates the mode of the values and selects an index (lines 69-72) into an array of the appropriate mode of value. This is a weak area of

Pascal: it has been necessary to specify maximum numbers of each mode of value (lines 3 - 6), declare subranges of type integer (lines 8 - 11) for indices to the arrays that will hold the values (lines 75 - 78). Other languages, such as C (Kernighan and Ritchie, 1978), handle this more satisfactorily; however Pascal has two advantages, compilers are available for a wider range of computers and the Pascal structures given here could be translated into C if required (whereas if the full power of C where exploited the reverse translation would be difficult).

The information required for a multiway structure (lines 58 - 61) is more complex. The element _access_ is similar to the element _vecaccess_ for a vector, but additional accessing forms are possible: we have included the diagonal access of a multiway structure here (line 20). Other methods, such as considering the values as the lower triangle of a symmetric matrix, would also be useful; however care would be required to decide how this should be construed for multiway structures with more than two dimensions. Integer _nodimensions_ gives the number of dimensions and _attributes_ points to an attribute block for the data. Information about each dimension is kept in a linked list of records of type _dimension_ (lines 28 - 34). The element _nextdimension_ (line 61) points to the information for the first dimension, which contains a pointer to the next dimension, and so on. The record for each dimension contains the number of levels of the dimension, pointers to header blocks of structures that give names for the levels and an accessing order for the levels, and the pointer for the linked list.

A compound structure is a collection of structures, each component of which may be a structure of any type, including compound structures themselves; thus allowing hierarchical structures to be handled. Each component has a reference block; for a simple component the reference block points to a data attributes block and data values block, as described above, whereas for a compound component it contains the first element of a doubly-linked list of elements of type _component_ . There is an element in this list for each of the components which form the compound component under consideration. Each element of type _component_ is a record (lines 35 - 41) containing the name of the component (line 37), a pointer to the reference block of the component (line 38), and a pair of pointers (lines 39 and 40) to link to the previous and next components. The reference block for a compound structure, or a compound component, contains _whichcompound_ (line 62) indicating the type of compound structure or component (line 19) and _base_ (line 63), itself a record of type _component_ which simply points to the first and last components of the structure or component. An example of a compound structure is the two-way table with margins, described above. This may itself be a compound component of a structure representing the full two-way analysis of variance of the variable concerned.

```
const
   maxalpha    =    8;
   maxintvals  = 1000;
   maxrealvals = 2000;
   maxnamevals =  500;
   maxptervals =  500;
type
   intvalpter   =  1..maxintvals;
   realvalpter  = 1..maxrealvals;
   namevalpter  = 1..maxnamevals;
   ptervalpter  = 1..maxptervals;
   ptheadblock  = ↑ headblock;
   ptrefblock   = ↑ refblock;
   ptattblock   = ↑ attblock;
   ptcomponent  = ↑ component;
   ptdimension  = ↑ dimension;
   valtype      = (ints,reals,names,pters);
   structtype   = (scalar,vector,multiway,compound);
   typecompound = (factor,ssp,other);
   accesstype   = (normal,reversed,diag);
   subaccess    =  normal..reversed;
   ident        = array [1..maxalpha] of char;
   fullid =       {   stores suffixed      }
      record      { structure identifier }
      alphaident : ident;
      suffix     : integer
      end;
   dimension =       {   stores information about one    }
      record         { dimension of multiway structure }
      nolevels       : integer;
      levelnames,
      acclist        : ptheadblock;
      nextdimension  : ptdimension
      end;
   component =       {   stores information about one    }
      record         { component of compound structure }
      compident      : ident;
      ptref          : ptrefblock;
      prevcomponent,
      nextcomponent  : ptcomponent
      end;
```

```
    headblock =     { stores header block information }
       record
       identifier    : fullid;
       ptref         : ptrefblock;
       lefttreelink,
       righttreelink : ptheadblock
       end;
    refblock =     { stores reference block information }
       record
       case strtype : structtype of
          scalar    : (value          : real);
          vector    : (vecaccess      : subaccess;
                       novalues       : integer;
                       levelnames,
                       acclist        : ptheadblock;
                       vecattribs     : ptattblock);
          multiway  : (access         : accesstype;
                       nodimensions   : integer;
                       attributes     : ptattblock;
                       nextdimension  : ptdimension);
          compound  : (whichcompound  : typecompound;
                       base           : component)
       end;
    attblock =   { stores data attribute block information }
       record
       novals    : integer;
       case mode : valtype of
          ints   : (intvals  :  intvalpter);
          reals  : (realvals : realvalpter);
          names  : (namevals : namevalpter);
          pters  : (ptervals : ptervalpter)
       end;
var
    intdata  : array [ intvalpter] of integer;
    realdata : array [realvalpter] of real;
    namedata : array [namevalpter] of ident;
    pterdata : array [ptervalpter] of ptheadblock;
    directory : ptheadblock;
```

3. Conclusion

The specification given above has been implemented on a MIDAS 3D (Z80 CPM) microcomputer at Rothamsted and procedures have been written to form and manipulate the various structures. The experience gained should be very useful in designing the next generation of statistical packages.

4. References

Albrecht, P.F., Garrison, P.E., Graham, S.L., Hyerle, R.H., Ip, P. and Krieg-Bruekner, B. (1980). Source-to-source translation: Ada to Pascal and Pascal to Ada. SIGPLAN Notices 15 , 183-193.

Alvey, N.G. et. al. (1977). Genstat: a general statistical program. Rothamsted Experimental Station, Harpenden.

Jensen, K. and Wirth, N. (1978). PASCAL User manual and report. Springer, New York.

Kernigan, B.W. and Ritchie, D.M. (1978). The C programming language. Prentice-Hall, New Jersey.

Knuth, D.E. (1968). The art of computer programming: Vol. 1 Fundamental algorithms. Addison-Wesley, London.

Payne, R.W. and Nelder, J.A. (1976). Data structures for statistical computing. Proceedings of the 9th International Biometric Conference, Vol. II, 191-208.

Lamacraft, R.R. and Payne, R.W. (1980). A new look at data structures for statistical languages. In: COMPSTAT 1980: Proceedings in Computational Statistics (ed. Marjorie M. Barritt and D. Wishart), 463-469. Physica-Verlag, Vienna.

Data Management and Tabulation in OPCS

J. Dixie, Office of Population Censuses and Surveys, Titchfield, G.B.

Summary

The Office of Population Censuses and Surveys is the largest social statistical office in the United Kingdom, and has an extensive computing requirement. Generalised statistical software has been used to an increasing extent over the past 20 years by support staff with specialist training. The move to generalised systems from applications programs was a major advance. We are now beginning a second major advance - the dispersal of computing facilities, and the provision of powerful software which is easy for subject matter specialists to use themselves. Three areas are of particular importance. We are using a data dictionary system for the first time to support the processing of the 1981 census of population. We are currently setting up a database management system (SIR) to hold all of the data from social surveys. And in September we are to launch a new tabulation and table manipulation system (TAU) which will handle the bulk of our analysis. As these systems become established the onus of specifying statistical analyses will largely be transferred from computer staff and clerical staff with specialist training to the subject matter specialists. This will enable us to further improve the service that we give to the government, and other users.

Keywords

Population Census, Social Surveys, Population Statistics, Data Dictionary, Statistical Database, Tabulation, SIR, TAU, Wafer Aggregation, Record Selection, Asymmetric Nesting, Table Arithmetic, Table Restructuring, Combining Tables.

1. The Work of OPCS

The United Kingdom Office of Population Censuses and Surveys, (OPCS), was formed in 1970 by the amalgamation of the General Register Office, and the Government Social Survey. It is now the biggest single office concerned with social statistics in the UK, with responsibility for conducting population censuses, and the production of demographic and medical statistics, and for carrying out both continuous and ad hoc social surveys.

2. Statistical Computing Requirements

We operate an ICL 1906S computer under George 3, and an ICL 2966 computer mainly under DME (George 3). Later this Autumn the 2966 will be switching to CME operation, so that George 3 and VME may be run concurrently. In due course we will be moving entirely to VME operation. Our users are located at two main sites, and in both cases the principal means of access to the machines is via video terminals, supported by line-printer facilities.

Our requirement for statistical computing facilities covers all stages in the processing of statistical data from acquisition through editing to tabulation, mathematical analysis, and publication. Our data sets vary in size from, say 3 million characters for a small survey, up to the 3,500 million characters of the 1981 census of population. Some of the data, particularly that concerning the registration of births marriages and deaths, has a simple structure, while the census data and that from most social surveys has a complex structure.

OPCS has used generalised software for statistical tabulation to an increasing extent over some 20 years. The move to generalised systems from applications programming was a major advance, and now nearly all tabulation is done with generalised systems. However, our current software is mainly used by staff with specialist training, and the subject matter specialists do not use the computer directly to any great extent. Editing and data management are done almost entirely by specialists in computing, albeit using utilities such as Filetab as well as applications programs. Only in the field of mathematical statistical analysis, where packages such as SPSS, Package-X, Clustan, Glim, and EDP (developed by OPCS) have been available for several years, does the end user normally specify the work himself.

3. The Dispersal of Computing Power

We are now at the beginning of a second major advance, the provision of communications equipment and suitable software for direct use by subject matter specialists. This may be expected to have as great an impact on our work as the previous move to generalised systems. The use of packages such as SPSS, Clustan and Glim is increasing where it is appropriate to use them, and will no doubt continue to grow. But the biggest changes will be in the use of computerised metadata systems, data management systems, and software for tabulation. It is on these three areas that I want to concentrate in this paper.

4. Meta Data Systems

The 1981 census of population saw the introduction of our first full scale data dictionary system, and this is having a substantial influence on all stages of processing the data. The system enables users to see easily what is available on the data files, and helps them in specifying their tabulations. It provides the foundation for the various administrative systems associated with servicing the needs of the wide range of customers for analyses of the census data. It is also used to generate automatically the descriptions of the data files required by the software used to analyse the data.

The census data dictionary was the forerunner to an enhanced system which was released for use earlier this summer, and onto which we are now loading the documentation for all of our major data files for population and medical statistics. This is a substantial task, and will be spread over some months. The data dictionary system is also likely to be valuable for aiding the processing of the regular social surveys, such as the General Household Survey and the Labour Force Survey.

For processing ad hoc social surveys we have adopted a slightly different approach. We are introducing a database management system to bring together and organise all of our survey data. This provides adequate computer held documentation for these generally smaller and self-contained data sets.

5. Data Management

Over the past year we have been considering the possible benefits from the use of a database management system (DBMS). We have reached the point where it has become clear that for our smaller data sets, particularly where they have a complex structure, (but without any penalty where they do not), the key to making data available to users without specialist computing knowledge is to employ a DBMS. We have selected the SIR system for social survey data. This has recently been implemented on our ICL 2966 computer, and will enable us to bring together data from diverse sources; for example personal interviews, administrative records and telephone follow-up interviews. It will also support the procedures for checking the data for validity, and for correcting it where necessary. The third, and vital stage, is to derive new variables and store them on the database before the data is ready for analysis. These usually form a large part of the total data volume for a survey. It will then be simple for users to produce extract files for tabulation, mathematical analysis, and for distribution direct to other government departments and associated bodies.

The facilities provided by a database system would be no less suitable for the users of our large data sets. The only serious drawback is the significantly higher use of computer resource that this would entail, over our present systems. Through the database group of the Statistical Computing Project of the United Nations Economic Commission for Europe we are investigating the RAPID database system. This was initially developed by Statistics Canada for statistical applications, and has a transposed file structure. It therefore offers the hope of greater computational efficiency. As an alternative it may be that in the future a dedicated database computer would provide a cost effective solution. In the meantime we make extensive use of the utility FILETAB, which is used by programmers. We are increasing our use of a "user friendly" version, USERTAB, which offers fewer facilities but is suitable for both programmers and non-programmers. We have enhanced USERTAB to produce output in the form of a magnetic tape, in addition to the (standard) line printer output. For Census data the data descriptions required by FILETAB and USERTAB are created automatically by a utility run off the data dictionary system. We plan to extend this procedure to other applications as more metadata is loaded onto the dictionary system.

6. Tabulation

This September we will be introducing into service a new tabulation system, TAU, which has been designed to meet the requirements of all parts of our Office. These are extensive. It is not surprising therefore that TAU has also been adopted by a number of other government departments, and installations are scheduled to begin in October. (In the UK the Government Statistical Service is dispersed to the various departments - eg Employment, Health, Industry, etc - and each department can choose the software that best meets its needs). In designing TAU we were seeking a combination of ease of use by those without specialist computing knowledge, powerful facilities, and an adequate degree of computational efficiency.

We started by considering the command language, and after careful study took that of TPL (developed by the United States Bureau of Labor) as the basis for TAU. It is a well tried language, and in our view strikes an excellent balance between conciseness and intelligibility, and has a clear and simple structure. These characteristics are essential in a system which is to be suitable for all classes of user. Like TPL, and various other systems, TAU requires the data file to be defined in a codebook. This removes from the user much of the burden of table specification, and is a second major factor in making the system easy to use. To this basis we added our experience in developing our present systems, particularly in the matter of efficient counting routines (which differ from those in TPL) and techniques for developing generative systems. So much for our starting point; what then is new about TAU?

We have added substantially to the facilities of TPL. The main extensions are:

i. Facilities to aggregate the wafers of a table to form new wafers of totals and sub-totals. Virtually any combination of wafers is possible; for example selecting wafers on the whole, or just a part, of the wafer identification code.

ii. The application of selection conditions to individual tables within a job and to subtables within a table. Also the application of selection conditions to the computation of new variables from those on the input data file, and to calculations performed on the cells of a table after the counting phase.

iii. The provision of the standard descriptive statistics: mean, standard deviation, percentiles, and percentages.

iv. The facility to nest variables asymmetrically. In the UK the most common retirement age for males is 65, while that for females is usually 60; the definition of "persons of retirement age" therefore must be based on two variables, sex and age.

v. A set of facilities for saving and retrieving tables for reprinting, or further processing. The latter we term secondary analysis, and the facilities provide for table arithmetic (eg dividing one table by another to produce a table of ratios), restructuring tables (eg by collapsing out a level of nesting, or transposing, or condensing the grouping of a variable), and for interleaving tables (eg four quarterly tables to produce an annual table) or placing tables with a common stub (or heading) side by side (or one above the other) on the page.

The benefit of storing the details of data files in codebooks for repeated or shared use is well proven. We have extended this idea to commands that are likely to be used repeatedly or by different users, by introducing a Statement Library System. This enables groupings of codes and the associated print labels, selection conditions, text for printing in tables, print formats, and wafer aggregation procedures to be named and stored. To make use of a stored command in a job it is only necessary to reference the appropriate name. The Statement Library System also includes interrogation facilities, which will help users to exploit it fully.

Although it is our intention to document all major data files on the data dictionary and to create codebooks for TAU automatically by means of a utility, there will always be some ad hoc files. These may not have a sufficiently long life to justify the effort of documenting them on the data dictionary. Therefore we have also developed a User Code Book System for creating ad hoc codebooks, which employs a command language compatible with that of TAU. The Codebook System also provides for details stored in an existing codebook to be copied to the codebook which the user is setting up where this will save effort. As a further refinement, it is only necessary for the user to define in a codebook the variables which are needed for the current analysis. The other variables may be defined if wished, (say in preparation for further analyses), or they may be ignored.

7. Conclusions

During the past year we have introduced more powerful hardware and more communications equipment into our computer system. We are now about to introduce new software for statistical processing, some bought in, and some which we have developed ourselves. Over the next few years the onus of specifying statistical analysis will be largely transferred from computing specialists to subject matter specialists. I am confident that this will help us to produce more timely statistics, and to offer our customers - the various levels of Government and related bodies and the general public - a better and more flexible service.

Archaeology, Data Analysis, Computer Science: How to Run Proper Treatment of Archaeological Data

F. Djindjian, Musée Antiquites Nationales, Paris, France
H. Leredde, University of Paris-Nord, France

Classical data analysis have been applied to archaeological data for 15 years. In the light of numerous experiments, we can try to answer the following question : using modern methods of information analysis, what is the most appropriate strategy for a given set of archaeological data ?

KEY-WORDS : *Archaeology, Data Analysis, Methodology.*

Modern methodologic and algorithmic developments are described for the resolution of classical archaeographical and archaeometrical problems like typometry, seriation, culture identification, etc..., for which are defined patterning and assignment constructions generalizing classical archaeographical and archaeometrical formal constructions, modelling excepted.

Data analysis methods used in quantifying, reducing description space, understanding and patterning reduced configurations, validating and assigning are illustrated with examples choosen in the field of archaeology.

1. - FIRST PART : *Two knowledge-making constructions in Archaeology*

Two constructions are defined :

1.1 - *Patterning construction*, for the discovery of patterns in the data, based on equivalence, order, or spatial distribution relations. At this level are distinguished :

1.1.1 - *Stricto sensu patterning construction*, is aimed at the discovery of a pattern model (partition, seriation, distribution) in an intrinsic exclusively meaning description (technologic, functionnal, symbolic, social, economic,...) of physical finds or physical finds-closed-group. These constructions bring together classical constructions of typometry, taxinomy, seriation, typological facies determination. A study is made by using data analysis algorithms applied to *elements* × *attributes* tables where the elements are physical finds or physical find-groups and the attributes are an intrinsic description.

1.1.2 - *Explanation construction*, for the discovery of a given extrinsic meaning pattern inside any intrinsic description of physical finds or physical find-groups. These constructions bring together classical chronological or geographical facies searching, climatic or ecologic interaction, specific working areas within a habitat, determination of exchange networks. Research is performed by applying data analysis patterning methods to *attributes* × *attributes* tables where the first attribute group is an intrinsic information and the second attribute group is an extrinsic information (by which is meant measuring interactions between physicals finds and their surrounding).

1.2 - *Assignment constructions*, for the assignment of a physical find or physical find-closed-group to a primarily obtained pattern from a patterning construction. These constructions bring together classical typological classification, cultural attribution, relation to dating sequences, raw material origin identification. Assignment is made by using data analysis assignment methods.

2. - SECOND PART : *Mathematical construction*

This second part describes the mathematical algorithms realised to execute each function.

2.1 - *Patterning construction* :

2.1.1 - *Quantifying* : This part concerns simultaneous observation of discrete and continuous attributes and the appropriate metric choice. For a both discrete and continuous attribute description, a coding is made labelled complete disjonctive coding : this is obtained by chopping continuous attributes into defined boolean attribute *modalities*.

The coding only allows one to work on attribute *modalities*. A complementary coding is then used, by an optimal scoring algorithm [TENENHAUS(1978), YOUNG(1976)] which transforms discrete variables into continuous variables by optimizing inertia in a fixed reduced factorial space. The coding allows one to work directly on attributes.

On a *elements* × *attributes* table, where attributes are continuous or are made continuous by optimal scoring, euclidian distance will be used. Whenever attributes are coded in a complete disjonctive form, χ^2 distance will be prefered. On a contingency table *attributes* × *attributes*, χ^2 distance will be used.

2.1.2 - *Reduction* : Reduction is obtained by factorisation and selection of pertinent factors, interpreted by attributes. With the choice of χ^2 distance, reduction is obtained by correspondance analysis. Further, treatments are made on element or attribute-weighted-factors in reduced space.

2.1.3 - *Comprehension of reduced configuration* : The comprehension is obtained by both simultaneous projection of elements and attributes in factors, optimized with correspondance analysis, and by interpretation-helping-ratios [BENZECRI(1973)].

With cluster analysis (agglomerative hierarchical method), clusters are interpreted by factors from correspondance analysis or principal components analysis and by attributes, allowing then to determinate attributes and factors leading to the clustering [JAMBU(1978)] .

2.1.4 - In the case of partition research, cluster analysis is used directly on weighted factors [JAMBU(1978)] ; cluster analysis algorithms are particularly fast agglomerative hierarchical methods with χ^2 distance and variance agglomerative criterion [BRUYNOOGHE(1978)] .

2.1.5 - *Patterning* : Patterning is seen here as searching for characteristic geometrical configuration upon a reduced space (clusters, serial polynomial curve, spatial distributions) by an interactive process, in making supplementary an element or an attribute (by which is meant a simple projection in factor space).

2.1.6 - *Modelisation* : Patterning may also be modelised by adjustment to the reduced configuration of a theoretically reduced configuration with a Schönnemann and Caroll algorithm which adjusts two multidimensionnal spaces [SCHONNEMANN-CAROLL(1970)] .

2.1.7 - *Validity* : The same algorithms used on a randomised permutation data table allows one to determinate the inertia level beyond which factors and clusters are no longer relevant [LEBART(1977)] .

2.1.8 - *Remark* : Description may be either a type description or an attribute description. Our preference is a hierarchical description from attributes to very useful types with the supplementary element possibility, which allows one to retain finally the only relevant description level.

2.2 - *Assignment methods* :

2.2.1 - *Assignment by factorisation* is made by a single projection in reduced space. In fact, it is a multiple factorial regression [BENZECRI(1973)] .

2.2.2 - *Assignment to the clusters of a cluster analysis* : At each level of the hierarchy, the element is assigned to the nearest cluster in the following manner : the ratios between the squared distances of elements at the centers of gravity of each cluster q and q' are compared to a level chosen in order to minimize possible errors in assigning new elements to the clusters q and q' [JAMBU(1978)] .

It is particularly interesting in the case of partition patterning to use simultaneously the two methods with the possibility of representing cluster gravity centers and cluster inertia ellipses obtained by cluster analysis upon reduced space.

It is also possible to estimate assignment validity by computing the assignment error percentage of classified elements in regard to their own cluster.

2.3 - *Discriminating* (only for partition pattern) :

Classical discriminant analysis is seen as either an explanation construction or an assignment construction. This property only means that correspondance analysis (on contingency tables) and discriminant analysis (without assignment) on a measurement table are two particular cases of generalized canonical analysis.

In the case of both discrete and continuous attribute description, an efficient method is to select stepwise more discriminant attributes by a test (derived fom χ^2), then to make a correspondance analysis on an *elements* x *discriminant attributes* table where the discriminant attributes are coded in a complete disjonctive form, and finally to make a classical discriminant analysis on factors [SAPORTA(1976)] .

3. - *THIRD PART* : *Examples*

The following examples will illustrate some of archaeological problems that we have raised and the method of treating them.

4.1 - *Example I* :

We shall illustrate the formalization of an explicative construction with the example of the determination of chronological facies for the beginning of the Upper Paleolithic in the roch-shelter of La Ferrassie near Les Eyzies, in Dordogne (France). This shelter contains more than a score of aurignacien and perigordian levels, rich in stone tools, and stratigraphically superposed. The essential information is conveyed here by a typology of 25 tools classified according to their technique and manufacture. The contingent information is given by the stratigraphical and hence chronological position of the artefacts.

By using together correspondance analysis and hierarchical cluster analysis we can define five facies which are chronological facies because the levels to which they correspond are, with one exception, stratigraphically superposed.

4.2 - *Example II* :

The illustration of an assignment construction is given by the dating of other sites in Perigord by assigning them to one of the chronological facies. The mathematical method used to this end is the assignment algorithm of a supplementary element - here it is one of the series of stone tools - to one of the classes of classification dendogram which results from the explicative construction.

4.3 - *Example III* :

This example deals with the study of a group of silver-inlaid plate-buckles from northeastern France, dating from the end of the Sixth to the beginning of the Eighth Century. The group is made up 59 plate-buckles on which we have observed the presence or the absence of 26 selected technical criteria, having to do with manufacture, form and decoration [LEREDDE(1979), LEREDDE-PERIN(1980), PERIN(1980)] .

The problems were to structure these data. To do this we carryied out a series of permutations of the lines and the columns of the data table in order to make clear the links between the criteria and the plates. We wanted to establish a typology of these plate-buckles in the first place and in the second place to go beyond this simple typology to discover whether this evolution was chronologically significant. In other words, we wanted to be able to answer the following question : does a seriation underlie this typology ?

We can safely conclude from the various treatments that they disclose a typology of plate-buckles which unmistakably reflects a technical and chronological evolution. Although the principal characteristics of these data, discovered by the treatments, were predictable, this was not true of the finer analysis into sub-groups, and the exact place of certain criteria in this evolution appeared to us only during the study of the results. Such observations would receive even stronger confirmation if we were dealing with a more voluminous group of data.

In conclusion, the fact that we experimented with three treatments of different natures in order to disclose an evolution within this typology has proved to be most fruitful. In fact, if these methods yield identical results for the major groups, in the definition of the sub-groups they reveal certain distinctions which are of great interest to specialists.

BENZECRI J.P. (1973), *L'Analyse des Données : la Taxinomie* (tome 1), *l'Analyse des Correspondances* (tome 2), Paris, Dunod.

BRUYNOOGHE M. (1978), *Classification ascendante hiérarchique de grands ensembles de données*, Les Cahiers de l'Analyse des Données, III-1, 7-73.

DJINDJIAN F. (1980), *Les faciès chronologiques aurignaciens et Périgordiens à la Ferrassie (Dordogne)*, Les dossiers de l'Archéologie, 42, 70-74.

DJINDJIAN F. (1981), *Construction de systèmes d'aide à la connaissance en archéologie préhistorique, structuration et affectation*, Thèse de doctorat de 3e cycle, Paris, Université Paris I.

DJINDJIAN F., LEREDDE H. (1980), *Traitement automatique des données en Archéologie*, Les dossiers de l'Archéologie, 42, 52-69.

DORAN J.E., HODSON F.R. (1975), *Mathematics and computers in Archaeology*, Edinburgh, Edinburgh University Press.

GARDIN J.C. (1979), *Une archéologie théorique*, Paris, Hachette.

JAMBU M. (1978), *Classification automatique pour l'Analyse des Données : Méthodes et Algorithmes* (tome 1), *Logiciels* (tome 2, with LEBEAUX M.O.), Paris, Dunod.

LEBART L. (1977), *Validité des résultats en Analyse des Données*,Consommation, 1, 42-79.

LEREDDE H. (1979), *La méthode des Pôles d'Attraction, la méthode des Pôles d'Agrégation ; deux nouvelles familles d'algorithmes en classification automatique et sériation*, Thèse de doctorat de 3e cycle, Paris, Université Paris VI.

LEREDDE H., PERIN P. (1980), *Les plaques-boucles mérovingiennes*, Les dossiers de l'Archéologie, 42, 83-87.

LERMAN I.C. (1981), *Classification et analyse ordinale des données*, Paris, Dunod.

PERIN P. (1980), *La datation des tombes mérovingiennes*, Genève, Droz.

SAPORTA G. (1976), *Stepwise discriminant analysis when all variables are nominal*, Spring meeting of psychometrica society, Murray Hill.

SCHONNEMANN P.H., CAROLL R.M. (1970), *Fitting one matrix to another under choice of a central dilation and a rigid motion*, Psychometrica, 35, 245-255.

TENENHAUS M. (1978), *Analyse en Composantes Principales d'un ensemble de variables nominales et numériques*, Revue de Statistique Appliquée, XXV-2, 39-55.

YOUNG F.W. (1976), *Optimal scoring with the variety of models*, Spring meeting of the psychometrica society, Murray Hill.

BLINWDR: Robust and Bounded Influence Regression

R. Dutter, Technical University Graz, Austria

0. Summary

BLINWDR is a single FORTRAN program which represents an extension of the widely used linear robust curve fitting program LINWDR. Besides options for different loss functions in order to handle the problem of distributional robustness in regression, the program offers the choice between several weight functions which should bound the influence of outliers in the factor space. It combines these sophisticated methods with the user oriented features of LINWDR which are important in applied regression analysis.

Keywords: Robust Regression / Statistical Software / M-estimators / Robust Procedures / Bounded Influence Regression

1. Introduction

The domain of robust statistics grasps methods which are relatively insensitive to small changes of the assumptions. With this we usually mean insensitivity to distributional changes. This problem in the linear regression case is solved in the program LINWDR (Dutter, 1978). Isolated points in the factor space, however, influence substantially estimates in regression problems or even make them absurd. Procedures for finding stable (robust) solutions are of course not unique. Many different proposals for weighting down influential observations can be found in the literature (Hampel, 1978, Krasker and Welsch, 1979).

Marazzi (198o) describes a subroutine library which offers FORTRAN subroutines with certain procedures for solving the bounded influence regression problem. Unfortunately the user has to write his own main program and links between the user's data and all the subroutines are not always easy to establish. Moreover, the user cannot immediately take advantage of a user oriented high-level computer program with usually many auxiliary statistics which facilitate data analysis.

In addition to the numerous possibilities of LINWDR the program BLINWDR offers most features of that part of the mentioned subroutine library which is concerned with regression. In this paper we describe shortly different methods of bounded influence regression, state their implementation in BLINWDR and resume other capabilities of the program.

2. Robust Regression

We consider the linear regression model

$$f(\underset{\sim}{\theta}) = \theta_o + \theta_1 x_1 + \theta_2 x_2 + \dots + \theta_k x_k$$

with $p = k+1$ parameters $\theta_o, \theta_1, \ldots, \theta_k$. Suppose that n observations $y_1, \ldots, y_n$ and corresponding values x_{ij}, $j = 1,\ldots,k$, $i = 1,\ldots,n$ of the independent variables are given. We write

$$f_i(\underset{\sim}{\theta}) = \theta_o + \theta_1 x_{i1} + \ldots + \theta_k x_{ik}, \; i = 1,\ldots,n.$$

Then the unknown parameters $\underset{\sim}{\theta}$ might be fitted to the data by essentially minimizing the function

$$g(\underset{\sim}{\theta},\sigma) = \sum_{i=1}^{n} \rho(\sqrt{w_i}\,\frac{y_i - f_i(\underset{\sim}{\theta})}{\sigma})\sigma$$

or by equating to zero its derivative

$$\text{(1a)} \qquad g'(\underset{\sim}{\theta},\sigma) = -\sum_{i=1}^{n} \sqrt{w_i}\,\psi(\sqrt{w_i}\,\frac{y_i - f_i(\underset{\sim}{\theta})}{\sigma})\,\frac{\partial f_i(\underset{\sim}{\theta})}{\partial \underset{\sim}{\theta}} = \underset{\sim}{0}$$

where ρ is an even function satisfying certain robustness conditions, $\rho(0) = 0$, $\rho(t) > 0$ for $t > 0$, with its derivative ψ, w_i are optional weights for the observations, and where σ denotes a scaling factor which is to be estimated from the data. σ could simultaneously be computed by the condition

$$\text{(1b)} \qquad \sum_{i=1}^{n} \chi(\sqrt{w_i}\,\frac{y_i - f_i(\underset{\sim}{\theta})}{\sigma}) = a = \text{const.}$$

The solution of the system of equations (1a-b) is called <u>Huber-type</u> (see Huber, 1973).

The function ρ respectively ψ should reduce or eliminate the influence of large residuals. In order to bound the influence of a point in the factor space, i.e. of a row of the design matrix $x_{i1}, x_{i2}, \ldots, x_{ik}$, weights v_i, $i = 1,\ldots,n$ are introduced which are functions of the relative distance to the center of the point cloud. We say that the solution of the system

$$\text{(2)} \qquad \begin{aligned} &\sum_{i=1}^{n} \sqrt{w_i}\,\psi(\sqrt{w_i}\,\frac{y_i - f_i(\underset{\sim}{\theta})}{\sigma\, v_i})\, v_i\, \frac{\partial f_i(\underset{\sim}{\theta})}{\partial \underset{\sim}{\theta}} = \underset{\sim}{0} \\ &\sum_{i=1}^{n} \chi(\sqrt{w_i}\,\frac{y_i - f_i(\underset{\sim}{\theta})}{\sigma\, v_i})\, v_i^2 = a = \text{const.} \end{aligned}$$

is of type <u>Schweppe</u> and the solution of

$$\text{(3)} \qquad \begin{aligned} &\sum_{i=1}^{n} \sqrt{w_i}\,\psi(\sqrt{w_i}\,\frac{y_i - f_i(\underset{\sim}{\theta})}{\sigma})\, v_i\, \frac{\partial f_i(\underset{\sim}{\theta})}{\partial \underset{\sim}{\theta}} = \underset{\sim}{0} \\ &\sum_{i=1}^{n} \chi(\sqrt{w_i}\,\frac{y_i - f_i(\underset{\sim}{\theta})}{\sigma})\, v_i = a = \text{const.} \end{aligned}$$

is of type <u>Mallows</u> (see Hampel, 1978). The constant a may be determined so that the estimate of σ is asymptotically unbiased for normal errors.

The distances in the factor space are found after having (robustly) calculated a covariance structure as follows. Denote the i-th row of the design matrix $x_{i1},\ldots,x_{ik}$ by $\underset{\sim}{x}_i$. Then calculate a $(k \times k)$-matrix A such that

$$\frac{1}{n} \sum_{i=1}^{n} u(||\underset{\sim}{z}_i||)\underset{\sim}{z}_i \underset{\sim}{z}_i^T = \beta I_k$$

where $\underset{\sim}{z}_i = A(\underset{\sim}{x}_i - \underset{\sim}{b})$ with a certain location $\underset{\sim}{b}$. (β and u are given below.) The distance between $\underset{\sim}{x}_i$ and $\underset{\sim}{b}$ is then $||\underset{\sim}{z}_i||$ and the i-th weight v_i is a certain function $f(||\underset{\sim}{z}_i||)$.

BLINWDR offers four options for u and f (see Marazzi, 1980, for more motivation):

(a) Hampel proposal (for the Mallows-type estimator):

$u(t) = \min\{1, \mathrm{BOUND}/t^2\}$, $f(||\underset{\sim}{z}||) = u(||\underset{\sim}{z}||)$, $\beta = 1$.

(b) Mallows-Maronna proposal (for the Mallows-type estimator):

$u(t) = \min\{1, \mathrm{BOUND}/t^2\}$, $f(||\underset{\sim}{z}||) = \sqrt{u(||\underset{\sim}{z}||)}$,

$\beta = (1/p)\, E[u(||\underset{\sim}{z}||)\, ||\underset{\sim}{z}||^2]$ where the expectation is computed for a p-variate standard normal argument $\underset{\sim}{z}$.

(c) Hampel-Krasker proposal (for the Schweppe-type estimator):

$u(t) = 2\Phi(c/t)-1$, $f(||\underset{\sim}{z}||) = 1/||\underset{\sim}{z}||$, $\beta = 1$,

Φ denotes the cumulative standard normal distribution.

(d) Schweppe proposal (for the Schweppe-type estimator):

$w_i = \sqrt{1-h_{ii}}$ where h_{ii} denotes the i-th diagonal element of the hat matrix $X(X^TX)^{-1}X^T$ with the design matrix X.

It remains to define the functions ψ and χ. Four approved forms are included in the program.

(a) Least-squares:

$$\psi(t) = t,$$
$$\chi(t) = t^2/2.$$

(b) "ψ bends at c" (Huber):

$$\psi(t) = \begin{cases} -c & t < -c \\ t & |t| \leq c \\ c & t > c \end{cases}$$

for some $c > 0$ to be specified.

$$\chi(t) = \frac{\psi(t)^2}{2}.$$

(c) "ψ bends at a, b and d" (Hampel):

$$\psi(t) = \begin{cases} t & |t| \leq a \\ a.\mathrm{sign}(t) & a < |t| \leq b \\ \frac{d-|t|}{d-b}\, a.\mathrm{sign}(t) & b < |t| \leq d \\ 0 & d < |t| \end{cases}$$

with $0 < a < b < d$. σ is estimated by a robust fit using the monotone function "ψ bends at a".

(d) "ψ has the form of sine" (Andrews):

$$\psi(t) = \begin{cases} \alpha.\sin(t/\alpha) & |t| < \alpha\pi \\ 0 & \text{otherwise.} \end{cases}$$

σ is estimated by "ψ bends at α ".

3. Other features of BLINWDR

The data $x_{i1}, \ldots, x_{ik}, y_i$ and eventually the attached weight w_i have to be entered case by case. If a different model with this data set is wished, variables (independent and/or dependent) may be transformed, omitted or newly generated. Several problems can be treated in one computer run. The data set for the first problem must be entered by punched cards or by another storage medium. Further problems may reuse the same data set (but probably transformed) or other sets must be entered. A data set may contain several dependent variables which are sequentially treated. For each dependent variable in turn, different fits can be requested, e.g. a least-squares fit, several robust fits, different sorts of weighting in the factor space, etc. Thus, a requested statistic will be calculated for each fit and for each dependent variable.

Options to direct the run stream and to be specified on a control card include
- different forms of data input (format, storage facility)
- transformation of data
- weighting of observations
- $\theta_o = 0$ fixed (no intercept)
- listing of input data, transformations, sum of variables, correlations etc.
- information cards for display in the printout
- names of variables to be read
- specifications about the sorts of fit
- figures for cross validation (saving and reusing the computed coefficients).

Besides many univariate statistics the program calculates "robustified" residual mean squares and other "robustified" statistics which are based essentially on modified observations. Comparison of the residual mean square with the robustified one throws light on the form of the distribution of the residuals.

Tables of the observations, fitted values, residuals and distances of the points in the factor space as well as their assigned weights v_i are supplied. Large residuals are marked. The variance of the error term may be checked by a calculation of the "standard deviation estimated from residuals of neighboring observations". Many different plots for the graphical analysis are provided. The search for smaller sets of influential variables is supported by the calculation of C_p-values which are preselected, tabulated and plotted.

4. Implementation, comments

The program BLINWDR has been written in FORTRAN following international guidelines in order to facilitate transportability, readability and reliability. It has been implemented and extensively tested on a UNIVAC 1100 computer. Because of the numerous features it is about 10000 lines long and copies in form of a computer tape are available on request from the author. Algorithmic and programming details as well as user's instructions are documented in the manual (Dutter, 1982). A paper reporting a detailed study of dust generation in cardboard production illustrating the application of BLINWDR is on the way.

5. Acknowledgements

The author is grateful to Alfio Marazzi for the kind disposition of his programs. The author also likes to thank H. Löw for programming assistance and I. Ganster for improving an earlier version of the manuscript. This research was supported by the "Fonds zur Förderung der wissenschaftlichen Forschung", project no. 4232.

6. References

Dutter R. (1978), Robust Regression: LINWDR and NLWDR. COMPSTAT 1978, Proc. in Comp. Statist., L.C.A. Corsten, ed., Physica-Verlag, Vienna.

Dutter R. (1982), Robust and Bounded Influence Regression Program BLINWDR. Res. Rep., Inst. Statist., Techn. Univ. Graz.

Hampel F.R. (1978), Optimally Bounding the Gross-error-sensitivity and the Influence of Position in Factor Space. Invited paper read at the ASA/IMS Meeting in San Diego, Ca., Aug. 16, 1978.

Huber P.J. (1973), Robust Regression: Asymptotics, Conjectures and Monte Carlo. Ann. Statist. 1, 799 - 821.

Krasker W.S. and R.E. Welsch (1979), Efficient Bounded Influence Regression Estimation Using Alternative Definitions of Sensitivity. Techn. Rep. TR-3, Center of Comput. Res. i. Econom. a. Manag. Sci., Cambridge, Mass.

Marazzi A. (1980), ROBETH: A Subroutine Library for Robust Statistical Procedures. COMPSTAT 1980, Proc. in Comp. Statist., M.M. Barritt and D. Wishart, eds., Physica-Verlag, Vienna.

An Analysis of Developers' and Users' Ratings of Statistical Software Using Multiple Correspondence Analysis

I. Francis, Cornell University, Ithaca, U.S.A.
N. Lauro, University of Naples Italy

Abstract

This paper is a contribution to the problems of evaluation and classification of the Statistical Software.

These problems, which are in many ways problems of statistical analysis, led us to propose multidimensional data analysis techniques as methodology for the evaluation and comparison of Statistical Software (Lauro Serio,1981).

A recent survey of Statistical Software (Francis,1981) in which the developers of the software as well as users gave their ratings of the software on many characteristics, both qualitative and ordinal, provides a case study for demonstrating the usefulness of several multidimensional statistical techniques (Lebart,Morineau,Tabard, 1977) in classifying packages and comparing the users' and developers' ratings.

Keywords: STATISTICAL SOFTWARE,CLASSIFICATION,MULTIPLE CORRESPONDANCE ANALYSIS

1 - Introduction

In describing and modelling the properties of any system, science traditionally uses the quantitative measures. The emerging science of software is not an exception. The problem of comparing and evaluating statistical packages is a complex one and standardization is considerably difficult, in fact the objectives and the criteria of evaluation are not themselves well defined within existing large literature because their continue evolution (Dixon,1975;Francis,1975).

The evaluation criteria sometimes need properly weighted in order to make them suitable to the particular aspect of a survey (Dixon,1975).In the end, problem of software evaluation becomes a statistical problem where the observed variables are indirect and approximate measurements of the structural features of the package(Lauro-Serio,1981).In order to give a contribution to solve these problems in this paper we discuss the use of several multivariate techniques for analyzing the multidimensional ratings of statistical software by developers and users.

These methods do considerable synthesis of the different variables even when heterogeneous and at same time they allow the introduction of a weighting system which achieve maximum efficiency in improving any a priori scaling.

2 - Materials and methods

The data comes from a survey of statistical software (Francis,1980,1981) in which programs were rated by the respective developers and users on their usefulness as represented by 55 characteristics, each being rated on a 4-point scale.

The characteristics examined in the survey covered a program's capabilities for processing and analyzing statistical data, its portability, its ease of use, and its maintenance.The precise meaning of the ratings for each characteristic must be obtai

COMPSTAT 1982 © Physica-Verlag, Vienna for IASC (International Association for Statistical Computing), 1982

ned from the questionnaire (Francis,1980).For example, on question 30 regarding a package's capabilities in multiple regression analysis, a "0" rating indicated "no capabilities";a "1" indicated that this was a minor purpose;a "2" indicated "moderate" capabilities or a "significant "purpose;and a "3" indicated complete coverage of all multiple regression procedures including stepwise regression, all possible regressions, ridge regression,wide variety of residual plots available, summary statistics (Durbin-Watson,etc.), and regression through origin.On the other characteristics the scale attempted to measure general usefulness; for example, concerning statistical training required,"zero" indicated that no training was needed, while "three" indicated a university degree in statistics was needed.

A general view of the survey using developers' ratings, is given in Tab.1 which displays a clustering of 117 programs and 10 variables concerning the "life-cycle" of a statistical analysis from building a data file to mathematical modelling.
The clustering of both programs and variables was done simultaneously, as discussed by Hartigan (1972).The selection of variables and the clustering was done so as to cluster the high ratings down the main diagonal.The "life-cycle" ordering of the capabilities therefore imposed a similar ordering on the programs.This table permits broad comparisons of programs.Programs which have unusual characteristics stand out. What cannot be seen from this table are any suitable associations between variables, the relationships between these 20 variables and others not used in the clustering, and the comparison between these developers' ratings and the users'.

To examine the effectiveness of more multidimensional techniques we limit our attention to the 26 relatively homogeneous general-purpose packages in the two groups "Survey Analysis" and "General Statistical Analysis" in Table 1.The "active" variables, using the developer ratings, include the 10 variables in Table 1 under "Statistical Analysis" and "Econometrics" plus two reflecting "Bayesian" and "Other Multivariate" capabilities.The relationship of this analysis with other"supplementary" variables and user ratings can then be examined.

The most well-known technique for this problem would be a principal component analysis of the 12 active variables.The 26 programs (developers' ratings) could be projected on to the plane of the first two principal axes and a clustering attempted.

When this analysis was perfomed on the Software ratings the first axis was seen to represent "size";packages with many strong features had high scores on this axis. A second axis put in light software specialization opposing the experimental data analysis features(i.e.Anova) to modelling ones (i.e.Econometrics).
Similar results were obtained by a binary correspondence analysis adopting the so called "doubling code" transformation in order to give same weights in the analysis for low and high ratings.

Because of space limitations these two analysis cannot be presented here in detail. We limit ourselves to presenting the results of a multiple correspondence analysis . To introduce this technique we first compare its features with those of principal component analysis:

i) Principal component analysis is a tool for summarizing multivariate data on a two-dimensional picture. Each variable is represented by one point in the projection on to the plane of the first two principal axes. It assumes that equidistant levels on

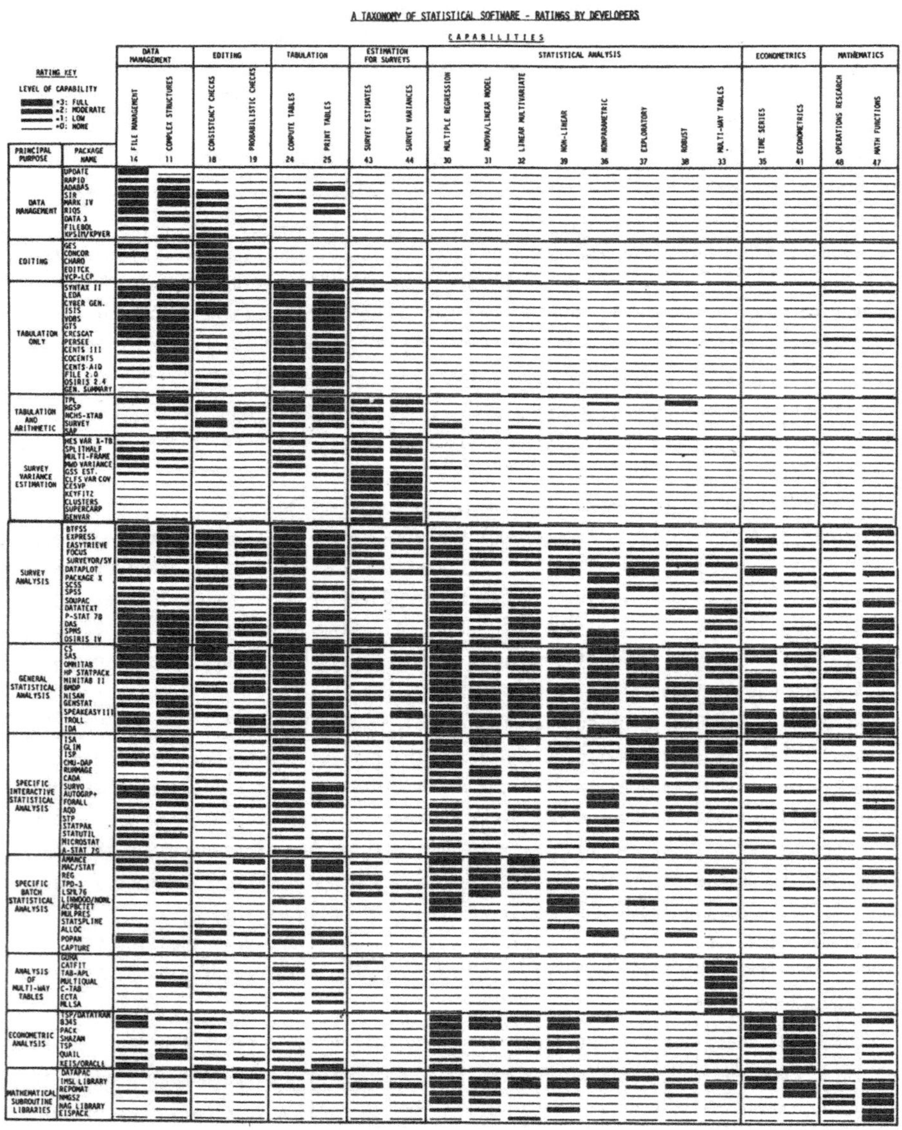

Tab.1 - Two way cluster analysis

the scale of a variable correspond to equal distances between concepts. It examines only the linear correlations among variables.
ii) Multiple correspondence analysis is a tool for explicating relationship between variables and individuals in multivariate data. Each level of the original scale is represented on the picture by a "disjunctive complete" coding. It does not assume linearity of scales or relationships. The metric used and transformation of the data are symmetric in levels of variables and individuals wich means that each level and individual can be plotted together on the same principal plane. Thus it is possible to examine the role that each variable plays in clustering the individuals. Finally

Multiple correspondence analysis (see Lebart, Morineau and Tabard, 1977) treats the data of a survey in the form of a matrix Z of "disjunctive complete" codes : each original variable is replaced by as many binary dummy variables as there are levels or modalities of the original variable ; the value of that dummy corresponding to the observed level is set equal to 1 and all others 0. The matrix Z'Z, referred to as Burt's matrix, is a partitioned block matrix in which the diagonal blocks are diagonal matrices containing the frequencies of each level of each vairable ; the off-diagonal blocks are two-way contingency tables crossing different variables.

In Figure 1 the first two principal axes of Z'Z are used to represent each modality (level) of the variables and, on the same picture, each program. The modalities for the developer' rating on the active variables, that is the data used to generate this two-dimensional projection are represented in capital letters using a three letter name plus a number from 1 to 4 for the levels "0" to "3" ; for example NOL4 and M/W1 represent the fourth level("3") of "Non-linear" and the first level("0") of "Multi-way tables" respectively. The programs are represented by their full name . A black point distinguishes developers' ratings from users' ratings; the former having been active in generating the projection, and the latter having been inactive and simply projected on to this plane after it was established. Similarly, inactive variables (small letters) can be projected on to this plane; for example fil3 represents the third level of "File Mnagement".
This picture allows us to see:(a) clusterings of programs, (b) the influence of different variables in determining the clusters, (c) in what way a developer differ from users, (d) progression of different modalities, (e) the effect of the chosen scale of a variable when it is qualitative,ordered,or quantitative.

3 - Results

The first axis in Fig. 1 is related, broadly speaking, to a progression from right to left in the levels of the variables, such as LIN1 to LIN4 (Linear multivariate); an exception is NOP (Non parametric). The co-ordinates on the first axis of the third and fourth levels for some of active variables are similar, wich might suggests that these two levels could be combined, for example OTH3 and OTH4 (Other multivariate). In some others variables their coordinates on the second axis may be sufficiently different so as to associate them with different clusters,for example ROB3 and ROB4.

For those programs to the left on first axis, i.e. generally high levels on the active statistical capabilities ,the second axis and the modalities associated with it characterize two kinds of program: (1) modelling programs; with high levels on ECO (Econometrics), BAY(Bayesian), NOL (Non linear fitting), ROB (Robust), EXP(Exploratory)

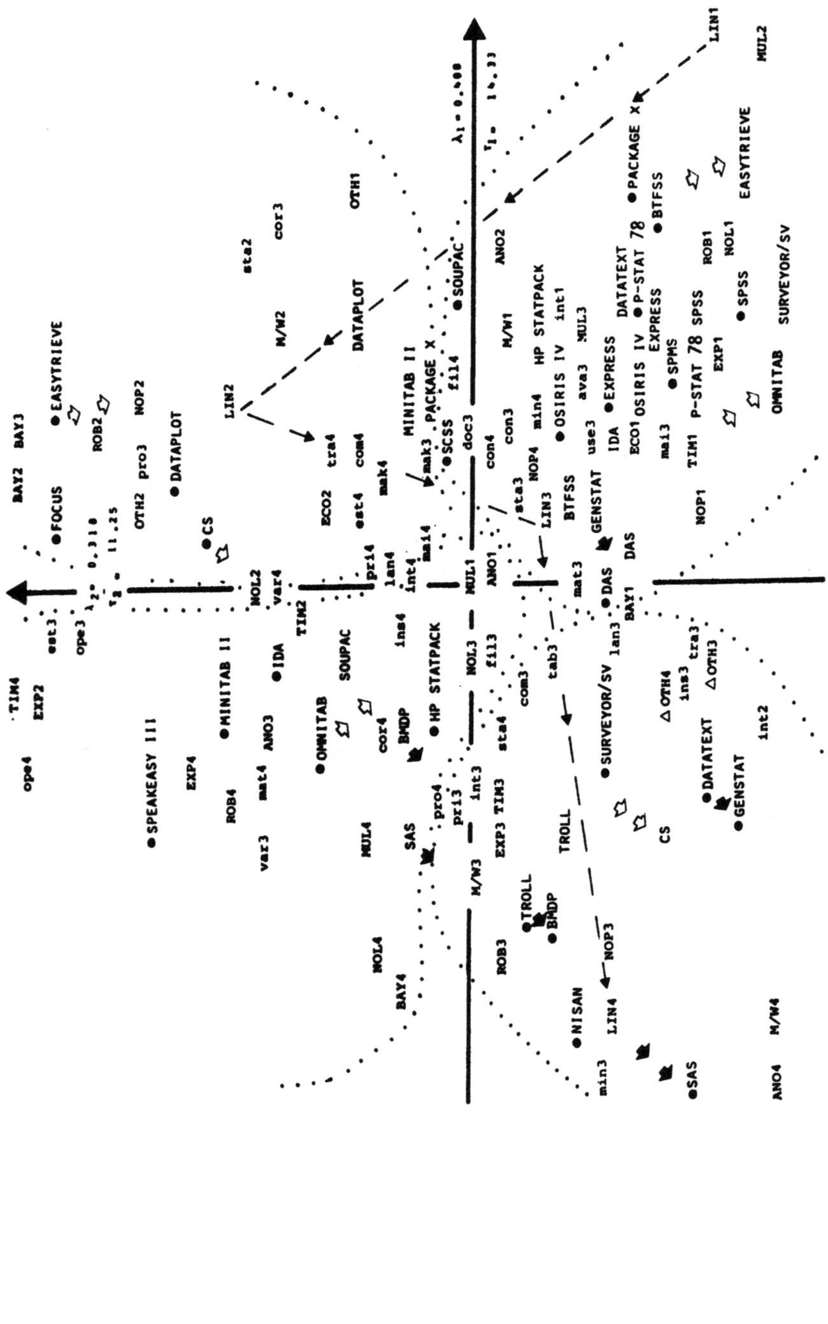

Fig.1 - Multiple correspondence analysis : Factorial plane 1-2 .

MUL(Multiple regression); (2) programs for analysis of experimental data,with high le vels on ANO(Anova). The programs in the lower right quadrant, characterized by a mixture of modalities such as NOP4,LIN3,MUL3, include generally packages wich are used for analysis of social data.

A dynamic clustering of the programs in this plane,as discussed by Diday(1971),suggests a typology of programs wich is more informative from the partition showed in Table 1. The lower-left cluster includes programs highly rated for the analysis of experimental data such as SAS,BMDP,GENSTAT,NISAN; the upper-left group includes programs oriented toward model building or interactive analysis;the upper-right group has low ratings for statistical analysis but appears to be related to displaying data, for example DATAPLOT,FOCUS,EASYTRIEVE; finally the lower-right cluster is oriented towards the social sciences including some well known packages like SPSS,OSIRIS,PSTAT.

The projection on to this plane of supplementary modalities,including computing and user-interface features, confirms this interpretration: ope4 (Operations Research) and mat4(Mathematical functions) are related to modelling; pri4 (Printing Tables), tra4 and com4 (Low training in Statistics and Computing) appear in the upper-right cluster; fil4 is necessary for social science applications.

By projecting the supplementary users' ratings of 21 programs and joining these points to the respective developers' ratings we can examine how users and developers differ. Some users place the programs in very different clusters, for example OMNITAB, EASYTRIEVE and CS. It can be seen(with exceptions of BTFSS,SOUPAC,and PACKAGE-X) developers claimed stronger statistical capabilities (first axis) than users. Developers of general-purpose packages tend to claim more special-purpose capabilities. To the extent that the second axis represents specialities, a big difference on this axis suggests that the user was not sufficiently familiar with programs.

4 - Conclusion

Multiple correspondence analysis has enable us to represent in a comprehensible way considerable details in this multivariate data and to explore the data for structure or explonatory hypotheses. While any conclusion must bear in mind the quality of the data, the generalizations which have been drawn from this analysis are in good general agreement with those of people who are familiar with this data.

Acknowledgements.

I. Francis received support for this work while a visitor at the Center for Operations Research and Econometrics, Université Catholique de Louvain. Table 1 is reprinted with permission of Elsevier North-Holland.

5. References

DIDAY E. (1971) - "La méthode des nuées dynamiques", Rev. Stat. App. 19, 19-34.

DIXON W.J. (1975) - "Statistical Software : touching on various parts of the elefant", Bull. of Int. Stat., 46, 566-571.

FRANCIS I. et al. (1975) - "Criteria and considerations in the evaluation of Statistical Program Packages", The Am. Stat., 29, 52-56.

FRANCIS I. (1980) - "A taxonomy of Statistical Software",in Compstat 80, Physica Verlag, Wien, 43-54.

FRANCIS I. (1981) - "Statistical Software : A Comparative Review", North Holland,N.Y.

HARTIGAN J. (1972) - "Direct clustering of a data matrix",J.Am. Statis. Ass.,67, 123-129.

LAURO N. and SERIO G. (1981) - "Criteria for evaluating and comparing statistical software : A multidimensional data analysis approach, in Multidimensional Data Analysis", INRIA - Rocquencourt, FRANCE.

LEBART L., MORINEAU A., and TABARD N. (1977) - "Techniques de la description statistique", DUNOD, Paris.

Real Time Graphical Techniques for Analyzing Multivariate Data

J.H. Friedman, J.A. McDonald and W. Stuetzle, Stanford University, U.S.A.

ABSTRACT

The Orion I workstation is an experimental computer graphics system, built at the Stanford Linear Accelerator Center in 1980-81. It is used to study applications of recent developments in computer graphics technology to statistics. Orion I is the newest of several descendants of an earlier system built at SLAC called Prim 9. The principal common feature of "Prim" systems is the use of real-time motion graphics to display three-dimensional scatter plots. We demonstrate some of the more basic methods for data analysis in our film "Exploring Data with Orion I".

SUPPORT

This research was supported in part by the Department of Energy under contracts DE-AC03-76SF00515 and DE-AT03-81-ER10843, the Office of Naval Research under contract ONR N00014-81-K-0340, and the U.S. Army Research Office under contract DAAG29-82-K-0056.

I. INTRODUCTION

This paper accompanies a film that demonstrates some programs written for the Orion I workstation. Orion I is an experimental computer graphics system, built in the Computation Research Group at the Stanford Linear Accelerator Center (SLAC) in 1980-81.

This paper introduces some basic tools in a new area of statistical research: applications of real-time interactive graphics. These basic tools are simple examples of new methods made possible by advances in computer graphics technology. Although graphical methods are among the oldest in statistics, interaction between a data analyst and a machine makes it possible to analyze data in ways that are qualitatively different from traditional methods.

1.1. DESCRIPTIVE STATISTICS

The goal of statistics is to analyze data.

How we analyze a given data set is very dependent on context. Sometimes we are given data that arises from an experiment designed to answer a particular question. In this case, we may have a great deal of prior knowledge about our data, and have confidence that we know what to expect in it. This is a situation in which inference may be appropriate.

An inference is a generalization from a given set of data to some larger population (real or hypothetical) from which the data set is presumed to be a sample. Statistical inference is based on a probability model of the mechanism by which the sample is drawn from the population. Our inferences will only be valid if the probability model is a reasonable approximation to the way the data is actually generated. A common and serious flaw in standard statistical practice is the use of inappropriate models.

If our prior knowledge of the data is limited, we need to first "just look at the data and see what is going on." This is called _description_ or _exploratory data analysis_ [18,15]. We concentrate on understanding features of the particular data set at hand; we do not draw (formal) conclusions about larger populations. There are two parts to description: _exploration_, to discover interesting features of our data, and _summary_, to report what we have discovered.

Description is a more primitive problem than inference. For inference to be appropriate, we need considerable prior knowledge about the data. Inference relies on models; good modeling requires looking at data.

1.2. INTERACTIVE GRAPHICS

Graphical methods are used primarily for description. Our research, which emphasizes interaction and real-time motion, concentrates on the exploratory part of description. We are developing methods that allow us to discover, understand, and summarize the "structure" of multivariate data. Since, in general, we have little reliable prior knowledge about our data, we are especially interested in methods that allow us to discover unanticipated kinds of structure and to respond to unexpected circumstances in an intelligent way.

This is where interactive graphics has much to contribute. First of all, computer graphics can quickly expose different views of data to human perception, so that the data analyst can detect many kinds of patterns in the data. The analyst can interpret apparent patterns using his knowledge of the context in which the data arose. Also, the analyst can select directions for further exploration using results so far. Interactive graphical methods combine human talents for perception of patterns and judgment using the full context of a problem with a machine's ability to do rapid and accurate computation.

1.3. AN EXAMPLE

What we mean by "structure" is any apparent pattern or interesting feature in a graphical (or numerical) description of data. Structure is a subjective perception. Quantitative measures of specific aspects of structure can be very useful, but any particular definition of structure would be too restrictive.

In developing new graphical methods for data analysis, it is useful to have some examples of structure in mind. One simple type of structure is illustrated in our film: "Exploring data with Orion I".

A typical thing that we would like to know about a set of data is the following: Does the data set separate in a natural way into distinct groups or clusters?

In the film, we look at a data set presented by Harrison and Rubinfeldt [14,2]. They measured 14 variables for each of 506 census tracts in the Boston Standard Metropolitan Statistical Area. They were interested in the dependence of housing value (represented by the median value of owner-occupied houses) on air pollution (represented by nitrogen oxide concentration). The remaining 12 variables measured other quantities thought to influence housing value, such as crime rate, average number of rooms per house, etc.

In the exploration movie, we do not consider explicitly the dependence of housing value on pollution. We concentrate instead on clustering.

If we were interested in the dependence of housing value, we might fit a model that predicts housing value as a function of the thirteen predictors. Before fitting any prediction rule, we should consider whether it is appropriate to fit one rule for all the data. If the data set separated in a natural way into distinct and internally homogeneous groups (or clusters), then we would want to consider fitting different rules for each group.

Most clustering algorithms rely on a notion of distance in the data space to partition a data set into isolated clumps. Observations are considered similar if they are close together and dissimilar if they are far apart. A cluster is a group of points that are close to each other and far from any other points.

With Orion I, we can use other criteria besides separation in a distance measure to partition a data set. In particular, we can use subjective perception of patterns to define natural groupings. A data set may naturally divide into two groups, which follow clearly distinct patterns. Yet the difference between the groups may not be easily summarized by any natural measure of distance.

In the film, we show how the Harrison-Rubinfeld data can be divided naturally into several groups. The major division turns out to be between urban and suburban-rural census tracts. However, the urban and suburban tracts do not form isolated clumps. Instead, in certain views, they lie in intersecting, perpendicular planes. Because the planes intersect, the groups are not isolated in most distance measures. But the separation in the two groups is obvious when seen.

2. LOOKING AT MANY-DIMENSIONAL DATA

The really challenging problem in multivariate data analysis is to discover and understand structure involving more than three variables at once.

2.1. THREE-DIMENSIONAL SCATTERPLOTS

The histogram is an effective tool for looking at one-dimensional data. A conventional, two-dimensional scatterplot is a successful method for looking at two-dimensional data.

To look at three-dimensional data, we draw a three-dimensional version of the scatterplot. Real-time motion graphics makes it possible to draw pictures that appear three-dimensional. We subject the data to repeated small rotations and display the projection of the rotated data as points on the two-dimensional screen. If we can compute and display rotations fast enough (> 10 per second), then we get an illusion of continuous motion. Apparent parallax in the motion of the points provides a convincing and accurate perception of the shape of the point cloud in the three-dimensional space.

In a conventional scatterplot, it is easy to read off approximate values of the two variables for any particular point. In moving, three dimensional scatterplots estimating the values of the three variables for any particular point is difficult, if not impossible. However, perceiving the shape of the point cloud is easy. Because we are interested in detecting patterns, we are much more interested in overall shape than in individual points.

2.2. MORE THAN THREE DIMENSIONS

There is no completely satisfactory method that lets us look at more than three variables at a time. Three basic approaches to viewing many- (more than three) dimensional structure are being studied with Orion I. They are:

1. Higher-dimensional views:
   We represent as many variables as possible in a single picture.
2. Projection Pursuit:
   We find a low-dimensional picture that captures the structure in the many-dimensional data space.
3. Multiple Views:
   We look at several low-dimensional views simultaneously; by making connections between the low-dimensional views we hope to see higher-dimensional structure.

2.3. HIGHER-DIMENSIONAL VIEWS

One way to see high-dimensional structure is to draw pictures that show as many dimensions at a time as possible.

One of the simplest ways to add dimension to a picture is through color. We start with a three-dimensional scatterplot. We can add a fourth variable to the picture by giving each point in the scatterplot a color that depends on the value of a fourth variable. With an appropriately chosen color spectrum, we can easily see simple relationships between the fourth variable and position in the three-dimensional space. Our ability to perceive distinctions in color does not compare to our ability to perceive position in space; we should expect to miss subtle or complicated relationships between a color variable and three position variables.

Color works best for a variable that takes on only a small number of discrete values.

Other tricks let us increase the dimension of a picture. For example, we can represent each observation in the scatterplot by a circle, rather than by a simple point. The radius of the circle can depend on the value of a fifth variable.

In a simple scatterplot, observations are represented by featureless points. We add dimension to the picture by replacing points with objects that have features, such as color, size, and shape. These features can be used to represent variables in addition to those represented by a point's position in the scatterplot. These "featurefull" objects, sometimes called glyphs, are well known [13].

With each new dimension, the glyphs become more complicated and the picture becomes more difficult to interpret. We need more experience to determine which are good ways of adding dimension to glyphs and to understand the limitations of each method.

Again, these methods seem to work best if the additional variables take on a small number of discrete values. So these techniques may be most suited to data sets that include a combination of continuous and discrete variables.

2.4. PROJECTION PURSUIT

The basic problem with adding dimension to a single view is that the picture quickly becomes impossible to understand. An alternative is to restrict our picture to a few (3 or less) dimensions and then select low-dimensional pictures that capture interesting aspects of the multivariabe structure in our data. This is the basic idea of projection pursuit. [6,12,10,8,9].

In general, we could consider any mapping from the many-dimensional data space to a low-dimensional picture. The projection pursuit methods that have been developed so far restrict the mappings considered to orthogonal projections of the data onto one-, two-, or three-dimensional subspaces of the data space. More general versions of projection pursuit would have a spirit similar to multidimensional scaling [13].

In general, no single view of the data will capture all interesting aspects of its many-timensional structure. So projection pursuit methods generally try to uncover many-dimensional structure by combining several low-dimensional views.

Until recently, work on projection pursuit has concentrated on automatic algorithms. In automatic projection pursuit, a numerical measure of interesting structure is chosen. Then an optimization algorithm chooses a projection to maximize the numerical measure.

The first projection pursuit algorithm [12] used a numerical optimizer to search for one- or two-dimensional projections that maximized a "clottedness" index, which was intended as a numerical measure of interesting structure.

Automatic projection pursuit methods have been developed for more well defined problems: non-parametric regression [10,8], non-parametric classification [8], and non-parametric density estimation [8,9]. In these problems, we build up a model that summarizes the apparent dependence in our data set of a response on some predictors. The model is constructed from several projections, using the projection pursuit paradigm [8]:

1. Choose an initial model.
2. Repeat
   (a) Find a projection that shows deviation from the the current model, indicating previously undetected structure (Projection Pursuit).
   (b) Change the model to incorporate the structure found in (2).
3. Until the current model is a sufficiently accurate summary in all projections.

Projection pursuit methods have a natural connection to graphics. As long as we project on a subspace of dimension no greater than three, we can look at a picture of the result. Also, an interactive graphics system lets the data analyst modify or take over some functions performed by machine in the automatic versions of projection pursuit. Some functions in the projection pursuit paradigm that can benefit from interaction are:

1. Searching for interesting projections.
2. Determining the model update.
3. Deciding when to end the iteration.
4. Interpreting the result.

For example, the system allows a user to manually imitate the Rosenbrock search strategy [17] used by the numerical optimizer in the automatic version. However, a human being can search to optimize subjective criteria, using perception and judgment, instead of having to rely on a single, numerical measure of what constitutes an interesting view.

2.5. MULTIPLE VIEWS

In the projection pursuit paradigm, we develop understanding of many-dimensional structure from a sequence of low-dimensional views. We would like to understand the relationships between the contents of two or more views in the sequence. The methods described in this section allow us to do this.

2.5.1. M-and-N-plots

This approach is inspired by M-and-N-plots of Diaconis and Friedman [4].

A two-and-two-plot is one kind of M-and-N plot; two-and-two-plots are used to display four-dimensional data. To make a two-and-two-plot, we draw two two-dimensional scatterplots side by side. The two scatterplots show different pairs of variables. We then connect corresponding points in the two scatterplots by lines. To get a picture that is not confusing, we will usually not draw all the lines. Diaconis and Friedman give an algorithm for deciding which lines to draw.

2.5.2. Color M-and-N-plots

Our idea is a modification of the above. Instead of connecting corresponding points by lines, we draw corresponding points in the same color.

Briefly, this is how the program works. On the screen there are two scatterplots, side by side, showing four variables. There is also a cursor on the screen in one of the two scatterplots. The scatterplot that the cursor is in is the active scatterplot. We can move the cursor by moving the trackerball. Points near the cursor in the active scatterplot are red. Points at an intermediate distance from the cursor are purple. Points far from the cursor are blue. Points in the non-active scatterplot are given the same color as the corresponding point in the active scatterplot. The colors are continuously updated as the cursor is moved. We can also move the cursor from one scatterplot to the other, changing which scatterplot is active.

It is useful to have other options for determining the colors in the active scatterplot. For example, instead of determining color by distance from the cursor, the colors can be determined by the vertical position of the cursor. Points above the cursor will be in one continuous range of hues, from blue through magenta to red. At the position of the cursor there is a discrete jump in hue, so that points below the cursor have hues from cyan through green to yellow.

Using color instead of line segments to connect points has a disadvantage; it is not as precise at showing us the connection between a pair of points representing the same observation. However, we are not usually very interested in single observations. More often, we want to see how a region in one scatterplot maps into the other scatterplot; the combination of local coloring and the moveable cursor is a good way of seeing regional relationships in the two scatterplots.

An advantage of color over line segments is that it is possible to look at more than two scatterplots at once. Connecting corresponding points with line segments in more than two scatterplots at a time would produce a hopelessly confusing picture. With color, on the other hand, it is no more difficult to look at three or more scatterplots at once than it is to look at two.

2.5.3. A variant for geographical data

Color raster graphics devices have seen considerable use in the representation of geographical data by colored maps. We can combine maps with two- or three-dimensional scatterplots by copying the idea in Color M-and-N-plots. A map is like a two-dimensional scatterplot; a region in the map, like a point in a scatterplot, represents an observation. If we draw a map and a scatterplot side by side, we can connect points in the scatterplot with the corresponding regions in the map, using color and a cursor, just as we connected corresponding points in the Color M-and-N plots.

For this to be effective, the colors of the points in the scatterplot and the regions in the map must be changed in real-time as the cursor is moved. This can be done on the present Orion I system only if there are no more than 255 regions in the map. However, a simple extension of the hardware (adding more color planes to the frame buffer) would make it possible to change the color of 1023 or more regions in real-time.

3.1. HISTORY

Orion I is the youngest descendant of a graphics system called Prim-9, which was built at SLAC in 1972 [6]. Prim-9 was used to explore 9-dimensional data. It used real-time motion to display three-dimensional scatterplots. Through a combination of projection and rotation, a user of Prim-9 could view an arbitrary three-dimensional subspace of the 9-dimensional data. Isolation and masking were used to divide a data set into subsets.

The computing for Prim-9 was done in a large mainframe (IBM 360/91) and used a significant part of the mainframe's capacity. A minicomputer was kept busy transferring data to the black and white vector drawing display. The whole system, including the 360/91, cost millions of dollars. The part devoted exclusively to graphics cost several hundreds of thousands of dollars.

Successors to Prim-9 were built at the Swiss Federal Institute of Technology in 1978 (Prim-S) and at Harvard in 1979-80 (Prim-H) [5].

Prim-S and Prim-H both used a large, shared computer for computations and expensive black and white vector drawing displays. Both systems cost more than several hundreds of thousands of dollars. Rotations were the most demanding computations that could be done in real-time.

3.2. ORION I WORKSTATION

There are two ways in which the Orion I hardware is a substantial improvement over previous Prim systems: price and computing power.

The total cost for hardware in Orion I in 1981-82 is less than $50,000. A preliminary version of Orion I cost less than $20,000. The increase in price of the final version was due to doubling the resolution of the graphics device, from a 640×480 pixel screen to 1280×1024. The higher resolution requires about four times as much memory and a more expensive color monitor. The higher resolution gives nicer pictures, but is not really necessary for statistical applications. A black and white system, with all the capabilities of the earlier Prim systems, can be had for about $10,000 [1].

The computing power of Orion I is equivalent to that of a large mainframe computer (say one-half of an IBM 370/168) and is devoted to a single user. This permits Orion I to execute much more demanding computations than rotation in real time.

REFERENCES

1. Bechtolsheim, A. and Baskett, F., "High-Performance Raster Graphics for Microcomputer Systems," Proceedings 1980 SIGGRAPH Conference, published as Computer Graphics 14 (3), July 1980.
2. Belsey, D. A., Kuh, E., and Welsch, R. E., Regression Diagnostics, Wiley, 1980.
3. Cleveland, W. A., "Robust Locally Weighted Regression and Smoothing Scatterplots," JASA 74, 1979.
4. Diaconis, P., and Friedman, J. H., "M and N Plots," Tech. Report #151, Dept.of Statistics, Stanford University, April 1980.
5. Donoho, D., Huber, P.J., and Thoma, H. The Use of Kinematic Displays to Represent High Dimensional Data, Research Report #PJH-5, Dept. of Statistics, Harvard University, March 1981.
6. Fisherkeller, M.A., Friedman, J.H., and Tukey, J.W., Prim-9, An Interactive Multidimensional Data Display and Analysis System. Proc. 4th International Congress for Stereology, 1975.
7. Foley, J.D. and Van Dam, C., Fundamentals of Interactive Computer Graphics, Addison-Wesley, 1982.
8. Friedman, J.H. and Stuetzle, W., "Projection Pursuit Methods for Data Analysis" Tech. Rep. Orion 001, Dept. of Statistics, Stanford University, June 1981.
9. Friedman, J.H., Stuetzle, W., and Schroeder, Anne, "Projection Pursuit Density Estimation," Tech. Rep. Orion 002, Dept. of Statistics, Stanford Univ.,July 1981.
10. Friedman, JH and Stuetzle, W., "Projection Pursuit Regression", JASA v. 76,1981.
11. Friedman, J.H. and Stuetzle, W., "Hardware for Kinematic Statistical Graphics", Tech. Rep. Orion 003, Dept. of Statistics, Stanford University, Feb. 1982.
12. Friedman, J.H. and Tukey, J.W., "A projection pursuit algorithm for exploratory data analysis," IEEE Trans. Comput. C-23 pp. 881-890, 1974.
13. Gnanadesikan, R. Methods for Statistical Data Analysis of Multivariate Observations, 1977.
14. Harrison, D. and Rubinfeld, D.L., "Hedonic Prices and the Demand for Clean Air," Journal of Environmental Economics and Management, 5, 1978.
15. Mosteller, F. and Tukey, J.W. Data Analysis and Regression, Addison-Wesley,1977.
16. Newman, W.M. and Sproull, R.F., Principles of Interactive Computer Graphics, 2nd ed., McGraw-Hill, 1979.
17. Rosenbrock, H.H., "An automatic method for finding the greatest or least value of a function," Comput. J. 3, 1960.
18. Tukey, J.W., Exploratory Data Analysis, Addison-Wesley, 1977.

About Fuzzy Discrimination

J.-M. Gautier, C.O.R.E.F., Boulogne-Billancourt, France
G. Saporta, Rene Descartes University, Paris, France

Il arrive souvent en analyse discriminante que l'appartenance des individus aux classes d'une partition de la population ne soit pas connue avec certitude, ou qu'il soit délicat d'attribuer strictement un individu à une catégorie lorsque la partition est définie à partir d'une variable numérique découpée en classes. Il semble alors plus raisonnable de se donner une distribution de probabilité sur les classes qu'une fonction booléenne d'appartenance surtout si un individu est proche de la frontière entre deux classes.

On établit alors les modifications à apporter aux techniques usuelles de discrimination (factorielle et décisionnelle) ainsi que les conséquences de ces modifications sur les indices usuels de qualité d'une discrimination.

Keywords :Fuzzy Discrimination
Discriminant Analysis

Discriminant analysis is the replicative or predictive study of a qualitative variable over a set of predictors that are generally numerical.

In certain situations, one cannot attribute with certainty a category of the variable to be explained to certain (even to all) individuals in the sample. This is particularly the case if :

- The classes are poorly defined, e.g., imprecise nomenclature or else discrete coding of a quantitative variable (1).

- The class to which an individual belongs cannot be determined with certainty, e.g., appearance of a symptom after absorption of a medicine. There is nothing to prove that the symptom could not have appeared naturally.

But it may also be that the determination of the class is too costly, and that one is content with an estimate.

From the formal viewpoint, we will therefore suppose that with each individual, there is associated a probability distribution P_j (i) for the classes j = 1,. . ., k of the variable to be explained, rather than a set of mutually exclusive indicative variables.

(1) In this case, the quantitative variable being divided up into classes, an individual in the neighborhood of the border between 2 classes probably cannot be attributed to a single one of the classes, if only because of possible measuring errors.

We will denote by X the matrix (n, p) of the p numerical predictors, n being the sample size

Hence we are going to describe how to extend to this situation the various research results of discriminant analysis, i.e., seek the best separation between the classes in the sense of a particular criterion (geometrical methods), or attempt directly to estimate the P_j (i) for every individual for which one sums up the p predictors (Bayesian research).

For geometrical research, this extension is made by introducing weightings into the calculations and by counting each observation i as an element of all of the classes j for which : $P_j (i) \neq 0$. The weighting of observation i in class j is then $P_j(i)$. We note $\alpha_j = \sum_{i=1}^{n} P_j(i)$ the weight of the class j.
It will be shown that in reality, it is not necessary to work on matrices of dimension k.n, but only n (on condition of having written the programs on an ad hoc basis).

The various quality measures of the discrimination are affected by the fuzziness of the classes of the variable to be explained, and for certain ones of these criteria, we will propose a limit calculation that will make it possible to judge the real discriminating power of a variable, or of a set of variables, relative to this limit.

As to the bayesian methods, if one excepts the case in which one considered the $P_j(i)$ as a sample of a random variable, the generalization is carried out in the same fashion as for the geometrical methods, since the only things affected by the fuzziness of the classification are the estimators of the parameters of the conditional distributions.

However, another approach is possible : a direct search for a formula for adjustment of the P_j by means of explanatory variables : logistic regression or linear regression with constraints.

I. GEOMETRICAL METHODS

I.1 Evaluation of the center of gravity of the classes

We will denote by P_j the diagonal matrix (n x n) of the P_j (i) associated with the group j, and $\underline{1}$ the vector of which the n components are equal to 1.

Since the classes are fuzzy, the centers of gravity g_j of each one of them are obtained by taking the average, weighted by the P_j (i), of the coordinates of the observations, hence :

$$\underline{g}_j = \frac{1}{\alpha_j}[X' \; P_j \; \underline{1}]$$

I.2 Expression of the matrices of variance

The total matrix of variance would then be :

$$V = \frac{1}{n} \sum_{j=1}^{k} X' \; P_j \; X = \frac{1}{n} \; X'X \cdot$$

if the data are centered.

The matrix of variance of the classe j is written :

$$V_j = \frac{1}{\alpha_j} \left(X' P_j X - \frac{1}{\alpha_j} X' P_j \underline{1}\, \underline{1}' P_j X \right)$$

The intra-class matrix of variance W is then :

$$W = \frac{1}{n} \sum_{j=1}^{k} \alpha_j V_j \quad \text{since} \quad \sum_{j=1}^{k} \alpha_j = n$$

Its current term is worth :

$$W_{l_1 l_2} = \frac{1}{n} \sum_{i_1 i_2} x_{i_1 l_1} x_{i_2 l_2} \left(\delta_{i_1 i_2} - \sum_j \frac{1}{\alpha_j} P_j (i_1) P_j (i_2) \right)$$

The inter-class matrix of variance B is therefore equivalent to :

$$B = \frac{1}{n} X' \left(\sum_j \frac{1}{\alpha_j} P_j \underline{1}\; \underline{1}' P_j \right) X$$

And its current term is :

$$b_{l_1 l_2} = \frac{1}{n} \sum_{i_1 i_2} x_{i_1 l_1} \left(\sum_j \frac{1}{\alpha_j} P_j (i_1) P_j (i_2) \right) x_{i_2 l_2}$$

I.3 Calculation of the distance from a point to the center of gravity of the classes

Let $\underline{e}$ be a point of R^p ; if the matrices V_j are not significantly different, one uses the Mahalanobis metric W^{-1} to calculate these distances :

$$d^2 (\underline{e} ; g_j) = (\underline{e}' - \underline{1}' P_j^* X) W^{-1} (\underline{e} - X' P_j^* \underline{1})$$

$$= \underline{e}' W^{-1} \underline{e}' + \underline{1}' P_j^* X W^{-1} X' P_j^* \underline{1}$$

$$- 2 \underline{e}' W^{-1} X' P_j^* \underline{1}$$

where $P_j^* = \frac{1}{\alpha_j} P_j$

If the V_j are significantly different, one applies V_j^{-1} instead of W^{-1} (Sebestyen's method).

We perceive that these formulas hardly from those of the classic case. Their main interest is that they supply a method of direct calculation not bringing in matrices of dimension kn.

I.4 Criteria of the quality of the discrimination

The first criterion that comes to mind is the one of the percentage of correct classification by the method of reassignment (about which it is known, incidentally, that it yields biased results). In the classic case, this percentage can reach 100 % ; here it is quite obviously limited by :

$$\sum_i \max_j P_j(i) \times \frac{100}{n}$$

In the case of 2 groups, the Mahalanobis distance D^2 between the two centers often serves as a criterion for separability, in particular for the selection of the variables, other criteria such as the F, can be deduced from it for a monotonic transformation :

$$D^2 = (\underline{g}_1 - \underline{g}_2)' \; W^{-1} \; (\underline{g}_1 - \underline{g}_2)$$

In the usual case, this distance is not bounded above and may theoretically be infinite if and only if each group is reduced to a point projected onto straight line $g_1 g_2$. Here this distance is always limited, and this limit can be calculated by the following procedure :

D^2 is a maximum if, in projection on straight line $g_1 g_2$, all the points such tha $p_1(i) > P_2(i)$ are confused in a point X_1, and all the others in X_2. Hence one is led back to a uni-dimensional problem on straight line $g_1 g_2$. Let us place ourselves in this case. D^2 being invariant for change of radius and of scale on the variables, one may suppose that on straight line $g_1 g_2$, the total variance is equal to 1, and that the variable is centered. From this, one deduces at the values of X_1 and X_2 and the value of σ^2, which is not zero, since the variances of each group cannot be zero if there is at least one i in which $P_j(i)$ is different from 0 or from 1,

whence $D^2 \leqslant \dfrac{(g_1 - g_2)^2}{\sigma^2}$

If we denote by P_1 the proportion of individual i affected in group 1, such that $p_1(i) > p_2(i)$:

$$x_1 = \sqrt{\frac{1 - P_1}{P_1}} \qquad x_2 = - \sqrt{\frac{P_1}{1 - P_1}}$$

The calculation of σ^2, g_1 and g_2 depends on the 3 parameters : p_1, $a_1 = \sum p_1 (i)$, $a_2 = \sum p_2 (i)$.

In the case of k groups, a usable criterion is the sum of the Mahalanobis distances of the groups taken two by two ; as previously, the sum of the $D^2 (g_j ; g_1)$ is maximum if in the space created by the k centers of gravity, all of the observations such that p_j (i) is a maximum are projected onto the same point $\underline{K}_j$. One may always suppose that the observations are centered and that the matrix of over-all variance is I_{k-1}, which leads to a single configuration of the $\underline{K}_j$ neglecting an isometry. One can then calculate the criterion that supplies the desired increase, which depends only on the distribution of the weightings.

One concrete calculation method consists in takings any set of k points $\underline{z}_i \dots \underline{z}_k$ and in carrying out the linear transformation that leads to the $\underline{K}_j$.

II. PROBABILISTIC METHODS

II.1. Bayesian methods with hypothesis of normality

See Aitchison and Begg (1976).

If one makes the hypothesis of a normal distribution $N_p (\mu_j ; \sum)$ in each class, the only problem consists in estimating the parameters of the model before applying Bayes' formula (cf. Anderson).

The estimators of the μ_j and of $\sum$ are precisely the $\underline{g}_j$ and W previously defined.

II.2. Direct estimation of the P_j

Since one has a sample of P_j and explanatory variables X, one may use the regression techniques in the broad sense, or :

a) Logistic regression

This method reduces to supposing that Log $\frac{P_j}{P_k}$ is a linear function of the explanatory variables. The coefficients of these functions being estimated then by the method of maximum likelihood (Cox's model).

b) Linear regression under constraint

One regresses each p_j on the explanatory variable while imposing the constraint $\sum\limits_{P_j \geq 0} p_j = 1$ (which is easy), and the constraints $P_j \geq 0 \quad \forall_j$, which leads to optimization programs on cones. In other words, it is a question of carrying out the canonical analysis between a convex cone and a vectorial sub-space.

REFERENCES

AITCHISON J. BEGG C.B (1976) Statistical Diagnosis when basic cases are not classified with certainty Biometrika, 63, 1-12.

ANDERSON T.W. (1958) "Introduction to multivariate statistical analysis" - Wiley , New-York.

ANDERSON J.A. (1972) Separate sample logistic discrimination, Biometrika 59, 19-35

COX D.R. (1970) "The analysis of binary data" - Methuen, London

MARTIN J.F. (1980) Le codage flou et ses applications en statistique. These 3è cycle Universite de Pau

SEBESTYEN (1962) "Decision making processis in pattern recognition" Mc Millan, New York.

Acknowledgement :

The authors would like to thank P. Besse for his comments and corrections.

A Fortran Preprocessor for Statistical Data Analysis

J.E. Gentle, IMSL Inc., Houston, U.S.A.

SUMMARY

Most software packages for statistical data analysis restrict the user's flexibility in data handling, in program control, and in access to the computations within the program. If the package is implemented as a preprocessor which allows intermixing of the applications statements with the statements of a procedural language, however, the user can have both the ease of use and power of the applications package as well as the flexibility of the procedural language. A family of problem-solving computer systems, called PROTRAN, built around a preprocessor is described, with emphasis on the specific package for statistical data analysis, STAT/PROTRAN. Various considerations and techniques of the implementation of the preprocessor are discussed.

KEYWORDS: Computer program; User interface; Fortran; Preprocessor

DESIGN OBJECTIVES AND GENERAL DESCRIPTION

One of the greatest disadvantages in the use of most applications software packages is the lack of flexibility for the user. In order for the package to be usable for a given problem, the problem must fit the mold imposed by the design of the procedure developed in the package for that particular kind of problem, and the input data must be structured to conform to the data handling capabilities provided by the package. If more than one problem is to be analyzed, often more than one computer run is necessary. This is especially true if the input of one analysis depends on the output of another analysis. Of the various methods utilized in applications programs, some offer more flexibility than others. The preprocessor approach, which has been discussed by Anderson (1980), has as one of its major advantages the flexibility it affords the user.

PROTRAN (PROblem TRANslator) is a family of problem-solving software systems built around a preprocessor. The first two members of the PROTRAN family, MATH/PROTRAN and STAT/PROTRAN, are intended for use in general mathematical and statistical application areas. Although the orientation of the systems are different, the syntax of PROTRAN is consistent over the individual members, and both can be used together. The choice of what procedures to implement in STAT/PROTRAN was made with a view toward frequency of usage and with consideration of what should constitute an initially small, but useful and extensible, general system for statistical analysis.

From the user's standpoint, the basic need in a statistical analysis system is for a method of bringing in data and performing simple manipulations on data. Statistical data sets can most often be conceptualized as rectangular arrays in which the columns correspond to variables and the rows correspond to observations. The data set itself needs a name for convenient reference, both within the analysis system and within the host computer system. The user must also be able to assign names to the variables (or columns) of the file, and some provision must be made for special codes representing missing or exceptional data. The overwhelming proportion of usage of existing statistical analysis systems is in the areas of computations of simple univariate or bivariate statistics and frequency tabulations.

Based on those considerations, the following areas were included in the first edition of STAT/PROTRAN:

1. Basic univariate statistics
2. Frequency tables and crosstabulation
3. Correlation analysis
4. Regression analysis
5. Analysis of variance for balanced complete designs
6. Data management and file handling

Within each of these general areas, several options are provided. In regression analysis, for example, ordinary least squares can be used, or stepwise or all subsets regression can be requested.

In addition to the general goals of flexibility and ease of use, the design objectives of STAT/PROTRAN also included:

1. Data structures must be general and not depend on the kind of analysis to be performed. The same data must be usable by different analysis routines.

2. Related areas of analysis must be accessed by a common procedure, with optional keywords to specify the particular analysis desired.

3. Computed statistics must be available to the user throughout the user's program.

4. Allowance must be made for missing data.

5. While size restrictions may be imposed by the host computer system, STAT/PROTRAN must not impose any restrictions on the size of the user's problem.

PROTRAN Syntax

Ease of use, flexibility, and compatibility with Fortran were the guiding factors in the development of PROTRAN syntax, which utilizes mnemonic keywords for the majority of its commands. PROTRAN statements (also referred to as PROTRAN procedure references) have the form

```
$     procedure_name procedure_argument (optional)
      keyword_phrases (optional)
```

The following is a complete STAT/PROTRAN program to compute basic statistics for a data set with two variables.

```
$     BUILDFILE MEASUREMENTS
      READ HEIGHT, WEIGHT
      DATA
      ====
list of data
      ====
$     STATS
      FILE MEASUREMENTS
      VARIABLES HEIGHT, WEIGHT
$     END
```

The program could aso be written as

```
      $BUILDF MEASUREMENTS;READ HEIGHT,WEIGHT;DATA
      ====
list of data
      ====
      $STATS
      $END
```

Note that the FILE and VARIABLES statements are not necessary.

Data Handling in STAT/PROTRAN

Data in STAT/PROTRAN are organized into PROTRAN files or arrays, which are essentially Fortran one- or two-dimensional arrays. The basic concepts are cases, or observations, and variables in either arrangement. PROTRAN files allow the user to choose the variable names and the names of the files. They also contain other information such as missing value codes. PROTRAN arrays are similar to Fortran arrays in which the rows are observations and the columns correspond to variables. Each PROTRAN array has a name which the user assigns to it, but the variables always have the names C1, C2, etc., corresponding to the first, second and sequential columns.

The procedures for data manipulation are:

BUILDARRAY	Building a PROTRAN array from a PROTRAN file
BUILDFILE	Building and modifying a PROTRAN file
LIST	Printing of PROTRAN files
PRINT	Printing of arrays of scalars

Statistical Analysis in STAT/PROTRAN

Users access broad types of analysis with simple procedure names. ANOVA, for example, is the procedure name for analysis of variance for one-way classifications and for balanced experimental designs. After the procedure name, users specify parameters of the problem with keyword phrases.

Analysis procedures are:

ANOVA	Analysis of variance for one-way classifications and for balanced complete experimental designs
CORR	Correlation analysis and bivariate statistics
FREQ	Frequency tables and crosstabulations
REGRESS	Regression analysis
STATS	Basic univariate statistics
TTEST	Student's t test

STAT/PROTRAN also includes several utility procedures:

ASSIGN	Assignment of values to arrays or scalars and arithmetic operations on arrays
GENRAN	Generation of random numbers
PLOT	Printing of plots of functions or data
SORT	Sorting of elements of a vector

The analysis procedures in STAT/PROTRAN can accept data from either a file or an array. For example, the program used above for computing basic statistics may be written as

```
$    DECLARATIONS
     REAL MATRIX MEAS(100,2)
     :
     bring in data by Fortran
     :
$    STATS
     ARRAY MEAS
     VARIABLES C1,C2
$    END
```

The variable names are C1 and C2, representing the first and second columns of the array. (In this case, the VARIABLES statement is unnecessary.)

The results of the computations could be saved in variables for use later in the program, for example, in the following segment the means are saved in a variable that has been declared to be a vector of length at least 2:

```
$    STATS
     MEANS XM
```

XM(1) will contain the mean of the first variable and XM(2) will contain the mean of the second.

IMPLEMENTATION

In addition to the objectives of flexibility and ease of use, the objectives in the development and implementation of PROTRAN are concerned with wide availability and with orderly growth of the system. The development objectives included:

1. The PROTRAN system must be transportable across a range of computer systems, including both the common "superminis" and large mainframes.

2. PROTRAN must be extensible at the development and implementation level. This extensibility relates both to the PROTRAN family, in which new packages may be developed, and to the individual members of PROTRAN, such as STAT/PROTRAN, in which new procedures may be implemented.

3. PROTRAN must utilize the existing large collections of tested Fortran software.

In order to achieve the various objectives, the PROTRAN system was implemented as a Fortran preprocessor with a large library of utility subroutines. There is a single preprocessor for all of the members of the PROTRAN family. The preprocessor is built around a fairly complete programming language designed for generating Fortran programs. Templates are written in this language and the preprocessor matches the user's input with the templates to generate Fortran code. The preprocessor does not process user Fortran code, which is indicated by the special procedure FORTRAN or by a $ character appearing as the final character in a PROTRAN statement. After the preprocessing step, the single Fortran program, consisting of code generated by the preprocessor and possibly by the user, is compiled, linked with the necessary PROTRAN utilities and with any subprograms that the user may have called, and finally executed.

The preprocessor language has a vocabulary of keywords that includes the keywords of the PROTRAN procedure, which the user may specify to define the options desired, and internal keywords useful for branching within the template. There is a facility for assigning values to the keywords but not for doing arithmetic on the values. Program control is provided by means of IF-THEN-ELSE, DO, and INCLUDE constructs. There are four portions of the Fortran program for which separate files are produced by the preprocessor: declarations, data statements and statement function definitions, body of the program, and complete Fortran subroutines. The code generated by the preprocessor is assembled so that the declarations are first, data statements are next, and so on, regardless of when the preprocessor had generated those statements. The preprocessor language is line-oriented, with columns 1 through 72 being used for template source code. A continuation line is indicated by a period in column 1. No other column has any special meaning, and the template statements can start in any of the first 72 columns. The dollar sign character, $, is used often to indicate template operations.

The preprocessor language has facilities for generation of labels and variable names which conform to certain conventions. The PROTRAN generated variable names have six characters and end in V7, and the statement labels are numbers between 30000 and 39999 (unless the user resets these default names and labels through an OPTIONS procedure). This allows the user to intermix Fortran with PROTRAN without undue concern for conflicts of variable names. When the preprocessor encounters the string $L(LOOP1), for example, it generates a Fortran statement label (as a number between

30000 and 39999) or else substitutes the previously generated statement number; the string $NI(JJ) causes the preprocessor, on first encounter, to generate (and declare) a Fortran variable of integer type, and on subsequent encounters, to substitute the same Fortran variable name. For example, a template may contain code such as

```
         DO $L(LOOP1) $NI(JJ) = 1,$NI(NVAR)
            $(MEANS)($NI(JJ)) = IMSLSR($NI(MNDX)+$NI(JJ))
   $L(LOOP1) CONTINUE
```

A sequence of PROTRAN for computing and saving basic statistics such as

```
   $     STATS
         MEANS XM
```

may result in the following Fortran being generated (in addition to code for setup and other things)

```
         DO 30021 IAAWV7 = 1, IAAQV7
            XM(IAAWV7) = IMSLSR(IAACV7+IAAWV7)
  30021  CONTINUE
```

Special functions are available for such things as determining the type of a user-defined variable, the dimensions or ranges of arrays, the number of arguments in a keyword phrase, and so on. For example, $$DC(A) and $$RC(A) give the column dimension and column range, respectively, of the array A. If A is a vector, the column dimension is one. The following sequence of template code illustrates some of the functions as well as some of the control structures. The purpose of this code is to determine the number of variables to be used in an analysis. That number is to be stored in a variable to be referred to as NVAR within the template. The PROTRAN procedure has an optional keyword VARIABLES for specification of the variables. If the keyword is not defined by the user, all variables on the data set (which may be a file or an array) are assumed.

```
         $IF(DEF(VARIAB))
            $NI(NVAR) = $ITEMS(VARIAB)
         $ELSE
            $IF(NOTDEF(ARRAY))
               $NI(NVAR) = $NI(KREAD)
            $ELSE
               $NI(NVAR) = $$RC(ARRAY)
            $ENDIF
         $ENDIF
```

PROTRAN maintains a stack for workspace using an allocation and deallocation scheme based on a linked list and indicators of data type. The method is similar to that discussed by Fox, Hall and Schryer(1978). The total size of the stack can be set by the user through the OPTIONS procedure.

If the preprocessor detects severe errors in the user's input, the code generation is bypassed from that point forward and the program terminates before entering the compilation step.

The templates have a modular organization and are selected by the preprocessor based on the keywords specified by the user and/or the nature of the input data. There are over 60 templates in STAT/PROTRAN. Templates vary in length from 10 or so lines to approximately 500 lines. The longer templates tend to be those which perform operations that are fairly standard from one procedure to another. There are approximately 18,000 lines of templates in the first edition of STAT/PROTRAN.

Templates are written by a number of different programmers and may even be written by users not involved in the general development. Hence, it is desirable to have some rules and guidelines in template programming. Some of these are stylistic and are intended to facilitate reading and debugging of template code. Others are to insure that user errors do not get into the compilation stage, where, because of mismatches of data types or similar problems, they may result in meaningless compiler or runtime diagnostics. All variables are checked for attribute consistency at the start of the body of the Fortran code. Direct use of user generated Fortran expressions is minimized, and a comment is placed before each instance of such usage.

The template programming language and various utility functions and runtime utility Fortran subprograms are described by Rice, McWiggins, and Aird(1981).

With the diversity of procedural languages available such as Pascal, C, PL/I, etc., some comments concerning the choice of Fortran as the language for PROTRAN may be in order. There are several reasons Fortran was chosen. The wide availability of Fortran and its relative standardization are perhaps the main reasons. Fortran is available in most environments in which statistical computing is done; and, although there are many dialects of Fortran, the age of the language and the interests of standards groups have endowed it with a degree of stability that may not be reasonably expected for many other languages. The large amount of existing scientific software written in Fortran is a major benefit to both the developers of PROTRAN and to the user of PROTRAN who may wish to incorporate procedural code in a PROTRAN program. Despite certain limitations of Fortran (especially pre-77 Fortran), Fortran is generally suitable for the kinds of computing problems encountered in scientific and statistical applications. The principal disadvantage in using Fortran arises from the slowness of the usual Fortran I/O routines. To overcome this problem, the PROTRAN preprocessor uses some specialized I/O routines.

When STAT/PROTRAN is compared to other statistical analysis packages, several unique features become apparent. The "package family" concept, for example, can allow the solution of a differential equation and a regression analysis with essentially the same user interface. STAT/PROTRAN's ease of use makes it readily available to the non-programmer, its flexibility makes it valuable to the experienced programmer. STAT/PROTRAN is extensible at two levels. At the user level, it offers direct incorporation of procedures written in FORTRAN, giving the knowledgeable Fortran user general procedural language plus the macro-statements of PROTRAN. At the implementation level, STAT/PROTRAN can be modified by use of the general template programming language.

REFERENCES

Aird, Thomas J., and Rice, John R.(1982). PROTRAN: Problem Solving Software. IMSL Technical Report No. 8201. IMSL, Inc., Houston

Anderson, A. J. B.(1980). The Use of Preprocessors in Statistical Software. COMPSTAT 1980. Proceedings in Computational Statistics, Physica Verlag, Wein, 490-496.

Fox, P. A., Hall, A. D., and Schryer, N. L.(1978). The PORT Mathematical Subroutine Library. ACM Transactions on Mathematical Software 4, 104-126.

Rice, John R., McWiggins, Mark, and Aird, Thomas J.(1981). Template Programming Guide. IMSL, Inc., Houston

An Analysis of Continuous Proportions

R. Gilchrist, Polytechnic of North London, G.B.

It seems strange that 'continuous proportions' have received little, if any, discussion in the literature as they can arise naturally in a variety of situations; for example, the proportions of land which have certain uses in different countries (agriculture, housing, industry etc.). Our interest was stimulated by the wish to analyse data on nurse activities; i.e. the proportions of time spent on certain activities by nurses in different hospitals.

The key difference between such data and the widely discussed 'discrete proportions' is that 'continuous proportions' are scale invariant (i.e. the units of measurement are irrelevant to the analysis). This is not the case with 'discrete proportions' as the 'Binomial denominator' plays an important role in determining the significance of some effect.

This paper discusses a method of analysing continuous proportions by the use of generalised linear models. Our technique can be readily implemented in GLIM (Baker and Nelder, 1978). However, any statistical package which can correctly analyse unbalanced designs can be used to give approximately the same analysis. We illustrate the method by analysing data on the activities of 81 nurses in three British hospitals.

KEYWORDS : CONTINUOUS PROPORTIONS; GLIM; GAMMA ERRORS; POISSON ERRORS; COMPOSITIONAL DATA; DIRICHLET; NURSES; HOSPITAL ORGANISATION

1. INTRODUCTION

We here discuss the use of generalised linear models to analyse data in the form of 'continuous proportions'. Our main practical example concerns the activities of trained and untrained nurses in open and divided wards of three British hospitals. 81 nurses were 'trailed' and changes of location and activity were noted so that, over three hour observation periods, continuous records were available of the activities of each nurse. The activities were divided into 12 groups, thereby giving 12 'continuous proportions', namely the proportion of time spent on each activity by a given nurse. The units of measurement (of time) are clearly irrelevant and any analysis must therefore be scale invariant. This differs from the more widely known 'discrete proportions', where the 'Binomial denominator' plays a key role in the likelihood function. Indeed, there seems to have been little treatment of continuous proportions in the literature; this

COMPSTAT 1982 © Physica-Verlag, Vienna for IASC (International Association for Statistical Computing), 1982

seems somewhat strange as such data can arise naturally in a variety of situations; for example, proportions of time spent on various activities or proportions of land area devoted to various uses. There is some discussion of the problem in Connor and Mossiman (1969) and in a technical report by Stephens (1980). Stephens' Method assumes that the vector of square roots of the proportions follows a Von-Mises distribution; however, the comparison of square roots may be thought to be somewhat unnatural and, moreover, Stephens' technique is unable to deal with cross classifications. More recently, Aitchison (1982) has considered this problem, which he refers to as the analysis of 'compositional data'. His technique uses transformed-Normal models. We here discuss a method using generalised linear models; in this respect, our technique can be carried out using the GLIM package (Baker and Nelder, 1978).

2. DESCRIPTION OF THE TECHNIQUE

In Gilchrist (1981) we suggested the use of Gamma errors in GLIM to analyse continuous proportions, much in the same way that Poisson errors can be used to analyse discrete proportions. It is well known that, for discrete proportions with Poisson errors, the act of fitting margins leads to an analysis equivalent to that obtained with an assumption of multinomially distributed data. Similarly, an assumption of Gamma errors suitably conditioned, can lead to an analysis equivalent to that obtained by assuming the data follows a Dirichlet distribution. However, we can also show that our particular Gamma assumption can be more easily and equivalently analysed using Poisson errors. The basis of the technique can thus be outlined as follows :-

Firstly, we must assume that the data ranges over a number of 'factors'. Moreover, one factor will be treated as replications of the data. We then use the replications to estimate a 'scale factor', assumed common to all proportions within a replication. To illustrate this, for our 12 nurse activities, we may choose to treat all trained nurses in each of the 3 hospitals as being replications, and similarly all untrained nurses as being replications. This gives 6 distinct 'individuals' and we estimate one scale factor for each 'individual'; i.e. all proportions for a given 'individual' are assumed to have a common scale. (There is nothing in theory against estimating a different scale factor for each separate proportion of an 'individual' but this does not work well unless there are many more replications than are likely in practice).

Estimation of the scale factor for each 'individual' is straightforward as it will be seen to be calculable as a measure of the variability of the observations about their averages over the replications; i.e. for a given 'individual', we first find the average time spent on each activity and then calculate a measure of variability of the different times about their respective averages. We then use the inverses of the 6 scale factors as prior weights and perform a weighted Poisson fit.

An intuitive explanation of the techniques might prove useful. Our analysis may be viewed as treating each proportion as made up of a number of 'phases'. All 'phases' are of equal duration for each individual, but differ between individuals. Our analysis really works upon the number of phases; in fact, it can be seen that our technique is very similar to that obtained by first estimating the scale factor and then treating the scaled data (proportions divided by scale factor) as being multinomially distributed.

3. ANALYSIS OF NURSE ACTIVITY DATA

The data on nurse activities was collected in 1980 by 'trailing' 81 nurses in fourteen wards at three hospitals. Each nurse was observed for 3 hours. The fourteen wards comprised of 8 divided wards (where all beds are in bays of 8 or less beds) and 6 open wards (where 50% or more of the beds are in one bay). The interest was in how the activities of the nurses related to ward physical variables and to organisational variables, such as the number of medical and nursing staff on duty, workload indices, number of occupied beds etc. The activities were arranged in 12 groups, as follows :-

A1 : Rounds; e.g. doctors' rounds, TPR rounds, medicine rounds.
A2 : Moving, transporting items.
A3 : Outside ward to collect, return, escort etc.
A4 : Seeking/and or standing.
A5 : Bed making, tidying up etc.
A6 : Working behind bed curtains.
A7 : Direct care excluding A6.
A8 : Communication activities.
A9 : Technical activities, e.g. drug preparation, sluicing, washing etc.
A10: Clerical duties.
A11: Personal time.
A12: Travel time.

The resulting proportions of time spent on the 12 activities by each nurse is referred to as the nurse's 'activity pattern'. Of particular prior interest was the effect of divided or open wards on the activity pattern.

In fact, a very large number of variables were recorded and the initial analysis consisted of screening these for possible 'main effects' (i.e. first order interaction with the factor LEV). Now, our analysis with estimated scale factors for each non-replicated 'individual' assumes we have a-priori evidence of these 'individuals'. As this was not available, we adopted the heuristic approach of using the 'standard' GLIM technique with Gamma errors as a means of determining a reasonable set of heterogeneous 'individuals'. (This technique is equivalent to assuming the proportions consist of the same number of phases, but of different lengths). It was clear that factors N and H were needed in any model (N representing whether the nurse was trained or not; H representing the 3 hospitals studied). We therefore used the N*H model to estimate the scale factors; there were therefore 6 such scale factors d_i. A Poisson fit with weights $1/d_i$ was then carried out, again screening a large number of variables. Some results are shown on Table 1.

TABLE 1 Deviances (-2logL) for various variables fitted to nurse activity data. The data have been 'scaled' using the 6 nurse/hospital combinations. (Nurses, N, trained or untrained, for 3 hospitals, H).
Factor LEV represents the 12 levels of the activity factor.

Model	d.f.	deviance
LEV	960	987
LEV*(N+H)	924	810
LEV*WOR	948	959
LEV*NUR	948	937
LEV*M	948	949
LEV*T	948	976
LEV*GE	888	916
LEV*(N+H+M)	900	764
LEV*(N+H+WOR)	912	779
LEV*(N+H+NUR)	912	757
LEV*(N+H+WOR+M)	888	721
LEV*(N+H+GE)	852	710
LEV*(N+H+NUR+M)	888	728
LEV*(N+H+NUR+WOR)	900	733
LEV*(N+H+NUR+WOR+M)	864	705
LEV*(H+N*WOR)	900	778
LEV*(N+H*WOR)	900	746
LEV*(N+H*(WOR+NUR))	888	724
LEV*(H*WOR+N*NUR)	876	713

The N and H effects clearly remain after allowing for the differing scales between the 'individuals'. The ward physical variable T determines whether a ward was open or divided; it does not affect the activity patterns. The number of medical staff on duty in the trail period does appear to explain some variation above that explained by N + H; similarly, the main location of a nurses work (determined by the factor GE) gives a further reduction in deviance, although GE is not itself significant. Other variables, NUR and WOR measuring the organisational practices of the wards, provided alternative acceptable models. The analysis here presented is not entirely complete. High-order interactions have not been tried because of computing restrictions, the difficulty being that the number of parameters to be estimated gets very large, due to the crossing with the activity variable LEV. However, the technique has been of value in an exploratory sense. We have been able to remove variables which appear not to be of interest and to produce several models which appear to give a satisfactory explanation of the data. In terms of presentation to the 'consumer', we provided activity tables and pie-charts for various possible models. The following tables are for the models N + H + M and N + H + WOR + NUR.

TABLE 2 Model N + H + M for predicting nurse activity profiles taking into account number of medical staff.

M = 1 for 0 to 9 medical staff; M=2 for 10 to 19; M=3 otherwise.

Table shows rounded % of time in each activity.

Hospital	1						2				3	
Nurse	Trained			Untrained			Trained		Untrained		Trained	Untrained
Medical Staff	1	2	3	1	2	3	1	2	1	2	1	1
A 1	8	8	10	4	3	4	15	14	6	6	8	4
C 2	1	2	2	3	4	4	2	3	4	5	3	6
T 3	0	3	2	0	4	2	0	0	0	0	2	4
I 4	2	2	3	2	2	3	2	1	2	1	3	3
V 5	3	5	8	6	10	16	3	5	6	11	3	6
I 6	22	16	10	38	28	17	16	12	27	20	16	27
T 7	9	10	11	11	12	13	14	15	16	17	9	11
I 8	18	18	19	4	4	4	15	16	3	4	14	3
E 9	11	10	7	17	14	11	8	7	12	10	9	13
S 10	7	6	10	2	1	2	11	10	2	2	12	3
11	8	8	12	10	10	15	12	13	15	16	12	15
12	8	10	9	8	10	8	6	7	6	7	9	8

TABLE 3 Model N + H + NUR +WOR for predicting nurse activity profiles taking into account ward organisation variables.

WOR = 1 if using 'patient care'; zero if using 'task and patient care'.

NUR = 1 if 'nursing process' in operation; zero otherwise.

Hospital	1						2		3			
Nurse	Trained			Untrained			Trained	Untrained	Trained		Untrained	
WOR	1		0	1		0	1	1	1	0	1	0
NUR	1	0	1	1	0	1	1	1	1	0	1	0
A 1	10	5	12	4	2	5	14	6	10	7	4	3
C 2	2	2	2	3	3	4	2	5	3	3	6	6
T 3	2	2	3	2	3	4	0	0	1	3	2	5
I 4	2	2	2	2	2	2	2	2	3	2	3	2
V 5	2	4	7	12	3	13	5	9	3	3	7	5
I 6	6	22	12	20	39	21	12	22	10	20	17	36
T 7	11	11	8	13	12	9	15	17	12	8	13	9
I 8	11	18	18	4	4	4	16	3	15	13	3	3
E 9	19	9	6	15	14	9	7	10	11	7	16	10
S 10	10	10	5	9	2	1	2	2	17	9	4	2
11	10	7	15	11	8	18	13	16	11	13	13	16
12	10	9	8	9	9	7	7	7	10	8	10	7

4. CONCLUSION

The technique here presented provides a comparitively simple means of 'cleaning-up' complicated continuous proportions data. More work is needed on the theoretical background before we would claim it to be a thoroughly respectable technique; for example, the use of χ^2 tests is somewhat ambitious, particularly bearing in mind the non-maximum-likelihood estimates of the scale factors. However, our emphasis is to attempt to use such tests in a somewhat vague fashion, avoiding fixed probability levels and choosing to remove only those effects which are clearly small.

This can result in several final competing models; nevertheless, in the cases considered by us, it has led to a significant understanding of the data and, in this respect, we can commend the technique as a useful exploratory tool.

REFERENCES

AITCHISON, J. (1982). The statistical analysis of compositional data. J.R.Statist.Soc B, to appear.

BAKER, R.J. and NELDER, J.A. (1978). The GLIM system. Generalised linear interactive modelling. Oxford: NAG.

CONNOR, W.S. and MOSSIMAN, J.E. (1969). Concepts of independence for proportions with a generalisation of the Dirichlet distribution. J.Amer.Stat. Ass.,64,194-206.

GILCHRIST, R. (1981). Using GLIM to analyse continuous proportions Proceedings of the Rome conference on the analysis of multidimensional contingency tables. Metron.

SEELYE, A. (1982). Ward layout and nurse staffing. Polytechnic of North London: Medical Architecture Research Unit 1/82.

STEPHENS, M.A. (1980). The von-Mises distribution in p-dimensions, with applications. Stanford Univ., Dept. Statistics. Technical Report No.290.

A Sampling Procedure for Historical Data

J. Gordesch, Free University of Berlin, F.R.G.

SUMMARY

For historical data, a sampling plan is designed meeting the following conditions:
(i) Clusters (in the sense of sampling theory) are not pregiven but have to be determined yet (clustering in the sense of automatic classification).
(ii) Sampling should be optimal in the sense of minimum variance estimators, taking into account restrictions on costs, time, and manpower.
(iii) Independence of scaling and projections has to be ensured.
A single-stage cluster sampling plan with unequal sizes of clusters is developed, using non-linear optimization techniques and a Cayley-Klein projective metric.

KEYWORDS

sampling, cluster analysis, projective geometry, optimization

INTRODUCTORY EXAMPLE

The origin (region, workshop, etc.) of the specimens of an archæological collection has to be ascertained. For this purpose, both external properties and the chemical composition of the clay may be used. The chemical composition (e.g. the content of Mg, Al, Si, Ca, Fe, and Ba) is analyzed by the method of 14-MeV-neutron activation and subsequent Ge(Li)-gamma spectroscopy. The composition as well as the external properties of the regions etc. are sufficiently known. The chemical analysis, though inevitable, is rather costly compared to a visual inspection. Thus clusters are formed according to external properties, guaranteeing some homogeneity. Then a sample of the resulting clusters is drawn, and each cluster in the sample is analyzed, either partly (two-stage sampling procedure) or totally (cluster sampling). A statistical procedure should be designed to yield an optimal, i.e. least costly and most precise result.

SAMPLING MODEL

A sampling plan has to be designed in the case of not yet determined sampling units consisting of smaller items. The sampling units ("clusters", e.g. regions) have to be determined according to optimal sampling conditions.Thus,a cluster sample of m units (clusters;regions) with $n = \Sigma m_j$ elements, m_j thereof being in the jth cluster,is drawn from a population of N elements divided into M groups.Various restrictions may be imposed:

(i) Clusters (in the sense of sampling theory) are not pregiven but have to be determined yet (clustering in the sense of automatic classification).

(ii) Sampling should be optimal in the sense of minimum variance estimators,taking into account restrictions on capital,time,manpower,etc. Most sampling procedures are optimized with respect to one (possibly composite) property (i.e.variable) only.

(iii) Any further structure may be prescribed in the sense that the sample must contain elements with certain properties only (i.e. merely elements lying in some subspace are sampled).

In the sequel,a single-stage cluster sampling plan with unequal sizes of clusters will be considered.

In this case,mainly two types of estimates are used:

(1) Ratio to size estimate.

$$\hat{x}_R = \sum_1^M m_i . \sum_1^m x_i / \sum_1^m m_i$$

$$V(\hat{x}_R) = \frac{M^2(1-m/M)}{m} . \frac{\Sigma m_i (\bar{x}_i - \bar{\bar{x}})^2}{M-1} \quad , \; i=1\ldots M$$

$$x_i = \sum_j x_{ij} \; , \; \bar{x}_i = x_i/m_i \; , \; \bar{\bar{x}} = \Sigma x_i / \Sigma m_i \; , \; i=1\ldots m \; , \; j=1\ldots m_i$$

For minimizing the sample variance,the m_i must be known or equal.We might use a two-stage optimization procedure,and first form clusters by the use of a clustering algorithm and then draw a sample.For various clustering results,the variances may be computed and compared to each other

(2) Unbiased estimate.

$$\hat{x} = \frac{M}{m}\Sigma x_i \; , \; x_i = \sum_j x_{ij} \; , \; i=1\ldots m \; , \; j=1\ldots m_i$$

$$V(\hat{x}) = \frac{M^2(1-m/M)}{m} . \frac{\Sigma (x_i - \bar{x})^2}{M-1} \; , \; \bar{x} = \Sigma x_i / M \; , \; i=1\ldots M$$

Various other estimators exist and may be used in particular cases. Their performance depends more or less on the composition of the clusters.As a rule,they are based on sampling with unequal probabilities, which complicates the model with respect to the application of cluster algorithms considerably.

The precision of the unbiased estimate is poor only when the $\bar{x}_i$ (means per element) vary little from unit to unit and the m_i vary greatly. By clustering, however, the difference between clusters is raised, and therefore a gain in the precision of sampling reached. The variance of the unbiased estimate is independent from the unknown cluster sizes and therefore easily connected with clustering procedures which determine also cluster sizes and yield better results than in the case of pregiven clusters.

In the sequel we modify an integer programming approach as it was suggested by Vinod (1969) and by Rao (1971) because the optimization procedures of variance minimization as well as optimal clustering can be combined in one algorithm.

ALGORITHM

Notation

a_{ij} coefficients of objective function. $a_{ij}=1$ if the ith element belongs to the jth cluster, otherwise zero. A is the matrix $[a_{ij}]$.

C available amount of capital

c_j amount of capital spent on the jth cluster

d_{ij} distance measuring the "cost" of assigning the ith object to the jth group. D is the matrix $[d_{ij}]$.

m_j elements per cluster

M total number of clusters

m number of clusters in the sample

n total number of elements in the sample

N number of elements in the population

P available amount of personnel

p_j amount of personnel spent on the jth cluster

r,s scaling or weighting factors greater than or equal to zero

T time available

t_j amount of time spent on the jth cluster

u_j $u_j=1$ if the jth cluster belongs to the sample, 0 otherwise

y_j $y_j=1$ if the jth object is group leader, 0 otherwise

$\overset{n}{\underset{i_o}{\forall}} i$ "for all i from i_o to n" (quantor)

As the number of clusters is not known in advance, the order of D and that of A is also not known in advance. To circumvent this arrangement we assume a total of n groups with n-m empty groups. For the identification of groups the jth object of the group is denoted as "group leader".

The "cost" or "distance" matrix is computed from the p-dimensional data vectors $x=(x_1,\dots x_p)$. The use of a projective space instead of a euclidean one as well as of a projective metric in the sense of Cayley and Klein (1928) to ensure invariance under projective transformation is possible. The clustering problem is to minimize the total cost of any resulting classification (partition) subject to certain constraints: minimize

$$\Sigma\Sigma a_{ij} d_{ij}^2 , \quad i,j=1\dots n ,$$

subject to the following constraints:

(1) $\sum_j a_{ij}=1$; $i,j=1\dots n$, stating that any object cannot be put into more than one group (non-overlapping partition).

(2) $\Sigma y_j = m$, stating that there are exactly m groups.

(3) $ny_j \geq \sum_j a_{ij}$; $i,j=1\dots n$. In the one-dimensional case this condition states that the jth object must be a leader before any objects can be placed in the group corresponding to it.

(4) The quantities d_{ij} in the jth column can be reordered in increasing order of magnitude:

$$d_{jj} \leq d_{i_1 j} \leq \dots\dots \leq d_{i_{n-1} j} \quad .$$

The string of ones in the jth column of $[a_{ij}]$ should follow the pattern given by $i_1, i_2, \dots i_{n-1}$ according to

$$d_{jj}=0 \leq d_{i_1 j} \leq \dots\dots \leq d_{i_{n-1} j} \quad .$$

This condition (a modification of Vinod's "generalized string property") is - particularly in the univariate case - intuitively appealing and may also accelerate the algorithm. It is not, however, a necessary or a sufficient condition for the existence of a minimum.

Another version of a "generalized string property" achieving much the same was given by Rao (1971): Each group consists of at least one object (group leader) such that the distance between the leader and any object that does not belong to the same group is not less than the distance between the leader and any object within the group.

Once the clusters are fixed, the minimization of the variance is - as one can see from the pertaining equation on p.2 - attainable only by a magnification of the size of the sample, subject to certain restrictions. Hence using the matrix $[a_{ij}]$ and the vector $[u_j]$, the objective function to be minimized reads

$$\Sigma\Sigma a_{ij} u_j \quad , \quad i,j=1\dots n \ .$$

The obvious restrictions on time,cost,personnel,etc. are

$$\Sigma t_j m_j u_j \leq T\ ,\ \Sigma c_j m_j u_j \leq C\ ,\ \Sigma p_j m_j u_j \leq P\ ,$$
$$j=1\ldots m\ .$$

The last step remains,the combination of the two optimization problems.First,the alteration of sign of the second objective function is necessary to obtain two minimization problems.Then,several combinations of the two objective functions make sense;however,a simple addition has often proved successful,and also reflects the fact that good clustering requires homogeneity,while good cluster sampling (the results are compared to other sampling techniques) depends on the heterogeneity of the clusters.The introduction of weighting (or scaling) factors r,s appears to be recommendable (r=0,s>0: sampling only; r>0,s=0: clustering only).Thus we reach at the following formulation of the optimization problem:

Minimize the sum

$$\sum_1^n \sum_1^n (ra_{ij}d_{ij}^2 - sa_{ij}u_j)$$

with respect to a_{ij},u_j under the constraints (1) to (8):

(1) $\forall_1^n i: \ \sum_1^n a_{ij}=1$

(2) $\Sigma y_j=m\ ,j=1\ldots n$

(3) $\forall_1^n i \forall_1^n j: \ ny_j \geq \sum_j a_{ij}\ ,j=1\ldots n$

(4) $\Sigma t_j m_j u_j \leq T\ ,j=1\ldots m$

(5) $\Sigma c_j m_j u_j \leq C\ ,j=1\ldots m$

(6) $\Sigma p_j m_j u_j \leq P\ ,j=1\ldots m$

(7) $\forall_1^n i \forall_1^n j: \ a_{ij},u_j,y_j \in \{0,1\}$

(8) Optional constraints on the x (e.g. Gx=b or other linear relations)

In certain cases - cf.Vinod (1969) and Rao (1971) - the quadratic programming problem can be reduced to a linear programming problem (assignment problem).In the general case,a branch and bound algorithm is recommended,particularly if the number of sampled units is small. For larger systems,however,heuristic techniques leading to suboptimal solutions only may be prefered for reasons of workability.

PROJECTIVE METRIC

Projective geometry had been designed to avoid the cumbersome distinction between many special cases but had failed to encounter also metric properties until the work of Cayley (in 1859) and Klein (1928).In a projective space over a formally real or an algebraically closed field,distance between two points as well as angle between two hyperplanes are introduced as invariants of a group of projective transformations.In the non-degenerate case e.g.,an arbitrary surface of dimension n-1 and order two (absolute conic section of the metric) is marked out.Two points x,y of the projective space determine a line which cuts the absolute conic section in a and b.Distance,then,is defined by the logarithm of the cross ratio,and is determined up to a scaling constant:

$$d(x,y) = c.\log(x,y,a,b) .$$

In the degenerate case,a surface of dimension n-2 and of order 2 is chosen as absolute conic section.By a limiting process,we again reach at a projective definition of distance.The main cases are euclidean and pseudoeuclidean distances.For further details refer to comprehensive text books on projective geometry as Klein (1928).

REFERENCES

Elsner E. (1973),Zur Optimierung der Stichprobenpläne bei Klumpenstichproben,Berliner Statistik,125-127.

Klein F. (1928,reprint 1968),Vorlesungen über nicht-euklidische Geometrie,Springer,Berlin.

Rao M.R. (1971),Cluster analysis and mathematical programming,J.Am. Stat.Ass.,66,622-626.

Vinod H.D. (1969),Integer programming and theory of grouping,J.Am. Stat.Ass.,64,506-519.

Use of Systemic Networks for Text Analysis

F. Grize, University of Neuchâtel, Switzerland
J. Bliss and J. Obgorn, University of London, G.B.

SUMMARY

Classical statistical methods are generally well suited for analysing numerical data. But methods of processing informal data, such as textual data, which could be: interview transcripts, free questionnaire responses, observational material, etc... are much less well developed. The method presented in this paper allows the analysis of such data in which the structure is arbitrarily complex.
The method is based on a concept of linguistics: the systemic network. Such a network combines elementary schemas leading to a general schema which formalizes elements of the structure of a text or discourse. Each particular text to be analysed is an instantiation of a general schema.
A computing system, presented here, makes the method utilizable on a computer.

Keywords: analysis of qualitative data, systemic networks, representation of a structure of possibilities, computing system for systemic networks.

1. WHAT A SYSTEMIC NETWORK IS?

Systemic linguistics (Berry 1975, 1977; Halliday 1978; Kress 1976) derives its name from the fact that it represents the language as a finite set of systems. It has been used by Turner (1972) in sociological studies and by Winograd (1972) in understanding natural language.

The term **system** is used because it is the fundamental concept of the grammar. The grammar itself is based on the notion of choice. Indeed, the user of a language may be regarded as someone who has simultaneously and successively to make a variety of distinct choices.

Let us take an example from the French language to illustrate the simplest case of a system. In French, a "déterminant" is either an "article" or an "adjectif". Such a set of exclusive choices would be represented by a **bar**:

```
               ┌─ ARTICLE
DETERMINANT ──┤
               └─ ADJECTIF
```

In this example, the choice is binary; of course, a system can have any finite number of options.

The elements of a system are called terms. Immediately we can make the distinction between a term which introduces a system such as DETERMINANT and others. In this case DETERMINANT is called an entry condition. Indeed, ARTICLE and ADJECTIF may be chosen if and only if DETERMINANT has been chosen. Notice the use of capital letters to write DETERMINANT which will be used when we refer to any formalized term of a network.

An important concept is the concept of delicacy, i.e. the possibility of constructing a network by stepwise refinement. For instance, as we know that in French an "article" is in addition either "défini" or "indéfini" we are able to say more about what it is. Delicacy is represented by linking terms in a **left-to-right tree**, therefore we have the more delicate network:

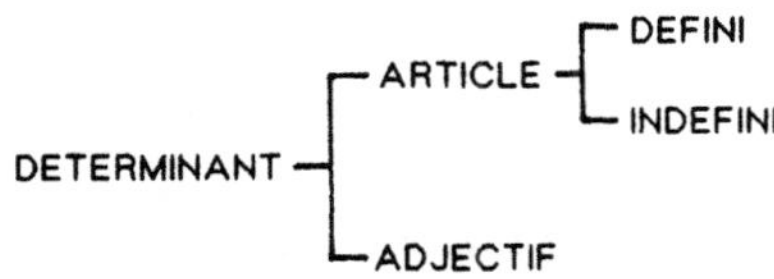

The next notion is that of simultaneity. Let us consider the systems of "genre" and "nombre". They would be represented by:

In French, the notion "genre" and "nombre" are always linked to the notion of "déterminant"; i.e. if one speaks of a "déterminant", simultaneously one must always consider its "genre" and its "nombre". To represent this simultaneity, we use a **right-facing bracket**:

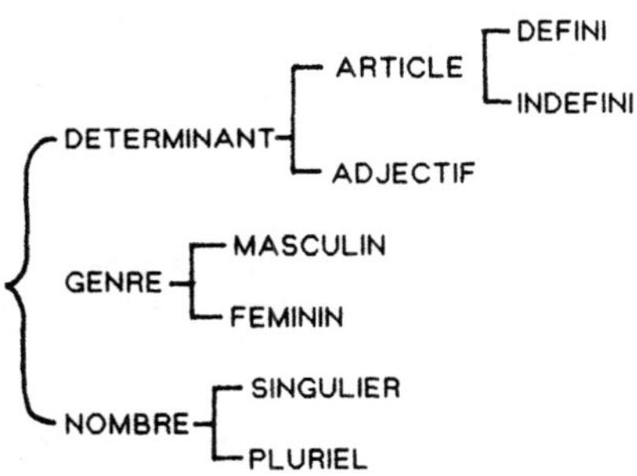

Such a combination of systems is a systemic network and must be interpreted as follows: after having chosen an option in the system DETERMINANT (ARTICLE or ADJECTIF, and if ARTICLE either DEFINI or INDEFINI), we must choose an option in the system GENRE (MASCULIN or FEMININ), and finally an option in the system NOMBRE (SINGULIER or PLURIEL). The set of all possible choices of a network is called the set of paradigms.

The above two notations alone suffice for many purposes and it is sometimes best to restrict oneself to them. However, a number of further notations are sometimes of use. They are defined in the following summary of the complete formal notation:

Network notation

if A then X_1

(1) 

if A then $X_1,\ldots,X_1$

(2) 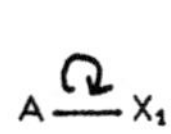

if A then X_1 or...or X_n

(3) 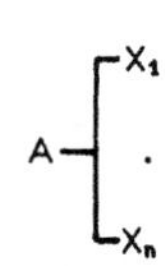

if A then X_1 and...and X_n

(4) 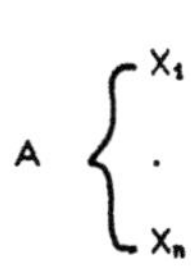

if X_1 and...and X_n then A

(5) 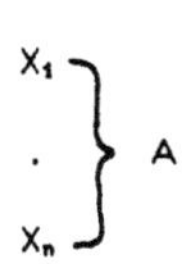

if X_1 or...or X_n then A

(6) 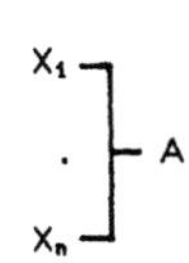

Remarks:

1) the X_i's may designate either a system (entry condition) or a terminal term.
2) each system has a name:
   1) **simple system** (for delicacy).
   2) **recursive system.** This system means: enter X_1, as many times as desired, but at least once. In fact, this system represents a **non-exclusive or.**
   3) **bar-system** (for exclusive choice).
   4) **bra-system** (for simultaneity).
   5) **con-system.** This system means: enter A only if all (i.e. on the only **condition**) of $X_1, \ldots, X_n$.
   6) **any-system.** This system means: enter A if one or more (i.e. **any**) of $X_1, \ldots, X_n$.

2. USE OF SYSTEMIC NETWORKS FOR NON-LINGUISTIC ANALYSIS

The method described in this paper originated in some empirical research by Bliss and Ogborn (1977). 115 undergraduate students were asked to talk about their feelings (**good** or **bad**) during their time of study. The interviews produced some 300 stories for analysis: a formidable bulk of complex material. Even on making global and general categories it was difficult to split the stories into well-defined categories. In consequence, they turned to systemic networks which offered the possibility of expressing interrelated classifications with any desired delicacy.
Later the method became somewhat more precise (Bliss and Ogborn 1979; Grize 1981) and was (or is) used in different fields. Ogborn (1980) used a systemic network for managing and describing an item bank. Mujib (1980) has developed a network to describe how the student's attention shifts from one aspect of a problem to another. Monk (1981) used the networks to study teacher-pupil interactions. To examine the attitude of people towards the army and more particularly their reasons for choosing a military career, Scherrer (1982) used the method.

The general me hod of analysing data with systemic networks may be divided into four steps:

- **conception of the network**
- **coding** the raw data using the terms of the network
- **checking** the completeness and coherence of the code with regard to the network
- **questioning** the code in terms defined by the network.

Conception of the network

This is the hardest stage, the rest following from and depending on it.
This step consists of formalizing, with the aid of a network, aspects of the structure of the data to be analysed. Suppose that people are asked to freely answer a question such as: "Do you smoke? In any case, say why you smoke or why you do not".
This question will produce as many texts - written in natural language - as people who are asked. All these texts should talk about the same topic ("the reasons why a person smokes or does not"). As soon as the analyst can isolate from these different texts all their common elements, with the aid of a systemic network, he can organize and relate them to one another. The answers being free, the network cannot translate all the elements contained in each text. In fact, the network will codify and fix, in an artificial description language, the perceptions of the analyst: what the analyst is interested in.

For lack of space, it is not possible to give such a network which might have more than 100 items. Instead, to illustrate our purpose, let us consider the following network:

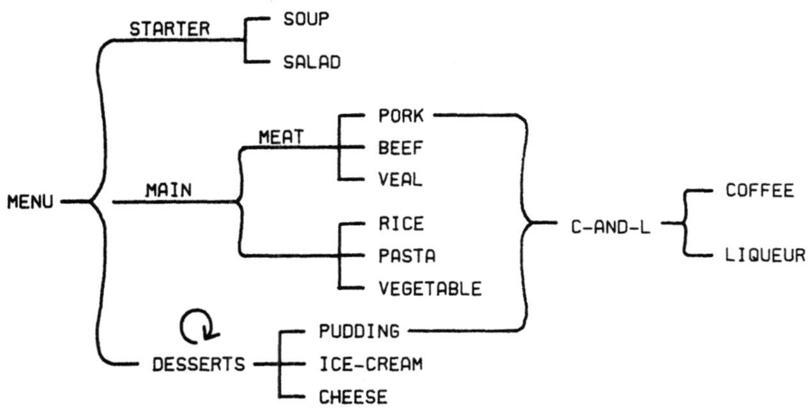

The left hand part of this network describes a menu. To the right, we find what could be a "hidden menu", perhaps noticed by an observer who found that the patron refused coffee and liqueur to all but those who had suffered both his pork and his pudding, but insisted on it for them!

Coding

This step consists of writing each particular text using the terms of the network. If we take again the example of the interviews on smoking, the analyst must decide, sometimes arbitrarily, how to interpret this or that idea appearing in the text. It is clear that the network does not describe the data, but gives an interpretation to the data. That is to say, what it is important is not the way that each term must be interpreted but the way each term is related with the others. In a sense, the code is what the analyst could say if he was asked to summarize what he understands in each text. The code can also be regarded as an artificial language, whose "grammar" is described by the systemic network.

Checking

After having coded each text, one have to verify that the code belongs to the set of the paradigms (a possible path in the network). Of course, this checking is purely mechanical and can therefore be automatically processed by a computer. In the case of the network-MENU, one must check that, if Mr. X chooses a meal, he has taken a STARTER then a MAIN then one or several DESSERTS; that is, in the STARTER, he has taken either a SOUP or a SALAD, etc...

During the validation of the code, it is necessary to store on the network all the traversed routes. To remember, for instance, that Mr. X has chosen SOUP and not SALAD, then PORK and not the three other meats, etc... This step is also purely mechanical and can be automatically realized by computer.

Questioning

This step consists of questioning the data related to the network. For instance, one could wonder: "Are there people who have taken PORK and RICE?", "Are there people who have taken three DESSERTS? If yes, who are they?", etc... That is, questions are formed from terms of the network combining with logical operators. Answering questions which use logical operators, or which require cross-tabulation, is also a purely mechanical step which can be realized by a computer.

3. THE COMPUTING SYSTEM

It is clear that a network may rapidly become very complex and that the code may have hundreds of items. In that case, it is practically impossible to handle this information by hand. For this reason and to minimize error, a computing programme, named BARBARA, has been developed (Grize, 1981). This programme processes everything which can be done automatically. It requires of the user little previous knowledge since it progresses under the form of a dialogue between the user and the computer. It is designed in three logically different parts:

- input of a systemic network
- input and checking of the code
- questioning of the data.

Input of a systemic network

Of course, BARBARA cannot conceive a network on the user's behalf, but it helps him in his work. It is able to accept any systemic network. It checks the coherence of the network: warning of possible accidental redefinition of any term, signalling the inaccessible terms, checking the connexity of the network, etc...

With the aid of a **help-function**, the user can, at any time, know what he needs to do or what he has previously done. For instance: which terms have already been used; whether they are entry conditions or not, and, if they are, which system they define; how many times each term appears in the network; which systems are not connected to the network, etc...

The formalism used to enter a network is quite simple. Each system is translated by a function which has the same name as the system itself, e.g. the bar-system will be translated by the function BAR(X_1 X_2 ... X_n). For instance, the input of the network-MENU, would be given by the set of the following statements:

```
MENU = BRA(STARTER MAIN DESSERTS);
STARTER = BAR(SOUP SALAD);
MAIN = BRA(MEAT BAR(RICE PASTA VEGETABLE));
MEAT = BAR(PORK BEEF VEAL);
DESSERTS = REC(BAR(PUDDING ICE-CREAM CHEESE));
CON(PORK PUDDING) = C-AND-L;
C-AND-L = BRA(COFFEE LIQUEUR);
```

Remark:
The **format** is free, i.e. a statement may lie on several lines. It is for this reason that each statement is ended by a semicolon.

Input and checking of the code

Once a correct network has been introduced, BARBARA is able to accept the corresponding code. It checks that the code is a possible path in the network and stores on the network all the traversed nodes. If there is an error, the programme signals its exact nature i.e. what was, at that moment, the expected system. Furthermore, the **help-function** can give other information to the user: which items are already defined, their definitions, etc...

Items are defined for input simply by the terms they use. For instance, one could have the three following items:

```
X = STARTER(SOUP) MAIN(MEAT(PORK)RICE) DESSERTS(ICE-CREAM CHEESE);
Y = SOUP VEAL PASTA PUDDING;
Z = STARTER(SALAD) MAIN(PORK PASTA) DESSERTS(PUDDING) COFFEE LIQUEUR;
```

Remark:
The **format** is also free. Moreover, one can notice that the entry conditions are specified or not (e.g. Y). Indeed, they are not compulsory, since the programme already **knows** the different relations between each term. They are specified only for purposes of readability.

Questioning of the data

This last step allows analysis of data. There are essentially three functions for this purpose. Each of these functions admits, as parameter, a logical expression formed with the operators **and, or, not** combined with terms of the network.

- LIST(L)
  gives the list of all the items for which the logical expression L holds. For instance, LIST(PORK **and** RICE) will give the list of all people who have chosen as main: pork and rice; in our example, there is only one: X.
- NUM(L)
  says how many items they are for which L holds.
- TAB(<list1> , <list2>)
  cross-tabulates all the items which have the features of <list1> against all the items which have the features of <list2>.

There are some other possibilities such as: to locate all the terms accessible from an entry condition; to abbreviate logical expressions, etc...

Finally we will mention that BARBARA maintains, manages and updates all data and network files.

REFERENCES

- Berry M. (1975,1977), Introduction to Systemic Linguitics Vol. 1: Structures and Systems, Vol. 2: Levels and Links, Batsford, London.
- Bliss J. and Ogborn J. (1977), Students' Reactions to Undergraduate Science, Heinemann, London.
- Bliss J. and Ogborn J. (1979), The Analysis of qualitative Data, Eur. J. Sci. Educ., 1, 427-440.
- Grize F. (1981), Barbara: Analyse de données informelles à l'aide de réseaux systémiques, Université, Neuchâtel.
- Halliday M.A.K. (1978), Language as Social Semiotic, Edward Arnold, London.
- Kress G. (Ed.) (1976), Halliday: System and Function in Language, Oxford University Press, Oxford.
- Monk M. (1981), Network Analysis of the "People in my Class" Questionnaire Data, Chelsea College, London.
- Mujib F. (1980), Students' Problem solving Behaviour in selected Problems related to undergraduate Physics, with specific Attention to the Analysis of Protocols, Chelsea College, London.
- Ogborn J. (1980), Use of a systemic Network for managing an Item Bank, Chelsea College, London.
- Scherrer P. (1982), De la perception du phénomène "armée" dans un milieu universitaire: étude empirique, Université/GPA, Neuchâtel.
- Turner G.J. (1972), Social Class and children's language of control at age 5 and age 7, in Bernstein B. (Ed.) (1972), Class, Codes and Control, Vol. 2, Routledge and Kegan Paul, London.
- Winograd T. (1972), Understanding Natural Language, Edinburgh University Press, Edinburgh.

SPASP – A Statistical Program for the Analysis of Stochastic Processes

W. Grossmann and G.Ch. Pflug, University of Vienna, Austria

Summary. We present the concept of a program system for the analysis of stochastic processes. The system is designed for the analysis of diffusion models, stochastic differential equations, regression models or Poisson models. The data structure is two dimensional i.e. we have for each case and each variable the possibility of repeated measurements in times.

Keywords: Software packages, analysis of stochastic processes

1. Purpose and scope of SPASP

Most of the statistical software packages have one of the following two basic structures. The first one is the case oriented structure or more precisely for the observed variables a general linear model is assumed (besides the possible exception of nonlinear regression routines). The other is the time oriented set up and by classical methods of time series analysis the identification of the structure of the underlying linear stochastic process is done. However in many applications the collected data are of the following two dimensional structure: for each case a set of time dependent variables is observed at different times. The times of observations as well as the number of times may vary from case to case. As an example for such situations consider data from medical observations where health data for individual patients at different times are collected. Similar situations occur also in econometrics or biology.

For the analysis of such data the statistical packages based on one of the structures defined above (BMDP, SPSS, SAS, GLIM and others) are not very well designed not only from the data structure point of view but also from statistical model building. Therefore for some time the need was felt to have a statistical program package which allows to analyse problems of the type described above, a package which makes a new start with respect to the data structure as well as with respect to the statistical model (c.f. Kubale and Schmidt (1981)). To the authors opinion Markov processes and regression processes with stationary error processes are together appropriate models for such problems. Because the models are rather general the analysis must include data analytic methods (together with graphic displays) as well as methods for parametric inference.

The paper presents first steps in these directions part of which has been done already whereas others are more or less tentative and will be completed in the future.

2. Data structure of SPASP

As it was pointed out the data structure of SPASP is two dimensional. On the one hand we have as usual cases as units of the observations. Each case $i(1 \leq i \leq n)$ is characterized by a set of time independent variables (CVARIABLES) and a set of time dependent variables (TVARIABLES). The CVARIABLES may be of any type (numeric or alpha numeric) and are global data for the case. The TVARIABLES are characterized by the times of measurements and the values of the variables. Consider as a typical example the following data set:

CVARIABLES: name, number, age, sex;
TVARIABLES: number, times, height, weight, blood-pressure;

That variable definition corresponds to the following typical data set for one case.

CVARIABLES:
name: maier number = 1 age = 14 sex = 0

TVARIABLES:
number: 1

times :	158	165	186	188	209	219	220
height:	165	--	--	167	--	--	171
weight:	58	58.5	58.3	58.2	58.5	58.2	58.2
blood pressure:	121	119	123	--	115	124	--

Observe that the number is necessary in order to give the correspondence between CVARIABLES and TVARIABLES. Furthermore the classical data structures are contained as special cases. If we have no TVARIABLES then the structure corresponds to the well known case structure of statistical packages. If there are only TVARIABLES then the model is a standard time series model.

3. Statistical Models and Analysis

We assume that for each case $i(1 \leq i \leq n)$ an independent stochastic process $X_i(t)$ can be observed at the times $t_{ij}(1 \le j \le m_i)$, together with some concomitant random or nonrandom processes $\underline{a}_i(t)$. For easier notation the models are formulated in terms of continuous time processes although the processes are only observed partially. Basically four different process structures are considered:

3.1. Diffusion models (D-models)

$X_i(t)$ is a diffusion process of the form

$$dX_i(t) = f(\underline{a}_i(t),t)dt + g(\underline{a}_i(t),t)dW_i(t)$$

where $W_i(\cdot)$ are independent copies of Wiener-processes.

3.2. Differential equations (DE-models)

$X_i(t)$ is a solution of the stochastic differential equation

$$dX_i(t) = f(t,a_i(t))dt + g(t,a_i(t),x(t))dW_i(t)$$

3.3. Regression models (R-models)

$$X_i(t) = F(t,a_i(t)) + Z_i(t)$$

where $Z_i(t)$ is a (stationary) error process.

3.4. Poisson models (P-models)

$$X_i(t) = P(f(t,a_i(t))$$

i.e. $X_i(t)$ is a Poisson process with intensity $f(t,a_i(t))$.

We have formulated the models in nonparametric form in order to allow more general structures.

The analysis of the models is done by performing the following two steps:

(i) analysis of the trend component
(ii) analysis of the second order properties

In the nonparametric set up trend components can be estimated by any of the following methods: spline estimates with variable knots, kernel estimates , nearest neighbor estimates (c.f. Collomb (1981)).

In the parametric set up, the functions defining the trend component depend on an unknown parameter θ. In order to estimate θ we use the classical methods of parameter estimation, e.g. M-estimates (including least squares). Furthermore confidence regions for the parameters are determined (Liptser, Shiryayev (1978)).

Analysis of second order is based on the residuals after removal of the trend components. According to the model we either analyse the diffusion constant or the variance-covariance structure of the process.

Based on this analysis classification and discrimination of curves can be done by various methods. Additional methods for the classification problem like nearest neighbor classification (Devroye,Wagner (1979)) or L_p-curve classification are provided.

4. SPASP-syntax and the structure of the package

The command language of SPASP is defined by a simple syntax diagram and hence the corresponding grammar is context free. The main structure of a SPASP-program is as follows

SPASP-program:

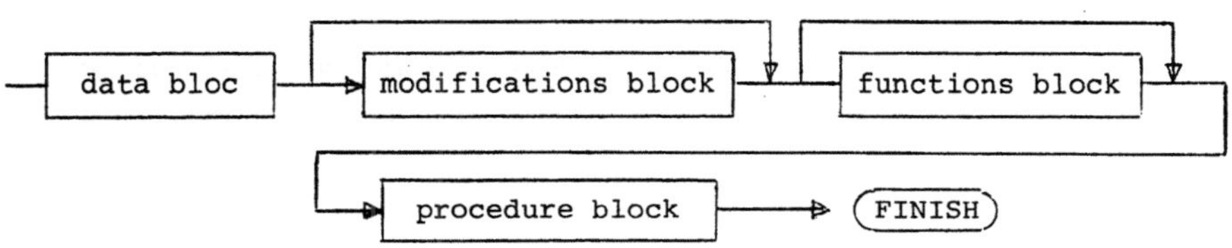

The data-block contains the definition (file, format, labels,etc.) of the CVARIABLES (CVARS) and the TVARIABLES (TVARS):

data-block:

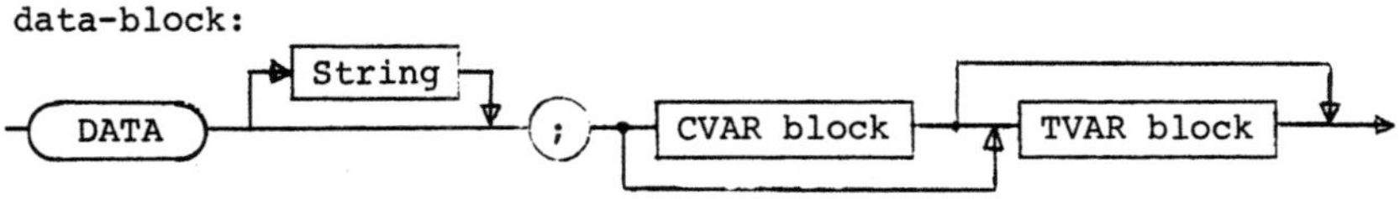

The modifications-block is devoted to the transformation of old variables and the definition of new ones. Within the functions-block the user can define arbitrary functions and procedures written in FORTRAN 77 language, which are automatically sent to compilation and linked by the system.

The procedure-block contains the keywords for the chosen analysis as well as the parameters and options for these procedures.

A typical SPASP command string looks as follows:

```
DATA :        health data;
CVARIABLES:   name, number, age, sex;
              TEXT: name;
              CASE ID: number;
              INPUT FILE: file 1;
              FORMAT: FREE;
              N OF TIMES: UNKNOWN;
TVARIABLES:   number, times, height, weight, blood-pressure;
              INPUT FILE: file 2;
              FORMAT: FIXED (3X, I3, 4F 7.2)
              TIMES: times;
PROCEDURES:   kernel-estimates;
              VAR LIST: blood-pressure;
              CASE LIST: ALL;
              OPTIONS: kernel = k1;
FINISH
```

Mainly for reasons of portability the computer code was written in FORTRAN 77. The lack of dynamic storage allocation in this language as well as the possibility of user-defined subroutines in SPASP made it necessary to process in two stages. A preprocessor was designed for the first analysis step. This preprocessor has the following tasks:

1. Syntactic and semantic analysis of the SPASP command input.

2. Counting and checking the data for correctness

3. Linking the different (source-) modules from the SPASP-library needed for the analysis

4. Calling for the FORTRAN 77 compiler

An error system was designed to cope with the different errors which can occur during these steps.

The following flow-chart represents the main features of the system:

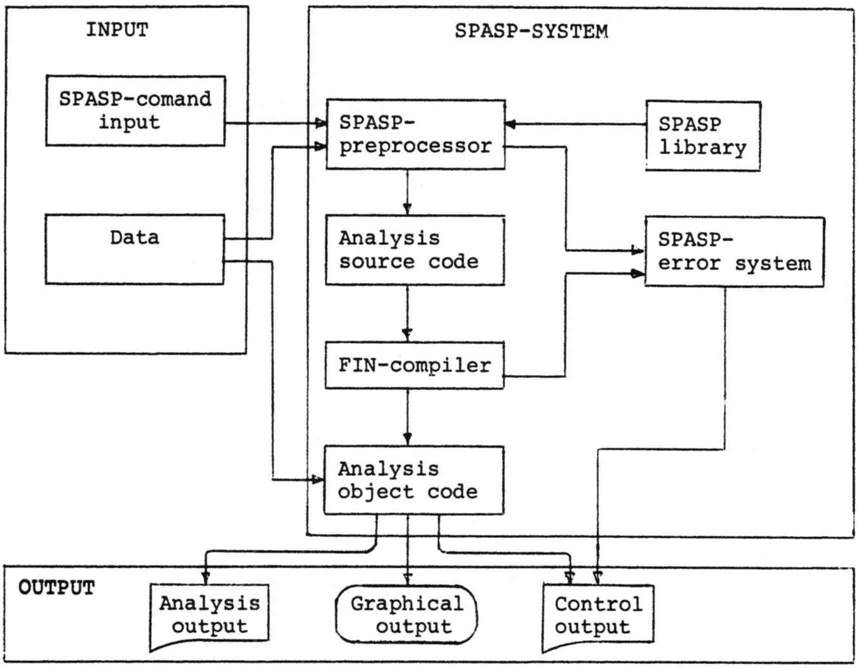

References

Collomb G., Estimation nonparametrique de la régression: Revue bibliographique, Int. Statist.Rev. 49, 75-93.

Devroye L.P., Wagner T.J. (1979), Distribution-free consistency results in non parametric discrimination and regression function estimation, Ann.Statist.8, 231-239.

Kubale R.,Schmidt F.W. (1982), Ergänzende Programme zur Dokumentation und Auswertung von Krankheitsverläufen, Statistical Software Newsletter 1, 1982.

Liptser R., Shiryayev A. (1978), Statistics of Random Processes Vol. 1,2, Springer, Berlin.

TOOLPACK: Improving the Programming Environment for Statistical Software

S.J. Hague, B. Ford and T.W. Lambert, NAG Central Office, Oxford, G.B.

Software tools can assist users to construct and refine applications software. The use of such tools by the Numerical Algorithms Group has increased our awareness of their benefits. The collaborative project entitled TOOLPACK will produce an integrated set of software tools which are relevant to developers of statistical software; some examples are cited.
Keywords: Software tools, portability, statistical software, machine independence, Fortran.

1. Introduction

Software tools are programs to assist in the development, testing, implementation, maintenance and distribution of computer software. These tools include:

dialect verifiers (e.g. PFORT)	formatters (e.g. POLISH)
static analysers	portability aids
data flow analysers	structuring tools
control flow analysers	language translators
program verification tools	probe insertors
test data generators	syntax-driven transformers

plus general text editors, string processors ...

Their use allows some phases of the software production cycle to be partially or wholly automated. If this mechanisation is successful it provides benefits such as:

- economy of programming effort, through relieving programmers from tedious and repetitious coding tasks
- greater flexibility by permitting the mechanical generation of software variants
- increased reliability, through more thorough testing
- more consistency, by the mechanical imposition of coding standards and layout conventions.

For further discussion of the potential benefit of automated program manipulation, and of the uses and availability of software tools, see Hague (1978) and Miller (1979).

2. A Summary of the Use of Software Tools by the NAG Central Office

The Numerical Algorithms Group develops and distributes a numerical and statistical algorithms library in FORTRAN, Algol 60 and Algol 68, the current FORTRAN version of which, NAG (1982), comprises over 460 user-callable routines. The total body of code including test software comprises over 200,000 source text records. The overall software management task is large; the Library is implemented on over 30 machine ranges and the advantages of a mechanised approach are apparent. See Ford (1979) for a description of the technical functioning of NAG as a whole - the processing aspects alone concern us here. Processing of FORTRAN code specifically is summarised in the rest of this section .

NAG code is received in the NAG Central Office from software authors working on diverse machines. Many of these have compiler extensions to the FORTRAN 66 standard; the PFORT verifier, Ryder (1975), is first used to detect use of these extensions. A program called DECS, developed in-house, is then used to insert explicit declarations of all variables in each segment - this is stylistically desirable, facilitates precision conversion, and reveals errors such as mis-spelt variable names. The POLISH program, Dorrenbacher et al. (1974), is used to impose a standard statement layout and to make such changes as statement label ordering, DO loop indentation and the collection together of FORMAT statements.

A DOUBLE PRECISION source code version is generated using the Automatic Precision Transformer APT, another in-house development.

3. The Toolpack Project

The use of a limited set of software tools has assisted NAG's software development and could also be applicable to software projects on a smaller scale, particularly if inter-machine range portability is of interest. However, these tools are not integrated in any sense. They are supplied or have been produced to meet a specific need, and not in an attempt to provide a software development environment. The provision of such an environment is the major motivation of Toolpack.

Toolpack is a collaborative project which began in 1979, see Cowell (1979) and involves six American institutions and NAG . Its aim is to produce a systematic collection of software tools to facilitate the development and maintenance of FORTRAN 77 programs. As such, it is of interest to many statistical software developers who are working with that language. The tools are being designed mainly for interactive use, and are intended to be transportable between host systems with minimal difficulty.

The following tool capabilities are regarded as constituting a sound minimal basis for a programming environment. Tool 0, the compiling/loading system is provided by the host environment. Tools 1 to 9 are now in various stages of development, and it is hoped to include most of them in the first public Toolpack release, planned for completion in 1983.

0. Compiling/loading system   5. Dynamic debugger

1. Fortran-intelligent editor
2. Formatter
3. Structurer
4. Dynamic tester and validator
6. Static error detector and validator
7. Static portability checker
8. Documentation generator
9. Program transformer

Further tools will be added in later releases; see Osterweil et al (1982) for current plans.

4. The Relevance of Toolpack to FORTRAN Statistical Software

Computational statisticians developing FORTRAN programs fall into three broad categories:

-individual programmers, writing software largely for their own use.

-algorithmic library developers, producing suites of subroutines which perform particular statistical tasks and are to be incorporated in other users' programs. The statistical chapters in the NAG Library are one such suite.

-package developers, constructing self-contained statistical systems which provide problem-orientated interfaces. GLIM, see Baker and Nelder (1978) and GENSTAT, see Alvey et al. (1977) are examples.

Toolpack will assist FORTRAN programmers in general, including these three categories of statistical computer users. We know statistical software development to which the use of currently available tools and the forthcoming facilities of Toolpack are particularly relevant.

4.1 Language Standardisation

Despite the advent of FORTRAN 77, a large proportion of statistical software is still written in some form of FORTRAN 66. Thse forms are usually extensions of the standard language. Statistical software writers have often used these manufacturer-specific extensions because FORTRAN 66 is lacking in several crucial respects, not least:

- no character data type
- no random access input/output
- no file handling facilities (e.g. dynamic open/close)

These defects have caused particular difficulties in statistical applications because such work is characterised by the treatment of large volumes of out-of-core data, of mixed numeric/character type.

FORTRAN 77 has remedied some of these shortcomings and despite other limitations such as the lack of a data structure facility is a more attractive language to the statistical programmer. Therefore, the advent of Toolpack will be an added incentive for such programmers to adopt FORTRAN 77.

However, there is wide divergence between the various extensions of FORTRAN 77 offerec by compiler suppliers. Thus the provision of a FORTRAN 77 equivalent of PFORT in Toolpack will be of importance in detecting non-standard coding, particularly in the context of software developments organised as collaborative activities between developers working on different machine ranges. In adition to validating program segments independently of one another, the verifier can also perform inter-segment checks which are often not performed by the compiling system. These checks include validation of actual parameter lists against dummy arguments, and consistency checks on the use of COMMON storage. Many statistical packages use large COMMON work areas for the storage of data and compiled or interpreted commands. Often, parts of these areas are EQUIVALANCEd to economise on storage on small systems. PFORT 77 will detect invalid or non-portable use of these language features.

The assertion that Toolpack will encourage users to adopt FORTRAN 77 begs the question of converting existing FORTRAN 66 programs. Currently there are no plans within Toolpack to build a "universal" FORTRAN 66 to 77 converter because of the diversity of FORTRAN 66 dialects. However, Toolpack will provide transformational tools which will enable users to construct their own language conversion utilities . For instance, if a suite of subroutines is found to be compatible with both language standards except for statements of the form:

```
DOUBLE PRECISION <name>
DATA <name> / 6H<six-character string> /
```

then an editing instruction such as the following (written in expanded notation for clarity) could be globally applied to the above statements so that the CHARACTER data type is used:

```
for:
   <double precision> <identifier> {1} <end-of-statement>
   <data> <identifier> {1} / <Hollerith>{length=6,value=<string>}/
substitute:
   <character> <identifier> {1} (6)
   <data> <identifier> {1}/<quoted>{length=6,value=<string>}/
```

producing pairs of statements of the form:

```
CHARACTER <name> (6)
DATA <name>/'<string>'/
```

4.2 Program Transformation

The Toolpack editor, Hague (1981), is intended to perform ordinary editing of textual data and more linguistically-significant tasks such as this conversion role. As a complementary facility to the editor, a programmable syntax-driven transformational tool is being developed for inclusion in Toolpack. This transformer, TAMPR Boyle et al. (1977) can apply grammar-preserving transformations and can provide fine control over formatting, as the transformed software is restored from parse tree form to source text form.

This offers to the software designer the possibility of coding at a generalised level, which for instance may include data structures not present in standard FORTRAN. It also allows the designer to be unspecific about how certain sub-algorithmic steps, such as forming sums of squares and products or updating a rank one matrix, are to be implemented. From the generalised form, specific realisations are produced by reliable transformations; one form perhaps emphasising an operational property over another (e.g. speed versus storage) or possibly being tailored for a special computing environment.

For regularly recurring program manipulation tasks specific "black-box" tools will be provided. The source text formatter is an example; it is also a transforming tool in that it can re-label the program uniformly, terminate DO-loops clearly and perform other tasks in its code standardising role. Another "black-box" transforming tool is the Automatic Precision Transformer, Hague (1980) which converts a program from single precision to double precision and vice versa. This is of interest to the statistical software writer because statistical algorithms often incorporate numerically sensitive operations. It is sometimes desirable to code part of an algorithm in DOUBLE PRECISION. For example, the computation of statistical distribution functions and their inverses may be required to high accuracy. This might be particularly true in the case of the Normal distribution, whose inverse is used as a data transformation in some applications.

4.3 Other Relevant Facilities in Toolpack

We have identified some tools which may be of special interest to the statistical software writer. No doubt the other tools to be provided will be found useful by the statistical computing community. Tools for version control and multi-variant storage may be of interest to package producers, assist in overcoming their software variant management problems. The lexer/parser generator facility of Clemm (1979) could also be of relevance. This will mechanically produce lexical analysis and parsing software for the language of the package, once the syntax rules have been devised.

A major design aim for Toolpack is extensibility; new tools can be added, whether they are written by users or provided through standard channels. Thus particular software communities, such as computational statisticians, may enhance the generally available Toolpack releases by facilities of particular relevance to them.

5. Conclusion

The provision of an integrated set of software tools as outlined here will be of benefit to statistical software developers and the users of their products.

References

Alvey N.G., et al. (1977), GENSTAT, a General Statistical Program, Rothamsted Experimental Station, Harpenden, Hertfordshire, U.K.

Baker R.J. and Nelder J.A. (1978), the GLIM System, Release 3, Numerical Algorithms Group, Oxford.

Boyle J.M., Dritz K.W., Arushanian O.B. and Duchevskiy, Y.U. (1977), A Program Generation and Transformation - Tools for Mathematical Software Development, Proceedings of IFIP 77 Congress, Toronto, Ed. Gilchrist B., North Holland.

Clemm G.M. (1979), FSCAN Report and User's manual, Department of Computer Science Technical Report CU-CS-166-79, University of Colorado at Boulder.

Cowell W.R. and Miller W.C. (1979), The Toolpack Prospectus, Technical Memorandum TM-34, Argonne National Laboratory.

Dorrenbacher J., Paddock D., Wisneski D. and Fosdick L.D. (1974), POLISH - A FORTRAN Porgram to Edit FORTRAN Programs, Department of Computer Science Report CU-CS-050-74, University of Colorado at Boulder.

Ford B., Bentley J., Du Croz J.J. and Hague S.J. (1979), The NAG Library "Machine", Software-Practice and Experience, **9**, 65-72.

Hague S.J. (1978), Software Tools, Proc. of Conference on "Numerical Software:Needs and Availability", Brighton 1977, Ed. Jacobs D.A.H., Academic Press, London.

Hague S.J. (1980), APT-X-Automatic Precision Transformer, Toolpack Memorandum SJH430, NAG Central Office, Oxford.

Hague S.J. (1981), The provision of Editors for the Manipulation of FORTRAN, Toolpack Memorandum SJH11112, Numerical Algorithms Group, Oxford.

Miller E. (1979), Tutorial Automated Tools for Software Engineering, Proceedings of Compsac 79, IEEE Computer Society, Long Beach, California, 90803.

NAG FORTRAN Library Manual, Mark 9 (1982), Numerical Algorithms Group, Oxford.

Osterweil L.S., Hague S.J. and Miller W.C. (1982), Toolpack Architectural Design : The User's Perspective, Applied Maths. Division, Argonne National Laboratory.

Ryder B.G. (1974), The PFORT Verifier, Software-Practice and Experience, **4**, 359-379.

A Programming Technique for Software in Statistical Analysis

R. Haux, University of Heidelberg, F.R.G.

SUMMARY: A new programming technique, called 'the stepwise reduction of data structures' is proposed for the development of software, especially in statistics. It will be demonstrated, using some well known examples (PEARSON's χ^2-test, KRUSKAL-WALLIS test), that the programming technique is easy to use. One of the advantages of this technique is that the user can call a program (that deals with statistical analysis) with different kinds of input data structures (and not only with one kind as conventional). In spite of this fact the effort for the programmer is usually not larger than that needed to develop a conventional program. It will be shown that the stepwise reduction of data structures can already be used within application oriented languages, like FORTRAN or PASCAL. It complements very well with WIRTH's stepwise refinement for the purpose of writing well structured programs. Finally, it will be outlined how to design statistical analysis systems so that this technique can also be used within those systems, for greater flexibility and better detection of errors.

KEYWORDS: stepwise reduction of data structures, statistical analysis systems, program packages, statistical computing, programming technique, method programming, data structures, χ^2-test, KRUSKAL-WALLIS test.

1. INTRODUCTION

As WIRTH (1971) stated, the creative activity of programming can be considered as a sequence of design decisions. In WIRTH's stepwise reduction of data structures and operations, the program construction consists of a sequence of refinement steps, where in each step, one or several instructions of a program are decomposed into more detailed instructions.

In statistical computing, various kinds of data structures have to be taken into account, for finding out design decisions. In programming an (analysis-) method, respectively procedure, it should be taken into account, that data do not always exist in a case-oriented form, but also in the form of tables etc.
A method programmer, who designs an analysis method for a statistical analysis system (like BMDP, SAS, SPSS etc.; for references see FRANCIS, 1977) only considers the possible case-oriented structures. So, if data is not stored in this way, it is often very troublesome and complicated for a user to transform the data into such a case-oriented form.

In this paper, I want to propose a programming technique, which is very simple, and which takes into account the variety of kinds of data structures in statistical computing. This technique, called the *stepwise reduction of data structures* has, in addition, the property that the effort in programming a method is usually not larger than in convenient programming.
After some basic definitions in section 2, the technique will be introduced and demonstrated in section 3. Discussion can be found in section 4.

2. BASIC NOTATION

A *statistical analysis system* consists of a *database* and a *method-base*. Both are created and managed by a program (or by a set of programs), called the *data- and methodbase management system (DMBMS)*. The DMBMS can activate *methods* (procedures). In this context, a method is a program-module (subprogram, main program, ...), which produces a certain performance, essentially: statistical analysis. It will be assumed, that a method has the possibility to call other methods. The database of a statistical analysis system contains a set of *data structures* (temporarily or permanently), which are not necessarily related. For a user and for a method programmer a statistical analysis system is characterized by its *interfaces,* among others: its language(s), usually the *command language* of the user (e.g. the language of BMDP,...) and the *language of the method programmer* (e.g. FORTRAN, PASCAL, PL/I). In the sense of SAMMET (1969), command languages will be called *special purpose,* languages like FORTRAN *application oriented.*
Note, that the term 'data structure' is used here in the logical, conceptual sense, not in its physical meaning (ANSI/SPARC, 1975).
A statistical analysis system will be called *integrated* if a DMBMS is given (like in BMDP, SAS, SPSS). It will be called *not integrated,* if the user has to do the data- and/or methodbase management (like in subroutine packages as IMSL, NAG (for references see also FRANCIS, 1977) or like in private libraries)).

3. THE STEPWISE REFINEMENT OF DATA STRUCTURES

3.1 INTRODUCTION

The design decisions for a stepwise reduction of data structures can be outlined in the following way; for programming a certain method one has to

- find all possible (and meaningful) input data structures, respectively groups of input data structures, as well as the (group(s) of) output data structure(s), which contain the results of the method (perhaps the test statistic, degrees of freedom, p-value, status or error messages etc.) then one has to
- try to find a 'path', that each input data structure (group) can reach the output data structure (group(s)) via one or several *reduction steps*. The paths of the input data structure (group(s)) to the output data structure (group(s)) should be minimized.

Reduction means in this article: There exists a definite rule, that a (group of) data structure(s) can be transformed into another (group of) data structure(s). This transformation rule can be realised as a program. To get a clearer idea of the stepwise reduction of data structures, the technique will now be demonstrated in designing methods for two well known tests: PEARSON's χ^2-test (see e.g. BISHOP et al., 1975) and the KRUSKAL-WALLIS test (KRUSKAL and WALLIS, 1952). It will be assumed, that the reader is familiar with the statistical model of the test as well as with the construction of the test statistic.

A more detailed description of the proposed programming technique can be found in HAUX (1982).

3.2 Example 1: PEARSON's χ^2-Test

One possible input data structure, DS1 say, for the χ^2-test is a sequence of records. Each record represents a 'case' and contains, e.g., three variables: the case-identification, the two grouping variables with domains $1,\dots,R$ and $1,\dots,C$, $R \geq 2 \wedge C \geq 2$. This input data structure is usually stored as a file (WIRTH, 1976), respectively data matrix (NELDER, 1969). It could also be stored as a relation (CODD, 1979).

A second data structure, DS2 say, is the $R \times C$ table, containing the frequencies. Usually, this data structure is stored as a two dimensional array of non negative integers. The output data structure for the test, DS3 say, can e.g. consist of the test statistic, the degrees of freedom, etc. and can be stored in a record.

One immediately recognizes, that DS1 can be reduced into DS2 (but DS2 not into DS1), and DS3 can be reduced from DS2. Figure 1, called *reduction diagram*, shows this connections.

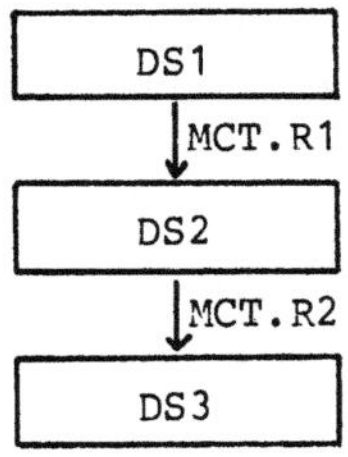

Figure 1: Reduction diagram for a χ^2-test method.

MCT.R1 and MCT.R2 are programs, respectively methods. For a user who has the data stored in DS1, a method 'χ^2-test' consists of MCT.R1 and MCT.R2. The user will call MCT.R1, this 'sub-'method will then call MCT.R2. For a user who has the data stored in DS2, a method 'χ^2-test' consists only of MCT.R2. Note, that a reduction from DS1 to DS3 is not necessary, because the program corresponding to this reduction can be replaced by MCT.R1 and MCT.R2.

3.3 Example 2: KRUSKAL-WALLIS Test

The first possible input data structure, DS1 say, is also the case-oriented one, perhaps consisting of a sequence of records each record containing an identification variable, a treatment variable and a variable for the criterion. It is also possible that the data are organized case-oriented, but that the criterion has already been replaced by its rank (DS21). Another possible data structure is a $K \times n_k$ table with the $K \geq 2$ treatments and the $n_k > 0$ realisations of random variables for the k-th treatment, $k = 1,\dots,K$ (DS22). The criterion in DS22 could also have been replaced by its rank (DS3). Lastly, there is, of course, an output data structure containing the results (DS4). The reduction diagram for the method 'KRUSKAL-WALLIS test' is given in figure 2.

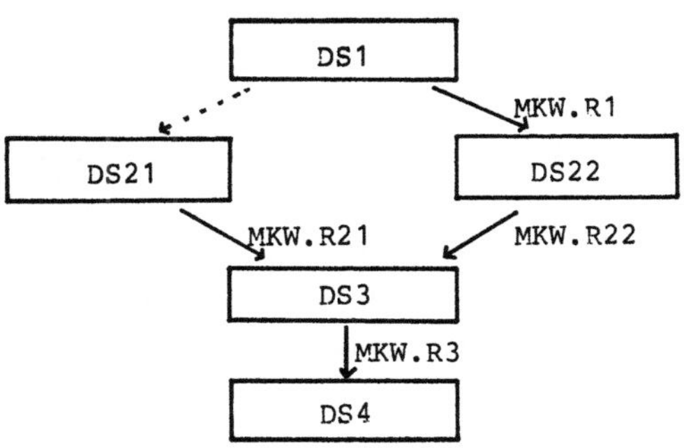

Figure 2: Reduction diagram for a KRUSKAL-WALLIS test method.

For a user who inputs DS1, the method consists of MKW.R1, MKW.R22 and MKW.R3. For a user who has the data stored as DS3, e.g., the method consists only of MKW.R3. A reduction of DS1 to DS21 is not necessary, because DS4 can be reached from DS1 via DS22. Of course, the paths DS1 to DS4 via DS21 or DS1 to DS4 via DS22 are arbitrary. The method programmer has to decide which way seems to be more appropriate.

4. OUTLOOK AND DISCUSSION

Of course, the examples are simplified. Taking the KRUSKAL-WALLIS example, one can see that there is an equivalent effort in programming MKW.R1, MKW.R22 and MKW.R3, or in programming a method, that inputs only the case-oriented data structure and computes the results, without modularizing the method through reduction steps. In the first case, three input data structures are considered in the second case only one. If the method programmer wants DS21 to be available, too, there is some additional effort: to program MKW.R21. Within each 'sub'-method the stepwise refinement can be used.

The technique of stepwise reduction can already be used within application oriented languages. A more complicated test has been programmed by the author in FORTRAN and is documented in HAUX, SCHUMACHER and WECKESSER (1981). There, each sub-method corresponds to one or several subroutines.

Within the commonly available integrated statistical analysis systems, there is no support to stepwise reduction, because a user can usually only define case-oriented data structures - data matrices - in the system's command language. And there is also commonly no possibility, that in integrated systems methods can call other methods (in example 2 MKW.R22 had to call MKW.R3, e.g.). So, for greater flexibility, a DMBMS of an integrated statistical analysis system should provide a sufficient variety of kinds of data structures and a better management of its methodbase, as well.
A better management of the methodbase and more modularized methods would, in addition, enable the DMBMS to undertake a better error detection. At each method call for example, it could check the contents of the input or output data structure(s), e.g. via schema definitions and/or specified integrity constraints (HAUX, 1981).

Finally, I want to remark, that the stepwise reduction of data structures is *not* dependent on a specific statistical analysis system. It is only depending on the characteristic of the language, either special purpose or application oriented, and on the variety of possible data structures. For the sake of the user it could be hoped that statistical analysis systems, especially the integrated ones, will be improved along these lines.

ACKNOWLEDGEMENTS: I would like to thank my colleagues from the department of Medical Documentation, Statistics and Computer Science (University of Heidelberg) and from the department of Clinical Documentation (University of Ulm) for many helpful discussions.

REFERENCES

ANSI/SPARC (1975). ANSI/X3/SPARC study group of database management systems - interim report. FDT Bull.ACM SIGMOD 7, No.2.

BISHOP, Y.M.M., FLIENBERG, S.E. and HOLLAND, P.W. (1975). Discrete multivariate analysis. Cambridge (Mass.).

CODD, E.F. (1979). Extending the database relational model to capture more meaning. ACM TODS 4, 397-434.

FRANCIS, I. (Ed.,1977). A comparative review of statistical software. IASC, Voorburg (NL).

HAUX, R. (1981). How to detect and prevent errors in computer supported statistical analysis. Paper, given at a meeting on 'design criteria for statistical analysis systems' at the University of Heidelberg (to be published).

HAUX, R. (1982). The use of complex data structures in statistical analysis systems. Doctoral thesis - in German.

HAUX, R., SCHUMACHER, M. and WECKESSER, G. (1981). A rank test for complete block designs. Preprint No. 134 of the 'Sonderforschungsbereich stochastische mathematische Modelle', University of Heidelberg.

KRUSKAL, W.H. and WALLIS, W.A. (1952). Use of ranks in one-criterion variance analysis. JASA 47, 584-621.

NELDER, J.A. (1969). The description of data structures for statistical computing. In: MILTON, R.C. and NELDER, J.A. (Eds.). Statistical computation.

WIRTH, N. (1971). Program development by stepwise refinement. CACM 14, 221-227.

WIRTH, N. (1976). Algorithms and data structures. New York.

A Comparison of Types of Database System Used in Statistical Work

G.R. Hext, Civil Service College, London, G.B.

SUMMARY

Eleven aspects of database systems as they apply to statistical work are discussed and categorised. They are applied to thirteen existing systems, some statistical, some more general; the reader can apply them to other systems.

KEYWORDS

Computer packages, computer systems, statistical analysis, database management.

1. Introduction

While early computer systems for statistical analysis such as BMD and SPSS could fluently analyse only the cases of one static rectangular data matrix at a time, more recent developments in survey, census, and business inquiry work use data that may be changing, and may have more complicated structure. This has led to the addition of database facilities to many statistical systems, the development of database systems especially for statistical work, and the use of more general systems in statistical applications.

This paper looks at a wide range of systems, typical of most that are in use for survey, census and inquiry work, and classifies their data handling features in a variety of ways, shown at the end of the paper.

2. The systems considered here

The chosen systems are mostly available on a variety of machine ranges, including ICL; they are either widely known and used in their own right, or are typical of a class of database systems. They are:

SAM	typical of systems developed on micro-computers
SPSS, BMDP, Genstat, Minitab	typical of systems primarily for statistical analysis
Package X, P-STAT, SAS	typical of systems for statistical analysis with considerable power to manipulate hierarchic and network (within each case) data held as data matrices.
SIR	a system specifically designed for handling statistical data of complicated structure within and between cases
RAPPORT	typical of general relational-like database systems
IDMS	typical of Codasyl database systems, and here also representing others where linkage is through pointers
RAPID	typical of systems using variable-by-variable or "transposed" files
CAFS	an example of specialised hardware.

With the advent of the cheap "chip" designers are developing hardware to remove from the central processor certain types of operation that hitherto were carried out there by software. For statistical processing an attachment to disc drives such as CAFS ("Content Addressing File Store") that will examine records and only send to the central processor those satisfying given logical criteria is of particular interest. For so often when handling large censuses or surveys the analyst may only need a sub-group such as pensioners (who in the United Kingdom must satisfy the relatively complex criterion of being the women aged 60 and over and men aged 65 and over). Hardware like CAFS may well complement and enhance statistical software, not displace it.

3. Database features

Features of the data handling parts of these systems, with notes and abbreviations for the table below, are as follows:

o. Data independence. The data should be stored and capable of access independent of any pre-specified use. This holds for all the systems considered here within the limitations of the structures they can handle, and so no further reference to it need be made.

a. Data structure. The basic structure for statistical work is the data matrix (DM), the rows of which hold information on the different cases or units of analysis, and each column the values of the same variable. The next structure, in order of complexity, is the hierarchy (H); next comes data of network structure within each case (N), and finally networks both between and within cases (NN). Here the concept of the case is extended to cover the major unit of analysis (for example the household, which may include lesser units such as cars, persons, their visits to a doctor, and the trips that people take in their cars).

A statistical database of relational structure, or a "relational database" is, in essence, a group of related data matrices (RDM).

b. Storage and linkage. To the user data of complex structure appear to be held either

i. case-by-case (CBC) where, for example, the information on a given household, its people, cars and trips, are all held together; or

ii. as a group of related data matrices, one or more matrices holding all the household information, other(s) all the people, other(s) the cars, and other(s) all the trips.

Under ii. the information on each case must be brought together for analysis, and methods of linkage are

- by position (POS), that is, the cases all occur in exactly the same order and position in data matrices that are being linked;
- by key (K) or other identifying feature, for example, the serial number of the household and, if necessary, of the person it;
- by pointers (POI) built by the system into each case as it is brought into the database, giving the computer addresses of the associated cases in other data matrices.

Pointer-based systems require more machine power to hold and maintain the pointers, but are fast in well-specified applications. Relational systems and others that link by key-matching at run-time offer greater flexibility in use and are easier to maintain. Matching by position is simple to implement but of limited use.

c. Data dictionary, schema. These describe the structure of the data to the system. At their simplest they may hold the list of VARIABLES in each file (V), perhaps supported by extended VARiable LABELS and for categoric variables their VALUE LABELS (L). More powerful systems may also hold information on keys (K), allowed values and ranges (R) and expected quantities of information (Q), while in Codasyl and other pointer-based systems these will also specify the linkages between different data matrices (P).

d. Languages. Those systems primarily or exclusively designed for statistical work probably have their own "very high level language" commands (COM) for operations such as REGRESSION, SCATTERGRAM, ANOVA. Their database manipulation commands will also be in this same language, for example, COLLATE, ADD CASES, SAVE. Within this framework there will probably be arithmetic and logical facilities looking much like those of Algol, Fortran or Cobol for deriving new variables, selecting cases and similar functions.

By contrast the "host language" systems (HLS) appear to a programmer as a high-level language (such as Fortran, Cobol) in which extra commands are available for accessing the database and navigating around it. For quick and not too complicated enquiries of such a data base a "query language" (QL) is usually provided, saving the user the need to write his own retrieval programs in the extended high-level language.

e. User interface. In the early days computers could only handle work as "Batch jobs" (B). As facilities for interaction between the user and the computer have developed so statistical systems have taken advantage of this. Users can give commands to their system (I); in some the system can also draw out their requirements by a "conversation" (C). Also batch systems have been made to appear interactive by giving the user an interactive command editor (E) for setting up and running groups of commands.

For systems using a host language the user interface will be specified by the user or his computer staff.

f. Data take-on, update, data vet, integrity. Many systems for statistical analysis assume that the data are static and "clean" (S), that is, no items will be added or deleted, and those that are there are checked and corrected as far as possible for interviewer and coding errors, logical inconsistencies and so on. A true database system should allow for its information being dynamic (D), and liable to change, and should provide for keeping it correct and consistent. It may trap errors "now" (N), as data are being submitted, or only later (L) in a run of a separate "data vet" program. The former would be particularly important if one or more users are allowed to update the data while others are analysing it.

g. Security, simultaneous access. Simple statistical systems will depend on passwords in the operating system to stop unauthorised access to groups of files or individual files (OS). Those for database use may give much more detailed protection, down to individual variables, to groups of cases, and different degrees of access to different people according to their needs (DB).

Most operating systems allow simultaneous access to people reading a given file, but only one person may access it to alter it (OS). Database systems will allow much more simultaneous access, and can be programmed to prevent simultaneous and/or inconsistent changes (DB).

h. Recovery. What happens if either the hardware or software breaks down? For a static file this will not matter provided a back-up copy is held - but what about databases that are being changed? Regular dumps (D), the keeping of a "journal" of transactions since the last dump (J) and a "recovery" utility (R) to recreate the latest state of the database from these two, are the facilities needed.

i. Portability. In a large organisation like the UK Government, computers of many manufacturers' ranges are used, and yet data and software should be widely available. So the ease with which these can be transported is very important.

i. Program portability (P) will be between machine ranges. This is achieved through a widely used high-level language (almost always Fortran), preferably with some tool for translating between the dialects and machine ranges.

ii. For data portability (D) we assume all systems can produce and read alphanumeric data matrices in fixed format. So we look for portability from one system to another in some internal format that carries the variable names, labels, and other descriptions, thus sparing the user from resubmitting them all. The SPSS system file is one very popular format. Such files must stay on the one machine range.

iii. One system, SIR, offers data portability across machine ranges (DR), the format being its own system file.

j. Statistical facilities. The simplest output from any of these systems would be a data matrix (M) for analysis elsewhere. In the host-language systems this facility, as well as any statistical analysis, must be programmed in. The statistical facilities in the other systems range from descriptive statistics and scattergrams (S) through tabulation (T) and regression (R) to many forms of advanced analysis (A).

4. Conclusion

Very broadly the systems are ordered in the Table by increasing power as database systems. As this power increases so the statistical facilities diminish. The most powerful are general-purpose systems, which are not specifically designed for statistical work, and use a host language (with all the programming work this entails), not a command language.

For many statistical purposes SIR offers a very good combination of database power and ease of use, on a wide range of computers. Next best, also widely available, may be P-STAT - with it some database power is lost, but more statistical facilities are included.

CAFS or similar hardware is well worth trying in conjunction with the other systems, particularly for the larger data sets. Such a combination could reduce the need for "transposed" file systems, as they give comparable speed of access without losing the advantages of case-by-case storage.

5. Disclaimer

The views expressed in this paper are those of the author, and not necessarily those of the Civil Service College.

TABLE: ASPECTS OF SOME DATABASE SYSTEMS USED IN STATISTICAL WORK

	Structures	Storage, Linkage	Dictionary	Languages	Interface	Data take-on	Security	Recovery	Portability	Statistics
Minitab	DM	-	-	COM	I, B	S (L)	OS	-	P	S, T, R, A
BMDP	DM	-	V, L	COM	B	S	OS	-	P (D)	S, T, R, A
SAM	DM	-	V	COM	I	D, L	-	-	P	S, T, R, A
SPSS	DM	POS	V, L	COM	B	(D) (L)	OS	-	P, D	S, T, R, A
Genstat	DM (a)	-	V, L	COM	B	S (L)	OS	D	P	S, T, R, A
Package X	RDM H, N	POS K	V	COM	I, C B	D, L	OS	D, J	P	S, T, R
SAS	RDM H, N	POS K	V, L	COM	B, E	D, L	OS	D	D	S, T, R, A
P-STAT	RDM H, N, NN	POS K	V, L K	COM	B, E	D, L	OS	D, J R	P, D	S, T, R, A
SIR	DM H, N NN	K	V, L K, R, Q	COM HLS	B, E	D, N, L	DB	D, J R	P, D, DR	M, S, T
RAPID	RDM	POS	V, L, K *	HLS	I, B ∅	D, L	DB	D, J R	-	M
RAPPORT	RDM, H N, NN	K	V, L, K, Q	HLS QL	∅ I, B	D, L	DB	D, J R	P	∅
IDMS	RDM H, N, NN	POI	V, P	HLS QL	∅ I, B	D N∅, L	DB	D, J R	-	∅
CAFS	DM (b)	-	V, L	QL HLS	I, B ∅	S	-	-	-	M

Notes * denotes exceptional strength

( ) denotes very modest presence

∅ denotes a facility to be specified by the user and programmed by his staff

(a) Genstat also handles an exceptionally wide range of structures, such as tables, correlation matrices

(b) CAFS also handles "header and trailer" records

ABBREVIATIONS USED IN THE TABLE

a. Structure

- DM Data matrix
- H Hierarchy
- N Network within
- NN Network between and within
- RDM Related data matrices

b. Storage and linkage

- CBC Case by case
- POS Position
- K Key
- POI Pointers

c. Data Dictionary

- V Variables
- L Labels
- K Keys
- R Values and ranges
- Q Quantities
- P Pointer links

d. Languages

- COM Command
- HLS Host-language
- QL Query language

e. User interface

- B Batch
- I Interactive - command
- C Interactive - conversational
- E Editor
- S As specified

f. Data take-on, etc

- S Static
- D Dynamic
- N Trap "now"
- L Vet "later"

g. Security, etc

- OS Operating system
- DB Database system

h. Recovery

- D Dumps
- J Journal
- R Recovery utility

i. Portability

- P Program
- D Data
- DR Data across machine ranges

j. Statistical facilities

- M Data matrix
- S Descriptive statistics, etc
- T Tabulations
- R Regression
- A Advanced analysis

References

Minitab Reference Manual, January 1980. T A Ryan, B L Joiner, B F Ryan. Statistics Department, Pennsylvania State University, PA, USA

BMDP Main Reference Manual (1981) Ed W J Dixon and Mary Ann Hill. University of California Press, CA, USA

SAM: User's Guide to SAM - Statistical Analysis for Micro-computers. Version 81.7 (1981) Ed R Mays, PO Box 160, Welwyn Garden City, England

SPSS: Statistical Package for the Social Sciences (2nd Edition, 1975). N H Nie et al. McGraw Hill

Genstat Manual (Release 4.01) (1977) N G Alvey et al. Rothamsted Experimental Station, Harpenden, England

Package X Practitioner's Handbook (Release 6). (1982) Central Statistical Office, London SW1

SAS User's Guide, 1979 Edition. SAS Institute, Cary, NC, USA

P-STAT 78 User's Manual. S and R Buhler, Princeton, NJ, USA

SIR (Scientific Information Retrieval) User's Manual, Version 2. (1980) B N Robinson et al. Evanston, IL, USA

RAPID Various guides and papers. Statistics Canada, Ottawa

RAPPORT-3 User Manual; Interactive Query Language Manual (1982). Logica, Ltd, London W1

IDMS Various guides. ICL, London SW15

CAFS Various guides. ICL, London SW15

The Frontiers of Statistical Analysis

B.L. Joiner, University of Wisconsin, U.S.A.

SUMMARY

This paper focuses on three of today's frontiers of statistical analysis. These are: developing better methods for data checking, being more creative in data plotting, and learning to understand how to format and document results so that understanding is enhanced. Progress on these frontiers will require a broader view of statistics and of statistical research as well as careful study of the statistical analysis process. We will also put much more of our research emphasis on the effective use of computers.

Key Words: Data checking, plotting, formatting or results, computers, statistical analysis process.

1. INTRODUCTION

Some of the most challenging and important research areas facing statisticians today are:

* Developing better methods for data checking
- Dealing with non-statistical errors in data
- Handling data with complex structures
- Analyzing data with complex statistical error structures
* Being more creative in data plotting
- Automating assumption checking as much as possible
- Documenting data sets
* Understanding how to format and document results
- Developing empirical models for data sets having many variables
- Analyzing very large data sets
- Analyzing data in Real-Time.

The emphasis in this paper is on the three starred topics: data checking, data plotting and the formatting of results. Data plotting and data checking are considered in the next section; the formatting of results is treated in Section 3; and some ideas on where we might go from here are in Section 4.

2. DATA PLOTTING AND CHECKING

This section consists of four examples; each is followed by lessons learned and implications for what we might do to improve matters. Three of these examples are drawn from the author's consulting experience.

Example 1. Efficiency of Nuclear Power Plants

Would nuclear plants be cost effective generators of electricity? Historically the efficiency of nuclear plants had been low but proponents argued that the newer reactors were more reliable, and further, that each reactor became more reliable as it got older. The data we had were triangular-shaped: the oldest reactor had been operating 16 years and the youngest had only one year of operation.

We worried about autocorrelations, within and between components of variance, year-to-year effects and many other technical sorts of issues. We worked hard to address these issues but on the witness stand the credibility of our analysis was sharply reduced because of numerous errors in the data base.

The main point illustrated by this example is the paramount need for concern about data quality. Our principle suggestion is that the analysts should pay close attention to data quality, and should spot check values against reliable sources whenever possible. Secondly, more research effort should be devoted to making it easier to fit complex models that take into consideration such factors as autocorrelations, within and between components of variance, etc.

Example 2: Oxygen Content of Steel Rods

Steel rods were to be sold by the U.S. National Bureau of Standards as a way for others to calibrate their instruments. It was hoped that the oxygen content was homogeneous; as a check 20 rods were haphazardly selected and two measurements were made on each. A one-way analysis of variance revealed no statistically significant inhomogeneity, but showed more variation in oxygen content than was expected. Part of the analysis focussed on the time order in which the measurements were made. The measurements were plotted, for example, versus measurement order, versus order with different coded symbols for different days, versus order with order-in-day coded, versus order-in-day and versus order-in day with day coded. One of these plots, measurement versus order-in-day, showed a sharp decrease in values as the day wore on, thereby pointing to the presence of a serious "lurking variable" (see Joiner, 1981, for details).

The points illustrated by this example include the advantage of the computer in making plotting simple, the desirability of making many different plots and the desirability of plotting data versus time order. We suggest that more emphasis be put on developing software that will automatically make a variety of plots, or at least suggest a menu of plots which we might want to consider.

Example 3. Anscombe's Data Sets

Anscombe (1973) constructed a series of data sets with a dependent and independent variable. The X-Y plots of these data sets show sharp differences in the patterns yet the regression output for each is exactly the same--the same coefficients, standard errors, variance-covariance matrix, R^2, and so on. These data sets provide a dramatic example of the pun, "Regression without plotting is _true_ regression."

Models should never be fit without plotting both the data and the residuals. In addition, we should seek to develop software that will automatically do as much assumption checking as practical. When a possible violation of assumptions is sensed, the computer should suggest a menu of plots for us to consider.

Example 4. Electricity Pricing Experiment

We analyzed an enormous data base--over 100 million values--from an electricity rate experiment. Numerous problems were found with the data including some we accidentally created. For example, we shifted the data the wrong way by one hour when daylight savings time changed and we somehow got one third of the customers off by 8 hours and one-third off by 16 hours at the time the test electricity rates started. The apparent effect of this accidental shifting was a flattening of the daily load curves--which the test rates were designed to achieve. Luckily the error also flattened the load curve for the control group, and so it was obvious that there were data problems. (See Pollack and Joiner, 1979, for more details.)

This example illustrates the compounding of problems that occurs with very large data sets. We suggest that key components of the data should still be plotted, even with enormous data bases; here computers are absolutely essential. One must expect that all data sets will have serious errors and further that more will be introduced in the analysis. Intermediate and final answers should be carefully checked. Finally, one should recognize that data preparation and data checking are very important and time-consuming aspects of the statistical analysis process and should not be frustrated by the amount of time and effort that must be devoted to these areas, especially with large and complex data sets. After all, "Fisher himself had said somewhere that the first responsibility of a statistician is to cross-examine his data" (Mahalanobis, 1964).

It seems possible to automate some aspects of data checking such as the following: range (are all values in bounds), outliers (in one or more dimensions), order (are all values of a variable in order except for a few), granularity (are all values except a few integers, halves), days (are all values except a few weekdays, weekends), change (maximum change between adjacent values, in 24 hour period), and runs (of identical values, up and down, etc.).

It may also be desirable to begin storing a veracity flag with each datum, such as: Missing, Known to be wrong, Lower bound--real value is at least this large, This one looks suspicious--remind me of this if it ever seems doubtful in some other way, Based on only 1/2 month's data, Imputed from other data, and so on. It is noteworthy that the 1980 U.S. Census tapes do have a flag for each datum indicating whether or not it was imputed.

3. FORMATTING OF RESULTS

The precise format in which results are presented to us can make a big difference in our understanding of what the data really have to say. Ehrenberg (1978,1981) has much to say on this point. An example is given by the stem-and-leaf display in Tukey (1977, p. 20). The displayed values are the highest elevations in each of the 50 states of the U.S. When the extreme points in the plot are labelled, our understanding is greatly enhanced. Such labels on extremes and clusters help us recognize patterns due to other, perhaps unrecorded, factors. Computer software should begin to do this labelling for us automatically.

Ehrenberg's rules have direct applicability to the design of computer software. As an example, the results from a stepwise regression have traditionally been printed much as the results of an ordinary regression analysis. In some cases, each step (model) results in a page or two of output. In other cases the intermediate results are ignored and the full output is printed only for the last model. In Minitab (Ryan, Joiner, Ryan, 1981) key results from each model are put in one table (see Exhibit 1). This seems to be a substantial improvement, but further refinement may still be possible. Exhibit 2 shows one possibility. The key changes are that the things we should compare most, e.g., how coefficients change across models, are now in columns rather than rows, and the t-ratios have been moved to a separate table. The R^2 values have also been rounded to two effective digits as Ehrenberg recommends. Note, however, that this would present a problem to some users who would prefer five or more digits for some purposes. One answer would be to give the user control over the number of digits with the default being along the lines prescribed by Ehrenberg.

One lesson seems to be that computer output should be desiged to encourage comparisons across models, across estimators and across data subsets.

4. CONCLUSIONS

Some specific analysis frontiers have been mentioned. It would appear that there are many more of a similar nature. Apart from these specifics it seems as if making progress on these frontiers will require

- A broader view of statistics and of statistical research,
- Careful study of the statistical analysis process as practiced by good statisticians,
- Much knowledge of computers and computer science

Exhibit 1

Stepwise Regression Output from Minitab

Key aspects of all four models are put in one table.

Step	1	2	3	4
Calcium	-0.738	- 0.614	-0.237	
T-ratio	-4.77	-12.62	-1.37	
Aluminum		1.44	1.45	1.47
T-ratio		10.40	12.41	12.10
Silicon			0.416	0.662
T-ratio			2.24	14.44
S	8.96	2.73	2.31	2.41
R-SQ	67.45	97.25	98.23	97.87

Exhibit 2

Improved(?) Stepwise Regression Output

Comparable values are put in columns rather than rows.

Step	Calcium	Aluminum	Silicon	%R SQ	S	t-ratios Ca	Al	Si
1	-.74			67	9.0	5		
2	-.61	1.44		97	2.7	13	10	
3	-.24	1.45	.42	98	2.3	1	12	2
4		1.47	.66	98	2.4		12	14

5. ACKNOWLEDGEMENT

Many thanks are due to Alison K. Pollack who contributed much to the ideas and exposition of this paper.

6. REFERENCES

Anscombe, F.J. (1973), "Graphs in Statistical Analysis," Amer. Stat., 27, 17-21.

Ehrenberg, A.S.C. (1978), Data Reduction: Analyzing and Interpreting Statistical Data (rev. reprint), John Wiley, New York.

——————(1981), "The Problem of Numeracy," Amer. Stat., 35, 67-71.

Joiner, B.L. (1981), "Lurking Variables: Some Examples," Amer. Stat., 35, 227-233.

Mahalanobis, PC. (1964), "Some Personal Memories of R.A. Fisher," Biometrics, 20, 368-371.

Pollack, A. and Joiner, B.L. (1979), "All Data Sets Have Errors--Well Almost All," Modelling and Analysis of Electricity Demand by Time-of-Day, Electric Power Research Institute, Palo Alto, California.

Ryan, T.A., Joiner, B.L. and Ryan, B.F. (1976), Minitab Student Handbook, Duxbury Press, North Scituate, Mass.

Tukey, J.W. (1977), Exploratory Data Analysis, Addison-Wesley, Reading, Mass.

Implementation of a Statistical Database System: HSDB

Y. Kobayashi, K. Futagami and H. Ikeda, Hiroshima University, Japan

Summary: This paper describes an implementation of a statistical database system with real-time response for interactive use, extended data dictionaries, data editing facilities, and special security features. Especially a new approach, use of summary tables, is proposed to improve the responsibility and data dictionaries are expanded to manage special information in statistical environment. The system enhanced interactivity to statistical analysis and maintainability of statistical data rather than a conventional statistical package or a statistical application system of a conventional data management system.

Keywords: statistical databases, derived statistical data, data dictionaries, interactive statistical analysis

1. Introduction

The statistical packages enable non-programers to analyze statistical data easily. It is not so easy to maintain the data in these packages. We have some statistical packages which use a conventional database management system (DBMS). But it is not still sufficient to manage the whole environment around statistical data. A new system or package is required to support the statistical environment effectively. Such a system is often referred as a statistical database system.

Our previous paper, Ikeda and Kobayashi (1981), discussed the following requirements of a statistical database system:

(1) real-time response to interactive analysis,

(2) data derivation by statistical computations,

(3) data aggregation features,

(4) special security features for statistical data, and

(5) preparation of various statistical functions.

There are already many systems or techniques which attain the requirements partially. RAPID of Statistics Canada (Hammond (1981)) adopted a transposed file organization to improve the responsibility for exhaustive searches, which often occurs in statistical analysis. SAS (Bragg (1981)) also succeeded to support interactive use by a special file organization. A data compression approach by Shoshani and Eggers (1980) and a random sampling approach by Burnett and Thomas (1981) also enhanced responsibility of interactive use. Especially the random sampling approach is an excellent property for responsibility, that is, the response time of data analysis is independent on the quantity of data. This property is referred as quantity independent responsibility. But the random sampling has a disadvantage, that is, difficulty of specification, because multistage sampling and stratified sampling have some strategic parameters to be specified by an analyst. Fault of the paramecer specification has a significant effect on the results of analysis.

This paper proposes a new approach to improve the responsibility. The approach has quantity independent responsibility and overcomes the disadvantage of the random sampling. The approach is to establish a temporary summary table corresponding to an interesting file. Interactive analysis is performed on a comparatively small summary table instead of the original data with large quantity. So the real-time response is realized. This approach gives an approximated result as well as sampling, but is easier to specify parameters of summarization than that of sampling.

Summary tables can establish securities independently on the corresponding original files. This solves the security problems of statistical data partially. More complicated security problems of statistical data have been referred to by Schwartz, Denning and Denning (1979).

The derivation and aggregation are usually treated as data editing/subsetting or user's view handling (Burnett and Thomas (1981)). One of the most important capabilities of statistical database system is feature/friendliness of view handling. Our system supports to produce various views interactively.

Data dictionaries (DD/D) have also to be expanded to manage the statistical environment effectively. Description of securities and prepared statistical programs/subprograms are also managed in the DD/D.

A diagram of our implemented system, HSDB, is given in Appendix.

2. Derived files

It is necessary to temporarily derive various files during the exploratory phase of data analysis while hypotheses are being formulated and tested (Burnett and Thomas (1981)). This phase is here called the pre-processing process of files in statistical data analysis. Some reseachers and practitioners call the process file/data subsetting or editing. The word "user's view" or "logical file" is also used if resulting files are not realized physically. In this paper, we refer to a resulting file by the pre-processing as a _derived file_. It is important for an analyst to define derived files easily. In our system, a derived file is defined as a combination of the following basic operations.

operation	function
SEL	selection of records and/or fields
UNION	merging of files
JOIN	concatenation of files
COMP	addition of new fields by computation
NORM	normalization of a repeating field
GROUP	creation of a repeating field
DIFF	difference of two sets
SORT	sorting of records

SAMPL	selection of records by random sampling
AGG	data aggregation
SAVE	save values of file and field attributes in DD/D

The system defines a derived file with a trial and error approach and produces an actual file according to a definition.

3. Expansion of DD/D

3.1 Expansion of file attributes

As file attributes, we add a discrimination mark between a derived file and an original file, a scheme of a derived file, a name of a person who collects or summarizes the data, and notes of a file, e.g., referred bibliographies.

3.2 Expansion of field attributes

The following attributes for each field are added to the ordinary field attributes.

1) established date
2) missing values
3) unit
4) precision
5) discrete domain or continuous domain
6) the number of actual records with a presented field
7) theoretical distribution
8) maximum
9) minimum
10) medium
11) mode
12) mean
13) variance
14) skewness
15) kurtosis
16) percent points

After computing the values of these field attributes, a user can refer to these values in a real time. He can also access statistics derived from these values, e.g., variance and range. This expansion enables the user to retrieve records which satisfy a condition such as $M_1 - 2SD \leq WEIGHT \leq M_1 + 2SD$, where WEIGHT is a field name and M_1 and SD are the mean and the standard deviation of WEIGHT respectively.

3.3 Dictionary of programs

The DD/D manages statistical programs, function subprograms deriving new fields and familiar statistical subprograms, in addition to ordinary application programs of the database system.

4. Statistical analysis

4.1 Interactive statistical analysis

Let X_i be a field in a file whose domain is a set Y_i for each i in {1, 2, ··, n}. If each Y_i is divided into m_i disjoint sets $C^{(i)}_1, C^{(i)}_2, \cdots, C^{(i)}_{m_i}$, then we call a set $\{ (\prod_{i=1}^{n} C^{(i)}_{j_i}, k): 1 \leq j_i \leq m_i \}$ a <u>summary table</u> of the file having parameters

$\{ C^{(i)}_{j_i} : 1 \leq i \leq n,\ 1 \leq j \leq m_i \}$, where $\prod_{i=1}^{n} A_i$ is Cartesian product of a collection of sets $\{A_1, A_2, \cdots, A_n\}$ and k is the number of records in $\prod_{i=1}^{n} C^{(i)}_{j_i}$.

Before starting of interactive statistical analysis (ISA), a user has to prepare a summary table of an interesting original file. During ISA, statistical programs process summary tables. The result can be obtained speedily, because a summary table has comparatively less records than its original file. There is a trade-off problem between precision of the result and response time of the analysis. Figure 1 shows an example of factor analysis after correlation analysis for ACCOUNT file which has the fields: CPU, REGION, , LPP. In Figure 1, notation "W=10" means a classification by intervals with length 10 and underlined strings of characters are commands or parameters inputted from a terminal.

<u>ISAP</u>	<u>ISAB</u>
INPUT FILE --------- <u>ACCOUNT</u>	PRELIMINARY NAME --- <u>ACTSUM</u>
PRELIMINARY NAME --- <u>ACTSUM</u>	*<u>CORR</u>
CPU ---------------- <u>W = 10</u>	*<u>FACTOR ACTFACT</u>
REGION ------------- <u>W = 100</u>	*<u>SAVE ACTFACT</u>
⋮	*<u>END</u>
LPP ---------------- <u>W = 10</u>	

Figure 1 Interactive statistical analysis

4.2 Detailed analysis in batch mode

After ISA by a trial and error process, more detailed analysis can be executed in batch mode. A detailed analysis processes the original file corresponding to a summary table analyzed during ISA. The system searches automatically both of the original file and the statistical analysis programs, and analyzes the data in batch mode.

We use an audit trail to implement detailed analysis. The trail records original file names, user identification numbers, processing dates and sequences of operations with parameters.

4.3 Interfaces to statistical packages

The system has interfaces of two famous statistical packages SPSS and SAS to our DBMS. Our future plan is to develop interfaces of other familiar packages and some statistical programs developed by users.

5. Presentation

The system has various commands for tabulation, Japanese KANJI reporting and graphical presentation such as horizontal bar chart, vertical bar chart and line graph. Each program of ISA presents the result in standard form, but the presentation commands above enable the user to make reports in various forms. Figure 2 illustrates a two-dimensional table presented in the following process.

HVIEWD	- view definition command
*AGG INPUT ---------- ACCOUNT OUTPUT ---------	to aggregate the file ACCOUNT and a work file will be created because of no OUTPUT parameter
FACULTY (C/O) -- C CLASSIFICATION - D = ENGINEER, * COMPUTER(C/O) -- C CLASSIFICATION - D = M-200H, M-180 MODE (C/O) -- C CLASSIFICATION - D = BATCH, TSS	specification how to classify the fields: FACULTY, COMPUTER, and MODE
FREQ ? (Y/N) --- Y	- do you want to set the frequency field ? (the answer is "yes")
*SAVE INPUT ---------- OUTPUT --------- AGGACT	to save the current work file and name it as AGGACT
DISP AGGACT, FACULTY, COMPUTER MODE	- to display AGGACT as two-dimensional table in the followings

<table>
<tr><th rowspan="4">FACULTY</th><th colspan="4">COMPUTER</th><th rowspan="4">TOTAL</th></tr>
<tr><th colspan="2">M-200H</th><th colspan="2">M-180</th></tr>
<tr><th colspan="2">MODE</th><th colspan="2">MODE</th></tr>
<tr><th>BATCH</th><th>TSS</th><th>BATCH</th><th>TSS</th></tr>
<tr><td>ENGINEER</td><td>18987</td><td>13735</td><td>49797</td><td>39818</td><td>122337</td></tr>
<tr><td>OTHERS</td><td>10475</td><td>10355</td><td>60237</td><td>99294</td><td>180361</td></tr>
<tr><td rowspan="2">TOTAL</td><td>29462</td><td>24090</td><td>110034</td><td>139112</td><td rowspan="2">302698</td></tr>
<tr><td colspan="2">53552</td><td colspan="2">249146</td></tr>
</table>

Figure 2 Two-dimensional tabulation

References:

Batory D.S. (1979), On Searching Transposed Files, ACM Transactions on Database Systems, Vol. 4 No. 4, 531-544.

Bragg A.W. (1981), Data Manipulation Languages for Statistical Databases -- The Statistical Analysis System (SAS), Proceedings of the First LBL Workshop on Statistical Database Management, 147-150.

The British Computer Society Data Dictionary System Working Party Report (1977), reprinted in SIGMOD Record, Vol. 9, No. 4.

Burnett R.A. and Thomas J.J. (1981), Data Management Support for Statistical Data Editing and Subset Selection, Proceedings of the First LBL Workshop on Statistical Database Management, 88-102.

Fry J.B. (1981), Data Manipulation in SPSS and SPSS-X, Proceedings of the First LBL Workshop on Statistical Database Management, 151-152.

Gey F. (1981), Data Definition for Statistical Summary Data or Appearances Can Be Deceiving, Proceedings of the First LBL Workshop on Statistical Database Management, 3-18.

Hammond R. (1981), Metadata in the RAPID DBMS, Proceedings of the First LBL Workshop on Statistical Database Management, 123-131.

Ikeda H. and Kobayashi Y. (1981), Additional Facilities of a Conventional DBMS to Support Interactive Statistical Analysis, Proceedings of the First LBL Workshop on Statistical Database Management, 25-36.

Lemon J.S. and Knowles J.S. (1980), Data Management Facilities in Statistical Packages, COMPSTAT '80, 108-114.
Merrit D. (1981), Volume Testing of Statistical Database Using Model 204 DBMS, Proceedings of the First LBL Workshop on Statistical Database Management, 380-389.
Mustonen S. (1980), Interactive analysis in SURVO 76, COMPSTAT '80, 253-259.
Reid A.J. and Lemon J.S. (1980), A review of interactive statistics packages in British Universities and Polytechnics, COMPSTAT '80, 260-268.
Schwartz M.D., Denning D.E. and Denning P.J. (1979), Linear Queries in Statistical Databases, ACM Transactions on Database Systems, Vol. 4, No. 4, 156-167.
Shoshani A. and Eggers S.J. (1980), Efficient access of compressed data, Proceedings of the Sixth International Conference on Very Large Data Bases, 205-211.
Turner M.J., et.al. (1979), A DBMS for Large Statistical Data Bases, Proceedings of the Fifth International Conference on Very Large Data Bases, 319-327.

Acknowledgement: The authors would like to thank the members of our group, Messrs M. Ohtomo, M. Fukushima, W. Ishihara and M. Kambe in Hiroshima University for their valuable discussions on the topic.

Appendix: Diagram of HSDB

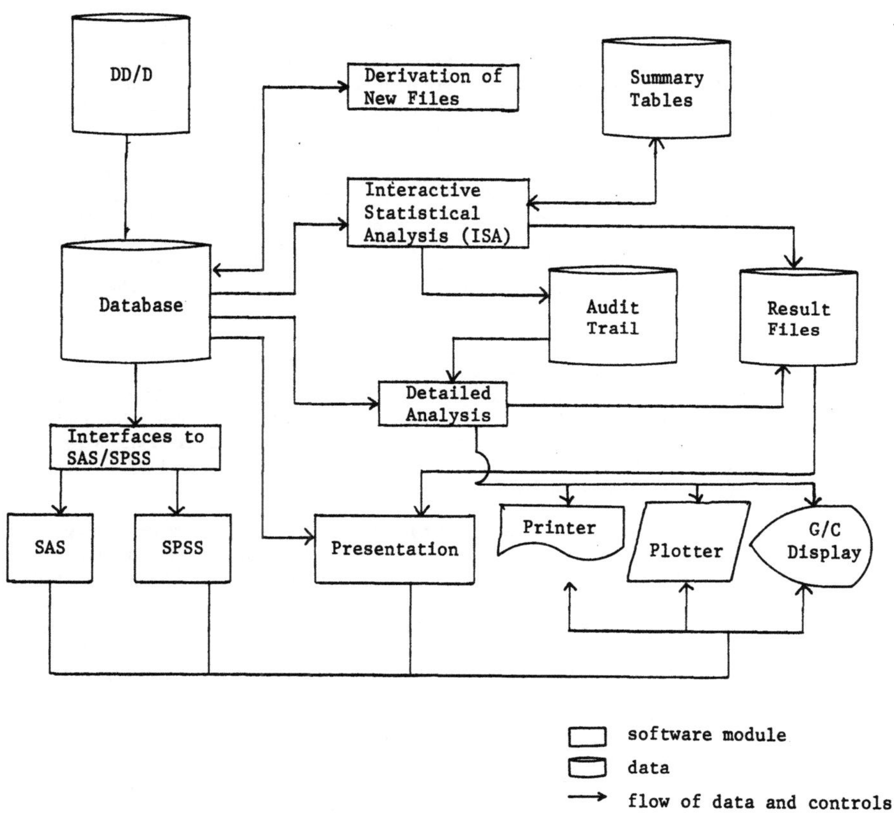

A Synthetic Approach to Multivariate Normal Clustering

P. Korhonen, Helsinki School of Economics, Finland
E. Blåfield, University of Jyväskylä, Finland

ABSTRACT

Two methods have been suggested for grouping together observations originated from the same multivariate normal distributions, both based on a maximum-likelihood (ML) estimation, but leading to different conclusions. In this paper we compare the use of Day's approach and that of Scott and Symons in clustering procedures from the theoretical and computational point of view. Based on this comparison, we suggest an approach unifying those approaches. The workability of the approach will be verified by numerical experiments.

Key words: Clustering, ML-approach, Normal Distribution

1. INTRODUCTION

Assume a set of observations, where each observation may arise from any one of g possible p-variate normal distributions, with the common, unknown, covariance matrix but different means. A clustering problem is to group together observations originated from the same multivariate normal populations. As a by-product it is desired that the parameters of the distributions be estimated.

For solving this well-known problem two interesting approaches have been proposed by Day (1969) and Scott & Symons (1971) (see also Symons 1981). Both are based on ML estimation, but they produce, in general, different results. Day was primarily concerned with estimating the parameters of the normal distributions, though he mentioned briefly the application of assigning sample observations to the populations. Corresponding considerations have been published by i.a. Wolfe (1970).

Scott & Symons (1971) adopted a different approach. Using the maximum likelihood technique they assigned individual observations on an 'all-or-nothing' basis to one of possible populations as part of the solving process. This is an actual clustering approach with an objective to estimate the 'true grouping' vector. Estimating the parameters of the component distributions is done after grouping. Related topics have also been discussed e.g. by Binder (1978). His approach is based on a Bayesian analysis and is very interesting.

Marriott (1975) has compared these two approaches and shown that Day's method leads to consistent estimators of the parameters of the component distributions, while that of Scott & Symons does not. Bryant & Williamson (1978) extended Marriott's results and showed that the approach of Scott & Symons may be expected to give asymptotically biased results for the parameters of the distributions quite generally. In addition, Marriott (1975) and Bryant & Williamson (1978) pointed out that Day's method is more preferable, too, when solving the actual clustering problem. In Day's method the allocation rule is consistent in the sense that the parameters for computing classification probabilities are consistently estimated. Instead, in the method of Scott & Symons the estimates of the 'true grouping' vector are not, in general, consistent, since estimation is based on the within-groups scatter matrix W, which is not a consistent estimator of the distribution dispersion matrix. Bryant & Williamson (1978) have also reported various bias effects in the estimate of the 'true grouping' vector.

Judging from those theoretical studies Day's approach seems to be superior to that of Scott & Symons. However, this is not the whole truth. In practical problems the method of Scott & Symons works fairly well (see e.g. Friedman & Rubin 1967, Wolfe 1970, and Binder 1978) and for it a computational efficient procedure has been developed (see e.g. Korhonen 1979). For instance the comparative studies on Fisher's iris data have given the same results with both the methods (Wolfe 1970, p. 345). In addition, the method of Scott & Symons is superior to that of Day from the computational point of view.

Both the methods may be extended to deal with unequal covariance matrices. However, this situation is more complicated than the restricted one, because infinite likelihood appears. Extra restrictions are needed to avoid this singular case and the best solution is not the global optimum of the likelihood function, but a local one. This extension will also be included in our theoretical considerations.

In the sequel, we shall compare basic formulae derived for those two approaches and suggest a synthetic approach, which utilizes the computational advantages of the method of Scott & Symons and the theoretical ones of Day's method. The paper consists of 5 sections. In Section 1 we have outlined the problem and considered the approaches from the theoretical point of view. In Section 2 we present some theories, and computational aspects are considered in Section 3. Numerical experiments are described in Section 4. Conclusions are drawn in the last section.

2. THEORY

In the following we give a brief unified theoretical presentation for the approach of Scott & Symons and that of Day. Bryant & Williamson (1978) called Day's approach the mixture maximum likelihood (MML) approach and the other one the classification maximum likelihood (CML) one. We adopt this terminology.

Let $\underline{x}_1, \underline{x}_2, \ldots, \underline{x}_n$ denote n p-dimensional observations. We denote the 'true grouping' vector by $\underline{\gamma} = (\gamma_1, \gamma_2, \ldots, \gamma_n)$, where $\gamma_i = k$ implies that the i th observation arises from the k th population, k=1,2,...,g. The number of populations g is assumed to be known. The component distributions are assumed to be normal $N_p(\underline{x};\underline{\mu}_k,\Sigma_k)$ with mean vectors $\underline{\mu}_k$ and covariance matrices Σ_k. Let π_k be the probability that the observation originates from the k th population. A clustering problem is to estimate $\underline{\gamma}$.

In the mixture maximum likelihood approach Day considers the finite mixture of the densities $N_p(\underline{x};\underline{\mu}_k,\Sigma_k)$. The estimate of parameters $\underline{\mu}_k, \Sigma_k$ and π_k are found by maximizing the likelihood function

$$(1) \qquad L_D = \sum_{i=1}^{n} \prod_{k=1}^{g} \pi_k N_p(\underline{x}_i;\underline{\mu}_k,\Sigma_k)$$

The grouping vector $\underline{\gamma}$ is estimated by means of the posteriori probability (see Table 1).

Scott & Symons (in their classification maximum likelihood approach) are primarily interested in estimating $\underline{\gamma} = (\gamma_1, \gamma_2, \ldots, \gamma_n)$ by maximizing the likelihood function

$$(2) \qquad L_S = \sum_{i=1}^{n} \pi_{\gamma_i} N_p(\underline{x}_i;\underline{\mu}_k,\Sigma_k)$$

The values of parameters μ_k, Σ_k and π_k will be estimated as a by-product, (see e.g. Symons 1981). (Actually, parameter π_k was not included in the original article of Scott & Symons, but we make this extension in our theoretical considerations.)

All parameters in both the methods can be estimated by using formulae presented in Table 1 (see Wolfe 1970, Scott & Symons 1971 and Symons 1981). For estimating the allocation parameter $\underline{\gamma}$ there exists no analytical expression in Scott & Symons' method. In Table 1 formulae have been presented without and with the equality constraint of the covariance matrices.

Table 1: Formulae for estimating parameters in Day's and Scott & Symons' methods

Param	Day	Scott & Symons
π_k	$\hat{\pi}_k = \frac{1}{n} \sum_{i=1}^{n} \hat{p}(k/\underline{x}_i)$	$\hat{\pi}_k = \frac{n_k}{n}$ $(n_k = \sum_{\gamma_i=k} 1)$
$\underline{\mu}_k$	$\hat{\underline{\mu}}_k = \sum_{i=1}^{n} \hat{p}(k/\underline{x}_i)\underline{x}_i / \sum_{i=1}^{n} \hat{p}(k/\underline{x}_i)$	$\hat{\underline{\mu}}_k(\gamma) = \sum_{\gamma_i=k} \underline{x}_i/n_k$
	$\hat{p}(k/\underline{x}_i) = \frac{\hat{\pi}_k N_p(\underline{x}_i; \hat{\underline{\mu}}_k, \hat{\Sigma}_k)}{\sum_{k=1}^{g} \hat{\pi}_k N_p(\underline{x}_i; \hat{\underline{\mu}}_k, \hat{\Sigma}_k)}$	
	Unequal Covariance matrices:	
Σ_k	$\hat{\Sigma}_k = \sum_{i=1}^{n} \hat{p}(k/\underline{x}_i)(\underline{x}_i-\hat{\underline{\mu}}_k)(\underline{x}_i-\hat{\underline{\mu}}_k)' / \sum_{i=1}^{n} \hat{p}(k/\underline{x}_i)$	$\hat{\Sigma}_k(\gamma) = \sum_{\gamma_i=k} (\underline{x}_i-\hat{\underline{\mu}}_k)(\underline{x}_i-\hat{\underline{\mu}}_k)'/n_k$
$\underline{\gamma}$	$\hat{\gamma}_i = k$, if $\hat{p}(k/\underline{x}_i) = \max_h \hat{p}(h/\underline{x}_i)$, $i=1,2,\ldots,n$	$\max_{\underline{\gamma}} [\ln(\prod_{k=1}^{g} n_k^{n_k}) - \frac{1}{2} \ln(\prod_{i=1}^{g} \|\hat{\Sigma}_k(\underline{\gamma})\|^{n_k})]$
	Equal covariance matrices:	
Σ	$\hat{\Sigma} = (\sum_{i=1}^{n} \underline{x}_i\underline{x}_i' - \sum_{k=1}^{g}\sum_{i=1}^{n} \hat{p}(k/x_i)\hat{\underline{\mu}}_k\hat{\underline{\mu}}_k')/n$	$\hat{\Sigma}(\gamma) = (\sum_{i=1}^{n} \underline{x}_i\underline{x}_i' - \sum_{k=1}^{g} n_k\hat{\underline{\mu}}_k\hat{\underline{\mu}}_k')/n$
$\underline{\gamma}$	(as above)	$\max_{\underline{\gamma}} [\ln(\prod_{k=1}^{g} n_k^{n_k}) - \frac{1}{2} n \ln\|\hat{\Sigma}(\gamma)\|]$

If parameter π_k is assumed to be known as Scott & Symons did, so the ML-estimate of the grouping vector will be found by minimizing the determinant of the covariance matrix. Hence forward, we use Scott & Symons' original version.

As we can see from Table 1, the formulae for estimators π_k, μ_k and Σ_k (or Σ) coincide, if the posteriori probabilities p (k/x_i) should be either 0 or 1. Scott & Symons' method may be regarded as an approximation of Day's method, where the greatest value of the posteriori probability is approximated by units and the others by zeroes. This feature will be utilized in our synthetic approach.

3. COMPUTATIONAL CONSIDERATIONS AND A SYNTHETIC APPROACH

As we pointed out in the introduction, Day's method suffers from many computational difficulties. The algorithms for Day's method are iterative by nature and seeking the optimum (not necessary global) is a very laborious task.

There seems to be basically two ways for finding the solution for Day's problem: 1) solving the equation system (Table 1) or 2) searching the greatest possible (finite) value of the likelihood function (1).

Equations in Table 1 define a straightforward and simplified iteration method, which can be accelerated in many ways, e.g. by applying a modification of Aitken's process, as Wolfe has proposed. A similar principle has been suggested by Day (1969), too. The method is quite laborious and Wolfe reports that "the amount of computation is more extensive than statisticians are accustomed to". Wolfe (1970) has developed two computer programs NORMAP (normal mixtures with common covariance matrix) and NORMIX (normal mixtures with different covariance matrices) which are widely known and available either as such or in statistical packages e.g. in CLUSTAN (see Wishart 1978).

An other way is to maximize the likelihood function. This approach has been applied by Quandt (1972) to estimating parameters in a switching regression problem, which is actually Day's problem with additional structural information. There are several methods available for maximizing the likelihood function. Quandt (1972) used Powell's conjugate gradient method, but he reported very high failure rates in his article. One of the most efficient methods used for solving the maximizing problem is a quasi-Newton method called the Davidon-Fletcher-Powell method (see, e.g. Blåfield 1980). Also the method of Hooke and Jeeves has been used for the problem. In any case, many difficulties appear when attempting to find satisfactory local maxima for the likelihood function. It is widely known that the likelihood function has local optima, sometimes very difficult to locate.

So, judging by the reports dealing with solving Day's problem there exists no efficient algorithm, which should converge in a reasonable number of iterations or should give results accurate enough.

On the contrary, it is easy to seek a local solution in the method of Scott & Symons. Several authors (e.g. Friedman & Rubin 1967) have described a widely known method called "the exchange method". The method generates a sequence of groupings by processing observations one at a time. The sequence converges to a grouping yielding a local optimum for the criterion function modified from the likelihood function (2) (see Table 1). Unfortunately, the use of the criterion in the exchange method has been quite laborious, not quite so laborious as in Day's method. This feature has prohibited its wide use in practical problems. However, Korhonen (1979) developed an efficient method for the case of equal covariance matrices. Thanks to the method the performance time of the exchange method has reduced drastically. The exchange method is applicable to the case of unequal covariance matrices, but for this case there exists no efficient method.

In addition to computational advantages, the method of Scott & Symons gives quite good results in practical problems (c.f. e.g. Wolfe 1970, Symons 1981). In order to utilize these computational advantages and the fact that it works quite well in practice, we suggest the following synthetic approach, which also utilizes the theoretical advantages of Day's method:

1. Assume equal covariances matrices and solve a clustering problem with the method of Scott & Symons. Repeat the clustering k times by using random initializations and choose the best solution (k is appropriately chosen). If the assumption about the equal covariance matrices is relevant, go on to Step 2, otherwise go on to Step 3.

2. If the solution is satisfactory, stop, otherwise continue with the mixture model. Use the Davidon-Fletcher-Powell's method for searching the optimal value of the likelihood (or loglikelihood) function. As initial values use the values of the parameters found in Step 1. Stop.

3. Relax the assumption of the equal covariance matrices and continue with the Scott & Symons' method, until the (local) optimum is found.

4. If the solution is satisfactory, stop, otherwise continue as in Step 2, now with unequal covariance matrices.

In the next section we shall consider the application of the first two steps to numerical data.

4. NUMERICAL EXPERIMENTS

In the following we briefly report a few explorative experiments on artificial data to demonstrate that the synthetic approach is workable. Specifically, we are interested in its performance as to clustering and execution times.

Both the methods are written in Fortran and implemented on the Univac 1100 computer. In the following text Procedure I will refer to the computer program for Scott & Symons' method and Procedure II for Day's one. Procedure I is based on Korhonen's algorithm and it starts with type observations, which can be given or randomly chosen. Procedure II uses the Davidon-Fletcher-Powell's method for maximizing the likelihood function and it starts with given initial values for parameters.

For test purposes we generated randomly four 6-variate normal samples, containing 10, 20, 30, and 40 observations, from distinct populations with equal covariance matrices, which was assumed to be an identity matrix. This assumption can be done without loss of generality, because the grouping vector is scale invariant. As the theoretical mean vectors were used (2,0,1,1,1,-2), (-3,-2,-1,1,2,3), (-1,1,-3,4,-2,0), and (0,3,-4,-3,0,-4). From the basic data matrix we formed four sub-data sets of 3 and 6 variables and of 2 and 4 groups. In the two-group data the groups of size 20 and 30 were used.

Each data set was grouped five times with both the procedures using the same type observations. In addition, in each case the grouping produced by Procedure I was used to specify the initial values for parameters for Procedure II (i.e. a synthetic approach). The results were quite interesting. Procedure II found either the global optimum or did not converge to any sensible solution. It always found the global optimum when following Procedure I. In that case it found the global solution in spite of the fact that the solution produced by Procedure I was sometimes the local one. These results together with average execution times have been summarized in Table 2.

Table 2: The ratio of the number of the optimal groupings to the number of all found groupings and the average execution times (s) (separated by ":").

	Procedure I		Procedure II		Synthetic Approach	
g p:	3	6	3	6	3	6
2	2/5 : 0	0/5 : 0	2/2 : 23	5/5 : 100	5/5 : 22	5/5 : 110
4	0/5 : 1	3/5 : 2	1/1 : 170	2/2 : 1200	5/5 : 120	5/5 : 900

From Table 2 we can see that the performance of Procedure II depends very heavily on good initial values for parameters. For instance, in case of 4 groups and 3 variables Procedure II succeeded only once to find a solution, which was optimal. Procedure I converged many times to a local optimum. Usually the local solutions are avoided by repeating the grouping with various initial values as we recommend in our synthetic approach. In this experiment we did not use the repetition, because we were interesting to see how Procedure II recovers from the local solution. As we have seen it seems to recover very well.

The average execution times between Procedure II and the synthetic approach were roughly equal. (The execution times of Procedure II does not include the unsuccessful cases.) It is worth noting that the execution times of the synthetic approach were lower than the average, when Procedure II started with the parameters of the optimal solution. From that result and the fact that the execution times of Procedure I are small fractions of the execution times of Procedure II it follows that it pays to search the global solution with Procedure I.

5. CONCLUDING REMARKS

The preliminary tests on the synthetic approach clearly indicate that the fusing of those methods is worth-while. The synthetic approach converged every time to the global optimum, in spite of the fact that Procedure I converged to a local solution. On the contrary, Procedure II did not always converge. The best execution times were obtained, when Procedure II started from the global optimum. Because the execution times of Procedure I are only small fractions of those of Procedure II, it pays to use Procedure I repeatedly by starting with different initializations and by so trying to find the global optimum.

Our results are very encouraging and we are currently working on thorough test problems including the consideration of parameters.

REFERENCES

BINDER, D.A. (1978), "Bayesian Cluster Analysis," **Biometrika**, vol. 65, N:o 1, pp. 31 - 38.

BLÅFIELD, E. (1980), **Clustering of Observations from Finite Mixtures with Structural Information**, Jyväskylä Studies in Computer Science, Economics and Statistics 2.

BRYANT, P. and WILLIAMSON, J. (1978), "Asymptotic Behaviour of Classification Maximum Likelihood Estimates", **Biometrika**, vol. 65, N:o 2, pp. 273 - 281.

DAY, N.E. (1969), "Estimating the Component of a Mixture of Normal Distributions", **Biometrika**, vol. 56, N:o 3, pp. 463 - 474.

FRIEDMAN, H.P. and RUBIN, J. (1967), "On Some Invariant Criteria for Grouping Data", **J. Am. Statist. Assoc.**, vol. 62, 1159 - 1178.

KORHONEN, P.J. (1979), **A Stepwise Procedure for Multivariate Clustering**, Computing Centre, University of Helsinki, Research Reports, N:o 7, (The Ph.D. thesis)

MARRIOTT, F.H.C. (1975), "Separating Mixtures of Normal Distributions", **Biometrics**, vol. 31, pp. 767 - 769.

QUANDT, R.R. (1972), "A New Approach to Estimating Switching Regressions", **J. Am. Statist. Assoc.**, vol. 67.

SCOTT, A.J. and SYMONS, M.J. (1971), "Clustering Methods based on Likelihood Ratio Criteria", **Biometrics**, N:o 27, pp. 387 - 397.

SYMONS, M.J. (1981), "Clustering Criteria and Multivariate Normal Mixtures", **Biometrics**, N.o 37, pp. 35 - 43

WISHART, D. (1978), **CLUSTAN User Manual**, PLU, Edinburgh University

WOLFE, J.H. (1970), "Pattern Clustering by Multivariate Mixture Analysis", **Mult. Behavioral Res.**, vol. 5, pp. 329 - 350.

Variable Selection in Generalized Linear Models

Ch. Kredler, Technical University of Munich, F.R.G.
L. Fahrmeier, University of Regensburg, F.R.G.

Summary.
Quadratic approximations of the log-likelihood function of multivariate generalized linear models are discussed and an upper bound for the truncation error is given. Then well known variable selection procedures are applied to the resulting linear models.

KEYWORDS: Multivariate generalized linear models, ML-estimation, variable selection, approximation of nonlinear by linear models.

1. Multivariate generalized linear models

Univariate generalized linear models were first discussed by Nelder/Wedderburn (1972) and several authors have given multivariate extensions. We follow the approach of Fahrmeir/Kredler (1982), where proofs not given here are to be found. Consider the arithmetic mean y_i of N_i independent responses of a random variable $Y_i \in \mathbb{R}^q$. Given regressor variables $x_i \in \mathbb{R}^p$, $i = 1,\dots,I$ the $(Y_i|x_i)$ have exponential family densities

$$f(Y|\theta_i,\phi) = c(Y,\phi)\exp\{(Y^T\theta_i - b(\theta_i))/a(\phi)\}\ , \tag{1}$$

where a, b, c are known real valued functions, $\theta_i \in \mathbb{R}^q$ and $\phi \in \mathbb{R}^1$ are unknown parameters. The density of $(y_i|x_i)$ is of exponential form (1) too, with $a(\phi)$ replaced by $a(\phi)/N_i$ and a slightly modified c. Moreover the arithmetic mean y_i is the maximum likelihood (ML) estimator for $\mu_i = E(Y_i|x_i)$. Suppose the $(y_i|x_i)$ are independent, and the x_i and unknown parameter vectors $\beta_{(j)} \in \mathbb{R}^p$, $j = 1,\dots,q$ are related to the μ_i by a known twice differentiable one to one link function $\gamma \in \mathbb{R}^q$

$$\gamma^T(\mu_i) = x_i^T(\beta_{(1)},\dots,\beta_{(q)}) = x_i^T B, \qquad i = 1,\dots,I. \tag{2}$$

Sometimes it is more convenient to arrange the parameters to a column vector β (here $\beta \in \mathbb{R}^{pq}$) and to define a design matrix X_i (here $X_i \in \mathbb{R}^{qp,q}$), which yields the equivalent representation

$$\gamma(\mu_i) = B^T x_i = X_i^T\beta, \qquad i = 1,\dots,I. \tag{3}$$

Assume the usual regressor matrix $X = (x_1,\dots,x_I)^T$ to have rank p and let $X^T = (X_1,\dots,X_I) \in \mathbb{R}^{pq,Iq}$ and y, μ, $g(\mu)$ be the $(Iq\times 1)$-column vectors of the y_i, μ_i and $\gamma(\mu_i)$. Then a multivariate *generalized linear model* is given by (1) and

$$y = \mu + \varepsilon, \quad g(\mu) = X\beta, \quad E(\varepsilon) = 0\ . \tag{4}$$

Among the various examples (probit analysis, McCullagh 1980) so called *natural* link functions play an important role. Let $\psi(\mu_i) = \theta_i$ be the canonical transformation. Then $\gamma \equiv \psi$ leads to a linear model for the canonical parameters

$$\theta_i = \psi(\mu_i) = \gamma(\mu_i) = B^T x_i = X_i^T \beta \ . \tag{5}$$

Examples are the usual linear model for normal, the log-linear model for Poisson and the logit model for multinomial distributed $(Y_i | x_i)$. For $\gamma \equiv \psi$ the following expressions simplify considerably. Let $\Sigma_i = \mathrm{cov}(Y_i | x_i) = \left(\frac{\partial^2 b(\theta_i)}{\partial \theta_i \partial \theta_i^T}\right) a(\phi)$, $G_i = \left(\frac{\partial \gamma(\mu_i)}{\partial \mu_i}\right)^T$, $V_i = G_i \Sigma_i G_i^T / N_i$, $G(\mu) = \mathrm{diag}\{G_i\}$, $V(\mu) = \mathrm{diag}\{V_i\}$, then gradient h and information matrix F of the log-likelihood function $L(\beta)$ according to model (4) can be given in terms of $\mu_i = \mu_i(\beta) = \nabla b(\theta_i)$

$$\begin{aligned} h(\mu) &= \nabla L(\beta) = X^T V^{-1}(\mu) G(\mu)(y-\mu) \ , \\ F(\mu) &= E[-H(\beta)] = X^T V^{-1}(\mu) X \ , \end{aligned} \tag{6}$$

where H denotes the Hessian of L. For natural $\gamma \equiv \psi$ we have $h(\mu) = X^T(y-\mu)/a(\phi)$, $F(\mu) = -H(\mu) = X^T \Sigma X$, where $\Sigma = \mathrm{diag}\{N_i \Sigma_i\} / a^2(\phi)$. Here $L(\beta)$ is strictly concave.

2. Computation of the ML - estimator $\hat{\beta}$

Assume that a maximizer $\hat{\beta}$ of $L(\beta)$ with $h(\hat{\beta}) = 0$ exists. Starting with an initial guess $\hat{\beta}_0 = 0$ or $\hat{\beta}_0 = \beta_{WLS}$ (which results from model (8)) approximations $\hat{\beta}_0, \hat{\beta}_1, \ldots$ for $\hat{\beta}$ are computed iteratively by

$$\hat{\beta}_{r+1} = \hat{\beta}_r - \lambda_r H_r^{-1} h(\hat{\mu}_r), \ \hat{\mu}_r = \mu(\hat{\beta}_r), \ r = 0,1,\ldots \tag{7}$$

The relaxation factor $\lambda_r \in (0,1]$ improves global convergence properties. $H_r = -F(\hat{\mu}_r)$ yields the widely used Newton method with Fisher scoring. For problems of higher dimension ($pq \geq 10$) the following optimally scaled quasi - Newton method reduces computation time considerably. Take $H_0 = -F(\hat{\mu}_0)$, which can be evaluated in an efficient manner for $\hat{\beta}_0 = 0$. The H_r are updated by *Broyden - Fletcher - Goldfarb - Shanno* rank - 2 - corrections envolving only the gradient h. The choice of λ_r is not crucial. Usually take $\lambda_r = 1$ and choose $\bar{\lambda}_r = \lambda_r/2$, if the first correction does not increase the value of L, and so on.

3. The weighted least squares estimator β_{WLS} and asymptotic properties

Analogously to the approach of Grizzle/Starmer/Koch (1969) β_{WLS} is defined as the LS - estimator of the linear model

$$g(y) = X\beta + \varepsilon_g, \ \ E(\varepsilon_g) = 0, \ \ \mathrm{cov}(\varepsilon_g) = V(y) \ . \tag{8}$$

If $N_i \to \infty$, $i = 1,\ldots,I$ the arithmetic mean y_i converges to its expectation μ_i and g(y) and V(y) are well defined almost surely. β_{WLS} then is the best asymptotic normal estimator and converges to $\hat{\beta}$ in probability.

Mixed data problems in logistic discriminant analysis (cf. Anderson 1972) lead to generalized linear models with multinomial response variables. In this and many other cases some (or all) of the N_i remain unity even if $N = \sum_{i=1}^{I} N_i$ approaches infinity. Unfortunately β_{WLS} is not defined in these cases. For $\gamma \equiv \psi$ the asymptotic theory of Haberman (1977) for $\hat{\beta} = \hat{\beta}_{(N)}$ requires only $N \to \infty$ and ensures that $\hat{\beta}_{(N)}$ exists for sufficiently large N. Further $\hat{\beta}_{(N)}$ is (strongly) consistent and asymptotic normal, i.e.

$$\hat{\beta}_{(N)} \sim_a N[\beta, F^{-1}(\mu(\hat{\beta}_{(N)}))] .$$

These desirable properties of $\hat{\beta}$ hold for a wide class of problems, for which β_{WLS} does not exist at all. In the following sections a linear model analogous to (8) is constructed for $\hat{\beta}$. Then well known variable selection procedures which are applicable to the linear estimator β_{WLS} can be applied to $\hat{\beta}$ in a modified way.

4. Quadratic approximation of the log-likelihood function

In this section only natural link functions are considered. Let $\mu = \mu(\beta)$, $\hat{\mu} = \mu(\hat{\beta})$ and $V(\hat{\mu}) = CC^T$ be a symmetric factorization. Using $h(\hat{\beta}) = 0$ we get the Taylor expansion

$$L(\beta) - L(\hat{\beta}) = -\tfrac{1}{2}(\beta - \hat{\beta})^T X^T C\{I + \tfrac{1}{3}E\} C^T X(\beta - \hat{\beta}). \tag{9}$$

According to Kredler (1981) for Poisson, gamma and multinomial response variables the bound

$$\|E\| \le (q+1) \max_{i;0\le j\le q} \left|\frac{\mu_{ij} - \hat{\mu}_{ij}}{\hat{\mu}_{ij}}\right| + o(\mu_{ij} - \hat{\mu}_{ij}) \tag{10}$$

is valid, where

$$\overset{(\wedge)}{\mu}_{io} = \begin{cases} 1 - \sum_{j=1}^{q} \overset{(\wedge)}{\mu}_{ij} , & \text{in the multinomial case} \\ 0 , & \text{else } \left(\dfrac{\mu_{io} - \hat{\mu}_{io}}{\hat{\mu}_{io}} = 0 \text{ in this case} \right). \end{cases}$$

Clearly $\|E\| = 0$ for normally distributed response variables. As $\hat{\beta}$ is consistent, $\hat{\mu}$ is consistent too. So for increasing N the quadratic form

$$Q(\hat{\beta},\beta) = (\beta - \hat{\beta})^T X^T V(\hat{\mu}) X(\beta - \hat{\beta}) \tag{11}$$

is a good approximation of

$$\chi^2(\hat{\beta},\beta) = -2\{L(\beta) - L(\hat{\beta})\} . \tag{12}$$

On the other hand Q characterizes the log-likelihood function of a linear model

$$\eta = X\beta + \varepsilon_\eta, \quad \varepsilon_\eta \sim N(0,V(\hat{\mu})). \tag{13}$$

By (4) $g(\hat{\mu}) = X\hat{\beta}$ and the LS - estimator

$$\hat{\beta} = (X^T V^{-1}(\hat{\mu})X)^{-1} X^T V^{-1} g(\hat{\mu}) \tag{14}$$

of this model with unobservable η can be written analogously to β_{WLS}. Note that β_{WLS} results from (14) with $\hat{\mu}$ replaced by y, the ML - estimator without model restrictions. We call (13) the *approximating linear model* of (4).

5. Variable selection

Denote by β_k^T the k-th row of B from (2). Testing

$$\begin{aligned} &H: \beta_k^T = 0, \quad k \in K \subset P = \{1,\dots,p\} \\ \text{against} \quad &\overline{H}: \beta_k^T \neq 0 \text{ for some } k \in K \end{aligned} \tag{15}$$

corresponds to a significance check of the regressor variables x_{ik}, $k \in K$. Denote by $\hat{B}_K$ the maximum of L under the hypothesis H, then

$$\chi_K^2(\hat{B},\hat{B}_K) = -2\{L(\hat{B}_K) - L(\hat{B})\}$$

is asymptotic chi square with q|K| degrees of freedom, assuming the assumptions ensuring asymptotic normality hold (see e.g. Mathieu 1981). Computation of χ_K^2 for all possible subsets K requires the solution of 2^p nonlinear optimization problems and is too expensive in most applications. Let $\hat{\beta}$ and $\hat{\beta}_K$ denote the column vectors corresponding to $\hat{B}$ and $\hat{B}_K$, $\hat{\mu} = \mu(\hat{\beta})$, $\hat{\mu}_K = \mu(\hat{\beta}_K)$.
Some manipulation of (9) - (12) yields

$$\chi_K^2(\hat{\beta},\hat{\beta}_K) = Q(\hat{\beta},\hat{\beta}_K)(1 + \rho) \ , \tag{16}$$

where $|\rho|$ is dominated by the bound of $\|E\|$ given in (10).
If the hypothesis H of (15) is true, for increasing N the leading digits of $\hat{\mu}$ and $\hat{\mu}_K$ will agree with high probability and χ_K^2 is well approximated by Q. Although these results are strictly valid only for natural link functions, the following variable selection procedure can be applied in the general case.

(i) At first $\hat{\beta}$ and a symmetric factorization of $F(\hat{\mu}) = X^T V(\hat{\mu}) X$ or $F^{-1}(\hat{\mu})$ are computed with methods mentioned in section 2.

(ii) Now $\hat{\beta}$ is viewed as the LS - estimator of the approximating linear model (13). For the ease of notation we describe, how approximations of $\hat{\beta}_K$ and $Q(\hat{\beta},\hat{\beta}_K)$ are obtained in a very simple way using $F^{-1}(\hat{\mu})$ explicity. A good computer program will perform the following operations on the factor R of $F(\hat{\mu}) = RR^T$. The LS - estimator $\tilde{\beta}_K$ of model (13) under the hypothesis H and the corresponding residual sum of squares RSS_K results from $q|K|$ Gauss - Jordan sweeps at those columns of the following tableau, which correspond to components of $\hat{\beta}$ annihilated by H of (15).

$$\begin{array}{|c|c|} \hline (X^T V^{-1}(\hat{\mu}) X)^{-1} & \hat{\beta} \\ \hline -\hat{\beta}^T & 0 \\ \hline \end{array} \qquad (17)$$

Having performed these steps, $\hat{\beta}$ is overwritten by $\tilde{\beta}_K$ and 0 by RSS_K. Now RSS_K is an approximation of χ_K^2 as well and error bounds can be given in the same way as described in (16). So the easily computed RSS_K can be used for the comparison of variable sets K.

(iii) Assume that $K_1,\ldots,K_r$ are some favourable variable sets, suggested by the method of Furnival/Wilson (1974) or other exploration strategies. The exact solutions $\hat{\beta}_{K_j}$ can then be computed with at most 2 iteration steps (7) starting at $\tilde{\beta}_{K_j}$ and $H_{K_j}^{-1}$ which can be found in suitable rows and columns of the tableau (17) modified by Gauss - Jordan sweeps.

6. Numerical results

A FORTRAN 77 program for the proposed variable selection procedure was implemented. Among others 2 examples of discriminant analysis with mixed variables were examined (cf. Fahrmeir/Kredler 1981 for a more detailed discussion). First 1000 credit clients had to be classified in classes with good and bad solvency. For every client C_i a vector x_i of 20 characteristic attributes was given. This problem leads to a generalized linear model with binomial response variables and the natural link-function $\gamma(\mu_i) = \log(\mu_i/1-\mu_i)$. In the second example 249 rheumatism patients were classified into 3 categories by means of 8 regressor variables and the logistic model of Anderson (1972), which can be viewed as a multivariate generalized linear model with natural link function. In both cases the 3 best variable sets K_{rj}, $j = 1,2,3$ with r variables $r = 1,\ldots,p$ were determined using the method of Furnival/Wilson (1974) in model (13). Comparison of variable sets K was accomplished by the above defined $RSS_K = RSS(K)$. In all cases (cf. Kredler 1981)

$$RSS(K_{ri}) < RSS(K_{rj})$$

yielded (18)

$$\chi^2(\hat{\beta},\hat{\beta}_{K_{ri}}) < \chi^2(\hat{\beta},\hat{\beta}_{K_{rj}}).$$

Consequently variable selection in the approximating linear model (13) leads to no loss of information in the examined examples.

7. Conclusions

Variable selection in approximating linear models was suggested e.g. by Lawless/Singhal (1978). Here in the special case of generalized linear models conditions can be given, which justify the quadratic approximation of the log-likelihood function. Numerical experience with multinomial response variables confirm the theoretical considerations. Similar results for the binary logit model were given by Nordberg (1981).

References

Anderson J.A. (1972), Seperate sample logistic discrimination, Biometrika, 59, 19-35.

Fahrmeir L., Kredler Ch. (1982), Verallgemeinerte lineare Modelle, in: Fahrmeir L., Hamerle A., Multivariate statistische Verfahren, de Gruyter, Berlin.

Furnival G.M., Wilson R.W. (1974), Regression by leaps and bounds, Technometrics, 16, 499-511.

Grizzle J.E., Starmer C.F., Koch G.G. (1969), Analysis of categorical data by linear models, Biometrics, 25, 489-504.

Haberman S.J. (1977), Maximum likelihood estimates in exponential response models, Ann. Statist., 5, 815-841.

Kredler Ch. (1981), Schrittweise Verfahren für multinomiale Regressionsmodelle, Dissertation, TU München.

Lawless J.F., Singhal K. (1978), Efficient screening of nonnormal regression models, Biometrics, 34, 318-327.

McCullagh P. (1980), Regression models for ordinal data, J.R. Statist. Soc. B, 42, 109-127.

Mathieu J.R. (1981), Tests of χ^2 in the generalized linear model, Math. Operationsforsch. Statist., Ser. Statistics, 12, 509-527.

Nelder J.A., Wedderburn R.W. (1972), Generalized linear models, J.R. Statist. Soc. A, 135, 370-384.

Nordberg L. (1981), Stepwise selection of explanatory variables in the binary logit model, Scand. J. Statist., 8, 17-26.

Approximation of Surfaces in (p+1)-Dimensional Spaces

H. Läuter, Academy of Sciences, Berlin, G.D.R.

Summary: We give a method for constructing a robust regression estimator by composing a classification procedure and a minimization problem. This two-step procedure has very good mathematical properties and is realizable on computers easier than other known proposals.

Keywords: Robust estimation, regression, classification procedure, two-step procedure

1. Introduction

Let $y_1,\ldots,y_n$ be independent random variables which satisfy the model

$$y_i = f(x_i,\beta) + \varepsilon_i \quad , \quad i = 1,\ldots,n. \tag{1}$$

Here $x_1,\ldots,x_n$ are points of the experimental region $\mathcal{X} \subseteq R^p$ and ε_i are identically distributed random variables with $E\varepsilon_i = 0$, var $\varepsilon_i = \sigma^2$. f is a known function and $\beta \in R^k$ is an unknown parameter which is to be estimated. The standard method for the identification of β is the least squares method. Here $\hat{\beta}$ is a solution of

$$\sum_{i=1}^{n} (y_i - f(x_i,\beta))^2 = \min. \tag{2}$$

One finds in some papers results on the asymptotic behaviour of $\hat{\beta}$ (R.J. Jennrich (1969), H. Bunke & W.H. Schmidt (1980)). J.W. Tukey (1960) remarked that this $\hat{\beta}$ is highly sensitive against gross errors in the observations. Many proposals exist to construct more robust estimators. Here we mention approaches connecting with trimming, Winsorizing, ect. Another type of methods for finding robust regression estimations is due to P.J. Huber. The basic idea in this approach is to find a solution of the minimization problem

$$\sum_{i=1}^{n} \rho\,(y_i - f(x_i,\beta)) = \min \tag{3}$$

for a given function ρ. Especially functions

$$\rho(t) = \begin{cases} \frac{t^2}{2} & \text{for } |t| \leq K \\ K|t| - \frac{K^2}{2} & \text{for } |t| > K \end{cases} \tag{4}$$

which depend on a parameter $K \geq 0$ were pointed out as very appropriate to obtain as efficient estimators as possible. With this ρ one gets estimators being not very sensitive to the presence of gross errors.

This effect increases if one works with a function ρ (depending on three parameters a,b,c) defined by $\psi(t) = \frac{d}{dt}\rho(t)$:

$$\psi(t) = \begin{cases} t & \text{for } |t| \leq a \\ a \cdot \operatorname{sgn} t & \text{for } a < |t| \leq b \\ \frac{a}{b-c}(|t|-c)\operatorname{sgn} t & \text{for } b < |t| \leq c \\ 0 & \text{for } c < |t| \end{cases} \tag{5}$$

Several proposals for the choice of the constants are given in D.F. Andrews et. al. (1972). Of course it would be appropriate to choose the a,b,c in dependence to the quality of the observations. This aspect we have kept in mind in the development of the preliminary-classification estimation (p.c.e.).

2. Preliminary-classification regression

The p.c.e. of β is a two-step procedure where the results of a given classification procedure are used in the following minimization problem. First we describe the classification procedure. Let sets $I_1,\ldots,I_k$ in R^1 be given. We assume that densities $g_1,\ldots,g_k$ on $\mathcal{X}$ are available with

$$g_j(x) = \frac{1}{\int_{\mathcal{X}} h_j(\xi)\,d\xi} \quad h_j(x)$$

where $P^{y|x}(I_j) = h_j(x)$ and $P^{y|x}$ denotes the distribution of the variable y at the point $x \in \mathcal{X}$ satisfying the model (1). The functions g_j characterize by construction the probability for observing values in I_j. So we have to consider k classes $K_1,\ldots,K_k$ where the elements of K_j are distributed with the density g_j. A classification procedure is described by a disjoint partition $W_1,\ldots,W_k$ of $\mathcal{X}$ where $x \in W_j$ implies the classification to K_j. We denote by

$$P(j|i) = \int_{W_j} g_i(\xi)\,d\xi$$

the probability for classifying an element from K_i to K_j with such a partition. Then as the classification procedure we use that one which minimizes the averaged expected probability of misclassification

$$\sum_{i=1}^{k} \sum_{\substack{j=1 \\ j \neq i}}^{k} p_i \, P(j \mid i)$$

where $p_1,\dots,p_k$ denote the a-priori probabilities for $K_1,\dots,K_k$. As it is known (T.W. Anderson (1958)) a solution for this minimization problem is given by the following rule: We classify x to K_j if $x \in M_j$, $j = 1,\dots,k$ where

$$\begin{aligned} M_1 &= \{x \mid x \in \mathfrak{X},\ p_1 g_1(x) \geq p_t g_t(x) \quad \forall t\}, \\ M_i &= \{x \mid x \in \mathfrak{X},\ p_i g_i(x) \geq p_t g_t(x) \quad \forall t > i\} \setminus (M_1 \cup \dots \cup M_{i-1}) \\ &\quad i = 2,\dots,k-1, \\ M_k &= \mathfrak{X} \setminus (M_1 \cup \dots \cup M_{k-1}). \end{aligned} \tag{6}$$

In such a way M_j contains those points of $\mathfrak{X}$ in which we observe values in I_j with the highest probability.

If we have to estimate the $g_1,\dots,g_k$ then we use the substitution method for the described classification procedure. Some properties of such a strategy under given assumptions one finds in D. Schmidtke (1982). We propose here the application of kernel methods.

For this we define the sets

$$J_s = \{x_i \mid y_i \in I_s,\ 1 \leq i \leq n\}, \quad s = 1,\dots,k.$$

We put $m_i = \# J_i$ and define to a given kernel $K(x \mid x_i, \lambda)$ the estimates

$$g_s(x,\lambda) = \frac{1}{m_s} \sum_{i: x_i \in J_s} K(x \mid x_i, \lambda)$$

for $g_s(x)$. A popular choice for K is the choice as a normal kernel,

$$K(x \mid x_i,\lambda) = \frac{1}{(2\pi\lambda)^{p/2}} \exp\left\{-\frac{1}{2\lambda} |x - x_i|^2\right\}.$$

In analogous way to $M_1,\dots,M_k$ we now define the sets

$$\hat{M}_1 = \{x \mid x \in \mathfrak{X}, p_1 g_1(x,\lambda) \geq p_t g_t(x,\lambda) \quad \forall t\},$$

$$\hat{M}_i = \{x \mid x \in \mathfrak{X}, p_i g_i(x,\lambda) \geq p_t g_t(x,\lambda)\ \forall t > i\} \setminus (\hat{M}_1 \cup \dots \cup \hat{M}_{i-1})$$

$i = 2,\dots,k-1$

$$\hat{M}_k = \mathfrak{X} \setminus (\hat{M}_1 \cup \dots \cup \hat{M}_{k-1}).$$

An interesting question is the influence of λ on $\hat{M}_1,\dots,\hat{M}_k$. Here it that the $\hat{M}_1,\dots,\hat{M}_k$ are good estimates for $M_1,\dots,M_k$ and the influence of λ is small. Only in the case of a small m_i and "extreme" points under these m_i points one has some troubles to pay attention to. In this case λ should not be chosen too small because the extreme points have an influence on $\hat{M}_1,\dots,\hat{M}_k$ which is not desired. Therefore λ must not be too small in comparison with $m_1,\dots,m_k$. The following inequality

$$\lambda \geq \frac{c}{\min_i \varphi(m_i)} \tag{7}$$

for a constant $c > 0$ and a monotone increasing function φ has to hold.

Under our assumptions on the model there exist many proposals in the literature for constructing a robust estimation for β. The different estimators arise from different proposals for ρ in (3). Here we use the idea that only such parameters are admitted for which $f(x,\beta) \in I_i$ with a high probability if $x \in M_i$. By construction of the M_i this seems to be a sensible demand. Therefore we define in R^q a set of admissible parameters. For this we choose in M_i resp. $\hat{M}_i$ a sequence of points z_{ij}, $j = 1,\dots,n_i$, $i = 1,\dots,k$ and define

$$B = \{ \beta \mid \sum_i \sum_j d(f(z_{ij},\beta), I_i) \leq \tilde{c} \}$$

for a distance function d between a point and a set. $\tilde{c}$ is a nonnegative constant to be chosen. Very often the special case $\tilde{c} = 0$ is of interest in which for many distances d the set B is of the form

$$B = \{ \beta \mid f(z_{ij},\beta) \in I_i, \quad j = 1,\dots,n_i, \ i = 1,\dots,k \}.$$

Estimations for β and σ we get by minimizing the function $h(\beta,\sigma)$ over $B \times R^+$, where

$$h(\beta,\sigma) = \sum_{i=1}^{n} \varkappa_i \, \rho\left(\frac{y_i - f(x_i,\beta)}{\sigma} \right) \sigma . \tag{8}$$

$\varkappa_i$ is chosen in dependence on x_i, y_i, M_j, I_j.

In the simplest case on takes $\rho(t) = t^2$ or $\rho(t) = |t|$ and for $\varkappa_i$ one puts often

$$\varkappa_i = \begin{cases} 1 & \text{for } x_i \in M_j \text{ and } y_i \in I_j \text{ for one } j \\ 0 & \text{otherwise.} \end{cases} \tag{9}$$

In similar way we can work with a solution of the equation $h'(\ ,\ ) = 0$ in $B \times R^+$. We see that the essential difference in comparison with the classical methods is the restriction for β on B and the proposal for the choice of $\varkappa_i$.

We have said that the choice of $\varkappa_i$ depends on the classification procedure, i.e. dependson the $M_1,\ldots,M_k$. A somewhat more general choice of the $\varkappa_i$ one gets with a randomized classification procedure. As it is known the posteriori probability for the correspondence of a x to a class K_e is given by

$$q_e(x) = \frac{p_e g_e(x)}{\sum_{v=1}^{k} p_v g_v(x)} . \tag{10}$$

Then one chooses $\varkappa_i = q_{s_i}(x_i)$ where s_i is determined by $y_i \in I_{s_i}$.

Here we assume for sake of simplicity that the sets $I_1,\ldots,I_k$ are disjoint.

In this chapter we have worked mainly with the true densities $g_1,\ldots,g_k$. If these densities are unknown then we substitute the estimated $\hat{g}_1,\ldots,\hat{g}_k$.

In our approach it is important to choose the sets I_i in an appropriate way. Very often it would be an interval and we have then by our weights $\varkappa_i$ the possibility to prevent gross errors. In such a way we could use $\rho(t) = t^2$ in (8) and have even in this choice the same qualitative behaviour of our estimator as in the case where one works with (5) in (3). The advantage of our method is the data oriented choice of the constants.

Example: We consider a linear regression model (1) with $f(x,\beta) = f'(x)\beta$. For intervals $I_1,\ldots,I_k$ we see that B is a convex set. For $\rho(t) = t^2$ and $\varkappa_i$ as in (9) or (10) we can compute the robust regression easily with standard programs. If one can work approximately for g_j with normal distributions then the M_i are bordered by hyperplanes. The points z_{ij} are chosen not too near these borders.

We will remark that the composition of the classification methods with nonparametric estimators for surfaces can be very fruitful. This is discussed in H. Läuter (1982).

References

Anderson, T.W. (1958), Introduction to Multivariate Analysis, J. Wiley, New York.

Andrews, D.F., P.J. Bickel, F.R. Hampel, P.J. Huber, W.H. Rogers, J.W. Tukey (1972), Robust Estimates of Location: Survey and Advances, Princeton Univ. Press.

Bunke, H., W.H. Schmidt (1980), Asymptotic results on nonlinear Approximation of regression functions and weighted least squares, Math. Operat.forsch.Statist., Ser. Statistics 11, 3-22

Hampel, F. (1973), Robust estimation: A condensed partial survey, Z. Wahrscheinlichkeitstheorie verw. Geb. 27, 87-104.

Huber, P.J. (1972), Robust Statistics: a review, Ann. Math. Statist. 43, 1041-1067.
Huber P.J. (1964), Robust estimation of a location parameter, Ann. Math. Statist. 35, 73-101.
Jennrich, R.I. (1969), Asymptotic properties of non-linear least squares estimators, Ann.Math.Statist. 40, 633-643.
Läuter,H. (1982), Data preparation and parametric inference, Compstat-Lectures (submitted)
Schmidtke, D. (1982), Classification procedures with given probabilities for misclassification, Doctoral thesis, Acad. of Sc. of GDR.
Turkey, J.W. (1960), A survey of sampling from contaminated distributions. In: Contrib. to Prob. and Statist. (ed. Olkin), 448-485.

Standardized Comparison of Nonlinear Model Fitting Algorithms

S. LeRoux, A. Messean and J.-P. Vila, C.N.R.Z., France

Summary : An orthogonality true convergence criterion has been implemented in several well-known nonlinear least squares algorithms. This authorized a normalized comparison of these algorithms on several typical nonlinear models.
A strategy of use of this criterion is proposed to improve convergence process, classically based on relative change termination criteria,

Keys words : Nonlinear least squares ; convergence criterion ; orthogonality.

I- INTRODUCTION .-

The interest of this paper is twofold :

- comparison of several nonlinear squares algorithms performances,
- implementation and use of a statistically meaningful "true" convergence criterion.

This work is the first step of a nonlinear least squares software project, in which several new statistical features and numerical tools, suitable to the handling of the statistical and numerical model nonlinearity consequences, will be implemented.

This project started with the choice of a fitting algorithm, sufficiently well adapted to the class of nonlinear problems encountered in the field of biological data, usually characterized by a few number of parameters, frequent high nonlinearity and algebraic as well as differential definitions

We included several nonlinear least squares softwares from different origins, covering the most renown algorithms, Interested mainly by the algorithms performances, we decided to start a series of "convergence-numerlized" comparisons,

II- STANDARDIZATION AND CONVERGENCE .-

Comparison of nonlinear least squares softwares is not a new operation for statisticians concerned with the numerical aspects of fitting nonlinear models to data (HIEBERT, 1981 ; MORE, 1979), The underlying numerical methods are few and well known : Gauss-Newton, Levenberg-Marquardt, gradient and Newton-type methods.

Generally, codes are compared from two points of views :

- general numerical performances : success in fitting, precision, speed of convergence...
- handling and outputs : easy-to-use facilities, account of the fitting process, statistical outputs...

Very often it is hardly possible to distinguish the algorithm-effect from the programming effect, when comparing performances.

So an intended performance comparison test must be preceeded by some normalization of the codes.

In the case of a nonlinear least squares software, users are generally first interested in the speed of convergence of the minimization process to the obtained estimates, and by their closeness to the true least squares estimates. The convergence criteria encountered in most cases are of the "relative change" type :

- relative change in the sum of squares between two successive iterations :

$$(S(\theta^{(k)}) - S(\theta^{(k+1)}) / S(\theta^{(k)})$$

- relative change in parameter values :

$$|\theta_j^{(k+1)} - \theta_j^{(k)}| / |\theta_j^{(k)}| \qquad j=1,\ldots,p$$

When one or both of these two quantities fall under some prescribed value δ, solution is claimed, and the minimizing process stopped. Due to possible prematured convergence determined by these criteria, some other criteria are sometimes recommended (size of the gradient for example) but till now "there is no known criterion that is absolutely satisfactory" (Kennedy and Gentle, 1980).

In a recent paper, Bates and Watts (1981) emphazised the fact that all these criteria would be better called termination criteria, since they only reveal, when satisfied, a particular process behaviour in the specific path generated by each algorithm, not always associated with convergence. They proposed a convergence criterion based on the orthogonality of the residual vector to the derivative vectors.

In order to compare performances of several nonlinear least squares programs, we preliminarily introduced in the original source code of each program this orthogonality convergence criterion. This common implementation, while insuring true convergence, contributed to reduce the programming effect.

III- THE TESTED ALGORITHMS .-

We intended to compare convergence performances for the following well distributed nonlinear least squares softwares : NL2SOL (4) - VA05AD (5) - GAUSHAUS (9) - NONLIN (10) - VA02AD (13).

The algorithms used by these programs are based on four types of methods : Gauss-Newton, Levenberg-Marquardt, quasi Newton and Powell. We give briefly their main characteristics. They each solve the unconstrained nonlinear least squares problem :

$$\text{minimise} \quad S(\theta) = \frac{1}{2}\sum_{i=1}^{n} w_i \,(y_i - f(x_i,\theta))^2 = \frac{1}{2}\, R'(\theta)\, R(\theta) \qquad (1)$$

by finding θ such that the following necessary condition is satisfied :

$\nabla S(\theta) = J'(\theta)\, R(\theta) = 0 \quad (2)$,with J jacobian of R. All the first three methods use variations of the Newton step $\Delta\theta$ to improve successive approximations of (2) :

$$\nabla^2 S \,.\, \Delta\theta = -\nabla S \quad \text{or} \quad (J'J + \sum_{i=1}^{n} r_i \nabla^2 r_i)\, \Delta\theta = -\, J'R \qquad (3)$$

They differ in their approximations of the second term in the parenthesis :

- the Gauss-Newton method uses the null approximation, justified in the vicinity of the solution,
- the Levenberg-Marquardt approximates it by λI, when I is the pxp unit matrix and λ a scalar varying during the process, in order to take successively advantage of the steepest descent direction (λ high valued) and of the Gauss-Newton direction (λ low valued),
- the Quasi-Newton methods augment the Gauss-Newton hessian by a symmetric matrix, generally such that the second-order approximant transforms the current θ-change into the observed first order change of ∇S,
- the algorithm of Powell (1968) is a Gauss-Newton like algorithm, in which the expression of the correction vector $\Delta\theta$ as a linear combination of particular independent directions d, permits the replacement of the computation of all the first derivatives by the updating of only one of the directional derivatives at each step,

NONLIN uses the Gauss-Newton algorithm, GAUSHAUS the Levenberg-Marquardt algorithm. The algorithm of VA05A is hybrid, combining features from the Newton, steepest descent and Marquardt methods. NL2SOL uses a sophisticated algorithm which adaptatively decides when to use a Quasi-Newton update of the hessian. VA02A uses the Powell algorithm.

Most of the codes offer sets of processing options, with high computing effects we do not discuss here.

We do not claim for the best setting for each program in each test problem. Generally the default settings provided by the authors are used.

IV- THE ORTHOGONALITY CONVERGENCE CRITERION .-

Bates and Watts provided this already known criterion (Chambers, 1973 ; Dennis and al., 1981) with a meaningful scale, based on inherent variability of the estimates. They also proposed an efficient and computationally stable procedure for their criterion based on a QR decomposition of the jacobian.

In the space of the n observations, let $y = (y_1, y_2, \ldots, y_n)'$ and $\eta(\theta) = (\eta_1(\theta), \eta_2(\theta), \ldots, \eta_n(\theta))'$ where $\eta_t(\theta) = f(x_t, \theta)$, be the observed and expected responses vectors respectively. When θ varies, $\eta(\theta)$ generates the response-surface (the solution locus of Box and Lucas (1959)). Let $\hat{\theta}$ be the least squares estimate. If near $\eta(\hat{\theta})$ the solution locus can be reasonably approximated by its tangent plane at $\hat{\theta}$, the usual confidence region for the parameters is taken as those values of θ such that

$$\| \eta(\theta) - \eta(\hat{\theta}) \|^2 \leq p\, S(\hat{\theta})\, F(p, n-p;\, \alpha)/(n-p)$$

This defines a disk on the tangent plane with radius proportional to $[p\, S(\hat{\theta})/(n-p)]^{1/2}$. The residual vector $e = y - \eta(\hat{\theta})$ is orthogonal to this tangent plane. This orthogonality property will not be verified if the obtained $\hat{\theta}$ is not exactly the true least squares estimate : the "confidence" disk will be increased in radius and offset by $\| e_T \|$, where e_T is the now non null residual vector projection. This leads Bates and Watts to base the orthogonality criterion on the offset relative to the radius of the confidence disk. They avoid dependence on the confidence level by removing the factor $F(p, n-p\, ;\alpha)$ and define a relative offset at $\theta^{(i)}$ as $P = \| e_T \| / (p\, S(\theta^{(i)})/(n-p))^{1/2}$.

To manage with experimental designs including replications, where the residual sum of squares is inflated by the pure error sum of squares $Srep$, while $\| e_T \|$ remains the same, they define the relative offset as

$$P = (\| e_T \| / p^{1/2}) / ((S(\theta^{(i)}) - Srep)/(n-p-\nu))^{1/2}$$

where ν is the number of degrees of freedom for replications.

The criterion tolerance level can now be assigned a value on a meaningful scale, protecting the validity of the confidence region, while preventing premature termination. Meaningful setting of this kind are difficult to conceive with the former change criteria.

V- THE CRITERION IMPLEMENTATION .-

The computation of $\| e_T \| = (e'J\, (J'J)^{-1}\, J'e)^{1/2}$ is based on an orthogonal triangular decomposition of J as $J = QR$, where Q is an n by p matrix with unit orthogonal columns and R is a pxp upper triangular matrix (Chambers, 1977). Because the Q columns form a basis for the J columns, we have $\| e_T \| = \| Q'e \|$. The decomposition was done using the Householder triangularization method.

When the fitting algorithm is of the Gauss-Newton type, the computation of $\| e_T \|$ and then P, can proceed without anymore computing, from the efficient and numerically stable evaluation of the parameter increment :

$$\delta = (J'J)^{-1}\, J'e = R^{-1}\, Q'\, e.$$

As the computation of this criterion is more expensive than the classical criteria, we used the following strategy :

1) the process first begins with the classical sum of squares relative change

criterion (δ_s).

2) when the value of δ_s falls under a sufficiently small critical arbitrary value (10^{-2} in the test problems), the process goes on till the end, with the orthogonality criterion, which then can be seen as a convergence checking.

The change of criterion occurs for different values of the P criterion in the different codes. And for each test problem the comparison of these first P values is highly significative of the respective behaviour of each code : slackening of the process more or less sooner.

VI- THE TEST PROBLEMS .-

Most of our test problems were chosen among well known biological models, characterized by algebraic or differential equations.

Here we present results for four typical case problems :

- high number of parameters. Osborne test function (1977) :

$$S(\theta) = \sum_{i=1}^{65} \Big[y_i - \theta_1 \exp(-\theta_5 t_i) - \theta_2 \exp(-\theta_6 (t_i-\theta_9)^2) + \theta_3 \exp(-\theta_7 (t_i-\theta_{10})^2) + \theta_4 \exp(-\theta_8 (t_i-\theta_{11})^2)\Big]^2 \quad \text{where } t_i=(i-1)/10 \quad \text{(test 1)}$$

$i=1,\dots,65$ and y_i are specified values.

- low nonlinearity - Logistic growth curve :

$$f(\theta,x) = \theta_1/(1 + \exp(-\theta_2(x - \theta_3))) \qquad \text{(test 2)}$$

- high nonlinearity - Two compartmental kinetics

$$f_1(\theta,t) = \frac{\theta_2\theta_3}{\theta_1-\theta_2}\Big[\exp(-\theta_2 t) - \exp(-\theta_1 t)\Big] \qquad \text{(2}^{\text{nd}}\text{ compartment concentration}$$

test 3).

$$f_2(\theta,t) = \frac{\theta_3}{\theta_1-\theta_2}\Big[\theta_1 \exp(-\theta_2 t) - \theta_2 \exp(-\theta_1 t)\Big] \qquad \text{(total concentration)}$$

(tests 4 and 5 : two systems of initial values).

- differential definition : the differential forms of the last two preceeding models :

$$\frac{dA}{dt} = -\theta_2 A$$

$$\frac{dB}{dt} = \theta_2 A - \theta_1 B$$

$$f_1(\theta,t) = B \qquad \text{(test 6)}$$

$$f_2(\theta,t) = A+B \qquad \text{(test 7)}$$

The test 4 was rerun with different sets of initial values for the parameters in order to compare the robustness of the codes.

The tests were performed on a CII-HB Iris 80 computer.

VII- RESULTS AND DISCUSSION .-

The table shows the comparative behaviour of the five codes on the test problems.

The second rows give the values of the P criterion, computed for the first time when the sum of squares change criterion attained a chosen critical value (CR = 10^{-2}). The stopping convergence P level was of course the same for the five codes (10^{-5}). It is worth noting the high differences between these "first P" value for the same test : in all cases, the smallest first P value is realized by the

GAUSHAUS code (Levenberg Marquardt algorithm) and the highest by the VA05AD code (hybrid Newton-Gradient-Marquardt).

These different first P values for the same level of the sum of squares change criterion, reveal the respective behaviour of the codes : GAUSHAUS slackens more lately than any other code, for which the progression towards the solution is less smooth, with successive phases of damped acceleration and deceleration : this is evidenced by the third row which gives the number of times, NO, the relative change criterion value goes up again after the computation of the first P value.

As a strategy, it would be interesting for some algorithms to compute the first P value only when the relative change criterion falls under its critical value for the second or the third time.

The first row gives the CPU time required by each code to satisfy the stopping convergence P level. In each test problem GAUSHAUS is significantly faster than other codes. NL2SOL seems to have the second more regular convergence after that of GAUSHAUS. However its augmented Gauss-Newton hessian procedure, based on adaptative quadratic modeling, appears high time consuming, and made it the slowest of the five codes.

ALGORITHMS PERFORMANCES COMPARISON

ALGORITHM	PERFOR-MANCES	TEST 1 Osborne	TEST 2 Logistic	TEST 3 Comp-1	TEST 4 Comp-2(a)	TEST 5 Comp-2(b)	TEST 6 Comp-1 diff	TEST 7 Comp-2 diff
NONLIN Gauss-Newton	CPU	202	2	5	22	10	22	26
	P	.124+00	.311-01	.253+00	.102+01	.101+01	.253+00	.153+00
	NO	5	0	0	8	5	0	1
VA05AD Marquardt, Raphson Newton, Steepest descent	CPU	414	4	15	8	23	19	38
	P	.751+01	.423E-01	.252+00	.411-02	.339+00	.252+00	.102+00
	NO	39	3	5	3	54	4	22
VA02AD Powell	CPU	No convergence	4	248	15	28	164	39
	P	.134+00	.287-02	.699-01	.159+01	.138+01	.699-01	.966-04
	NO	-	1	12	5	6	1	0
GAUSHAUS Gauss-Marquardt	CPU	20	1	1	4	3	4	5
	P	.294-04	.86/-03	.157-05	.114-02	.285+00	.154-05	.520-01
	NO	0	0	0	0	0	0	0
NL2SOL Gauss-Newton augmented	CPU	1175	3	13	20	22	73	52
	P	.404-02	.592-03	.183-02	Overflow	.172+01	.183-02	.857-02
	NO	0	0	0	-	0	0	0

CPU : Computer time requested (sec) ; P : Orthogonality criterion value.
NO : Number of relative change criterion oscillations from the critical 10^{-2} value.

VA02AD (Powell algorithm) seems to have a rather good convergence. Its required CPU time is damaged by the necessary auxiliary evaluation of the jacobian for the P evaluation, as its algorithm was precisely designed to dispense with this derivatives calculation. The implementation of the P criterion in VA02AD seems less damaging for few parameters problems and makes this program preferable to NONLIN (Gauss-Newton algorithm). VA05AD is the least efficient on these test problems, with reported no progression and oscillations in decreasing the sum of squares.

CONCLUSION .-

At least for the class of problems we dealt with, the supremacy of the Levenberg-Marquardt algorithm seems well established. The computation of the relative offset orthogonality convergence criterion benefits from the algorithm itself and does not appear too much time consuming, when restricted to a vicinity of the least squares solution detected by a classical relative change criterion. This succeeding criteria policy seems to us the most economical strategy insuring true convergence to the least squares estimates.

Fortran subroutines of the jacobian QR decomposition and convergence criterion computation are available from the authors. A Basic version of the Levenberg-Marquardt algorithm can also be found in Jolivet (1982).

REFERENCES .-

[1] BARD Y. (1974). *Non linear parameter estimation*. New-York. Academic Press.

[2] BATES D.M. ; WATTS D.G. (1981). A relative offset orthogonality convergence criterion for nonlinear least squares. *Technometrics*, *23(2)*, 179-183.

[3] CHAMBERS J.R. (1973). Fitting nonlinear models : Numerical techniques. *Biometrika*, *46*, 77-90.

[4] DENNIS J.E. ; GAY M. ; WELSCH R.E. (1981). An adaptive nonlinear least squares algorithm. *ACM Trans. Math. Soft.* *7(3)*, 348-368.

[5] Harwell Subroutine Library-Catalog of Sûbroutines, AERE R 9185. *Comput. Sci. and Syst. Div.*, AERE, Harwell, Oxfordshire, U.K.

[6] HIEBERT K.L. (1981). An evaluation of mathematical software that solves nonlinear least squares problems. *ACM Trans. Math. Soft.* *7(1)*, 1-16.

[7] JOLIVET E. (1982). *Introduction à la modélisation en Biologie*. Masson,

[8] KENNEDY W.J. Jr. ; GENTLE J.E. (1980). *Statistical Computing*. New-York. Marcel Dekker.

[9] MEETER D.A. (1968). *Program GAUHAUS*. Numerical analysis laboratory, University of Wisconsin. Madison, Wisconsin,

[10] METZLER C.M. ; ELFRING G.L. ; Mc EWEN J.A. (1976). A user's manual for NONLIN and associated programs. *Research Biostatistics*. The Upjohn Company. Kalmazoo. Michigan 44001,

[11] MORE J.J. (1979). *Implementation and testing of optimization software*. DAMTP Rep 79/NA4, Cambridge Univ. Cambridge, U.K.

[12] OSBORNE M.R. (1972). *Some aspects of nonlinear squares calculations*. in Numerical methods for nonlinear optimization. F. Lootsma ed. Academic Press.

[13] POWELL M.J.D, (1965). A method for minimizing a sum of squares of nonlinear functions without calculating derivatives. *Comp. Journal* *7(4)*, 303-307,

Numerical Approach to the Optimal Design Problem for Regression Models with Correlated Errors

A.v.d. Linde, Free University of Berlin, F.R.G.

SUMMARY A numerical approach is presented for solving the problem of optimal design for linear regression models with correlated errors given a finite maximal sample size. The problem is posed as optimization problem for a nonlinear, non-convex objective function with linear constraints and an algorithm using numerical differentiation is applied to solve it. The method is applicable to find optimal designs with respect to the usual criteria related to statistical procedures of estimating and testing in the linear model as well as to some prediction problems.

KEYWORDS: Optimal design, linear regression with correlated errors, reproducing kernel spaces, optimization with linear constraints, optimization without evaluating derivatives.

1. INTRODUCTION

Consider a linear model

$$E(X(t)) = \Sigma_{r=1}^{R} a_r(t)\beta_r \quad , \quad t \in T = [a_o, b_o]$$

with regression functions a_r , $r = 1...R$, forming a Chebyshev-system, β a vector of unknown parameters. The dependency structure of measurements is given by a continous, positive definite covariance function

$$K: T\times T \ni (s,t) \rightarrow \mathrm{cov}(X(s),X(t)) \in \mathbb{R} \quad , \quad K(s,t) \neq c \cdot \delta_{s,t} \quad .$$

Let K be an element of the class of covariance functions satisfying

$$K(s,t) = u(s)v(t) \quad s < t$$

$$u,v \in C^2(T) \quad , \quad \frac{u}{v} \text{ positive and strictly increasing}$$

including the examples $K_\lambda(s,t) = e^{-\lambda|s-t|}$, $K(s,t) = \min(s,t)$. Observations are not repeatable and a design d is described by a subset $T_d = \{t_1...t_N\}$ or equivalently by a n-tuple $d = (t_1...t_N)$, $a_o \leq t_1 < t_1 <...< t_N \leq b_o$, given a sample size N.

The optimal design problem - to minimize the covariance matrix of the Gauss-Markov estimator $\hat{\beta}^*$

$$\mathrm{cov}\, \hat{\beta}^*(X_d) = A_d^T \Sigma_{K,d}^{-1} A_d$$

$$X_d = (X(t_1)...X(t_N))^T , \quad A_d = ((a_r(t_n)))_{\substack{r=1...R \\ n=1...N}} , \quad \Sigma_{K,d} = ((K(t_i,t_j)))_{i,j=1...N}$$

in terms of some functional (determinant, trace or a quadratic form) means to optimize the wellknown statistical procedures of estimating and testing in the linear model with resprect to the choice of observation points.

Usually an asymptotic approach (Sacks/Ylvisaker, 1966; Bickel/Herzberg, 1979) is made to solve the problem because of two reasons: to circumvent the problem of existence of an exactly N point optimal design and to avoid difficulties in computation. In this paper a solution to the first problem is proposed by defining an appropriate region of admissable designs. With respect to the second problem investigations about the objective function in the optimal design problem are presented that prove the applicability of optimization algorithms.

2. EXISTENCE OF OPTIMAL DESIGNS

The region of admissable designs Δ is to satisfy two conditions:
(i) it has to be compact, the objective function has to be continous in Δ;
(ii) the parameter vector β has to be estimable for every $d \in \Delta$. With the given assumptions, (ii) is valid for every design d with at least R distinct points.

a) $N = R$

In case $N = R$ the naturally given region

$$\Delta_R = \{d | d = (t_1 \dots t_R),\ a_o \le t_1 < \dots < t_R \le b_o\}$$

can be modified to achieve compactness by introducing a parameter ε , preserving the distinctness of observation points:

$$\Delta_R^\varepsilon = \{d | d \in \Delta_R\ ,\ t_{r+1} - t_r \ge \varepsilon\} \quad .$$

This parameter does not imply any practical restriction because anyway measurements can be taken according to the precision of the technical apparatus only.

b) $N > R$

First consider the region $\Delta_N^{cl} = \{d | d = (t_1 \dots t_N)\ ,\ a_o \le t_1 \le \dots \le t_N \le b_o\}$. By a result of Sacks/Ylvisaker (1966) an interpretation of boundary points of Δ_N^{cl} as designs is derived from the continuity of the objective function: a boundary point δ is interpreted as a design d which contains only the distinct components of δ. This result is valid for a class of covariance functions including the preassumed one that are not differentiable in the diagonal of $T \times T$. It is proved using a representation of elements of the information matrix - the inverse of the covariance matrix of $\hat{\beta}^*(X_d)$ - as inner products in the reproducing kernel Hilbert space $H(K)$ corresponding to the covariance function K. This interpretation of boundary points of Δ_N^{cl} yields two consequences: (i) The estimability of β is not garantueed in all points of Δ_N^{cl}; of course this can be treated as above introducing ε. (ii) There are designs in Δ_N^{cl} not exhausting the sample size N. Indeed it turns out to be reasonable to deal with a maximal sample size N instead of a fixed sample size N.

There are examples showing that - if there is a correspondence between regression functions and covariance functions - there exists an optimal design concentrated in less than N points. (For instance, consider the linear regression model $E(X(t)) = \beta_o + \beta_1 t$, $K(s,t) = \min(s,t)$. For any $N > 2$ the covariance matrix of $\hat{\beta}^*(X_d)$ depends on d only through t_1 and t_N.) Therefore it seems to be appropriate to choose $\cup_{R\leq L\leq N}\Delta_L$ as region of admissable designs in case $N > r$, respectively the corresponding compact region

$$\Delta^{\varepsilon}_{N,R} = \{d|d = (t_1\ldots t_N);\ a_o\leq t_1<\ldots<t_R\leq t_{R+1}\leq\ldots\leq t_N\leq b_o;\ t_{r+1}-t_r\geq\varepsilon\ ,\ r = 1\ldots R-1\}\ .$$

Such a choice also allows to evaluate numerically the gain of information by additional observations.

In this way the domain of the objective functions of the optimal design problem is constructed as a convex, compact subset of $\mathbb{R}^N$ characterized by linear constraints.

Also by this construction a dependency of the optimal design on the artificial parameter ε has been introduced. It can be shown however, using results of Evans/Gould (1970) and Dantzig/Folkman/Shapiro (1967) that the optimal design varies continously with ε, more precisely that the mapping

$$\varepsilon \to \inf\{\emptyset(d)\,|\,d \in \Delta^{\varepsilon}_{N,R}\}$$

is continous and the set-valued mapping

$$\varepsilon \to \{d^* \in \Delta^{\varepsilon}_{N,R}|\emptyset(d^*) = \inf_{d\in\Delta^{\varepsilon}_{N,R}} \emptyset(d)\}$$

is closed, $\emptyset$ denoting the objective function.

3. DIFFERENTIABILITY OF OBJECTIVE FUNCTIONS

To apply a usual minimization algorithm to the optimal design problem it is necessary that the objective functions are twice partially differentiable in a neighbourhood of the optimization region. As the objective functions in the optimal design problem are defined only in Δ an extension argument is used to prove differentiability. Essentially the proof of differentiability can be reduced to showing differentiability of the component functions of the information matrix

$$i_K(a_r,a_{r'}):\ \Delta^{\varepsilon}_{N,R} \ni d=(t_1\ldots t_N) \to (a_r(t_1)\ldots a_r(t_N))\ \Sigma^{-1}_{K,d}(a_{r'}(t_1)\ldots a_{r'}(t_N))^T \in \mathbb{R}$$

by showing that there is a continous extension of the partial derivatives from the interior of $\Delta^{\varepsilon}_{N,R}$ to the boundary. Using a representation of the component functions in terms of the corresponding reproducing kernel Hilbert space $H(K)$ this can be done by elementary analytic arguments. For the class of covariance functions considered here such a representation is given by

$$i_K(a_r,a_{r'})(d) = \Sigma_{n=1}^{N-1} \frac{[(\frac{a_r}{v})(t_{n+1}) - (\frac{a_r}{v})(t_n)][(\frac{a_{r'}}{v})(t_{n+1}) - (\frac{a_{r'}}{v})(t_n)]}{[(\frac{u}{v})(t_{n+1}) - (\frac{u}{v})(t_n)]} + \frac{a_r(t_1)a_{r'}(t_1)}{u(t_1)v(t_1)}$$

for $d = (t_1\ldots t_N)$, $t_1 <\ldots< t_N$.

4. EXTENSIONS

a) The arguments developed above are as well applicable to the prediction problem: The mean square prediction error of the best unbiased linear predictor $\hat{Y}^*(X_d)$ for a random variable $Y \notin \text{span}\{X(t_1),\ldots,X(t_N)\}$ given var Y and cov $(X(t_n),Y)$, $n = 1\ldots N$, depends on d only through quadratic forms in $\Sigma_{K,d}^{-1}$ to which investigations of analytic properties of the objective function can be reduced. Special applications are prediction of $Y = X(t_o)$, $t_o \notin T$ or

$$Y = \frac{1}{b_o - a_o} \int_{a_o}^{b_o} X(t)dt \; .$$

b) Everything said above is valid for the class of covariance functions $K_\lambda(s,t) = 1-\lambda|s-t|$, $\lambda \in \mathbb{R}\setminus\{0, \frac{2}{b_o - a_o}\}$. For some covariance functions of higher order of differentiability on the diagonal of T×T an analogous argument can be applied. (For example: $N = 2$, $R = 1$, $a(t) = 1$, $K(s,t) = 1 - (s-t)^2$) . In this case L_2-derivatives of the stochastic process $\{X(t),\ t \in T\}$ occur as observations.

c) Analogous to the stability of the optimal design with respect to a variation of the parameter ε, a continous dependency of the optimal design on a family parameter of covariance functions like $K_\lambda(s,t) = e^{-\lambda|s-t|}$ can be established.

5. NUMERICAL RESULTS

Because of the complexity of the objective functions an algorithm without evaluating derivatives was chosen to perform the optimization. Bräuninger (1979) introduced a quasi-Newton method for minimization under linear constraints with good convergence properties. This algorithm was applied to the optimal design problem after slightly adjusting the approximation of derivatives according to the restricted domain Δ of the objective functions.

EXAMPLES:

The table below shows some results for the polynomial and exponential regression with $K(s,t) = e^{-\frac{1}{2}|s-t|}$. The domain of the objective function $\emptyset$ is $\Delta_{N,R}^{\varepsilon}$ with $N = 4$, $R = 2$, $\varepsilon = 10^{-8}$. As objective functions the determinant of the information matrix (DET $\sim$ det (cov $\hat{B}^*(X_d))^{-1}$) , the variance in the extrapolation problem for $t_o \notin T$, (E(t_o) $\sim$ $(a_1(t_o)\ldots a_R(t_o))$ cov $\hat{B}^*(X_d)(a_1(t_o)\ldots a_R(t_o))^T$) and the prediction error of the predictor of $X(t_o)$, $t_o \notin T$, (P(t_o)) , are considered. The starting point is denoted by d_o, the optimal design by d*.

T	$a_r(t)$	$\emptyset$	d_o	$\emptyset(d_o)$	d*	$\emptyset$(d*)
[0,1]	t^{r-1}	DET	(0.01,0.5,0.55,0.6)	0.7814	(0.,0.329,0.671,1.)	1.5878
			(0.,0.5,1.)	1.5868		
			(0.,1.)	1.5820		
[-2,2]	$e^{-(r-1)t}$	DET	(-1.5,-0.8,0.5,1.7)	27.5478	(-2.,-1.039,0.334,2.)	85.4652
			(-2.,2.)	53.5982		
[-1,1]	$e^{-(r-1)t}$	E(2)	(-0.2,-0.1,0.1,0.2)	2.6376	(-1.,-0.464,0.201,1.)	0.9962
		P(2)		2.5794	(-1.,-0.473,0.189,1.)	0.8182
			(-0.99,-0.5,0.5,0.99)	1.0077		
				0.8276		
			(-1.,0.,1.)	1.0192		
				0.8232		
			(-1.,1.)	1.1375		
				0.8485		

The next table shows the variation of the optimal design d* depending on the covariance function. For T = [-1,1], $d \in \Delta^{\varepsilon}_{N,R}$ with N = 4, R = 2, $\varepsilon = 10^{-8}$, $\emptyset(d) = \det(\text{cov}\ \hat{B}^*(X_d))^{-1}$, the exponential regression $a_r(t) = e^{-(r-1)t}$ and $K_n(s,t) = e^{-n|s-t|}$, $K_\infty(s,t) = \delta_{s,t}$ the following results are obtained, $d^*_n = (t^*_{1,n},\dots,t^*_{4,n})$ denoting the optimal design with respect to K_n. (Note that in the uncorrelated case when repeated observations are allowed, the optimal design is given according to the equivalence theorem by the probability measure $\delta^* = \frac{1}{2}\varepsilon_{-1} + \frac{1}{2}\varepsilon_1$, the corresponding value of the determinant of the information matrix being 22.0975.)

n	$t^*_{1,n}$	$t^*_{2,n}$	$t^*_{3,n}$	$t^*_{4,n}$	$\det(\text{cov}\ \hat{B}^*(X_{d_n}))^{-1}$
1	-1.	-0.3333333	0.33333	1.	7.12702696
xc 5	-1.	-0.35915	0.463712	1.	12.26096098
10	-1.	-0.7437582	0.6297497	1.	14.76566943
50	-1.	-0.9226841	0.887618	1.	19.45418632
100	-1.	-0.954826	0.936253	1.	20.53291098
200	-1.	-0.974069	0.964476	1.	21.19922880
300	-1.	-0.981392	0.974919	1.	21.45496049
∞	-1.	-0.99999999	0.99999999	1.	22.097565

7. REFERENCES

Bickel, P.S. / Herzberg, A.M. (1979), Robustness of Design against Autocorrelation in Time I: Asymptotic theory, Optimality for Location and Linear Regression, Ann.Stat., 7, 77-93.
Bräuninger, J. (1979), A Quasi-Newton-Methid for Minimization under Linear Constraints without Evaluating Derivatives, Computing, 21, 127-141.
Dantzig, G.B. / Folkman, J. / Shapiro, N. (1967), On the Continuity of the Minimum Set of a Continous function, J.Math.Anal.Appl., 17, 519-548.
Evans, J.P. / Gould, F.J. (1970), Stability in Nonlinear Programming, Operations Research, 18, 107-118.
Parzen, E. (1967), Time Series Analysis Papers, Holden Day.
Pazman, A. (1974), A Contribution to the Time Series Design Problems, Transactions of the Seventh Prague Conference and of the European Meeting of Statisticians 1974, Vol. A.
Sacks, J. / Ylvisaker, D. (1966), Design for Regression Problems with Correlated Errors, Ann.Math.Statist., 37, 66-89.

An Experimental Design and Analysis Package for Microcomputers

J.W. McNicol and S.C.M. Ng, Scottish Crop Research Inst., G.B.

Summary

This paper describes a package which provides statistical design and analysis facilities for plant breeders' variety trials. The programs are user friendly and the package is mounted on a Apple II microcomputer.

Keywords: statistical computing; variety trials; alpha designs; UCSD pascal.

1. Introduction

Much of the routine statistical work at the Scottish Crop Research Institute, Pentlandfield (SCRIP) involves the design and analysis of variety trials which are part of the forage brassica, spring barley and potato breeding programmes. The objective in these plant breeding programmes is to produce new varieties which compare favourably with established varieties. A typical brassica variety trial might have three replicates each with thirty entries of which six are controls i.e. established varieties. The variety trial designs used at SCRIP are either randomised complete blocks or, when there are twenty or more entries per replicate, alpha designs as described by Patterson,Williams and Hunter (1977).

The Variety Trial Library (VTL) described in this paper allows users to design and analyse such trials on a microcomputer. The package forms part of an integrated data capture and data processing system based on Apple II microcomputers which was introduced at SCRIP in 1980 and is outlined in McNicol, Ng and Kidger (1982). The aims of this system are to allow data to be transferred from hand held data terminals and electronic balances directly to file on a microcomputer, and to allow such data sets to be analysed on the micro or to be sent to a local mainframe for processing. The link between the microcomputer and the mainframe is achieved through software provided by Edinburgh Regional Computing Centre, This software, described by Tate (1981), allows the microcomputer to act as an interactive terminal to the mainframe. By having the VTL package as part of this system users can analyse for example, brassica plant dry matter yields on the microcomputer immediately after the sample fresh and dry weights have been recorded from the electronic balance and the plot fresh weights appropriately adjusted.

2. General features of the package

The emphasis throughout the writing of this package has been to make it as easy to use as possible. It is intended that staff with little or no computing experience should be able to design and analyse their routine trials using these programs. No attempt has been made to provide the general data base or statistical facilities found in comprehensive mainframe packages such as described by Talbot and Robinson (1980).

The package consists of six programs, namely

1. Enter/edit variety names
2. Generate a randomised complete block (rcb) design
3. Generate an alpha design
4. Enter/edit data for analysis
5. Analyse data from an rcb experiment
6. Analyse data from an alpha experiment

The software is written in University of California San Diego (UCSD) Pascal for an Apple II microcomputer, and as such can be transferred to any micro which uses the UCSD operating system. The minimum Apple configuration is 48k of memory, a Pascal language card and two floppy disk drives.

All six programs are contained on the one turnkey disk and are menu driven. Input via the keyboard is made in response to fully explanatory screen prompts and is checked for errors and inconsistencies. Where possible the user is asked to retype any mistakes. This approach ensures that program crashes and the consequent retyping of input are almost completely eliminated. If a program is unable to complete a run successfully, for example because of a data file corruption, the appropriate message is written to the screen and the user is returned to the package menu.

Each trial involves at least three disk files, one for the variety names, one for the design details and one for each variate which is to be analysed. A summary of the results may also be saved on disk. Character files are used throughout. Although this slows down reading and writing to disk, such files can be examined and edited by means of the UCSD editor. This has two advantages, firstly; it enables naive users to see their design and data files, thus reducing some of the mystique of computing and secondly it allows users to set up their own design files and analyse data from trials which were not generated by this package.

3. Trial Design

The variety names are saved on disk using program 1 in the package. Then, for both randomised complete block designs and alpha designs, the user is prompted for the trial name, site, crop, the number of replicates, the number of entries per replicate and a random number seed. A maximum of 4 replicates and 100 entries is permitted.

For alpha designs further information is required. Each replicate is divided into small blocks, each usually though not necessarily, having the same number of plots. The entries must then be allocated to the small blocks in such a way that the number of concurrences over the small blocks of any two entries is 0, 1 or 2. Where possible concurrences of size 2 are avoided. To achieve this allocation the program must have access to a generating array. For the purposes of analysis the average efficiency of the design must also be made available. The set of generating arrays given in Patterson, Williams and Hunter (op cit) and their corresponding average efficiencies are contained within the design program. If the appropriate array and average efficiency for a particular design are not held within this set they may be entered by the user from the keyboard. The program will then check that the concurrences of all pairs of varieties are less than or equal to 2, and the run will be terminated if this is not the case.

The design program also allows the generation of field plans. These field plans are listed on the printer and show the layout of the trial as it will appear in the field. The user must consider the trial as a rectangular collection of plots viewed from its bottom left hand corner. Each column of plots is referred to as a bed. The user is first asked whether the replicates lie across the field, down the field or, in the case of four replicates, two across and two down. Next the number of beds per replicate must be entered. Finally, for alpha designs, the user is asked whether the plots within small blocks run across beds or with beds. An indication of the kind of output produced for a four replicate trial with twenty entries in five small blocks of size four is given in Figure 1. The corresponding field plan is also shown. In this layout the four replicates are laid out in a square, there are four beds per replicate and the plots within small blocks run across the beds.

code	variety	rep one plt	rep one vty	rep one bk	rep two plt	rep two vty	rep two bk	rep three plt	rep three vty	rep three bk	rep four plt	rep four vty	rep four bk
1	Lair	1	19	4	21	4	4	41	19	2	61	17	1
2	Canard	2	9	4	22	10	4	42	15	2	62	8	1
3	Emerald	3	4	4	23	11	4	43	6	2	63	1	1
4	Bishop	4	14	4	24	17	4	44	2	2	64	15	1
5	Samo	5	20	5	25	9	3	45	18	1	65	3	3
6	Nevin	6	15	5	26	15	3	46	1	1	66	12	3
7	B15	7	5	5	27	16	3	47	14	1	67	19	3
8	B20	8	10	5	28	3	3	48	10	1	68	10	3
9	B17	9	18	3	29	1	1	49	11	3	69	14	5
10	B2	10	3	3	30	13	1	50	20	3	70	16	5
11	C7	11	13	3	31	7	1	51	3	3	71	5	5
12	A5	12	8	3	32	19	1	52	7	3	72	7	5
13	G2	13	2	2	33	2	2	53	5	5	73	6	4
14	D1	14	12	2	34	14	2	54	13	5	74	20	4
15	G1	15	7	2	35	8	2	55	17	5	75	13	4
16	B5	16	17	2	36	20	2	56	9	5	76	4	4
17	C8	17	16	1	37	12	5	57	16	4	77	9	2
18	D5	18	1	1	38	5	5	58	8	4	78	18	2
19	E1	19	11	1	39	6	5	59	4	4	79	2	2
20	E4	20	6	1	40	18	5	60	12	4	80	11	2

*** field plan ***

10	16	8	4	12	9	18	2	11
9	9	17	13	5	4	13	20	6
8	11	20	3	7	14	16	5	7
7	10	14	1	18	10	19	12	3
6	19	15	6	2	17	8	1	15
5	16	1	11	6	12	5	6	18
4	17	7	12	2	20	8	14	2
3	18	3	13	8	1	13	7	19
2	10	5	15	20	3	16	15	9
1	19	9	4	14	4	10	11	17
bed	1	2	3	4	5	6	7	8

Figure 1.

4. Trial Analysis

Each variate to be analysed is held on a separate file on disk. The data for any one variate may be entered either by using program 4, or by using the UCSD editor. In either case individual data values must be entered one per line but otherwise in free format. Missing values are denoted by an asterisk. Transformations allowed are exponential, inverse, angular, natural log and square root. The analysis programs will accept data in standard order or in field plan order.Thus for the example shown in Figure 1, the data could be in variety order 19, 9, 4, 14, 20, 15, 5,.... or in variety order 19, 10, 18, 17, 16, 19, 10, 11, 9, 16, 9, 5,.... These are the two most commonly used scoring orders used as SCRIP.

At the analysis stage only the names of the design and data files need be entered. These are read and the appropriate analysis is performed. The full analysis includes the analysis of variance table, standard errors and least significant differences, and a table of entries together with their rank, percentage of the mean of the controls, mean, adjusted mean for alpha designs, and the plot data. Again missing values are denoted by asterisks. The controls as defined in the analysis do not necessarily correspond to those implied in the list of entry names. They are designated at the analysis stage and such designations are valid only during the current run of the program.

Full account is taken of missing values. Should a complete replicate be missing, as has happened when part of a field is flooded, the analysis will ignore this replicate altogether, and treat the trial as having one replicate fewer than the original design. The same is true of entries for which all plots are missing. A maximum of 16 missing values apart from those which come into the above categories can be handled by the complete blocks program. The only constraint in the alpha analysis program is that the residual degrees of freedom is positive.

Acknowledgements

We are indebted to the Agricultural Research Council Unit of Statistics, Edinburgh, which provided the alpha design and analysis algorithims, and which made available a comprehensive list of generating arrays and average efficiencies.

References

McNicol J.W., Ng S.C.M., and Kidger R. (1982), A data-capture system for agricultural research based on a microcomputer, Expl. Agric., 18.

Patterson H.D., Williams E.R. and Hunter E.H. (1977), Block designs for variety trials, J. Agric. Sci., 90, 395-400.

Talbot M. and Robinson D.L. (1980), CVT - a statistical computing and data management system in agricultural crop research, Compstat 1980, 81-87.

Tate A. (1981), The UCSD p-system at Edinburgh University, USUS News No. 4, 63-65.

Propositions for Fuzzy Characteristic Functions in Data Analysis

J.L. Mallet, Ecole Nationale Superieure de Geologie, Nancy, France

SUMMARY

The author proposes a data coding technique using fuzzy characteristic functions, the construction of which depends on initial data. Application of a linear data analysis method to these coded data is (approximately) equivalent to the corresponding non linear method applied to the initial data.

KEY-WORDS : Fuzzy-sets, Parzen's Kernels, Coding

ABBREVIATIONS :

R.V. = Random Variable
C.F. = Characteristic Function

1 - INTRODUCTION

1.a - Set $D^n = D^n(\Omega, A, P)$

Definition

We will designate $D^n = D^n(\Omega, A, P)$ the set of all R.V. X defined on (Ω, A, P) having their values in $\mathbb{R}^n$ and having a density f_X which satisfies the three following conditions :

1°) f_X has the following form where ν is a finite number and $\{x(1), \cdots, x(\nu)\}$ a set of ν points in $\mathbb{R}^n$:

$$f_X(x) = \sum_{i=1}^{\nu} p_i \cdot \Delta(x|x(i)) \qquad \forall\, x \in \mathbb{R}^n$$

2°) each function $\Delta(\cdot|x(i))$ is a probability density having $x(i)$ for mean value and such that :

$$\begin{cases} \Delta(\cdot|x(i)) \in L^2(\Omega, A, P) \\ 0 \leq \Delta(x|x(i)) \leq \Delta(x(i)|x(i)) < +\infty \qquad \forall\, x \in \mathbb{R}^n \\ \lim\limits_{\|x\| \to \infty} \|x\| \cdot \Delta(x|x(i)) = 0 \end{cases}$$

3°) coefficients p_i are non-negative

Justification

Using measure theory, it may be demonstrated that the probability law P_X of a R.V.X having its values in $\mathbb{R}^n$ always may be approximated by the probability law $P_{\tilde{X}}$ of a R.V. $\tilde{X}$ belonging to D^n. For this reason, we can always assume that each R.V. X with values in $\mathbb{R}^n$ approximately belongs to $D^n(\Omega, A, P)$.

Furthermore, Parzen-Rosenblatt's approximation theory for probability density functions gives us an useful technique for the approximation of f_X using an N-sample of X. (Cf. Tapia and Thompson, 1978 ; Hermans and Habbema, 1976).

1.b - Properties of D^n

The following implication is always satisfied :

$$X = [X_1, \cdots, X_n]^t \in D^n(\Omega, A, P) \Rightarrow X_i \in L^2(\Omega, A, P) \qquad \forall i$$

Furthermore, for all R.V. $X \in D^n$, we have the following relations where $x(i)$ and $C(i)$ are respectively the mean and covariance of the density function $\Delta(\cdot|x(i))$:

$$E(X) = \sum_i p_i \cdot x(i) \quad ; \quad E(XX^t) = \sum_i p_i \cdot \{C(i) + x(i) \cdot x^t(i)\}$$

1.c - Fuzzy characteristic functions

Definition

For all point s in $\mathbb{R}^n$, we call "fuzzy characteristic function" of s, all function $\varphi_s(\cdot)$ such that :

$$0 \leq \varphi_s(x) \leq \varphi_s(s) < 1 \qquad \forall x \in \mathbb{R}^n$$

Generally, we choose $\varphi_s(\cdot)$ in such a way that the value $\varphi_s(x)$ may be interpreted as a "proximity index" between the point x and the reference point s (cf. Kaufmann, 1971).

Fuzzy C.F. induced by a R.V. X

Let X be a R.V. defined on (Ω, A, P) taking its values in $\mathbb{R}^n$ and let $\varphi_s(\cdot)$ be a $\mathcal{B}^n$-measurable fuzzy C.F. of the point $s \in \mathbb{R}^n$. If we let ...

$$\varphi_s^x(\omega) = \varphi_s\{X(\omega)\} \qquad \forall \omega \in \Omega$$

... then, φ_s^x is a R.V. such that :

$$0 \leq \varphi_s^x(\omega) \leq \varphi_s^x(\omega_s) \leq 1 \qquad \forall \begin{cases} \omega \in \Omega \\ \omega_s \in X^{-1}(s) \end{cases}$$

If the elementary events ω are known only by their effects $X(\omega)$, then by induction, it is logical to interpret $\varphi_s^x(\omega)$ as a proximity index of ω to the A_x-measurable set $A_s = X^{-1}(s)$.

By definition, we shall say that φ_s^x is a fuzzy C.F. induced by X at the point s.

1.d - Fuzzy disjunctive and generating C.F.

Introduction

Let X be a R.V. belonging to D^n. For each point $x(i)$ used in the definition of f_x, there are an infinite number of ways to define an associated fuzzy C.F. $\varphi_{x(i)}(\cdot)$. Among these, there are two in particular that we have studied :

Fuzzy generating C.F.

We propose to call "fuzzy generating C.F." associated with X, the ν functions $J_i(\cdot)$ defined by :

$$J_i(x) = \frac{\Delta(x|x(i))}{\Delta(x(i)|x(i))} \qquad \forall \begin{cases} x \in \mathbb{R}^n \\ i = 1, \cdots, \nu \end{cases}$$

The qualifying "generating" is proposed to indicate that the knowledge of the functions $J_i(\cdot)$ and of the coefficients p_i allow us to reconstitute the distribution of X. In this paper, we shall say nothing more about this kind of fuzzy C.F. for the reason that we have already published some results about it (cf. Mallet, 1980).

Fuzzy disjunctive C.F.

We propose to call "fuzzy disjunctive C.F." associated with X, the ν functions $I_i(\cdot)$ defined by :

$$I_i(x) = p_i \cdot \frac{\Delta(x|x(i))}{f_X(x)} \qquad \forall \begin{cases} x \in \mathbb{R}^n \\ i = 1, \cdots, \nu \end{cases}$$

Fuzzy disjunctive C.F. induced by X will be noted I_i^X and we write :

$$I^X(\omega) = [I_1^X(\omega), \cdots, I_\nu^X(\omega)]^t$$

The qualifier "disjunctive" attributed to these fuzzy C.F. will be justified by the first property 2.a.

REMARK

Let us assume that densities $\Delta(\cdot|x(i))$ are of the following type where $\{B_i\}$ is a partition of $\mathbb{R}^n$:

$$\Delta(x|x(i)) = \begin{cases} \delta_i \text{ if } x \in B_i \\ 0 \text{ else} \end{cases}$$

Under these condition, we can easily verify that we have :

$$I_i(x) = 1_{B_i}(x) = \begin{cases} 1 \text{ if } x \in B_i \\ 0 \text{ else} \end{cases}$$

Therefore, in this particular case, fuzzy disjunctive C.F. can be identified with the notion of classical disjunctive C.F.

2 - FUZZY DISJUNCTIVE C.F.

2.a - Properties

Disjunction

For all R.V. X belonging to $D^n(\Omega, A, P)$, R.V. I_i^X are A_X-measurables, and if we let ...

$$A_i^X = \{\omega \in \Omega : I_i^X(\omega) > 1/2\}$$

... then we have the following disjunctions :

$$A_i^X \cap A_j^X = \emptyset \qquad \forall i \neq j$$

Invariance

Let Φ be a bijection from $\mathbb{R}^n$ to $\mathbb{R}^n$ such that Φ^{-1} be continuously differentiable, and let X and Y be R.V. such that :

$$Y = \Phi(X)$$

Under these conditions, if X belongs to $D^n(\Omega, A, P)$, then so does Y, and we have :

$$I^y \equiv I^x$$

2.b - Fundamental Proposition

Notations

In that which follows below, we shall write :

$$\left| \begin{array}{lcl} 1_\Omega(\omega) & = & 1 \quad \forall \ \omega \in \Omega \\ E^x(\cdot) & = & \text{mean operator conditioned by X} \end{array} \right.$$

Furthermore, we shall use the following notation concerning R.V. equivalence classes :

1) $\dot{U}$ = equivalence class containing the R.V. U :

$$\left.\begin{array}{l} U' \\ U'' \end{array}\right\} \in \dot{U} \iff U'(\omega) = U''(\omega) \qquad P\text{-}p.s.$$

2) For all $\dot{Z}$ such that :

$$\dot{Z} = [\dot{Z}_1, \cdots, \dot{Z}_n]^t \quad \text{with} \quad \dot{Z}_i \in L^2(\Omega, A, P) \quad \forall\, i$$

.... we shall designate by H_z, the $L^2(\Omega, A, P)$ subspace defined by :

$$H_z = \left\{ \dot{U} \in L^2(\Omega, A, P) \mid \exists\, u \in \mathbb{R}^n : \dot{U} = u^t \cdot \dot{Z} \right\}$$

3) For all $\dot{Y}$ belonging to $L^2(\Omega, A, P)$, we shall let :

$$\left| \begin{array}{l} \Pi_z(\dot{Y}) = \text{orthonormal projection of } \dot{Y} \text{ on } H_z \\ \mathcal{E}^z(\dot{Y}) = \text{linear regression of } \dot{Y} \text{ as a function of } \dot{Z} \end{array} \right.$$

Proposition

For all R.V. X belonging to $D^n(\Omega, A, P)$, we have :

$$\left\| \begin{array}{l} \left.\begin{array}{lcl} I^x_i \in L^2(\Omega, A, P) & & \\ E(I^x_i) & = & p_i \\ E(X \cdot I^x_i) & = & p_i \cdot x(i) \end{array}\right\} \forall\, i \\ \sum_i \dot{I}^x_i \quad = \quad \dot{1}_\Omega \\ \dot{X} \notin H_x \quad \text{in general} \\ \mathcal{E}^{I^x}(\cdot) = \Pi_{I^x}(\cdot) \end{array} \right.$$

Furthermore, for all R.V. Y with values in $\mathbb{R}^{n_y}$ and such that the pair (X,Y) belongs to $D^{n_x + n_y}$ and admits a "factorized" density of the following kind ...

$$f_{xy}(x,y) = \sum_i p_i \cdot \Delta_x(x|x(i)) \cdot \Delta_y(y|y(i))$$

... we have :

$$\Rightarrow \quad E^x(\dot{Y}) = \mathcal{E}^{I^x}(\dot{Y}) = \Pi_{I^x}(\dot{Y})$$

2.c - Application to data coding

Fuzzy disjunctive coding

Let X be a R.V. assumed to belong (at least approximatly) to D^n :

$$X = [X_1, \cdots, X_j, \cdots, X_n]^t$$

In data analysis, basic informations concerning X is always constituted by values $X(\omega_i)$ taken by X over a finite set of elementary events $\{\omega_1, \cdots, \omega_k, \cdots, \omega_N\}$; these values are stored in a data table D_x as indicated in figure 1-a. Starting from this table, it is always possible to construct an approximation f_x of the density of X. Therefore, we can construct a fuzzy disjunctive C.F. I^x associated with X :

$$I^x = [I_1^x, \cdots, I_i^x, \cdots, I_\nu^x]^t$$

As indicated in figure 1-b, we shall call "fuzzy disjunctive coding" of D_x, the data table D_{I^x} generated by I^x :

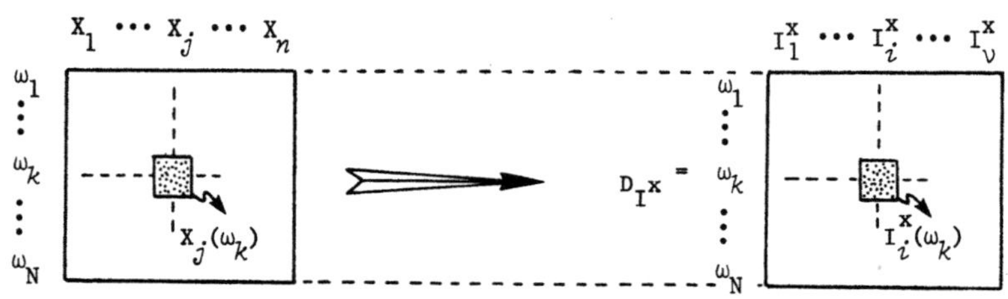

Fig. 1.a *Fig. 1.b*

Figure 1 - Fuzzy disjunctive coding of a data table

The transformation $X \rightarrow I^x$ may be considered as a change of variables taking into account initial data D_x through the approximated density f_x.

Corollary

Taking account proposition 2.b, it is clear that the linear regression operator $\mathcal{E}^{I^x}(\cdot)$ is a non-linear approximation of $E^x(\cdot)$. Consequently, all linear data analyses of D_{I^x} may be considered as approximations to the corresponding non-linear data analyses of D_x. The invariance property defined in paragraph 2.a guarantees a certain stability of results with respect to initial coding X of phenomena being studied.

CONCLUSION

Notions presented in this paper must be considered only as possible techniques to define fuzzy C.F. in data analysis. Other techniques are possible as for example those developed by Martin (1980). The common purpose of all these techniques is to give us tools for the development of non-linear data analyses.

REFERENCES

Hermans J. and Habbena J.D.F. (1976), The Alloc package multigroup discriminant analysis programs based on direct density estimation.
Compstat 1976.

Kaufmann A. (1971), sous-ensembles flous.
Tome 1, Masson, Paris.

Mallet J.L. (1980), Proposition pour un coefficient de corrélation entre individus.
Bulletin de l'Association des Statisticiens Universitaires, Vol. 1, pp. 19-31.

Martin J.F. (1980), Le codage flou et ses applications en statistique.
Thèse 3ème Cycle, Université de Pau , France.

Tapia R.A. and Thompson J.R. (1978) - Nonparametric probability density estimation.
Johns Hopkins, Baltimore.

Integrated System for Modelling Multivariate Time Series

M. Marti, A. Prat and J.M. Catot, Polytechnic University of Barcelona, Spain

SUMMARY

After a description of the evolution of the statistical software to model time series in recent years, we present a system that allows for an easy use of existing packages for modelling univariate and multivariate time series using parametric ARIMA models, as well as providing an interface with general purpose statistical software.

It provides an environment that allows a group of researchers to test new developments and interchange easily each one's results. It also simplyfies the use of existing software to treat situations with a great number of series to model, and without the requirement of people with a great knowledge or experience in using computers.

The system is conceived with a data adquisition and manipulations subsystem, a model base subsystem (that includes the existing packages), and a user subsystem interface, following the ideas developed in the context of Decision Support Systems.

Keywords: Software for time series modelling, multivariate ARIMA time series.

1. INTRODUCTION

The usage of the most part of the statistical models in the practical applications is almost always conditioned by: (i) the existence of an appropriate computer software; (ii) an easy access to the data to be analysed and modeled; (iii) an easy interaction between the user, the software models, and the stored data.

To describe the particular use of the time series analysis we can list the stages of the evolution of the ARIMA parametric methods for the treatment of univariate time series analysis. These stages can be summarized as follows :

a - Formulation of statistical models to represent stochastic stationary sequences.
b - Development of a methodology (Box, Jenkins (1970)) to identify, estimate and validate the model best suited for a sequence of real data (stationary or not).
c - Implementation in an academic environment, of the first versions of the specific software.
d - Marketing and trading of ad-hoc versions for particular installations (public or private).
e - Conversions to interactive software.
f - Incorporation to general purpose statistical packages.

In the case of ARIMA parametric methods for multivariate time series, the evolution is actually in the stage c, Tiao et al. (1979). In fact this evolution is not a particular one, and had been similar for the most part of advanced statistical methods.

Our experience using the ARIMA methods in their different levels -univariate analysis, transfer function models, and multivariate approach (see Martí et al. (1978); Prat et al. (1979); Prat, Solé (1980); Prat, Martí (1981); Prat, Martí (1982))- shows that the existing software is mainly thought of for use in an academic environment by highly specialized people, and it does not support any type of data manipulation and access.

In section 2, we describe the system structure and their main functions.

In section 3, we illustrate through an example the proposed system. Section 4 contains some conclusions.

2. SYSTEM STRUCTURE

To facilitate the usage of the ARIMA methods in the univariate case using the program PACK, we developed a first system (Catot (1981)), that generates the control cards and the parameters to run the program with very few instructions. This first system incorporated some aditional facilities, as the possibility to obtain the residuals of the model and analyse them with statistical packages as BMDP.

Nevertheless, this is not the only problem that the user faces when trying to use effectively a statistical model, or more generally a mathematical model, in a real situation; as we pointed out in the introduction it is necessary to access and manipulate easily the data.

In the last 10 or 12 years a great number of Decision Support Systems (D.S.S.) have been developed that faced this problem for a more or less general type of decisions. Those systems had been analysed and classified, see for example Alter (1979).

Sprague (1980) proposes a framework for the D.S.S. and for their development. The scheme generally accepted is that of the Exhibit 1.

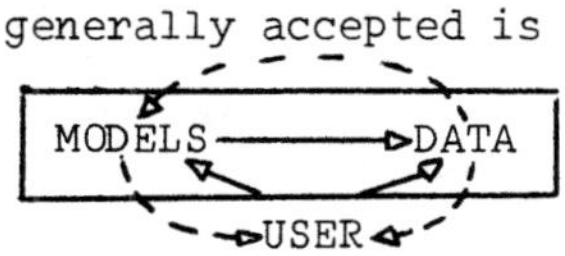

Exhibit 1

Haseman (1977) describes the GPLAN system whose conception and structure are well suited for our purposes, and are formalised in Bonzeck et al. (1981), distinguishing: (i) the language to communicate with the user, (ii) the models or functions available, (iii) the data and information stored.

In our system the user language is in format "menu", the models are, for the moment, the program PACK and the WMTS-1 (for univariate and multivariate analysis respectively), and also some auxiliary functions to obtain graphics, for example, and the data are the stored time series, and also the parameters of each particular model identified as the most suitable to represent a series or the interaction between two or more series in the multivariate case.

3. THE PROPOSED SYSTEM

After the conection, the system presents to the user the list of usable functions. Those included in the first version are presented in exhibit 2, as it appears in the screen.

```
FUNCIONS POSSIBLES =

          DIR . . . . DIRECT. DE SERIES

          LIST  . . . LIST. POR PANT. DE UNA SERIE

          GRFS  . . . GRAFICO DE UNA SERIE

          GRFR  . . . GRAFICO RESIDUOS DE LA
                      ULT. ESTIM.  REALIZADA

          GRFP  . . . GRAFICO DE LOS VALORES REALES,
                      LOS AJUSTADOS Y LAS PREVISIONES

          IDEN  . . . IDENTIF. UNIV. PROGRAMA PACK

          ESTU  . . . ESTIMAC. UNIV. PROGRAMA PACK

          PREL  . . . IDENT. MULT. PROGRAMA PRELIM

          STEP  . . . IDENT. MULT. PROGRAMA STEPAR

          ESTM  . . . ESTIMAC. MULT. PROGR. WMTS-1
FUNCIO ?
       ?  = DIRECTORI DE FUNCIONS
      FIN = ACABAR
```

Exhibit 2

If the user types DIR the system displays the directory of files that contains each one of the series. These files are managed, at this moment, by the editor of the Operating System or by a special program to incorporate data coming from a data base.

```
NUMERO DE DADES:     68

   100.0000    120.0000     90.0000    150.0000    155.0000    155.0
   200.0000    150.0000    135.0000    125.0000    100.0000    215.0
   200.0000    195.0000    280.0000    230.0000    225.0000    120.0
   130.0000    250.0000     80.0000     90.0000     70.0000     90.0
    70.0000    100.0000    100.0000    100.0000    170.0000    165.0
   100.0000    130.0000    165.0000    130.0000    200.0000    200.0
   170.0000    120.0000    200.0000    210.0000     70.0000    110.0
   200.0000    200.0000    130.0000    150.0000    200.0000    110.0
   170.0000    215.0000    200.0000    180.0000    175.0000    120.0
   150.0000    140.0000    120.0000    100.0000    100.0000    160.0
   175.0000    140.0000    225.0000    325.0000    260.0000    125.0
   140.0000    130.0000
FUNCIO ?
```

Exhibit 3

If the user types the function LIST the system asks for the name of the serie lo list, and then displays it, as it is presented in exhibit 3.

If the user specifies the function IDEN (univariate identification by the program PACK), the system will ask the name of the series, and after checking if this one exists, asks for the constant, the Box-Cox transformation to apply, and the stationarity of the data (see exhibit 4).

Only with this response the system constructs a first line with the parameters appropiate to the program that we will use, and a second file with the control cards required to this execution that runs automatically. The results are left in a file that can be printed or displayed. The control returns to the user.

```
IDEN
SERIE  ?
ARPRE0
NO EXISTEIX CAP SERIE AMB AQUEST NOM
SERIE  ?
ARPRE2
 CONSTANTE   TRANSFOR.   EST
 ----.----   ----.----   ---
      .0          .0      12
```

Exhibit 4

In a similar way we can request the function ESTU (univariate estimation by program PACK) as it is presented in the exhibit 5. We can see that it is only necessary to specify the constant, the transformation, and the degree of the autorregressive polynomial (AR), moving average polynomial (MM) and differencing (DIF). Afterwards the system runs in a similar way as in the function IDEN.

```
ESTU
SERIE  ?
ARFRE2
 CONSTANTE   TRANSFOR.
 ----.----   ----.----
      .0          .0
   PARTE REGULAR:
 AR  DIF  MM
 --- --- ---
   1   1   2
 PARTE   ESTACIONAL:
 AR  DIF  MM  EST
 --- --- --- ---
```

Exhibit 5

For the multivariate case, using the WMTS-1 program, the system asks for the number and names of the series to be modelled, for example in the case of estimation (ESTM function) we show the procedure in exhibit 6. The autorregressive and moving average matrices are requested in conventional presentation.

The functions that are previous to the estimation, PREL (PRELIM) and STEP (STEPAR) need only to be called with the name of the series and a few parameters.

```
ESTM
FUNCIO ?
             ?  = DIRECTORI DE FUNCIONS
           FIN  = ACABAR
ESTM
DONA NUM. DE SERIES A TRACTAR
2
ESPECIFICA NOM. DE LA SERIE NUM.  1
SERIE  ?
ARPRE2
ESPECIFICA NOM. DE LA SERIE NUM.  2
SERIE  ?
ARPRE3
ENTRA NUM. DE MATRIUS AUTORR. REGULARS
2
ENTRA ELEMENTS MATRIU AUT. ORDRE 1
       PHI(I,1)  PHI(I,2)
       ---.----  ---.----
I  = 1 >   .0100
I  = 2 >              .0210
ENTRA ELEMENTS MATRIU AUT. ORDRE 2
       PHI(I,1)  PHI(I,2)
       ---.----  ---.----
I  = 1 >
..........
```

Exhibit 6

Other functions allow the user, for example, to obtain a plot of the data of one series, the residuals of a model, etc.

It is easy to introduce a new function in the system to allow the researchers to incorporate new service functions or new complete programs.

In a second version the system will incorporate an interface with general purpose statistical packages, to provide more sophisticated analysis of some results of the models or data.

4. CONCLUSIONS

Although the proposed system is in its first release, we think that it achieves the following objectives that guided our work:

First the system takes existing software without modifications and makes easy to use those complex programmes to non highly expert users, and also simplifies the modelling of a great number of similar nature time series as we can find in a banking organization for their branches.

In addition, the user needs only to specify what kind of analysis he wants, without having to worry about how to get it, so making the user independent of the computer's operating system, and of the specific programmes used.

The incorporation of new functions, as result of research work, new releases of the existing software, new programmes and new service functions are easily performed in a transparent way for the user. Finally, the system is self contained and doesn't need a user's manual.

5. REFERENCES

- Alter, S.A. (1979); Decision support systems: current practices and future challenges; Addison-Wesley, Reading Mass.
- Bonzeck, R.H.; Holsapple, C.W.; Winston, A.B. (1981); Foundations of Decisions Support Systems; Academic Press.
- Box, G.E.P.; Jenkins, G.M. (1970); Time Series Analysis Forecasting and Control; Holden Day.
- Catot, J.M. (1981); La informática como soporte de los métodos de previsión. Procedimiento de ayuda para la utilización de programas de ordenador; Tesina de grado; Facultat de Ciències, Universitat Autònoma de Barcelona.
- Haseman, W.D. (1977); GPLAN: An operational DSS, edited by Carlson, E.D., on Data Base, Quarterly of SIGBDP, Vol 8, num. 3.
- Martí, M.; Prat, A.; Hernandez, C. (1978); Aplicación de la metodología de Box-Jenkins a la previsión de la punta mensual de carga de una empresa eléctrica; QUESTIIO, 2, 4.
- Prat, A. (1980); Análisis multivariante de la relación temperatura media - consumo diario de energía eléctrica; QUESTIIO, 4, 1.
- Prat, A.; Martí, M. (1981); Time series modelling: transfer function or multivariate model. A comparative case study; Workshop on Time Series Analysis in Management; European Institute for Advanced Studies in Management; Bruxelles.
- Prat, A.; Martí, M. (1982); Advantages of the multivariate time series approach versus the transfer function model; International Symposium on Forecasting; Istambul, July (accepted).
- Prat, A.; Bassols, J.; Tort, X.; Solé, I. (1979); Modelado dinámico estocástico de la influencia de la temperatura en el consumo diario de energía eléctrica. Cuadernos Económicos del I.C.E., Madrid.
- Prat, A.; Solé, I. (1980); Influencia de la temperatura en el consumo diario de energía eléctrica. Función de transferencia según la metodología Box-Jenkins; QUESTIIO, 4, 1.
- Sprague, R.H. (1980); A framework for the Development of Decision Support Systems, MIS Quarterly, December, p. 1-26.
- Tiao, G.C.; Box, G.E.P.; Hillmer, S.C.; Hudak, G.; Bell, W.; Chang, I. (1979); The Wisconsin Multivariate Time Series (WTMS-1) Programm. A Preliminary Guide; Department of Statistics. University of Wisconsin.

Software for Time Series Analysis

G. Mélard, Free University of Brussels, Belgium

Summary : This paper is devoted to the presentation of (a) a general approach in univariate time series analysis, and (b) the corresponding computer software called ANSECH. The methodology is similar to that of Box and Jenkins(1970) but the class of models is wider. The main algorithms for estimation and forecasting are briefly described.

Keywords : time series, Box and Jenkins method, statistical algorithms.

1. Introduction

Given a time series $w = \{w_t \; ; \; t=1,\ldots,n\}$, a parametric model belonging to the class of stationary and invertible autoregressive-moving average (ARMA) models is searched for. The model is defined by

$$w_t = \phi_1 w_{t-1} + \cdots + \phi_p w_{t-p} + a_t - \theta_1 a_{t-1} - \cdots - \theta_q a_{t-q} \, , \tag{1}$$

$t \in \mathbb{Z}$, $a = \{a_t \; ; \; t=1,\ldots,n\}$ being a white-noise process. Since the book of Box and Jenkins(1970), the tools of the time-series analyst have been improved, see the review by Newbold(1981). Many authors, including Box and Jenkins themselves, have admitted that the lack of stationarity of many time series implies that extensions of ARMA models should be used. Most computer programs on the market have few of these extensions. In this paper, we will describe a comprehensive class of models supported by a software called ANSECH (ANalyse des SEries CHronologiques), various estimation procedures and some aspects of the computer program.

Hipel and McLeod(1980) and Khabie-Zeitoune(1982) have had similar concerns.

2. The class of models

We will use the following notations :

(a) y_t^I, y_t^D, y_t^W, y_t^M and y_t^S are arbitrary functions of t ;
(b) f_t and g_t are positive functions of t ;
(c) μ_t and μ_t' are polynomial functions of t ;
(d) m_t and m_t' are periodic functions of t ;
(e) ϕ_{it} and θ_{jt} are either constants (ϕ_i, θ_j), or functions of t ;
(f) C(.) defines a transformation which is invariant with repect to t ;
(g) $\bar{\phi}_i$ and $\bar{\theta}_j$ are constants.

We denote by d, D, p, q, P, Q, s, non-negative integers, $s \neq 0$. Let $\nabla = 1 - B$ and $\nabla_s = 1 - B^s$, where B is the backshift operator ($B \cdot_t = \cdot_{t-1}$). Finally, $z = \{z_t \; ; \; 1-d-Ds \leqslant t \leqslant n\}$ is a time series and $a = \{a_t \; ; \; 1 \leqslant t \leqslant n\}$ is seen as a realization of a white-noise process with mean 0 and variance σ^2.

The following equations, called respectively the variable sub-model, the extended ARMA sub-model and the innovation sub-model, yield a_t from the z_t :

$$w_t = \nabla^d \nabla_s^D [C(z_t - y_t^I)/f_t] - (y_t^D + \mu_t + m_t) \tag{2}$$

$$b_t = \frac{w_t - \Sigma_{i=1}^{p} \phi_{it} w_{t-i} + \Sigma_{j=1}^{q} \theta_{jt} b_{t-j} + y_t^W}{1 + \Sigma_{i=1}^{P} \tilde{\phi}_i w_{t-i} - \Sigma_{j=1}^{Q} \tilde{\theta}_j b_{t-j}} \tag{3}$$

$$a_t = \exp(-y_t^S)[b_t/g_t - (y_t^M + \mu_t' + m_t')] . \tag{4}$$

The subscript t will be deleted when the context is clear. The parameters of the model are σ^2 and those included in (a-g). The analytical form of the functions (a-f) must be specified. The following remarks are intended to justify (2-4) :

1° Functions (a) can describe the effects of interventions. In particular, y^I has been used by Box and Tiao(1975) and y^S is concerned with the innovation variance. Instead of relying on transfer function models, a parametrization via piece-wise linear functions is used (Mélard,1981). This is much more efficient in all respects, e.g. when there are several outliers.

2° Several options exist for (b) in order to model a trend on the scale parameter of the time series (Mélard,1977).

3° Functions (c-d) can be used when deterministic trend or seasonal are better than the corresponding stochastic components provided by differencing. The seasonal component can be parcimoniously parametrized in terms of seasonal coefficients, Fourier coefficients (Abraham and Box, 1978) or a mixture of both.

4° The introduction of μ' and m' in (c-d) has been examined by Mélard(1982a).

5° In the extended ARMA sub-model, the coefficients ϕ_{it} and θ_{jt} can vary linearly or exponentially with t (Mélard and Kiehm, 1981). It is also possible to handle the natural parameters of various short-term forecasting methods (Mélard and Rouland, 1979).

6° The transformation (f) is based on the power family as in Box and Cox(1964) and Ansley et al(1977) :

$C(x_t) = \lambda^{-1}(x_t^\lambda - 1)$ for $\lambda \neq 0$, $C(x_t) = \log x_t$ for $\lambda = 0$.

7° The non-linear model including (g) has been used by Lentz and Mélard(1981).

The most useful elements among these extensions appear to be : (a) y^I, y^D and y^S ; (b) g and (f).

3. Estimation of the parameters of the extended model

Let α be the vector of all the parameters except σ^2, and assume that the a_t are n.i.d. (0 , σ^2). The likelihood function, conditional on a_t and z_t , $t \leqslant 0$, is given by

$$L(\alpha,\sigma^2;z) = (2\pi)^{-n/2} J(\alpha)\, \sigma^{-n} \exp\{-\frac{1}{2\sigma^2} \sum_{t=1}^{n} a_t^2(\alpha)\} , \tag{5}$$

where $a_t(\alpha)$ is derived from (2-4) and the Jacobian $J(\alpha) = |(\partial z_t/\partial a_s)|$ is

$$J(\alpha) = \prod_{t=1}^{n} \{f_t g_t \exp(y_t^S) \, |1 + \Sigma_i \tilde{\phi}_i w_{t-i} - \Sigma_j \tilde{\theta}_j b_{t-j}| \, (z_t - y_t^I)^{\lambda-1}\} . \qquad (6)$$

By letting $\sigma_* = \sigma(J(\alpha))^{1/n}$, the concentrated likelihood is

$$(2\pi)^{-n/2} \sigma_*^{-n} \exp\{ - S(\alpha)/(2\sigma_*^2)\}, \text{ where } S(\alpha) = \Sigma_t \{a_t(\alpha)/(J(\alpha))^{1/n}\}^2 . \qquad (7)$$

Maximizing (5) is equivalent to finding $\hat{\alpha}$ such that $S(\hat{\alpha})$ is minimum. Then an estimate $\hat{\sigma}^2$ of σ^2 is deduced from $\hat{\sigma}_*^2 = S(\hat{\alpha})/n$. In practice, the constraints $\Pi f_t = \Pi g_t = \Pi \exp(y_t^S) = 1$ are imposed and the Box-Cox transformation is normalized.

Brubacher(1977) has noted that if λ is included in the model, the information matrix may not be estimated by

$$\hat{\sigma}^2 \, \Sigma \frac{\partial a_t}{\partial \alpha'} \frac{\partial a_t}{\partial \alpha} \quad \text{but by} \quad \hat{\sigma}^2 \, \Sigma \{\frac{\partial a_t}{\partial \alpha'} \frac{\partial a_t}{\partial \alpha} + a_t \frac{\partial^2 a_t}{\partial \alpha' \partial \alpha}\} . \qquad (8)$$

The same is true when f, g or y^S contains parameters.

The algorithm for computing a by the conditional method is divided in 14 steps. The steps 1 to (r-1) are executed only once, where the r-th step is the first in which parameters are involved. The optimization routine calls steps r to 14 with the current parameters values, as many times as necessary. Only three auxiliary arrays X1, X2 and X3, with length not greater than (n+d+Ds), are needed. Let X4 contain a. The algorithm is summarized in Table 1.

Table 1.- Algorithm for computing a by the conditional method

Exit : Y(es) if parameters first encountered at the next step, N(o) otherwise

Entry : Y if parameters first encountered at the current step, N otherwise

Parameters : indicates where they are included

X1, X2, X3, X4 : operations performed (by default : on the same array)

List of operators : (op)ϵ { $=,+,-,.,/,\nabla^d,\nabla_S^D$,ARMA }

An operation in column Xi should be understood as an assignment Xi =: Xi(op) operand, except for "=" which is : Xi=:operand.

Step	Exit	Entry	Parameters	X1	X2	X3	X4
0	Y	Y		$= z$			
1	Y	Y	y^I	$- y^I$			
2	Y	Y	C(.)	=C(.)			
3	N	Y	f	/ f		= f	
4	N	N		∇^d		↓	
5	Y	N		∇_S^D		↓	
6	Y	Y	y^D, y^M	$- y^D$	= X1	↓	
7	Y	Y	μ , μ'	$- \mu$		↓	
8	Y	Y	m , m'	- m		↓	
9	Y	Y	$\phi,\theta,\tilde{\phi},\tilde{\theta}$	↓		↓	
10	N	Y	g	↓		= g	
11	Y	N		↓		↓	
12	N	Y	y^W, y^S	↓		↓	
13	N	N		↓	$= -y^W$ ARMA(X1)	↓	
14	Y	N		↓	↓	↓	$= X2/X3 - (y^M + \mu' + m') / \exp(y^S)$

Forecasting can be performed in two stages :

1° recovery of intermediary arrays, see Table 2 ;

2° computation of the forecasts, see Table 3. Four other arrays are necessary, denoted by X5, X6, X7 and X8.

Table 2.- Algorithm for forecasting : stage 1.

The macro-instruction [1,m](X1,X2,X3,X4) means that steps 1 to m of the algorithm of Table 1 are executed.

Instruction	Result
[1,11](X1,X2,X3,X4)	X1 = w
[1, 6](X4,X2,X3,X4)	X2 = $w + y^D + \mu + m$ and X3 = f
X2 =: X2 - X1	X2 = $y^D + \mu + m$
[1, 1](X1,X4,X4,X4)	X1 = $z - y^I$
X4 =: z - X1	X4 = y^I
[1, 3](X1,X3,X3,X3)	X1 = $C(z - y^I)/f$

Table 3.- Algorithm for forecasting : stage 2.

The distinction between the observation and the forecast intervals is made by the concatenation operator ";"

Instruction	Result
[1,14](X5,X6,X7,X8)	X5 = w
X8 =: w + X2	X8 = $w + y^D + \mu + m$
X8 =: X1 ; X8	X8 = $C(z - y^I)/f$; $w + y^D + \mu + m$
X8 =: X8 ; $(\nabla^d \nabla_s^D)^{-1}$X8	X8 = $C(z - y^I)/f$
X8 =: C^{-1}(X8.X3) + X4	X8 = z

Three other estimation methods can be used in the program :

1° The unconditional method, with the backforecasting procedure (Box and Jenkins, 1970), iterated on request. The algorithm is more complex than the algorithm for the conditional method because of the time-dependency of the coefficients ϕ and θ (Kiehm et al,1980) and presence of μ' and m' (Mélard,1982a).

2° The exact likelihood method for ARMA models with constant coefficients, without y^S, g, μ', m', $\tilde{\phi}$, $\tilde{\theta}$. The algorithm (Mélard,1982c) is faster than the approximate unconditional method.

3° The exact likelihood method for ARMA models with time-dependent models, without μ', m', $\tilde{\phi}$, $\tilde{\theta}$ (Mélard,1982b).

We will not give here the corresponding modifications to Table 1. In all cases, a sum of squares $S(\alpha)$ is minimized with respect to α by the Marquardt algorithm.

4. General description of the software

In the sequel of this paper we will mention a number of features provided by the computer program ANSECH (Mélard, 1975), written in Extended Fortran C.D.C. for a Cyber 170/750 system. Each problem is decomposed into sections, as shown in Table 4. The problem is described by a command sentence using separated commands of the form "identifier" or "identifier=option". Each identifier is associated with two default option values, one to be supplied when the identifier is absent from the command sentence, the other, when the first form of the command is used. Examples of identifiers are shown in Table 4.

Table 4.- Decomposition of a problem into sections, with some examples of commands

Section	Purpose	Examples of commands
A	input validation	ESTIM must be present if estimation is requested
B	access to data	KEY = name of data set (in case of random access)
C	model definition	DIFF = degree of regular difference (value of d)
		BOXCOX for normalized logarithmic transformation
D	parameter meaning	HOLT = 1 or 2 Holt or Winters model (instead of ARMA)
E	estimation	ITERATIONS = maximum number of iterations (Marquardt)
		LIKELIHOOD = estimation method
F	plots, diagnostics	CHECKING = a string of code letters defining the requests (A for autocorrelation, L for Ljung-Box,...)
G	forecasting	LEADTIME = number of steps ahead for which forecasts are asked for

Further lines of directives are required for

- preliminary transformations,
- selection of parameters : type, order, initial value,
- description of interventions (Mélard, 1981).

There are fourteen types of parameters, plus the various parameters in the intervention directives. For example, the letters M and N identify respectively the regular and seasonal moving average polynomials. The following two lines can be used to fit the airline passengers data (Box and Jenkins, 1970, Ch 9) by the model

$$\nabla \nabla_{12} \log z = a_t - \theta_1 a_{t-1} - \Theta_1 a_{t-12} + \theta_1 \Theta_1 a_{t-13}$$

```
ESTIM,READ,AUXIN=1,KEY=BJ-SERG,BOXCOX,DIFF,DIFFS
M 1  0.5  N 1  0.6
```

("READ,AUXIN=1" specifies reading from input file # 1, DIFFS corresponds to ∇_s).

The command language is easy to learn, as observed with about a hundred students and researchers, many of them without any experience in data processing. However, a program has been written recently that questions the user who wishes to enter new data or solve simple problems. Command sentences are composed, shown to the user, who can supplement them.

It is not possible to detail other features of ANSECH such as : data management, dating facility, extended residual tests, calculation of the likelihood on grids of points, transformations, Monte Carlo processor (which can yield a ready-to-use SPSS input file).

References

Abraham, B. and Box, G.E.P.(1978), Deterministic and forecast-adaptative time-dependent models, J.Roy.Statist.Soc.Ser.C Appl.Statist.,27,120-130.
Anderson, O.D. and Perryman, M.R. (eds)(1981), Time series analysis, North-Holland, Amsterdam.
Ansley, C.F., Spivey, W.A. and Wrobleski, W.J.(1977),A class of transformations for Box-Jenkins seasonal models, J.Roy.Statist.Soc.Ser. C Appl.Statist.,26,173-178.
Box, G.E.P. and Cox, D.R.(1964), An analysis of transformations, J.Roy.Statist.Soc. Ser. B,26,211-243.
Box, G.E.P. and Jenkins, G.M.(1970), Time series analysis, forecasting and control, Holden-Day, San Francisco(revised edition : 1976).
Box, G.E.P. and Tiao, G.C.(1975), Intervention analysis with applications to economic and environmental problems, J.Am.Statist.Assoc.,70,70-79.
Brubacher, S.R.(1977), Applications and extensions of the Box-Jenkins model, Communication at the Public Utilities Forecasting Conference, Browness-on-Windermere.
Hipel, K.W. and McLeod, A.I.(1980), Perspectives in stochastic hydrology, in O.D. Anderson (ed), Time series, North-Holland, Amsterdam,73-102.
Khabie-Zeitoune, E.(1982), Identification, estimation and prediction of linear and nonlinear multivariate models in time series, Proceedings of the 4th International Time Series Meeting, Valencia, June 1981, O.D. Anderson(ed), North-Holland, Amsterdam, to appear.
Kiehm, J.-L., Lefèvre, C. and Mélard, G.(1980), Forme rétrospective d'un modèle FARIMAG, Cahiers du C.E.R.O.,22,375-383.
Lentz, J.-R. and Mélard, G.(1981), Statistical analysis of a non-linear time series model, in O.D. Anderson and M.R. Perryman(eds)(1981),287-293.
Mélard, G.(1975), ANSECH-Programmes pour l'analyse des séries chronologiques, Presses Universitaires de Bruxelles, Bruxelles(4th edition : 1981).
Mélard, G.(1977), Sur une classe de modèles ARIMA dépendant du temps, Cahiers du C.E.R.O.,19,285-295.
Mélard, G.(1981), On an alternative model for intervention analysis, in O.D. Anderson and M.R. Perryman(eds)(1981),345-354.
Mélard, G.(1982a), On a deterministic sub-model for the innovation process in ARIMA models, 1981 Proceedings of the Business and Economic Statistics, American Statistical Association, Washington (to appear).
Mélard, G.(1982b), The likelihood function of a time-dependent ARMA model, in O.D. Anderson and M.R. Perryman(eds) Applied time series analysis, North-Holland, Amsterdam, in press.
Mélard, G.(1982c), A fast algorithm for the exact likelihood of autoregressive-moving average models (submitted).
Mélard, G. and Kiehm, J.-L.(1980), Spécification d'un modèle ARIMA évolutif : approche par décomposition, Cahiers du C.E.R.O.,22,265-277.
Mélard, G. and Kiehm, J.-L.(1981), ARIMA models with time-dependent coefficients for economic time series, in O.D. Anderson and M.R. Perryman(eds)(1981),355-363.
Mélard, G. and Rouland, O.(1979), Une méthode de sélection d'un algorithme de prévision à court terme, Communication aux Journées de Statistique de l'Association des Statisticiens Universitaires, Paris.
Newbold,P.(1981), Some recent developments in time series analysis, International Statistical Review,49,53-66.

Choice of Methods and Algorithms for Statistical Treatment of Large Arrays of Data

A. Morineau, CEPREMAP, Paris, France

SUMMARY

Exploratory analysis for large arrays of data requires specific statistical tools and computer procedures. Concerning the methods (descriptive factorial analysis and exploratory clustering), we must elaborate algorithms with a reasonable compromise between computer time and central core storage. The amount of data requires the use of adapted aids for interpretation of the results : graphical representations and numerical indicators. Finally, the conception of these programs and their arrangement must satisfy rules which allow the system to adapt to the user's needs, and to the evolution of practice and knowledge. The paper describes the main solutions adopted in the construction of the package SPAD "Système Portable pour l'Analyse des Données" (portable system for data analysis).

KEYWORDS : Data Analysis Package, Large arrays, Principal Component Analysis, Correspondence Analysis, Multiple Correspondence Analysis, Clustering Methods.

THE STATISTICAL METHODS.

The aim of the analyses contained in the program SPAD (8000 FORTRAN statements, 102 subroutines), is to provide the users with descriptive exploratory methods available for large arrays of data.

By "large arrays" it is meant : data files with several thousand rows (ie. cases), and several hundred columns : mixed data (binary, nominal, ordinal, continuous) as are provided by socio-economic surveys. By "descriptive exploratory methods" it must be understood as a statistical treatment of multivariate data leading to summarization by graphical displays in which associations are expressed in terms of proximities, together with numerical indicators to help their reading and the interpretation. These methods are described in detail for example in Lebart-Morineau-Tabard (1977).

a) Principal Component Analysis (PCA), Correspondence Analysis (CA).

These classical linear reduction methods describe the dispersion of an array of n points in p-dimensional space where coordinates are continuous variables (PCA), or display rows and columns of a two-way contingency table as points in low dimensional spaces (CA). The ability to display supplementary rows and/or columns in the space of principal axes is used to facilitate the interpretation of proximities. The main features of this implementation lie in the possibilities of quick selection and visualization :

- beforehand selection of active, supplementary and nominal illustrative variables (centroïds of observations) from a unic basic file,

- systematic sorting of the labels of all the elements (up to 60 characters) according to each axis,

- simultaneous interpretation of a set of axes with the help of a clustering of observations described by their principal coordinates (see below),

- various graphical displays of variables, attributes and observations.

b) Multiple Correspondence Analysis

Correspondence Analysis may be extended to the display of a wide range of non-negative data matrices (apart from contingency tables), demonstrating its versatility as an exploratory statistical tool. The most important application is Multiple Correspondence Analysis, which is an analogue of Principal Component Analysis applied to large sets of multistate nominal characters (in order to use data on all variables in a global analysis, the quantitative variables may be reduced to qualitative form). This technique produces a graphical display of all categories involved in the cross-tabulation of all couples of nominal variables (Burt's table).

The ability to make the projection of as many "supplementary categories" as wanted allows us to grasp easily the main characteristics and structures of the data. A glance at the graphical display replaces the reading of numerous two-way contingency tables, while taking into account their networks of relationships.

c) Clustering

To remedy the insufficiencies of the visual investigation of data (sequences of factorial two-dimensions graphic displays), a global description can be obtained by producing clusters. If redundancy among variables is suspected, the original data may first be transformed to a reduced dimensional space by an appropriate eigen-analysis. In this case data are "smoothed" in the subspace of factorial coordinates, and a large part of random noise is eliminated.

Clustering techniques results here in several forms of graphical outputs : dendrograms for hierarchical techniques, and graphical displays of group centers upon the factorial axes (this links both kinds of exploratory data analysis). The diagrams show the principal relations between the clusters in a concise and easily interpreted form. A strategy consisting of 3 steps which proves very useful, is employed to indicate how real and distinct the clusters are (the method is suitable for clustering several thousand units).

First step. The algorithm (which is an adapted version of the k-means method) produces a quick partition of the objects (in the initial space or in a factorial space). In fact, two or more independent partitions are produced. The final clusters are groups of objects always classified in the same class when comparing the sequence of partitions. There may be as many as 30 to 50 such "stable groups" (see Diday, 1971).

Second step. A hierarchical clustering method operates on the centers of stable groups considered as individuals. The centers are weighted according to the number of objects in the group (Wald's formula for fusion). This step ends with the dendrogram describing proximities between the centers.

Third step. A visual inspection of the dendrogram (or an a priori choice of the number of classes) leads to the choice of the right place where to cut the hierarchical tree (if this right place does exist). Each object is ascribed to its class, and a statistical processing leads to the description of the clusters in terms of the initial variables (quantitative and categorical characteristics, ranked in order of importance).

CHOICE OF ALGORITHMS

An acceptable compromise between efficiency and memory storage must be made in the choice of algorithms. In any case, the data array is never stored in the central core of the computer ; it is read and treated one row at a time.

Different types of algorithms are available according to the memory capacity of the computer, and the dimension of the problem. For correspondence analysis a program parameter allows the user to compute eigenvalues and eigenvectors without computing the matrix to be diagonalized. The algorithm is based upon stochastic approximation. It allows the determination of the first eigenvectors by direct reading of the data (Lebart, 1976).

When nominal data array is in a complete disjunctive form, the amount of computation can be limited to the number of variables, whatever the total number of modalities. Considerable savings in computing costs occur if the transformation to binary coding is implicitly performed within the programs. The same reduction is used for spectral analysis by stochastic approximation in case of multiple correspondence analysis, and for the clustering k-means algorithm operating on individuals characterized by nominal variables.

Other technical particularities allow computation savings. For example, the core diagonalization of Burt's table (by QL-modified method) is performed on the reduced matrix obtained after transformation of the table (in case all responses are binary, the dimension of the table to be diagonalized is reduced by half).

GRAPHICAL AND NUMERICAL RESULTS

Graphical pictorialization is one of the most important tools in data analysis. The package must provide powerful and flexible facilities, and make the common graphical operations easy to perform :

- treatment of coïnciding points : the hidden points are listed with their coordinates ;
- ability to label each point with a concise label of 4 characters, or with a complete name ;
- possibility of density scatter plots, each individual being labelled with the digit of its category in a given categorical variable ;
- possibility of reversing the direction of an axis ;
- selection in the scatter plot of noteworthy points (influent points in the determination of axes) ;
- elimination of outliers (returned to the margin).

MANIPULATION

In developing data analysis package, facility and efficiency must be balanced against one another. For convenience of use of the package, an analysis will be a sequence of well-defined "steps" (some of them are common to any analysis while others are specific). The communication of the results from one step to the other is automatically realized by means of external storage, without any intervention by the user. Modularity contributes to the flexibility of the package : modules can be used in unforeseen sequences for particular applications without change. Furthermore the modular package may easily grow : new methods can be added easily. (SPAD actually uses 14 steps).

Physically each step is called by using a keyword and by defining a short sequence of parameters (read in free-format). Here is an example demonstrating how to execute a multiple correspondence analysis with the package SPAD. We give the sequence of steps with some commentaries.

EXAMPLE OF A SEQUENCE OF STEPS

1. LILEX : This step allows selection of the observations and variables wich will be used in the analysis. We choose the active and illustrative variables among the categorical variables, as well as continuous supplementary variables.

2. MULTC or MULDI : This is the main step for computation, with in-core (MULTC) or direct-reading (MULDI) diagonalization. Categories with too few members are temporarily removed in order to insure robustness of principal axes. Among the results in the output, we get the eigenvalues and percentages of variance. The

coordinates and "contributions" of each point are listed for the first six axes. The correlation coefficients between each continuous variable and the first six axes are printed. We can obtain a graphical display of each axis, one at a time, in which most influent categories are ranked and printed with full labels (so that interpretation of the axis emerges at a glance from the listing). This step creates the main file of numerical results, which will be used throughout all subsequent steps.

3. GRAPH : We may obtain many different types of graphical factorial displays, on any two axes : active categories only, active and illustrative categories, observations (with four characters labels, or density display), plots mixing rows and columns, and graphical display of continuous variables in which coordinates are the correlation coefficients with the axes.

4. SEMIS : The analysis can proceed with a global description of the sample in terms of homogeneous clusters. The step SEMIS executes the "mixed algorithm" as describes above, including the graphical display of the dendrogram (computations are performed on factorial coordinates, one of the parameters being the number of axes).

5. TAMIS : The final partition is chosen after visual inspection of the dendrogram (otherwise a priori chosen). The step TAMIS provides a synthetic description of the classes with the basic continuous and categorical variables. We can obtain the contingency tables which cross-classify the partition with the categorical variables (and the CHI-2 statistics), and the comparison of the global means with the sample means in the classes. Then, the classes are characterized by the most significant categories (and continuous variables) which are ranked according to importance.

6. GRAFK : We obtain a graphical display of cluster centers on any factorial plan, as well as density plots in which observations are identified by the digit of their cluster.

The program SPAD will be a component of the forthcoming package "MODULAD" which intends to gather the contributions of several institutes or laboratories in the field of exploratory data analysis.

REFERENCES

BENZECRI J.P. (1973) . L'Analyse des Données, Tome 1 : La Taxinomie ; Tome 2 : L'Analyse des Correspondances. Dunod. Paris (2ème éd. 1976).

DIDAY E. (1971) . La Méthode des Nuées Dynamiques. Rev. Stat. Appl. Vol. 19, n° 2, pp. 19-34.

LEBART L. (1976) . Sur les Calculs Impliqués par la Description de Certains Grands Tableaux. Annales de l'INSEE, n° 22-23, pp. 255-270.

LEBART L., MORINEAU A. (1982) . S.P.A.D., Système Portable pour l'Analyse des Données. CESIA, 82, rue de Sèvres 75007 PARIS.

LEBART L., MORINEAU A., TABARD N. (1977) . Technique de la Description Statistique, Méthodes et Logiciels pour l'Analyse des Grands Tableaux, Dunod, Paris.

New Computing Tools and New Statistical Packages

B.P. Murphy, University of Western Australia, Nedlands, Australia

0. Summary

The lack of appropriate software for the new ranges of microcomputers gives rise to the fear that, as in the current mainframe era, professional statisticians are in danger of yielding to amateurs, or worse, control of statistical computing initiatives. We discuss here the challenges and problems involved in making the new computing tools, specifically microcomputers and the hardware and software they are generating, available to statisticians, and indicate incentives which may encourage new statistical program production.
Keywords: Microcomputers, Statistical Software, Screen Editors, Data Entry.

1. New computing hardware.

The inexpensive ('personal') microcomputer is clearly the major new computing tool of the decade, together with its associated peripherals such as high speed terminals, printers, plotters and communications couplers. The high speed response, immediacy of access and control, low running costs, and finally actual power, in terms of job sizes possible,have not only led to its complete acceptance in the business world, but also caused large social changes.

Business has clearly proven a Z80-type machine with 48-64K memory and substantial floppy disk backup to be a powerful workhorse in single and multi-user systems. Such a truly interactive system, given appropriate software, provides virtually all the practising statistician could desire, yet its impact so far seems marginal Since much future computing growth will take place in terms of expanding micros rather than shrinking minis, partly due to greater adherence to 'standards' for both software and hardware development in the area, statisticians must urgently take an interest.

2. New Ancillary Software

Some of the current business software is quite useful to the statistician. The screen editors, word processors, and data entry programs readily available from commercial sources (WordMaster, WordStar, DataStar (Lifeboat Assoc.,1981; or see any computing monthly)) are vastly superior, in terms of higher user efficiency and lower frustration levels, to similar mainframe facilities; few mainframes can expect to improve sufficiently to compete. Thus the micro is used as a superior edit-submit station for a mainframe by many statisticians, which must be considered as an absolute minimum capability for a serious micro. Much more is both desirable and possible.

Other commercial packages already found useful by statisticians are the table (financial modelling) programs (e.g. CalcStar, TMaker (Lifeboat,1982)), which permit many common statistical computations. These classes of commercial software illustrate to the statistical programmer the high level of user convenience which must be offered by any new software hoping to be widely accepted.

It seems certain that statisticians will have to reproduce some old programs, as well as produce new ones. Fortunately, the class of compilers now appearing (e.g. Lifeboat Associates, 1981) for micros should make future programming less frustrating. This is particularly true of compilers for the structured languages C, Ratfor, APL, Pl/1, Pascal and Ada, which beside conforming fairly well to standards, include many extra features, such as in-line editors, debuggers, linkers and thoughtful extensions, to maximise the new criterion of programmer efficiency. Extra effort in the case of Pascal and Ada reflects the strengthening view (Gilbreath,1981) of the micro programming community, over the protestations of the purists, that they are doomed to survive (Hoare,1980), with only a few conflicts in extensions.

3. New Statisticial Software.

Unfortunately the little statistical software which exists for micros is most unimpressive and reinforces the fears expressed above - superficial programs produced by non-statisticians using unsound calculation formulae for small word machines. New work is required.

An alternative to producing new programs is converting proven older packages. Most of these are heavily overlaid, batch oriented, Fortran monoliths, Murphy (1982) discusses possible methods of conversion, which centre of overlay facilities, absolute or compiler enforced, but recent experience has been mixed. Work continues in this area, but it is not clear however how much of this type of conversion will be accepted even when successful, since few (any?) current statistical packages, despite their enormous statistical value, reach the standards of user convenience exhibited by current editor software. These standards will soon be demanded by the statistician, and the future relevance of most old monoliths is doubtful.

4. Future Statistical Packages

Our concern here is not with statistical content of future packages, as theoretical developments and market forces will dictate these, but with aspects of program design and structure. Nelder (1978) lists many desiderata, all still highly relevant, and it will be argued here that many come naturally with the new programming languages, particularly Pascal and Ada.

In this Unit, experience has led to adoption of the certain points in our new programs. The major experiences were a 6000 line Pascal program (Murphy, 1981) for fitting log-linear models in contingency tables (LOLITA), and a table construction program now being written. Some facts are :

a) The LOLITA program has 75 procedures and is overlaid on a Cyber 172 to run in 30000 words. A 4000 line subset runs, under Pascal/M, on any Z80 micro. The conversion required 8 parameter resettings, 4 new lines of code to open files, and some reorganisation of the procedure order to assist overlaying on the micro; it took two days to set up on the first occasion. But the resulting micro code (with original parameters and deleted 4 lines) runs on the Cyber!

b) Others have read and modified the code, with little more than some explanation of the basic flow and statistical objective, the self documentation forced by the structure principles of Pascal being vital. Consequently, it is now easily supported by the part time efforts of one man, and code and documentation are maintained by a (micro) word processor program.

The lessons were inescapable:

i) New programs must be conceived in essentially interactive terms; batch modifications later are simple.

ii) Language design must be a first priority; the users' potential needs and VDU capabilities must be carefully assessed. It is then as easy to implement a sound language (Wirth,1976,Ch.5) as one with shortcuts reflecting programmer ignorance or nonchalance. Indeed it is not too extreme to suggest that a protototype User Manual should be complete before a single line of code is cut! A recursion based language seems necessary for language analysis which is not clumsy for user or programmer.

iii) Graphics will be increasingly wanted; as these (and other) new techniques develop,facility will be needed for convenient addition to the package - modular construction is an obvious necessity.

iv) Only the fully structured languages, which currently means Pascal, can provide the tools for the programmer to make possible satisfactory implementations, at lowest cost, at the level now demanded.

5. Illustration - User Facility

To illustrate the user and programmer facility structured languages encourage, consider the interactive program TABGEN for table generation from survey data currently under development in this Unit. The user has, as for other tabulation programs, a set of commands for data and table description. On invocation the program prompts the user for commands and parameters and will display a relevant tableau, as is common in word-processor and financial modelling packages. The terminal screen is divided into three areas: a Help/Menu display using the top few lines, a large 'window' on the instruction tableaux/files being created or modified and a 'prompt'/ 'status' area at bottom displaying current activity and options. The user responds to the prompts in the lower area and the resulting action is seen in the window, which can be alternatively accessed by a screen editor. This dual access approach allows for 'big' instructions (e.g. giving a long series of variable names) as well as individual item editing. Such tableaux are converted into 'compiled code' and associated directories, to be accessed when the data is actually read, either immediately, or in a later (batch) run.

The display below shows the Variable-Attributes tableau after execution of commands

```
HD      Stomach Surgery -- Feb. 1982                 !heading
V_N     age   sex  · quet ·  sbp   sugar opern?      !variable names
FMT     (f3,1xa1,f5.1,f5.0,a2)                       !format
U_L     99 - 200 350 20 -                            !upper limits
M_V     all  * ;  ht  to  sug 0,-1                   !missing values
C       agegp =age div 10  + 1                       !compute
R       sbp    40 60 .. 140 170.. 300 +              !recode
```

and the user is editing the Upper_Limit for 'sbp'. He will go on to enter lower limits (L_L), for further Recodes/Computes, and finally Table_Descriptors.

```
Menu Level 2                                                                (C)TABGEN
HeaDing   VarName    ForMaT     UpLimit    LoLimit     Group      Recode     Compute
MissVal   ASSgnMiss  LevlNames  Parallel   Tables      EditData   EditAttr   Files
Lists     Process    ForMaTOut  Space      VARiableATT

      VARiables ATTributes   for job headed:- 'Stomach Operations -- Feb.'
 #  VARNAME  LOLIM  UPLIM    COLS      DEC    MISSVL   LNAMES    REC    COMP
 1  age               99     1-3        3     *           n       n      y
 2  sex               -      4-5        a     *           y       n      n
 3  quet             200     6-10       1     * 0 -1      n       n      n
 4  sbp             <350>   11-15       0     * 0 -1      n       y      n
 5  sugar             20    16-20       0     * 0 -1      n       n      n
 6  opern?            -     21-22       a     *           n       n      n
 7  agesq     *       *                       *           n       n      n
 8  agegp     *       *                       *           n       n      n

  Space Used (Avail.) :- Var 240 (4000)    Comp/Rec 188 (2000)    Text 204 (2000)
  Current Command :-   SV                                   (Substitute in Variable)
          At  :- UL    For :- sbp      Value :- 450_____
```

Other tableaux and windows exist for further setup information. This genuinely interactive approach with user and programmed checks ensures that the main run has an excellent chance of immediate success, and also makes possible a mainframes version. Efficient run-time computation is separately ensured.

6. Illustration - Programmer Facility

Pascal/Ada are obvious languages for the program above. The tableau/file information is held in SETS and RECORDS constructs, and recursion is essential in syntax analysis and table building. Further illustrations of these features are seen in the procedure below, connecting log-linear model representations.

A log-linear model for a contingency table can be expressed by an array of elements, giving the highest order mutually exclusive interactions in the model. The 'graphmodel' below corresponds to this usual sufficient configurations representation Alternatively, it has been found convenient to have a set representation indicating every interaction in the model; this 'utermodel' formulation involves a further function (POSNU) defining a suitable mapping of the interactions onto the integers. The procedure builds a utermodel from a graphmodel; the template is given in Wirth (1976,Ch.3), for such recursive procedures.

```
{  TO CONVERT A GRAPH_MODEL TO A U_TERM_MODEL.  Assume defined :

        const      nfac      = 8 ;
        type       factor    = 1..nfac ;
                   interactn = set of factor  ;
                   graphmodl = record
                               size : integer  ;
                               intnvect  : array [1..70] of interactn  ;
                               end ;
                   utermindex = 0..255;
                   utermmodel = set of utermindex;
          function posnu(int:interactn ; i:integer) : utermindex ;  forward ; }

procedure grtouset(var grmodl:graphmodl;var umodl:utermmodl;var nfac:integer);
var  i,j,lastcliqno : integer ; intn : utermmodl ;
     procedure moreuze(first:factor) ; {Recursive proc to extend found intn }
     var i,j : integer ;
     begin                                  {Extend intn by factor and check}
     lastcliqno := grmodl.size + 1 ;
     for i := first to nfac do begin
          intn := intn + [i] ;
          grmodl. intnvect[lastcliqno] := intn  ;         { Placing sentinel}
          j := 0 ; repeat j:= j+1 until intn <= grmodl.intnvect[j]  ;
          if j < lastcliqno then begin                  {Intn found in grmodl}
               umodl := umodl + [posnu(intn,i)]   ;     {Build new formulation}
               if i<nfac then moreuze(i+1) ;            {Try next factor      }
               end ;
          intn := intn - [i]  ;
          end ;
     end ;                         {Of moreuze}
begin                              {    grtouset     }
umodl := [0]  ; intn := [] ; moreuze(1) ;
end ;                              {   of grtouset    }
```

7. Incentives

Statisticians should not view the move to new languages as negating twenty years of statistical software development. The necessary older work will be recoverable, and now is an appropriate time to reconsider our computing activity (Hoare,1980). The portability facility of new languages over new machines, and the market thus opened for software written on/for the micro, the refound elegance of the art of programming, and the facility to rapidly extend such programs in accord with recent statistical research, should be great incentives for statistical programmers towards regaining control of statistical computing direction.

8. References

Clarke M.R.B. (1981). A Givens algorithm for moving from one linear model to the next without going back to the data. Appl.Statist.,30,198-203.

Gilbreath J. (1981). A high level language benchmark, Byte,Sept.,180-199.

Hoare C.A.R. (1981). The Emperoro's Old Clothes. Comm.Assoc.Comp.Mach.,24,75-83.

Lifeboat Associates. (1981). Catalogue of available software, N.Y.

Murphy B.P. (1982). Production and reproduction of statistical software for micro-computers. Contemp.Maths. (in press).

Murphy B.P. (1981). LOLITA User Manual. Technical Memo RMSU-81-4. Raine Medical Statistics Unit, Univ.West.Aust.

Nelder J.A. (1978). The future of statistical software. Proc.Compstat'78 Conference. Springer-Verlag, Vienna.

Wirth N. (1976). Algorithms+Data Structure=Programs. Prentice Hall, N.J.

Statistical Computing Based on Text Editing

S. Mustonen, University of Helsinki, Finland

SUMMARY

Editorial mode is a new form of interactive computing enabling the user to perform mathematical, statistical and graphical operations in connection with normal text editing. Various aspects of the editorial approach are considered through examples.

Keywords: interactive analysis, statistical graphics, text processing, editorial mode, touch mode

1. INTRODUCTION

Interactivity in statistical computing is often considered almost synonymous with conversational use of the computer. Many of the friendly interactive statistical systems are controlled by a dialogue between the user and the system. Originally also SURVO 76 (Mustonen 1977,1980) had all the functions provided in conversational mode. When developing this system we soon realized that there are other forms of interactivity which on certain occasions may be still 'more interactive'.

In some areas of statistical data management and computing the conversational mode is surpassed by a new approach which we call editorial mode. Especially tasks closely related to report writing, such as editing of results, making of statistical graphs, manipulation with multiway tables, arithmetical and statistical calculations etc. seem to be very conveniently performed in editorial mode.

Our editorial approach is based on a general text editing program SURVO 76 EDITOR which is one of the SURVO 76 modules. It is planned primarily for report generating tasks in co-operation with the conversational programs.

All the operations of this editor are performed in an edit field which typically has 100-250 lines and 70-120 columns. The edit field is always partially visible on the screen, which acts like a window. The user may easily scroll that window in any direction and he may enter text, data and operations in any part of the edit field. Various data files and results may be loaded to the edit field and the contents of the field may be stored for subsequent use.

Simple text editing is done using the soft keys. These are labelled for typical word processing tasks like insertion and deletion of characters and lines, moving of the cursor etc. The leading principle in all SURVO 76 EDITOR operations is that a minimal effort, i.e. the least possible number of touches on the keyboard is needed. In this sense this editor is comparable with the most efficient text processing systems.

In our editor, however, standard text processing is merely a solid background for advanced work related to data management and to mathematical, statistical, and graphical applications.

In more complicated tasks the general idea is to enable the user to specify the job by typing some control information among the text and data and then let the system continue the work automatically. Thus the SURVO 76 EDITOR is normally in typing mode and it accepts and displays any text which is entered from the keyboard. The system may, however, be activated by pressing the key CONTINUE. What then happens, depends solely on the situation in the edit field and on the position of the cursor.

2. ARITHMETICAL AND STATISTICAL CALCULATIONS

For example, when the user has typed (as we are doing just now when preparing this manuscript)

```
p=0.837 log(p/(1-p))=_
```

and presses CONTINUE, the system responds by giving the value

```
p=0.837 log(p/(1-p))=1.6360738697
```

and it immediately returns to the typing mode thus permitting the user, for example, to round the resulting value, to alter the expression, to take a printout of specific lines or to continue typing etc.

Similarly, activation of

```
DER (X+A)↑X X_
```

by pressing CONTINUE leads to the display

```
DER (X+A)↑X X_
 Derivative of (X+A)↑X with respect to X is
    (X+A)↑X*(LOG(X+A)+X/(X+A))
```

Here DER is an editing operation which produces analytic derivatives of elementary functions.

More extensive mathematical and statistical calculations may be presented as computation schemes. For example, when comparing means of two samples by doing a standard t test we could have the following situation in the edit field:

```
35  *
36  *   Comparing means of two normal populations
37  *               size   mean     std.dev.
38  *   Sample 1:   N1=16  M1=38.5  S1=13.3
39  *   Sample 2:   N2=20  M2=30.0  S2=15.4
40  *
41  *   The test statistics is t=(M1-M2)/(S*sqr(1/N1+1/N2)),
42  *   where S=sqr((F1*S1↑2+F2*S2↑2)/(F1+F2) and F1=N1-1, F2=N2-1.
43  *
44  *   Assuming that the samples are from the same normal distribution,
45  *   the probability that t exceeds the observed value
46  *   t:=              is 1-t.F(F1+F2,t):=_
47  *   where t.F(n,t) is the cdf of the t distribution.
48  *
```

Now, if this scheme is activated by keeping the cursor either in the position after 't:=' or after '1-t.F(F1+F2,t):=' as shown above and CONTINUE is pressed, the last lines will be redisplayed in the form:

```
44  *   Assuming that the samples are from the same normal distribution,
45  *   the probability that t exceeds the observed value
46  *   t:=1.710791283   is 1-t.F(F1+F2,t):=0.0482531105
47  *   where t.F(n,t) is the cdf of the t distribution.
```

Observe, that to make a computation scheme of this kind corresponds to programming, but the form of presentation is more natural. It is like teaching a human being how to do a t test from the data presented. When the scheme has been activated the system itself has to find all the constituents needed for evaluating the expressions. To facilitate the specification of the computation schemes several mathematical and statistical functions, like 'sqr' and 't.F' above, are available. The user may also define more functions and use them in the computation schemes (Mustonen 1981a,c).

Another alternative for arithmetic operations with single numbers and numerical arrays is provided by a technique, which we call the touch mode. While working with a certain edit field the user can pass from normal typing mode to touch mode by pressing a specific key. In touch mode the cursor may be moved as in typing mode, but when the cursor touches a number in the edit field, this number may be taken into calculation simply by pressing any of the keys +,-,*,/ etc. and the expression formed and evaluated so far will be displayed on the last line of the screen. The touched numbers will be displayed in inverse video and the expression evaluated may be put to any place in the field by moving the cursor to that place and by pressing =.

In touch mode the operations defined as chains of consecutive key strokes may automatically be repeated from various starting points. This feature enables the user to perform rather complicated calculations in a very natural fashion.
For example, to compute a five term moving average from a time series in the edit field, we enter the touch mode and move the cursor to the first observation:

```
TOUCH MODE                                                   SURVO 76 EDITOR
 28    *
 29    * Mean temperature (C°) in July, Helsinki 1944-1978
 30    * year       temperature
 31    * 1944       18.7
 32    * 1945       19.5
 33    * 1946       18.8
 34    * 1947       17.9
 35    * 1948       17.9
 36    * 1949       17.3
 ..    . ....       ....
```

Now the chain for computing the desired average is defined by first calculating the mean of the first 5 observations as a sequence of key strokes

key	comments
BEGIN	start of a definition
+	+18.7
↓	cursor one step downwards
+	+18.7+19.5
↓	
+	+18.7+19.5+18.8
↓	
+	+18.7+19.5+18.8+17.9
↓	
+	+18.7+19.5+18.8+17.9+17.9
S	the present expression is evaluated (92.8)
C	enter a constant
5 RETURN	constant=5
/	92.8/5
↑↑→→→→→→→	cursor 2 steps upwards and 7 steps to the right
=	display 18.56 (=92.8/5) in the current position

After these key strokes (an experienced user needs 20 seconds at most to perform them) we shall have the following display:

```
TOUCH MODE                                                   SURVO 76 EDITOR
 28    *
 29    * Mean temperature (C°) in July, Helsinki 1944-1978
 30    * year       temperature
 31    * 1944       18.7
 32    * 1945       19.5
 33    * 1946       18.8    18.56
 34    * 1947       17.9
 35    * 1948       17.9
 36    * 1949       17.3
 ..    . ....       ....
```

To activate an automatic repetition, we merely move the cursor to touch the second value (19.5) and press CONTINUE. Then, the moving average will immediately be computed for all successive observations and we shall have the display

```
TOUCH MODE                                                        SURVO 76 EDITOR
 28    *
 29    * Mean temperature (C°) in July, Helsinki 1944-1978
 30    * year       temperature
 31    * 1944       |18.7|
 32    * 1945       |19.5|
 33    * 1946       |18.8|    18.56
 34    * 1947       |17.9|    18.28
 35    * 1948       |17.9|    17.54
 36    * 1949       |17.3|    16.9
 ..    . ....        ....     .....
```

3. OPERATIONS ON DATA MATRICES

Various statistical data sets may be loaded from SURVO 76 data files into the edit field. Small data sets can also be entered directly in the field. In addition to various editing operations for data matrices and multiway tables (which can be presented in a natural nested form), several statistical operations (for analysis of variance, log-linear models, linear and nonlinear regression etc.) are available.

Each data matrix to be handled by statistical operations has to be labelled with a DATA specification which indicates the lines used for the observation values and the line of the column labels. In the following display we have a small data set consisting of 12 observations and 5 variables presented in the edit field:

```
  1    *                 Consumption of various beverages
  2    *DATA COUNTRIES,A,B,C
  3    C                 Coffee  Tea  Beer   Wine Spirits
  4    A Finland           12.5 0.15  54.7    7.6  2.7
  5    * Sweden            12.9 0.30  58.3    7.9  2.9
  6    * Denmark           11.8 0.41 113.9   10.4  1.7
  7    * Norway             9.4 0.19  43.5    3.1  1.8
  8    * France             5.2 0.10  44.5  104.3  2.5
  9    * Ireland            0.2 3.73 124.5    3.8  1.9
 10    * Italy              3.6 0.06  13.6  106.6  2.0
 11    * Holland            9.2 0.58  75.5    9.7  2.7
 12    * Portugal           2.2 0.03  27.5   89.3  0.9
 13    * Switzerland        9.1 0.25  73.5   44.9  2.1
 14    * Spain              2.5 0.03  43.6   73.2  2.7
 15    B England            1.8 3.49 113.7    5.1  1.4
 16    *                          11111
 17    *SORT COUNTRIES,16_
 18    *
```

To sort the data according to beer consumption, we have typed a SORT operation on line 17. SORT refers to the data set (COUNTRIES) and to an image line (16) having a mask '11111' indicating the sort key. By now activating line 17 by keeping the cursor on that line and pressing CONTINUE, the data set will be sorted and we have the display:

```
  1    *                 Consumption of various beverages
  2    *DATA COUNTRIES,A,B,C
  3    C                 Coffee  Tea  Beer   Wine Spirits
  4    A Italy              3.6 0.06  13.6  106.6  2.0
  5    * Portugal           2.2 0.03  27.5   89.3  0.9
 ..    .........           .... .... .....  .....  ...
 14    * Denmark           11.8 0.41 113.9   10.4  1.7
 15    B Ireland            0.2 3.73 124.5    3.8  1.9
 16    *                          11111
 17    *SORT COUNTRIES,16_
```

To estimate linear and nonlinear regression models, an ESTIMATE operation is available. In the following display the preceding example has been continued by typing a regression model and an ESTIMATE operation on lines 17 to 20.

```
14     * Denmark          11.8  0.41 113.9  10.4  1.7
15     B Ireland           0.2  3.73 124.5   3.8  1.9
16     *
17     *MODEL Beer1
18     *log(Beer)=constant+coeff*log(Tea)
19     *
20     *ESTIMATE COUNTRIES,Beer1,21_
```

The ESTIMATE operation on line 20 refers to the data set (COUNTRIES) and to the model (Beer1) defined by the MODEL specification. The third parameter (21) is the first line for the results. When ESTIMATE is activated by pressing CONTINUE, the model 'Beer1' will at first be formally analyzed. It is converted into standard notation LOG(X(3))=A(1)+A(2)*LOG(X(4)) where A(1) and A(2) stand for those words 'constant' and 'coeff' which are not recognized as names of variables and are thus interpreted as parameters to be estimated. Also, the first and second partial derivatives of this model function will be formed analytically and this information is employed in selecting a proper computation algorithm. In this case the second derivatives vanish indicating a linear model with respect to the parameters, and so one iteration of the Newton-Raphson method (which is one of the standard alternatives) will be performed. Finally, the results are listed from line 21 onwards as follows:

```
20     *ESTIMATE COUNTRIES,Beer1,21_
21     * constant=4.488964       (0.1565749)
22     * coeff=0.3276288         (0.0752709)
23     * RSS=1.553654  R↑2=0.6545
```

The estimation process may be controlled in various ways by extra specifications typed in the edit field. For example, the estimation criterion may be specified by CRITERION=Lp, where CRITERION=L2 (i.e. ordinary least squares) is default. Similarly the observations may be weighted by entering WEIGHT='weight function', where the weight function is any function of variables in the data set, etc. (Mustonen 1981b).

4. STATISTICAL GRAPHICS

A similar approach is used when making plots of data sets, curves and surfaces (Mustonen 1982). Each extra specification has a default value depending on the data and on other specifications. Thus, a PLOT operation with a few additional keywords usually creates a decent graph. Since all the information needed for a certain plot remains in the edit field, it is easy to improve the result by editing the extra specifications. An extensive inquiry facility is also available for prompt information about details.

For instance, to make a scatter diagram for 'Beer' vs 'Tea' in the preceding example the following PLOT operation, assisted by an image line (23) and one extra specification (YSCALE=LOG,...) implying a logarithmic scale on the Y axis, will be sufficient:

```
21     * Denmark          11.8  0.41 113.9  10.4  1.7
22     B Ireland           0.2  3.73 124.5   3.8  1.9
23     * SSS                    YYYY XXXXX
24     *PLOT COUNTRIES,23_
25     *YSCALE=LOG,0.01,0.02,0.05,0.1,0.2,0.5,1,2,5
26     *
```

On the image line 23 the mask XXXXX refers to the X variable, the mask YYYY to the Y variable and SSS indicates the labels of the observations in the diagram. Thus, activation of PLOT on line 24 will produce the graph shown in Fig.1.

Similarly, to make a pie chart of the following table, a PLOT operation with two extra specifications (HEADER, TYPE) is used as follows

```
31 *
32 *HEADER=Number of males (1000) in various age groups
33 *
34 *DATA MALES,A,B,M
35 M           0-14 15-24 25-44 45-64  65-
36 A Sweden     841   571  1188   930  585
37 * Denmark    564   385   731   537  308
38 * Finland    506   399   727   468  202
39 * Iceland     32    22    29    20   10
40 B Norway     474   316   534   442  251
41 *
42 * PLOT MALES_
43 * TYPE=%PIE
44 *
```

and activation of PLOT on line 42 produces Fig.2.

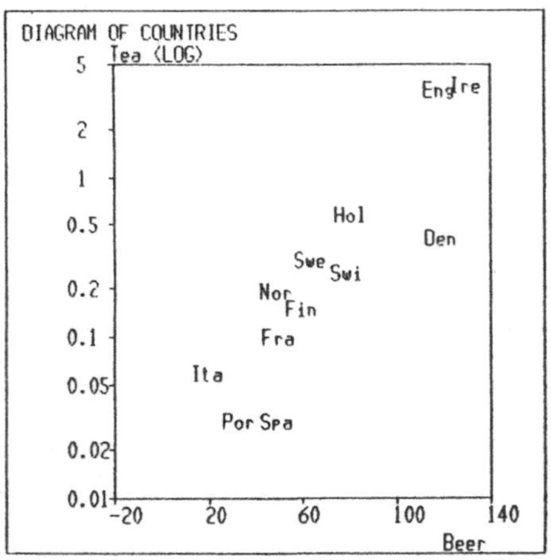

Fig.1

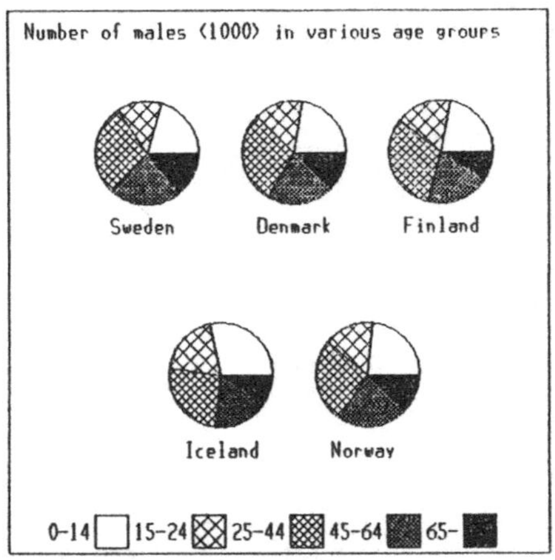

Fig.2

REFERENCES

Mustonen, S., SURVO 76, a statistical data processing system, Research Report No.6, Dept. of Statistics, University of Helsinki, 1977

Mustonen, S., Interactive analysis in SURVO 76, Proceedings in Computational Statistics, ed.by M.M.Barritt and D.Wishart, 253-259, Physica-Verlag, Wien, 1980

Mustonen, S., SURVO 76 EDITOR, a new tool for interactive statistical computing, text and data management, (RELEASE 2), Research Report No.24, Dept.of Statistics, University of Helsinki, 1981a

Mustonen, S., SURVO 76 EDITOR, Estimation of regression models, Research Report No.29, Dept. of Statistics, University of Helsinki, 1981b

Mustonen, S., Statistical computing with a text editor, Computational Statistics, ed.by Herbert Büning and Peter Naeve, 327-348, Walter de Gruyter, Berlin, 1981c

Mustonen, S., Statistical graphics in SURVO 76 EDITOR, Research Report No.33, Dept. of Statistics, University of Helsinki, 1982.

Cluster Analysis Based on the Maximization of the RV Coefficient

G. Nin, C.R.I.G., Marseille, France

Résumé

Nous utilisons un cosinus généralisé pour ajuster une partition à un opérateur de données; l'origine et les détails de cette approche sont précisés dans les deux premiers paragraphes. Le problème d'optimisation qui en découle est de type discret, mais nous lui associons un problème continu qui permet d'utiliser un algorithme classique de la théorie de l'optimisation: l'algorithme du gradient réduit; c'est ce qui est développé dans le troisième paragraphe. Les solutions obtenues permettent alors d'associer à tout individu de la population initiale un degré d'appartenance à chacune des classes de la partition cherchée. Dans le dernier paragraphe, en assouplissant notre formalisme, nous montrons que le critère étudié permet de retrouver celui de l'inertie inter-classes.

Keywords: Linear multivariate analysis; RV-coefficient; Cluster analysis Non concave optimization; Reduced gradient method.

1. The RV-coefficient.

It is well known since P.Robert and Y.Escoufier(1976) that some of the main methods in linear multivariate analysis can be presented from the following optimization way:

Maximize $RV(A,B)$, where A is a given operator which is calculated $B \in \mathcal{D}$ from the data, B an unknown operator which must lie in some domain $\mathcal{D}$ and

$$RV(A,B) = \mathrm{Trace}AB / (\mathrm{Trace}A^2)^{1/2} (\mathrm{Trace}B^2)^{1/2}.$$

More precisely, if the data are arranged into a $p \times n$ matrix X (p numerical variables, n individuals), if a positive semi-definite matrix Q defines the scalar product between two variables and a diagonal positive matrix T defines the scalar product between two individuals, the solution of the problem:

Maximize $RV(X'QXT, Y'YT)$

$$\begin{cases} Y \in \mathcal{D} \\ \mathcal{D} = \{ Y,\ k \times n\ / YY' = I_k \} \end{cases}$$

(X' is the transposed matrix of X, and I_k the matrix identity of $\mathbb{R}^k$)

is given by the k first principal components of X.

As the problem of maximization the RV-coefficient have been shown to be interesting from the point of view of linear multivariate analysis, we have applied this method to cluster analysis.

2. Cluster analysis.

Suppose that the matrices X, Q, T are given as in the first paragraph and k is a strictly positive integer. If P_k denotes a partition in, at most, k classes of individuals, it is customary to associate a matrix U, $n \times n$, to the partition P_k:

$U=(U_{ij})$, $1 \leq i \leq n$, $1 \leq j \leq n$, with U_{ij} equal to 1 if the individuals numbered i and j are in the same class, and equal to 0 otherwise. It is easy to see that $U=Y'Y$ where Y is the matrix in which each row Y_i contains the values of the characteristic function of the class numbered i $(1 \leq i \leq k)$. We can call such a function an "indicatrice variable". In order to find a partition, in at most k classes, of the individuals which is the closest, for the RV-coefficient, to the operator of data, we have to solve:

(*) $\underset{Y \in D}{\text{Maximize}}\ RV(X'QXT, Y'YT)$, where D is the domain of all matrices made of indicatrice variables with k rows and n columns.

This problem is a discrete one and rather than trying to solve it, we propose to associate a continuous problem the solutions of which will be considered as "fuzzy" solutions of our initial clustering problem.

Note C the set of the matrices $Y=(Y_{ij})$, $1 \leq i \leq k$, $1 \leq j \leq n$, which satisfie the following linear inequalities:

$$\begin{cases} Y_{ij} = 1, & 1 \leq j \leq n. \\ Y_{ij} \geq 0, & 1 \leq i \leq k, 1 \leq j \leq n. \end{cases}$$

We associate the following problem to problem (*):

(* *) $\underset{Y \in C}{\text{Maximize}}\ RV(X'QXT, Y'YT)$

C is a convex set and it is shown (G. Nin (1981), for instance), that the set D is constituted by the extreme points of C, so, by a classic result which can be found in T. Rockafellar (1970), if the function of Y that we maximize was concave, then the solutions of the problem noted (*) would be obtained by solving (* *).

In fact this function is not concave but a solution $\overline{Y}$ of (* *) is easy to interprete in a fuzzy way: $\overline{Y}_{ij}$ defines a degree of membership of the individual j to the class i. This notion less strict than the ordinary one seems to be appreciate by the users of the method.

3. Reduced gradient method

A) Generalities

This method was presented by P. Wolfe (1963), it is inspired from the simplex one and it is used to solve non concave maximization problems under linear constraints like in problem (* *).

We identify the matrix Y to the kn-dimensionnal vector in which the coordinates are $(y_{11},y_{12},\ldots,y_{1n},y_{21},\ldots y_{kn})$, then the function defined by: $Y \mapsto RV(X'QXT, Y'YT)$, is noted F and the problem (* *) can be written:

Maximize $F(Y)$, where $B=(I_n \vdots I_n \vdots \ldots \vdots I_n)$ is obtained by justaposition of k matrices equal to the matrix identity of $\mathbb{R}^n$, and $\underline{1}$ is the n-dimensionnal vector in which the coordinates are all equal to 1.

$$\begin{cases} BY=\underline{1} \\ Y \geqq 0 \end{cases}$$

A subset I of the index set $\{1,2,\ldots,nk\}$ is said to be a basis of the linear system $BY=\underline{1}$ if the matrix noted B^I formed with the columns of B indexed by I, is regular. Suppose now that I is such a basis then it is possible to reduce the number of independent variables in the problem (* *). The equality $BY=\underline{1}$ may be written

$$(B^I, B^{\bar{I}})\binom{Y_I}{Y_{\bar{I}}} = \underline{1}$$

($\bar{I}$ is the complementary subset of I relative to $\{1,2,\ldots,kn\}$), and we have

$B^{\bar{I}}Y_{\bar{I}} + B^I Y_I = \underline{1}$, so $Y_I=(B^I)^{-1}(\underline{1}-B^{\bar{I}}Y_{\bar{I}})$. We denote g, deduced of F, the function of the independent variables, and the problem (* *) is written as follows:

Maximize $g(Z)$, this problem is a reduced one and we are going to use a gradient method to solve it

$$\begin{cases} Z \geqq 0, \\ (B^I)^{-1}(\underline{1}-B^{\bar{I}}Z) \geqq 0 \end{cases}$$

It is important to notice that the special form of matrix B avoids the inversion of B^I since this matrix is formed of the same columns as the matrix identity of $\mathbb{R}^n$, when I is a basis.

B) Description of the algorithm.

Suppose that at the end of any iteration, an Y in the set C and a basis I, such that $Y_I > 0$, are obtained.

1°) Determine an "ascending" direction for the reduced problem, noted $c(Y_{\bar{I}})$, i.e. a $k(n-1)$ dimensional vector such that the function g increases along the half line directed by $c(Y_{\bar{I}})$.

Choose an "ascending direction h for problem (* *) in the following way:

if $i \in \bar{I}$, then $h_i=(c(Y_{\bar{I}}))_i$ and if $i \in I$, then $h_i=-(B^I)^{-1}B^{\bar{I}}c(Y_{\bar{I}})_i$

Test the condition $\|h\|<\epsilon$, where ϵ is a given "small" value, If $\|h\|<\epsilon$, stop; if $\|h\| \geq \epsilon$, go to the next step.

2°) Determine the length θ of the step in the direction h.

3°) Define $Y'= Y+\theta h$ and proceed to a change of basis to exclude basis variables which became equal to zero. Denote I' the new basis and return to 1° with Y' and I' instead of Y and I.

Some variants can be found in J.Denel(1980), where a theorem of convergence to a stationnary point of the function F is established. This stationnary point is not necessary a global optimum of F, but it gives, generally, a "good" value to the criterion. The question of global optimum is a classical difficulty in non concave maximization, and we have not yet found a satisfactory solution.

4.Connections with others criteria

It is customary to compare a new criterion to the currently used. This is why we have to modify our problem (*) by introducing a diagonal metric M between the indicatrice variables $Y_1,\dots,Y_k$; we obtain:

$$\text{Maximize } RV(X'QXT, Y'MYT)\quad \begin{cases} Y \in D \\ M \text{ diagonal metric on } \mathbb{R}^k . \end{cases}$$

If the data are normalised, i.e. if $(X'QX)_{ii} = 1$ for $1 \leq i \leq n$ if $T=(1/n)I_n$, and if the diagonal values of M are noted m_i, it is easy to show that :

$$RV(X'QXT, Y'MYT) = \frac{1}{n^2}\,\frac{\sum_j n_j m_j (n_j - It_j)}{(\mathrm{Trace}X'QXT^2)^{1/2}\,(\sum_j m_j^2 n_j^2)^{1/2}}$$

where n_j is the cardinal of the class Cl_j of the partition P_k, and It_j is the inertia of this class ($It_j = \sum_{x_i \in Cl_j} dist^2(x_i, g_j)$, g_j center of gravity of the class Cl_j).

If we define the metric M by $m_j = \frac{1}{n_j}$ (M depends on P_k):

$$RV(X'QXT, Y'MYT) = \frac{1}{n^2}\,\frac{n - \sum_j It_j}{(\mathrm{Trace}X'QXT^2)^{1/2}\; k^{1/2}}$$

This calculation proves that, for this choice of M, the maximization of the RV is equivalent to the minimization of the sum of the inertias of the classes; this criterion is actually underlying in many clustering methods as it is noted by M.R.Anderberg(1973).

From the point of view of hierarchical clustering it is shown by G.Nin(1981), that the maximization of the numerator of the RV leads to the Ward's algorithm.

To conclude we mention that other algorithms can be used to maximize our criterion, for instance discrete ones like the exchange algorithm mentionned by I.C.Lerman(1970), but it seems to us that it is better to solve the continuous problem, because it enables the use of methods of mathematical programming which is a part of applied mathematics very active and where performant codes have already been written.

Bibliographie.

M.R.Anderberg(1973),Cluster analysis for applications,Academic Press, pp.156-175.

F.Caillez and J.P.Pages(1976),Introduction à l'analyse des données, SMASH. Paris.

J.Denel(1980),Convergence de la méthode du gradient réduit en l'abscence de concavité,Bulletin de la D.E.R.,E.D.F.,série C,n°2, pp.55-62.

Y.Escoufier and P.Robert(1976),A unifying tool for linear multivariate analysis methods:the RV-coefficient, Journal of the Royal Statistical Society, vol 25,n°3,pp.257-265.

I.C.Lerman(1970),Les bases de la classification automatique,Gauthier-Villars,pp.66-73.

G.Nin (1981), Maximisation de la trace et du RV, applliquée à la classification automatique, Thèse, Université de Provence.

R.T.Rockafellar(1970),Convex analysis,Princeton University Press, pp.342-346.

P.Wolfe(1963),Methods of non linear programming,Mac Graw-Hill Book Co,pp.67-86.

G.Zoutendijk(1960),Methods of feasible directions,Thesis,Amsterdam, pp.64-80.

Statistical Approach to System Parameter Synthesis

M. Novák, Czechoslovak Academy of Sciences, Prague, Czechoslovakia

Abstract:

The synthesis of almost every technical system involves these fundamental parts:

a/ the design of the system structure,

b/ the determination of the system parameters,

c/ the determination of the acceptable system parameter tolerances.

For all these parts the necessity to find an optimal solution is typical. In this contribution the attention is concerned to the problems of the parts b/ and c/ supposing that an optimal system structure has been already found. After discussion of the respective problematics and presenting of the importance of the statistical approach to their solution the principle of the so called statistical derivative method is described and the basic numerical algorithm for the statistical optimized synthesis of the system parameter nominal vector and tolerance vector is shown.

Keywords: System synthesis, optimization, tolerances, statistical derivatives, design centering.

1. Introduction

Designing any technical system which has to be manufactured in repeated realization one has to consider several important technical and economical aspects. These are before all the system function, production yield, application reliability and the cost minimization. Few years ago practically all the synthesis and design procedures have terminated in finding of the suitable system structure and determining of the values of system parameters so that the requirements on the basic system properties /system functions/ had been reached. However, the increase in the size and complexity of the designed systems and the more severe requirements on the manufacturing and application economy has caused the necessity to find some new approaches to the system synthesis allowing to respect the whole set of these additional aspects.

In general, the synthesis procedure can be divided in the following three subsequent steps: approximation, realization and optimization. During the approximation period of the synthesis procedure, one has to find a set of the physically realizable system functions, so that the difference between this set and the required system properties is minimal. In the realization period, the system structure is to be constructed corresponding to the chosen system functions. During the last, optimization period the results of the previons two has to be modified with respect to the additional requirements of the cost, yield and reliability.

2. Problem formulation.

The actual values of any system parameter x_i of any realized system S oscillate around the nominals x_{io} in some region Δx_i with some density of probability $\varphi_i(x_i)$. Suppose, that the system S has N parameters x_i, $i = 1, 2 \ldots\ldots N$. The parameter vectors $X = \{x_i\}$ of all the M realized systems S_m, $m = 1, 2 \ldots\ldots M$ lie in the parameter space $\{x\}$ in certain region only, which we shall denote as the region of deviations R_E. This is projected on the axes x_i in the intervals of deviations Δx_i. If there are no correlations between $\varphi_i(x_i)$, R_E has the form of a hypercube. In the case when $\varphi_i(x_i)$ are correlated, its form is more complicated. The properties of the system S are represented by the set of the system functions $F = \{F_j\}$, where $j = 1, 2 \ldots\ldots J$. Let us denote the set of the nominal properties by $F_0 = \{F_{jo}\}$, to which corresponds the nominal parameter vector $X_0 = \{x_{io}\}$. The variance of vector X of actual parameters values in the region R_E causes the variance of F. Suppose that $X_0 \in R_E$ and that the maximal allowable deviation of F with respect to F_0 is ΔF. All the functions F satisfying the condition

$$F - F_0 \leqq \Delta F \qquad /1/$$

form the body F_A of the acceptable system functions. In the exp. /1/ the algebraic operators $-$ and $\leqq$ has to be considered in symbolic meaning in general; they can be taken in a presented form only if F is a scalar value. To each $F \in F_A$ corresponds one X inside the region of acceptability R_A, which is a part of the N-dimmensional space $\{x_i\}$. For a good design $X_0 \in R_A$. The correspondence between F_0, ΔF, X_0, R_E and R_A is shown schematically for N = 2 and scalar F in Fig. 1.

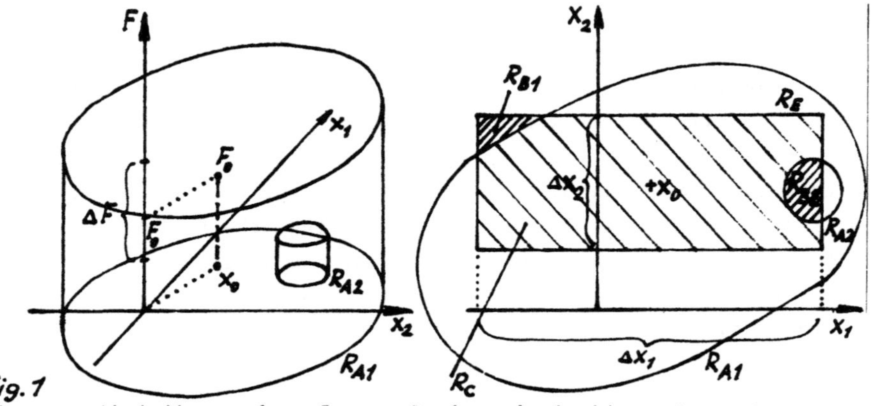

Fig.1

Suppose that the surface F over R_A is univaluable and continuous. During the manufacturing, all the produced systems are realized so that their X_s' fill out the R_E with respect to $\phi(X) = \{\varphi_i(x_i)\}$. If all the $R_E \in R_A$, the production is with the 100% yield å priori /Y = 1/. If certain part of R_E overtap R_A /as it is shown in Fig.1/ certain part of the production has to be thrown out and $Y < 1$. Therefore R_E seems to be reasonable to allow maximally so large, that it does not cross the boundaries of R_A. Being interested to minimize the production cost C, one has to consider that it is inverse proportional to R_E according the following:

$$C = C_c + \sum_{i=1}^{N} c_i , \qquad /2/$$

where C_c is the constant, tolerance independent production cost of the system and

$$c_i = \alpha_i + \frac{\beta_i}{\gamma_i^{\,\gamma_i}} , \qquad /3/$$

where $\alpha_i, \beta_i, \gamma_i$ are empirical real constants and

$$\gamma_i = \frac{x_i - x_{i0}}{x_{i0}} \qquad /4/$$

are the parameter tolerances. So, we are interested in finding an optimal compromise between the decrease of the cost C caused by the increase of the allowable tolerances and it increase caused by the decrease of the yield

$$Y = \int_{R_c} \phi(X)\, dX \left[\int \phi(X)\, dX \right]^{-1} , \qquad /5/$$

where $R_C = R_A \cap R_E$, below the value 1.

Preliminary, we have supposed that the nominal design $X_0 \in R_A$ is given. Trying to optimize C, we must allow that X_0 can be changed during the optimization procedure, if it appears that for $X_c \neq X_0$ certain lower values of C can be found. The problem can then be formulated as follows :

find X_c corresponding to minimum C subject to F_0, ΔF and $\phi(X)$.

3. Approach to problem solution

Let us try to analyze R_A at first for being able to evaluate the integrals in /5/. Beause the boundaries of R_A can be determined in a closed analytical form in very rare cases only, a statistical approach to their analysis is used usually. Here one generates the samples X inside R_E according to $\phi(X)$ and for every sample answers the question, wheather /1/ is fulfilled or not. Naturally, such a procedure can be very laborious and computer time consuming. Therefore, several methods vere proposed, which allow to fasten the way to find the points X, satisfactory close to the boundaries of R_A /see [1,2,3] e.g./ . Notwithstanding that the knowledge of R_A is very interesting and useful for learning the system properties, it is not necessary in principle for answering the mentioned fundamental problem. Suppose, that we approximate the integral expression /5/ in a stochastical way by

$$Y_p = \frac{P_p}{P} , \qquad /6/$$

where P is the number of all the samples generated in R_E and $P_p \leqq P$ is the number of samples when /1/ holds. The number P has no straight forward relation to the number M of the produced systems, it depends on the system properties and on the density $\phi(X)$. Evidently, for $P \longrightarrow \infty$ the statistical yield Y_p approaches Y. The practical experience has shown, that when analysing the mediumly complicated /N about 10/ electronic systems, $10^2 < P < 10^3$ is a reasonable choice. The approximation of the partial stochastical derivatives $\partial Y_p / \partial x_i$ can be found by the use of the values $Y_p(x_1, \ldots x_i + \Delta x_i, \ldots x_N)$. The size of the step Δx_i has to be chosen very carefully with respect to the danger to destroy the detailed structure of the multidimensional space $\{x_i\}$ through the integral character of Y_p . For the optimization of the tolerance vector $\tau = \{\tau_i\}$ can be used the following algorithms :

a/ for given x_{0i} and Δx_i , Y_p is determined,

b/ in the point $(x_{01}, x_{02} \ldots x_{0i} + \Delta x_i, \ldots x_{0N})$ are determined the values $\partial Y_p / \partial x_i$

c/ the components $$\frac{\partial \tilde{Y}_p}{\partial x_i} = \frac{\frac{\partial Y_p}{\partial x_i} - Y_p}{\sqrt{\sum_{i=1}^{N} \left(\frac{\partial Y_p}{\partial x_i} - Y_p\right)^2}}$$ of the normalized stochastical derivatives $\partial \tilde{Y}_p / \partial X$ are determined, d/ the nominal point X_0 is moved in the direction $-\frac{\partial \tilde{Y}_P}{\partial X}$ in X_1 so, that Y_P (X_1) remains higher than Y_P (X_0),

e/ the value $\partial \tilde{Y}$ (X_1) / ∂X_1 is calculated and a new point X_2 is chosen in the direction $-\partial \tilde{Y}$ $(X_1)/\partial X_1$.

The procedure is repeated untill the values Y_P increase /if not some higher number P is to be chosen/. When searching by the use of the described algorithm for the optimal centered nominal point X_C, one has to move the subsequent nominals in the directions opposite to $\partial \tilde{Y}_P$ / ∂X. However, each such value is only approximative. The accuracy of the calculation can be improved by the increase of P, which is in the contradiction to the tendency of computing time minimization. Therefore a reasonable compromise has to be chosen.

Suppose, that the calculation of the statistical approximation Y_P of the yield for J different points on the line going through X_0 in the direction $-\partial \tilde{Y}_P$ / ∂X_0 is made /see Fig. 2/

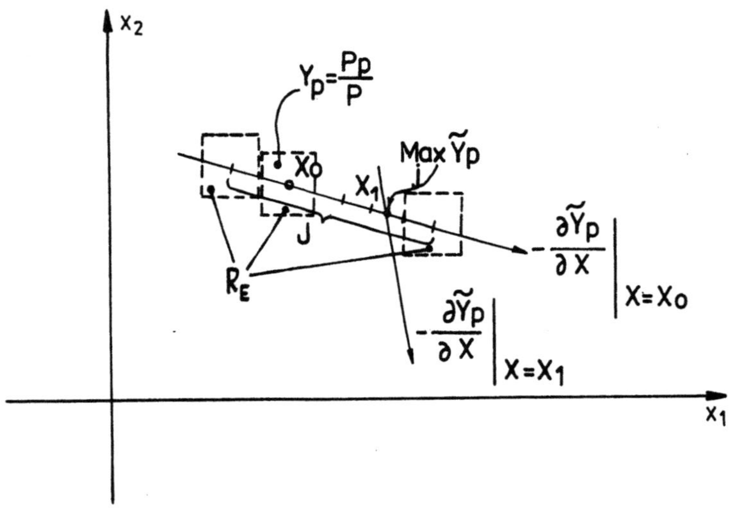

Fig. 2.

The point X_1 of the maximal Y_P is found as the center of the interval on the mentioned line, where the sum of Y_P in K adjacent points is maximal, when K << J. In X a new direction $-\partial \tilde{Y}_P$ / ∂X_1 for the further move of the nominal is determined.

4. Results and conclusion

The centering procedure mentioned above allows to save a considerable large amount of computing time. The practical experience obtained dealing with optimization of electronic selective networks shows that for N=5 and $0{,}2 \leqq \gamma_i \leqq 2\%$ the number P = 200 is satisfactory. If $\gamma_i > 2{,}5\%$, $P \geqq 600$ had to be taken. The decrease of the computing time for this procedure compared with the usual methods based on the analysis of R_A is more than 50%. Because no assumption about the behaviour of the surface F over $\{X_i\}$ was made, the procedure can be used also in cases of the discrete parameter space $\{x_i\}$ and discrete tolerances. The only need is to have a sufficiently fast analytical procedure for determining F (X) at disposal.

5. Refferences

[1] Director S. W., Hachtel C. D./1977/, The Simplicial Approximation Approach to Design-Centering, IEE Transactions on Circuits and Systems, CAS-24, No 7, 363 - 372,

[2] Mayne D. Q., Polak E., Voreadis A. /1982/, A Cut - Map Algorithm for Design Problems with Parameter Tolerances, IEEE Transactions on Circuits and Systems, CAS - 29, No 1, 35 - 45,

[3] Novák M. /1978/, The Role of Design Centering in the Synthesis of Tolerances, Proceedings of the 6th Colloquium on Microwave Communication, Budapest,

[4] Novák M. /1981/, Simulation der statistischen Eigenschaften von massproduzierten Systemen, INFO 81, Neubrandenburg, GDR.

Statistical Match of Samples Using Additional Information

G. Paass, GMD, St. Augustin, F.R.G.

Summary:

Statistical matching methods combine records from two samples with different but overlapping sets of variables in order to construct a new sample that comprises all variables. From theoretical reasons the two samples do not contain information about all statistical interactions present in U. Therefore every classical statistical matching method implicitly presupposes the conditional independence of specific variables. However often this assumption is violated and this causes distortions of the resulting sample. In this paper information from additional samples is used to replace the conditional independence assumption. The EM-algorithm is used to estimate population parameters from different sources. Using these estimates several criteria for the combination of records are constructed.

KEYWORDS: File merging, statistical match, matching criteria, additional samples

1 Introduction

Let us consider a population U, whose elements are described by three vectors of variables x,y and z. Suppose there are two samples containing different sets of variables. The first sample comprises the x- and y-variables while the second sample contains the x- and z-variables. Both samples are assumed to be error-free, come from the same population and the common variables in both samples are defined in the same way.

In a statistical match each case (e.g. person, individual, household) in the (x,y)-sample is assigned to one or more cases from the (x,z)-sample (Radner, et al. 1980). The assignment is based upon the similarity of the cases, usually expressed by some sort of distance function. Then from these selected cases a new case is formed consisting of the (x,y)-values from the (x,y)-sample and of the z-values from the (x,z)-sample. In contrast to other procedures of imputation (e.g. regression) statistical matching methods can be applied to nearly all sorts of distributions and thus have features of nonparametric methods.

As in statistical matching methods all information about the marginal (y,z)-distribution has to be communicated by the common vector of x-variables, every matching method implicitly postulates the conditional independence of the y- and the z-variables given x (Sims, 1972). This conditional independence assumption obviously is very restrictive. Empirical investigations of the performance of matching methods indicate (Paass, Wauschkuhn, 1980; Rodgers, DeVol, 1982; Paass, 1982) reasonable quality of the matched files unless the conditional independence condition is violated. The only way to improve statistical matching results if the conditional independence condition does not hold is to collect additional information about this relation and introduce it into the matching process. This information in general consist of additional samples or tables with different sets of variables.

The matching procedures described in this paper contain three major elements:
- the estimation of parameters of a common distribution of all variables from all available samples;
- the derivation of a matching criterion;
- the actual assignment of cases according to the matching criterion.

The estimation of a common distribution may be accomplished by the EM-algorithm or the iterative proportional fitting algorithm which are described in section 2. For the derivation of matching criteria two major approaches are developed. The first approach starts with a given distance measure between the cases of the two samples. The missing variables in either sample are estimated using the population parameters. This approach is discussed in section 3 for normal and multinomial populations.

The second approach starts with the relation of the elements of the samples after the assignment has been performed according to some distance measure. The distribution of the assigned cases (neighbors) is modelled approximately and used for the formulation of a matching criterion. It gives more weight to the similarity of x-values than to the estimates of unobserved variables and is liable to reconstruct the conditional variance more correctly. It is described for normal populations in sections 4 and 5. The last section gives a short discussion and summary of the results.

2 Estimation of Population Parameters

Assume we have three samples where the vectors $(x',y')'$, $(x',z')'$ resp. $(y',z')'$ are observable. Then observations on z, y, resp. x are missing. We assume that the distribution of the underlying population is known with the exception of population parameters which have to be estimated. As in this setup ML-estimates cannot be calculated in closed form, we have to employ an iterative algorithm, e.g. the EM-algorithm (Dempster, et al. 1977). It is applicable to a large class of distributions (normal distribution, multinomial distribution, etc.; c.f. Paass, 1982b), has very desirable convergence properties and can be applied to any number of samples with any pattern of missing data.

The normal theory of EM-algorithm is complicated if linear dependencies occur in the population. But from the matching aspect linear dependencies contain valuable information as exact linear relations imply logical dependencies in the data. Therefore we have developed a version of the EM-algorithm that exploits the information from linear dependencies and calculates covariances reflecting these singularities (Paass, 1982b). The EM-algorithm copes with different samples from the same population and tries to reconstruct the joint distribution. If marginal tabulations from the same sample are available we can estimate the joint distribution with the iterative proportional fitting algorithm which converges to the maximum likelihood estimate of the cell counts (Bishop, et al., 1975, p.93ff).

3 Distance Minimization

We assume that from theoretical or heuristical considerations we have selected a distance measure $d_0(w^1,w^2)$ between the full data vectors $w^1:=(x^{1'},y^{1'},z^{1'})'$ and $w^2:=(x^{2'},y^{2'},z^{2'})'$. (For a discussion of distance measures see Bock, 1974, p.35-73; Clifford, Stephenson 1975, p.49-82). We want to combine two cases w^1 and w^2 if $d_0(w^1, w^2)$ is minimal. As z^1 and y^2 cannot be observed they have to be estimated using the distribution $p(x,y,z)$ which has been determined by the methods of section 2.

$$(1) \quad d_1 \left(\begin{bmatrix} x^1 \\ y^1 \end{bmatrix}, \begin{bmatrix} x^2 \\ z^2 \end{bmatrix} \right) = \int d_0 \left(\begin{bmatrix} x^1 \\ y^1 \\ z \end{bmatrix}, \begin{bmatrix} x^2 \\ y_2 \\ z^2 \end{bmatrix} \right) p(z|x^1,y^1)\, p(y|x^2,z^2)\, dz\, dy$$

For discrete distributions the integral reduces to a sum. If we apply (1) to quadratic distances (Euclidean distance, Mahalanobis distance)

$$(2)\quad d_0\,[w^1,w^2] := (w^1-w^2)'\ M(w^1-w^2)$$

(M symmetric) we get

$$(3)\quad d_1\left(\begin{bmatrix}x^1\\ y^1\end{bmatrix},\begin{bmatrix}x^2\\ z^2\end{bmatrix}\right) = \bar{g}'M\bar{g}$$

with

$$\bar{g} := \begin{bmatrix}x^1-x^2\\ y^1-E(y|x^2,z^2)\\ E(z|x^1,y^1)-z^2\end{bmatrix}$$

The conditional expectation for normal variables is given by the linear regression.

One of the first who discussed statistical models for matching was Kadane (1978, p.166f). He estimated the missing y^2- and z^1-values by regression, calculated the covariance matrix of the pooled sample of estimated and observed values and proposed to use the Mahalanobis distance. If no additional samples are available his approach will give nearly the same results as above. Sims (1978, p.176) argued that the conditional variance of z given x would be biased downwards and therefore the x-variables should get more weight than the y- and z-variables, as the latter are compared with estimates and not with observed values. Such a criterion is derived in the next section.

4 Distribution of Neighbors

We define

$$E(w) = \begin{bmatrix}E(x)\\ E(y)\\ E(z)\end{bmatrix} \qquad S := \begin{bmatrix}S(xx),S(xy),S(xz)\\ S(yx),S(yy),S(yz)\\ S(zx),S(zy),S(zz)\end{bmatrix}$$

We assume that we have two independent identically l-dimensional normal distributed samples of size k with variables $w = (x',y',z')' \sim N(E(w),S)$. The parameters $E(w)$ and S could have been estimated by the EM-algorithm and are assumed to be known.

First we suppose that all variables x, y, and z can be observed in both samples. Assume we want to assign to each element of the first sample an element from the second sample such that the sum D of Mahalanobis distances is minimal

$$(4)\quad D = \sum_{i=1}^{k} (w_i-w_{j(i)})'S^{-1}(w_i-w_{j(i)})$$

If the permutation j is selected randomly, D/2 is distributed as a central chisquare with k*l degrees of freedom. As we have k! permutations j on the average we have to select k! permutations at random until we get the minimal distance. Hence we have for the corresponding distribution P_0 of the minimal distance

$$(5)\quad P_0\ (w \geq \beta) = P(w \geq \beta)^{k!}$$

where P corresponds to the distribution of D. This relation can be used to calculate any desired quantile of P_0.

Now we state an approximate model that gives us the relation between the neighbors after the distance minimization for the complete data vector has been performed. We suppose that any two neighbors w^1 and w^2 from the two samples come from a common origin w^0 and define

$$(6)\quad w_i^0 \sim N(E(w),(1-c)S) \qquad i=1,\dots,k$$

where the constant c is specified later.

$$(7)\quad w_i^1 \sim N(w_i^0,cS) \qquad w_i^2 \sim N(w_i^0,cS) \qquad i=1,\dots,k$$

Both distributions are supposed to be independent. Obviously w^1 and w^2 are distributed as N(E(w),S) which conforms with the distribution of the observed samples. We have

$$(8)\quad w_i^1 - w_i^2 \sim N(0,2cS)$$

As these vectors are assumed to be neighbors, their Mahalanobis distance should have the same distribution as derived above in (5). Because P_0 is not normal (6)-(7) will only be an approximate model. As the comparative magnitude of the variances (1-c)S and cS is essential for the relative weight of variables, we require that both distributions have a common parameter, say the median. This gives us the condition to fix the unknown constant c. If the Mahalanobis distance of neighbors is denoted by D_0 we have from (4) and (8) that $D_0/2c$ is distributed as chisquare with k*l degrees of freedom. Hence we can fix c in such a way that D_0 has the same median as P_0 in (5).

5 Neighborhood Matching Criterion

We denote the vectors of observable variables by $q^1 := (x^{1'},y^{1'})'$ and $q^2 := (x^{2'},y^{2'})'$ and arrive at

$$(9)\quad w_i^0 \sim N(E(w),(1-c)S) \qquad S_0 := (1-c)S$$

$$(10)\quad q_i^1 = H_1 w_i^0 + v_1 \qquad v_1 \sim N(0,R_1) \qquad i=1,\dots,k$$

with

$$H_1 = \begin{bmatrix} I, & 0, & 0 \\ 0, & I, & 0 \end{bmatrix} \qquad R_1 := Var(v_1) = c \begin{bmatrix} S(xx), S(xy) \\ S(yx), S(yy) \end{bmatrix}$$

and

$$(11) \quad q_i^2 = H_2 w_i^o + v_2 \qquad v_2 \sim N(0,R_2) \qquad i=1,\dots,k$$

with

$$H_2 = \begin{bmatrix} I, & 0, & 0 \\ 0, & 0, & I \end{bmatrix} \qquad R_2 := Var(v_2) = c \begin{bmatrix} S(xx), S(xz) \\ S(zx), S(zz) \end{bmatrix}$$

Using the Kalman filter (Schweppe, 1973, p.136) we can calculate the conditional moments

$$(12) \quad E(w_i^o|q_i^1) = Var(w_i^o|q_i^1) \left[H_1' R_1^{-1} q_i^1 + S_o^{-1} E(w)\right]$$

$$(13) \quad Var(w_i^o|q_i^1) = \left[H_1' R_1^{-1} H_1 + S_o^{-1}\right]^{-1}$$

The conditional distribution of q^2 given q^1 is normal with

$$(14) \quad E(q_i^2|q_i^1) = H_2 \, E(w_i^o|q_i^1)$$

$$(15) \quad Var(q_i^2|q_i^1) = H_2 \, Var(w_i^o|q_i^1) \, H_2' + R_2$$

The corresponding logarithmic density actually is a Mahalanobis distance between the observed q_i^2 and the expected value $E(q_i^2|q_i^1)$.

$$(16) \quad \log p(q_i^2|q_i^1) \sim (q_i^2 - E(q_i^2|q_i^1))' Var(q_i^2|q_i^1)^{-1} (q_i^2 - E(q_i^2|q_i^1))$$

Hence we should match a q^1 with a q^2 that gives the largest logarithmic density (16). This choice approximately will reproduce the neighborhood relations of samples generated by (9)-(11). Hence the solution imitates the distribution of neighbors from an optimal assignment for samples where all variables can be observed. If appropriate transformations of the variables are performed the actual calculation of (16) only involves the evaluation of an Euclidean distance (Paass, 1982b).

6 Discussion

The assignment problem can be transformed to a special linear programming task (Barr, Turner, 1981). In the solution the sum of distances (16) is minimized and the marginal (x,y)- and z-distributions are preserved. The information about the critical y-z relations is contained in the covariance matrix S which can be calculated with the EM-algorithm. As S is used in (16), the y-z relations will be reconstructed in the matched sample within the sampling variability.

In general samples of socioeconomic data will not follow a normal distribution. However, in an empirical comparison of discriminant techniques with missing variables by Titterington, et al. (1981) the EM-algorithm in conjunction with the Mahalanobis distance was 'a single method which performs very well for each variable set, whereas with the discrete models the choice of the appropriate model can be critical' (Titterington, et al., 1981, p.158). Discriminant procedures as well as matching methods try to assign an observation to a set of cases respectively to a single case, and hence the results seem to be transferable to the matching context. A full discussion is given in Paass (1982b).

The actual performance of the proposed matching criteria has still to be investigated. At present experiments are conducted to test whether the matched samples conform to the underlying distribution and the y-z relations are preserved. In any case, however, the use of additional information will improve the quality of matched files and probably the technique of statistical matching may be utilized in a number of new applications.

References

Barr, R.S., and Turner, J.S., (1981), Microdata File Merging through Large-Scale Network Technology, Mathematical Programming Study, 15, 1-22.

Beale, E.M.L., Little, R.J.A. (1975), Missing Values in Multivariate Analysis, Journ. Roy. Stat. Soc. Ser. B, 129-145.

Bishop, Y.M.M., Fienberg, S.E., and Holland, P.W. (1975), Discrete Multivariate Analysis: Theorie and Practice, MIT-Press, Cambridge, Mass.

Bock, H.H. (1974), Automatische Klassifikation, Vandenhoeck und Ruprecht, Göttingen.

Clifford, H.T., Stephenson, W. (1975), An Introduction to Numerical Classification, Academic Press, New York.

Dempster, A.P., Laird, N.M., and Rubin, D.B. (1977), Maximum Likelihood from Incomplete Data via the EM Algorithm, Journ. Roy. Stat. Soc. Ser. B. 39, 1-38.

Fellegi, I.P. (1978), Discussion, 1977 Proc. of the American Stat. Assoc., Social Science Section, 762-764.

Kadane, J.B., (1978), Some Statistical Problems in Merging Data Files, Compendium of Tax Research, S.159-179, Washington, D.C.: US Government Printing Office.

Paass, G. (1982), Comparison and Evaluation of Statistical Matching Techniques, Interner Bericht IPES.82.0202, GMD, St. Augustin, FRG.

Paass, G. (1982b), Statistical Match with Additional Information, Interner Bericht IPES, GMD, St. Augustin, FRG.

Paass, G. and Wauschkuhn, U. (1980), Experimentelle Erprobung und vergleichende Bewertung statistischer Matchverfahren, Interner Bericht IPES.80.201, GMD, St. Augustin, FRG.

Radner, D., Allen, R., Gonzales, M.E., Jabine, T.B., and Muller, H.H. (1980), Report on Exact and Statistical Matching Techniques, Statistical Working Paper No.5, U.S. Department of Commerce, U.S. Government Printing Office, Washingthon, D.C.

Rodgers, W.L. and DeVol, E.B. (1982), An Evaluation of Statistical Matching, Survey Research Center, Instiute for Social Research, Univ. of Michigan, Ann Arbor, Michigan.

Schweppe, F.C., (1973), Uncertain Dynamic Systems, Prentice Hall, Englewood Cliffs.

Sims, C.A. (1972), Comments, Annals of Economic and Social Measurement, 1, 343-346.

Sims, J.B., (1978), Comment, Compendium of Tax Research, S.172-177, Washington, D.C.: US Government Printing Office.

Titterington, D.M., Murray, G.D., Murray, L.S., Spiegelhalter, D.J., Skene, A.M., Habbema, J.D.F. and Gelpke, G.J. (1981), Comparison of Discrimination Techniques applied to a Complex Data Set, J.roy. Stat. Soc. A, 145-175.

An Exploratory Program Package for Non-Linear Data-Smoothers

W. Polasek, University of Vienna, Austria

1. Introduction and Summary

A special algorithm for running medians, called "TABLE" is developed and compared with the usual available sorting-algorithms. The algorithm is particularly useful for repeated sorting-algorithms of nearly sorted list. Time comparisons with other algorithms show the superiority of the new algorithm. The program package INDSMO (interactive non-linear data smoothers) is designed to allow a quick and comfortable application of such problems in exploratory data analysis.

Finally we demonstrate in an economic example the input and output of this program package.

2. Non-linear Data-Smoothers

Recent developments in applied statistics lead to the concept of exploratory data analysis (abbreviated by EDA, see e.g. Tukey (1977)). One useful exploratory method are non-linear data-smoothers, which can be applied to all sorts of data-smoothing (time series, histograms, etc.). Non-linear data-smoothers are mainly based on medians, rather than averages. For short series and short-span smoothers they can even be calculated by hand, but for longer series and especially for compound smoothers involving the running median principle, calculations on the computer are necessary.

According to the principle of moving averages we define by "moving medians" the smoothing operation

$$\text{(2.1)} \qquad Sm^{(\nu)}\{x_t\} = \text{med}\{x_{t-n},\ldots,x_t,\ldots,x_{t+n}\},\ \nu = 2n+1,\ n \in N,$$

where "med" stands for median. For even span medians (ν=2n) we define as smoothing value the arithmetic mean of the smoothers with span 2n-1 and 2n+1. Compound non-linear smoothers are defined by "smoothers in series" and are abbreviated by writing the corresponding smoother symbols side by side.

$$\text{(2.2)} \qquad Sm_1Sm_2\{x_t\} = Sm_2\{Sm_1\{x_t\}\}$$

COMPSTAT 1982 © Physica-Verlag, Vienna for IASC (International Association for Statistical Computing), 1982

and means smoothing first by $y_t = Sm_1\{x_t\}$ and then by $z_t = Sm_2\{y_t\}$. Finally we define the running median principle by defining formally '3R', i.e. the running median of span 3:

$$3_{k+1}\{x_t\}_t = \text{med}\left[3_k\{x_t\}_{t-1},\ 3_k\{x_t\}_t, 3_k\{x_t\}_{t+1}\right], k \in N,$$

(2.3)

$$3R\{x_t\} = \lim_{k\to\infty} 3_k\{x_t\}$$

where 3_k denotes the k-th replication of the moving median of span 3; The initial condition is given by $3_0\{x_t\} = \{x_t\}$ and the limit is assumed to exist.

3. A Fast Running Median Algorithm

Recent developments in computer science lead to various sorting algorithms for special purposes (see Knuth (1978) and Wirth (1975)). There are also optimal algorithms available for the calculation of a median (see Schönhage et al. (1976) and Hopcroft (1974)). Some of these are also concerned with finding the median (or any ordered element) for almost sorted lists (Cooke and Kim (1980)). But all these algorithms deal with the problem of finding the median of a large data set once. For the calculation of running medians we need an optimal algorithm for finding a series of medians, but for small data sets. This problem stimulated the development of a new algorithm specific for running medians, and the comparison with various other median search algorithms based on different sorting algorithms.

3.1 Outline of the New Running Median Algorithm

The algorithm is explained for odd-span moving medians, but can be easily extended to even-span medians, where we need two invocations of the program. Let us denote the time series with $z_1,\ldots,z_N$ and assume a k-span median (k odd). A sub-set of k elements $y_1,\ldots,y_k$ of the time series $\{z_t\}$ is ordered (by the INSERT-algorithm, because it is the fastest for small k's) and split into two parts $s = (s_1,s_2)$, where s_1 has (k+1)/2 elements and s_2 has the remaining (k-1)/2 elements. Because the set s is ordered the last element of s_1 is simply the median, which we will denote with $\tilde{y}_{0.5}$. If we move on to the next set of k elements, the element y_1 is replaced by the element y_{k+1}. For the calculation of the new median there are now four cases possible, as is shown in Figure 3.1. The 4 different types of actions have to be carried out depending on the relative size of the old element y_1, which has to be eliminated, and the new element y_{k+1} which has to be added either to s_1 or s_2.

Fig.3.1 Median Search Algorithm for Running Medians:

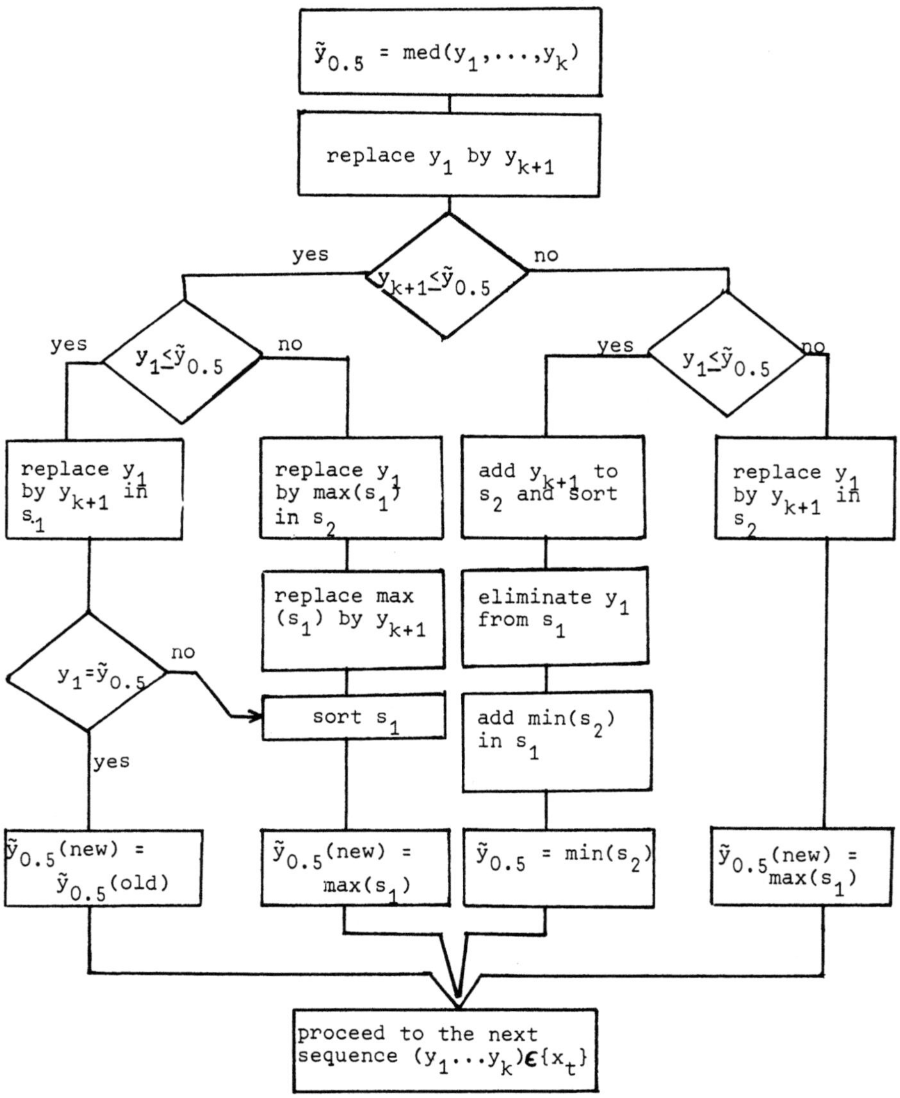

Note, that case IV ($y_{k+1} > \tilde{y}_{0.5}$ and $y_1 > \tilde{y}_{0.5}$) need no further sorting, while case I ($y_{k+1} \leq \tilde{y}_{0.5}$ and $y_1 \leq \tilde{y}_{0.5}$) only for the case when the old value y_1 is equal to the median. But also for the cases II and III where there has to be a sorting anyway, only half of the values ($y_1, \ldots, y_k$) have to be sorted.

3.2 CPU-Time Comparison for Different Algorithms

To evaluate the performance of the new algorithm in practice, several other sorting algorithms have been compared in an artificial example. 256 observations have been created and compared with running medians of increasing span. As can be seen from table 3.1 the new algorithm is substantially better than the other ones for this running median problem. As the length of the span increases the savings are about 2/3 CPU-time. Running Medians with span 13 or 25 are important for smoothing monthly and seasonal time series and are used in seasonal adjustment procedures.

Table 3.1: Time Comparison for different computer algorithms for running medians (milli-seconds)

	5R	7R	13R	25R	42	-24-12-
BUBBLE	518	735	1679	4752	84	1153
QUICK	647	899	1751	3407	102	903
FIND	423	529	873	1349	120	576
INSERT	416	502	749	1182	73	359
TABLE	238	275	364	523	58	242

The first four columns show odd-span running medians in ascending order, while the last two columns show the computing time for even span medians. From the Table 3.1 we can draw the following conclusion: BUBBLE-sorting is the worst procedure for large span medians. But for short-span medians BUBBLE-sort turns out to be better than QUICK-sort. QUICK-sort is second worst for large span medians, which is a little bit surprising.The reason lies in the still small amount of data which have to be sorted in a running median problem. The FIND-algorithm which evolved from the QUICK-sort algorithm needs approximately half of the computing time of the QUICK-sort algorithm. Only for even span medians the algorithm can be worse. (This is because the sorting has to be performed two times). The best of the available sorting algorithms for the running median problem is the INSERT-algorithm. The improvement is small for short-span medians, but the savings are considerably better for larger spans and even-span medians. The special newly developed median search algorithm "TABLE" is substantially better for all cases. This shows that skilful use of the structure of the sorting problem can substantially improve the computation times.

4. The Interactive Program System INDSMO

The program package INDSMO (Interactive Nonlinear Data Smoothing) was developed at the Institute of Statistics and Computer Science at the University of Vienna to get a quick and comfortable descriptive analysis for time series and to compare different data-smoothers. The program system was written in FORTRAN IV on a Cyber 170 computer by Hoffinger (1982) and incorporated the subroutine package from Vellemar and Hoaglin (1980). Besides the new running median algorithm of section 3 several graphic output routines have been implemented. In order to find a balance between a most flexible system and an easily handlable system five different input commands have been designed:

1) Number of observations of the time series (max. 999).
2) Variable format for the input of the time series (max. 80 alphanumeric characters: FORTRAN-format in parenthesis).
3) The time series data in the specified format.
4) The name (title) of the time series (max. 80 characters).
5) The type of non-linear data smoothers (max. 80 characters). For every different smoother an extra command (card) is necessary. For the syntax see section 4.1.

The program stops if there is no further smoother specified. For every smoother a graphical output is generated.

4.1 Types of non-linear data smoothers

All types and combinations of non-linear data-smoothers are allowed (as long they are less than 80 characters). The syntax is straightforward and corresponds to the usual abbreations in the literature (Tukey (1977), Mallows (1980) or Velleman (1980)). One-digit numbers refer to medians, two digit smoothers have to be separated by a hyphen (-). 'R' indicates a running median, SS a splitting-, H- a Hanning, and E an end-value procedure.

5. Example: Business Cycles of Gross National Products (GNP)

For the GNP-growth rates between 1955 and 1981 the quarterly time series is smoothed by an 53R smoother. The end points of the peak- or through-plateau phases can be used for the dating of turning points. This method reveals interesting theoretical and interpretative aspects and is fully discussed in Polasek (1982).

Fig.5.1 GNP (real) for FRG (West Germany): 53R-smooth:

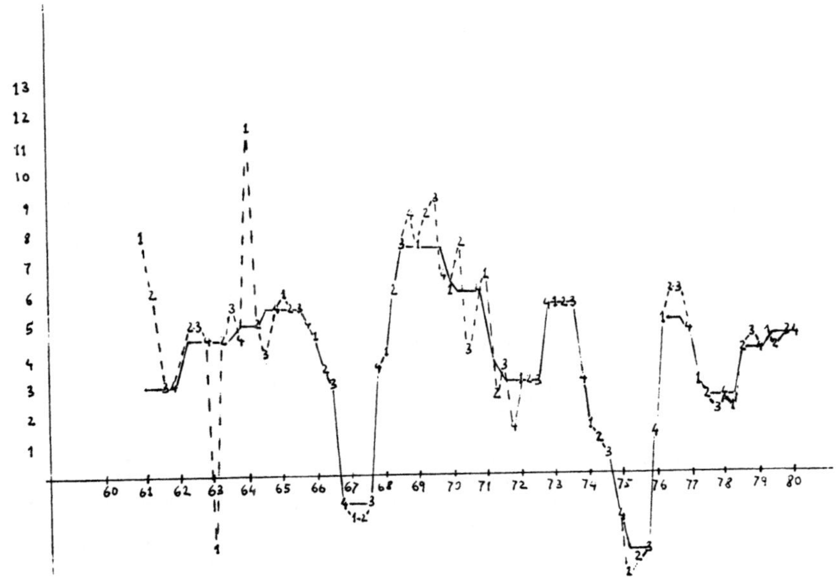

References:

Cooke C.R. and D.J. Kim (1980), Best Sorting Algorithm for Nearly Sorted Lists, Comm.ACM, Vol.23, No.11, 620-624.
Hoffinger H. (1982), Nichtlineare Datenglätter: Algorithmenvergleich und Computerprogramme, Diploma at the Institute of Statistics.
Hopcroft J.E. (1974), Complexity of Computer Computations, Information Processing 74, North-Holland Publ.Co. 620-626.
Knuth D.E. (1978), The Art of Computer Programming, Vol.3, Sorting and Searching, Addisson-Wesley, Reading MA (2nd ed).
Mallows C.L. (1980), Some Theory of Non-Linear Smoothers, Annals of Statistics, Vol.8, No.4, p.695-715.
Polasek W. (1982), Exploratory Business Cycles Analysis, to appear in Empirica, Austrian Institute of Economic Research.
Schönhage A., M. Paterson and N. Pippenger (1976), Finding the Median, Journal of Computer and System Sci. 13, 184-199.
Tukey, J.W. (1977), Exploratory Data Analysis, Addison Wesley.
Velleman P.F. (1980), Definition and Comparison of Robust Non-Linear Data Smoothing Algorithms, JASA, Vol.75, No.371, 609-615.
Velleman P.F. and D.C. Hoaglin (1980), Implementation of Some Resistant Non-Linear Smoothers, Cornell University, No.878/018, Preprint Series.
Wirth, N. (1975), Algorithms and Data Structures, Prentice Hall, Englewood Cliffs.

Applying Parallel Processing to Data Analysis: Computing a Distance's Matrix on a SIMD Machine

M. Raphalen, I.R.I.S.A., Rennes, France

SUMMARY : Nous étudions un algorithme de classification ascendante hiérarchique pour une mise en oeuvre sur calculateur tableau. Une partie de l'algorithme ainsi que sa parallélisation est décrite dans ce papier. Une évaluation de l'algorithme parallèle est effectuée, permettant de comparer les durées d'exécution de l'algorithme sur calculateur parallele et sur calculateur séquentiel.

KEYWORDS : SIMD computers, parallel processing, data analysis.

I. SIMD (Single Instruction, Multiple Data stream) computers

Figure 1 gives the overall structure of a SIMD computer. An array computer [Flynn 1966] consists of N processing elements (PE's) indexed from 0 to N-1, capable of performing arithmetic and comparison operations. Each processor is connected to a local memory. There are two ways for communicating the data to the PE's : Broadcasting scalars to all PE's or shifting a vector among the PE's. The PE's operate in a blocked manner under the control of a single control unit. It is possible to inhibit a subset of PE's by using a mask. At a given time, all enable PE's perform the same instruction broadcasted by the C.U. The PE's communicate via an interconnection network, which performs permutations.

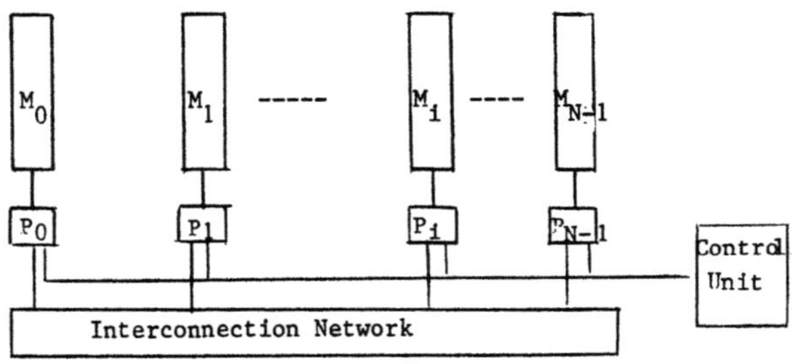

Fig.1

To compute on a SIMD machine, the operations must be expressed in terms of vector operations to use the internal parallelism of the machine and so to increase the speed. Suppose we have to compute the inner product of two N vectors A and B, using N processors.

A sequential program to compute IP is :

```
IP:=0
for i:=0 to N-1 do
    IP:=IP+A[i]*B[i]
```

The array elements are accessed one at a time.

A parallel program to compute IP could be (it is written in the Actus Language [Perrott 1978]) :

```
IP:=sum (A[0:N-1]*B[0:N-1])
```

(The notation "i:j" means that only the processors i to j execute the instruction. The others are idle).

The ith PE computes A[i]*B[i](i=0, N-1).Then a summation is performed on the N terms A[i]*B[i]. To compute IP, 2N operations (N+,N*) are necessary in the sequential manner, as only $\log_2 N + 1$ ($\log_2 N$+,1*) are performed in a parallel manner.

Definitions

If T_N is the time necessary to compute a given algorithm using N processors ($N \geq 1$) we define

- the speed-up ratio S_N

$$S_N = \frac{T1}{T_N}$$

- the efficiency E_N

$$E_N = \frac{T_N}{N}$$

II. An example of parallelisation

We study the parallelisation of an hierarchical clustering algorithm [Lerman 1981].

1) Statement of the problem

The data concerning individuals are contained in a table T :

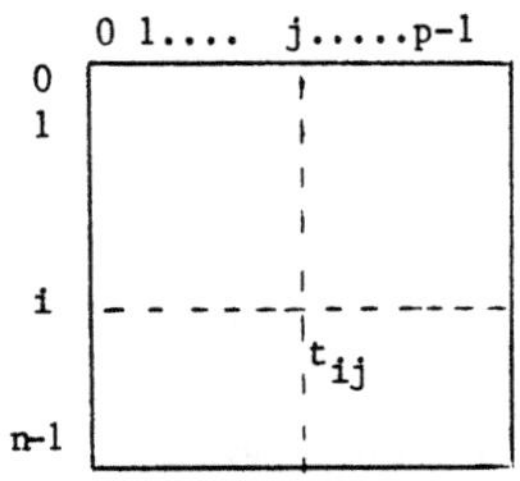

$I=\{0,1,\ldots,n-1\}$: set of individuals
$J=\{0,1,\ldots,p-1\}$: set of variables
t_{ij} : measure of the jth variable on the ith individual

The set E to classify is either I or J ; $|E|=e$ (n or p). The algorithm consists of three major steps :

s1 : computation of the distances (or indices) s[k,l] between the elements (k,l) of E. k = 0,e-1, l=0,k.

s2 : the pairs of elements of E are sorted according to the distance s.

s3 : aggregation of the E's elements.
Only the first step depends on the type of T.
We describe here the step 1 and we will just consider the case of E=J and t_{ij} taking the value 0 or 1 (i=0, n-1, j=0, p-1).

Step 1 then consists in computing a matrix S :

$$\text{※ } s[j,j'] := \frac{n \sum_{i=1}^{n} t_{ij}\, t_{ij'} - t_{.j}\, t_{.j'}}{\sqrt{n\, t_{.j}\, t_{.j'}}} \qquad \text{where } t_{.j} = \sum_{i=1}^{n} t_{ij} \text{ for } \begin{matrix} j=0,p-1 \\ j'=0,j \end{matrix} \quad (1)$$

As S is symmetric, we only need to compute the lower half of S.

$$\text{※ } MOY = \sum_{j=0}^{p-1} \sum_{j'>j} s[j,j']/(p(p-1)/2) \text{ : average of the indices}$$

$$\text{※ } VAR = \sum_{j=0}^{p-1} \sum_{j'>j} s^2[j,j']/(p(p-1)/2) - MOY^2 \text{ : variancy of the indices}$$

$$s[j,j'] := \frac{s[j,j'] - MOY}{\sqrt{VAR}} \quad (2)$$

※ To evaluate the neutrality of a variable in the formation of the clusters, we compute :

$$\nu(j) = \frac{1}{p-1} \sum_{j' \neq j} s^2[j,j'] - \left(\frac{1}{p-1} \sum_{j' \neq j} s[j,j']\right)^2. \quad (3).$$

for j=0 to p-1

2) Parallel evaluation

2.1. Step 1

There are two obvious ways to compute the $s[j,j']=\sum_{i=1}^{n} t_{ij}t_{ij'}$

- Algorithm 1

The n elements of a column of T are accessed in parallel :

```
For j:=0 to p-1 do
  For j':=0 to j do
    s[j,j']= sum (t[0:n-1,j]*t[o:n-1,j'])
```

- Algorithm 2

The elements of a row of T are accessed in parallel :

```
S:=0
For l:=0 to n-1 do
  For m:=0 to p-1 do
    s[m:p-1,m]:=s[m:p-1,m]+t[l,m:p-1]*t[l,m]
```

The operation counts are :

	Algorithm 1	Algorithm 2
*	$p(p+1)/2$	np
+	$p(p+1)/2.\log_2 n$	np
	cyclic shifting	

The choice of the algorithm depends on the values of n and p, and on the respective cost of the involved operations. But we may notice that in algorithm 1 the elements s[j,j'] are computed one at a time, while in algorithm 2, they are computed vector by vector. The further use of matrix S and machine considerations lead us to choose algorithm 2.

These two algorithms are suitable for a number N of processors larger than the size of the manipulated vectors (N>n in alg.1 and N>p in alg. 2). When this is not the case, we must cut the rows or columns in k blocks of at most N elements. In algorithm 2, J is cut in k=[p/N] subsets ($J=\bigcup_{l=1}^{k} J_l$). This leads us to build the lower matrix S in blocks of size at most (NxN). In a block S^{lm} (l=1,k;m=1,l) we compute the elements s(j,j') where $j\varepsilon J_l$, $j'\varepsilon J_m$.

The operation count for algorithm 2 is then :

* n.N. k(k+1)/2

\+ n.N.k(k+1)/2

In a S^{lm} block, we get the s[j,j'] of (1) vector by vector. If g[i] denotes the i element of the diagonal of S^{ll}, the ith PE computes :

```
for j'εJm :
  X[i]:=g[i]*s[j',j']
  s[(l-1)*N+i,j']:=(n*s[(l-1)*N+i,j']-X[i])/√(n*X[i])
```

The computation of the matrix S in (1) needs $O(N\frac{k(k+1)}{2})$ operations.

To compute MOY we build a vector X with N components :

$$X[i]:=\sum_{l=1}^{k}\ \sum_{m=0}^{(l-1)*N+i-1} s[(l-1)*N+i,m],\quad MOY:=sum(X[1:N])$$

VAR is obtained by the same process. $O(N.k(k+1)/2+\log_2 N)$ operations are necessary to compute MOY and VAR. The final matrix S of (2) is obtained in $O(N.k(k+1)/2)$ operations from (1).

The speed-up ratio S_N is of order (p+1)/(p/N+1) to compute S. When p grows larger, S_N gets very close to N and we can hope an efficiency of nearly 1.

We now examine the computation of (3). In the memory of PE's we have the following configuration.

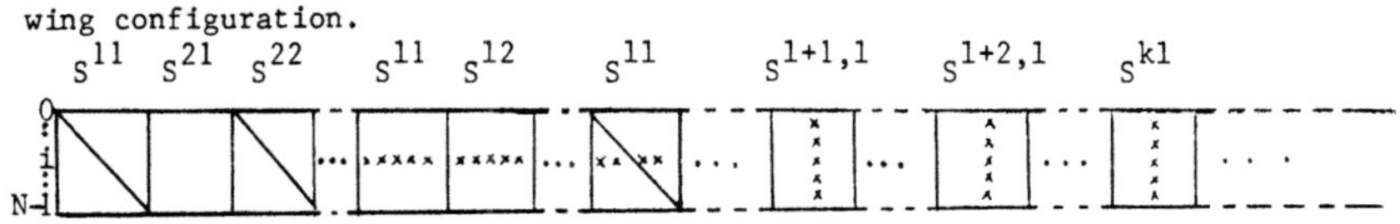

If j is the ith element of J_l ($j=(l-1)*N+i$), to compute $\upsilon[j]$ we need to access the pointed elements, i.e. the elements of the ith row of the blocks S^{lm} ($m\leq l$) (exepted the diagonal element of S^{ll}), and the elements of the ith column of the blocks S^{ml} ($m=l+1,k$). We denote :

$$E_l = \bigcup_{1\leq i\leq l} J_i - \{j\};$$

As we indiffenrently can't access simultaneously the rows or the columns or any set of N elements of a (N,N) matrix, we shall process in two steps :

(i) Computing two vectors h1 and h2 with N components. Each PE i computes :

```
h1[i]:=0 ; h2[i]:=0
for j' in El do
  h1[i]:=h1[i]+s[(l-1)*N+i,j']
  h2[i]:=h2[i]+(s[(l-1)*N+i,j'])²
```

Comment:h1[i] contains the partial sum : s[j,0]+s[j,1]+...+s[j,l*N-1]

(ii) Building two vectors V1 and V2 from scalars ;

For each j in J_l, PE i computes:

```
w1[i]:=0 ; w2[i]:=0
for m:= l to k do
  w1[i]:=w1[i]+s[l*N+i,j]
  w2[i]:=w2[i]+(s[l*N+i,j])²
```

We then obtain the scalars : v1[j]:=sum (w1[1:N]) and v2[j]:=sum(w2[1:N]).
Comment : v1[j] contains the partial sum s(lN,j)+s(lN+1,j)+....+s(p,j)

By setting these scalars in vectors and adding them to the vectors h1 and h2 we obtained before, we compute N elements $\upsilon[j]$ for $j\in J_l$ simultaneously. The algorithm has a complexity of $O((k-l)(kN+\log_2 N))$

The speed-up ratio S_N is of order $\frac{p(p-1)}{(p/N-1)(p+\log_2 N)}$

When p is large beside N, S_N isclose to N and the efficiency tends to 1.

CONCLUSION

We have seen that the algorithms for building the lower matrix S and computing the $\upsilon[j]$ ($j=0,p-1$) are asymptotically efficient (they have an efficiency close to1). This is a very theoritical result. In fact, the efficiency is reduced to c, c<1 because of the computer architecture and the constraints ondata communication due to the interconnection network [Tahé 1979]. An implementation of the algorithms on the array processor Propal II gives an efficiency of 0.5.

REFERENCES

ENSLOW P.H. (1974), multiprocessors and parallel processing, dans Wiley interscience publication.
FLYNN M.J. (1966), very high speed computing systems, proc. IEEE. Vol. 54.
LERMAN I.C. (1981), classification et analyse ordinale des données, Dunod.
PERROTT R. (1978), a languague for array and vector processors, Queens's University - Belfast - N. IRELAND.
TAHE S. (1979), transfert des données dans un calculateur parallèle vectoriel, thèse 3ème cycle, Université de Rennes I.

Proceedings of the 1st european conference on parallel and distributed processing. Toulouse, France, February 14,16 - 1979.

The Balance between Teaching Computing and Statistics

R.A. Reese, University of Hull, G.B.

Summary. Many research workers from a great variety of disciplines now make use of computers but, for obvious personal and philosophical reasons, such use must not be allowed to dominate or dictate the course of their work. This paper describes some attempts to provide relevant and acceptable training.

Keywords. Teaching statistics. Computer use. Data Analysis. SPSS. GLIM.

Introduction.

In 1964 the RSS held a meeting to consider trends in the teaching of statistics. Yates and Healy wrote, "The introduction of electronic computers should have a considerable effect on statistical courses. A good deal of dead wood, in the form of elaborate routines for desk calculation and inadequate approximations now rendered unnecessary, can be excised" (JRSS, 1964). Nine years later, Barnett (1973) could still write, "What seems most important is that we make every effort to maintain a proper balance in emphasis in teaching statistics. The student statistician ... needs to recognise the computer for what it is - a sophisticated tool, not a substitute for thought." Pukkila et al (1980) discuss the use of a computer as a teaching tool in statistics. Here, I discuss the simultaneous teaching of statistics and computing as research tools. This paper describes three courses with different balances and explains the design of each course and the reactions of students. If there is an overall message, it is that there can be no such thing as a purely computing or purely statistics course.

What is taught and the preferred methods of teaching will of course vary depending upon the audience - their background knowledge, ability, motivation and need. In a university computer centre our formal brief is to provide and maintain a computing facility; we are responsible for hardware and software. It is assumed that our customers - staff and research students - know what they want and need only technical (programming) advice. In running service (i.e. not examined) courses over several years, I have found this to be far from true. The need (and desire) for basic statistical advice is reflected in the problems that users bring, the way they choose solutions and the solutions then adopted. Written manuals are not enough; the choice of software can be considerably influenced by the attitudes and knowledge of advisors. The growth in use of, particularly, SPSS has followed the provision of courses explaining why it might be used as well as how.

Choice of syllabus.

One of the boards setting public examinations for school states as an aim of its A-level course (taken normally at age 18) "that the subject may be appreciated as an important component of modern applied mathematics and as a useful practical tool". It then emphasises, "Sound application of statistical methods requires a proper understanding of basic concepts" (JMB, 1981). These precepts will be accepted by most teachers of statistics at all levels.

However, many people within universities who need (or believe they need, or are told they need) to carry out statistical data analysis have neither the mathematical background to undertake a course in statistics nor the time to make good this lack. At COMPSTAT 80 I asked Prof. Tukey for his initial reaction to the user who presents himself thus, "I've got my data, what can I do with it?" Tukey's reply was that he would introduce this person to a student of statistics but it is extremely unlikely that there can ever be enough skilled applied statisticians to meet the total research demand on a one-to-one basis. The scheme also fails because university research is seen as an individual and competitive, rather than a team activity.

In courses, I discuss this last point by reference to Fig 1. Whatever the subject of the research - which I will call Xology - there will be some overlap with both statistics and computer programming but there will also be areas in all three subjects that are unrelated. One of the dangers of introducing intelligent and inquisitive people to novel, powerful tools is that they can become preoccupied with the method and forget the purpose of the exercise - to obtain results valid and interesting in Xology. It can be unrewarding to try to persuade some of our users that they should not worry about a compiler bug but should rewrite part of their program so as to avoid it.

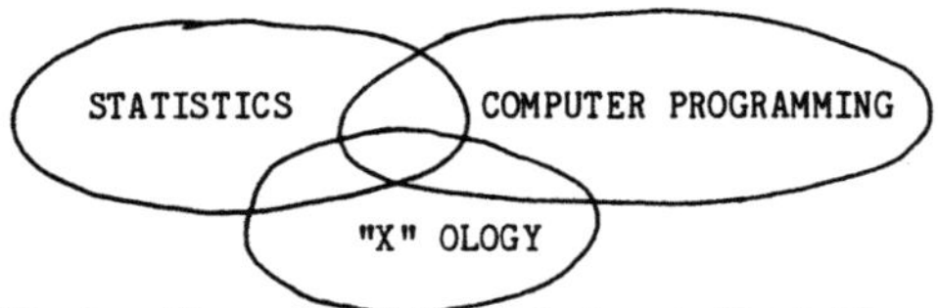

Fig 1 - The relationships between disciplines.

What "basic concepts" must be communicated to the would-be data analyst? A general consensus exists for the content of a first course in statistics. Examples are the JMB O- and A-level syllabuses (op. cit.), Austwick et al. (1971) and the contents' lists of many textbooks with titles of the form "Elementary Statistics for Xologists"; though the existence of so many such books is witness to the lack of consensus between the disciplines using statistical methods regarding terminology. The topics covered are: the concept of probability and axioms for combination; random events and random variables; description and summary of data; probability distributions exemplified by the Binomial, Poisson and Normal; sampling, estimation and hypothesis testing. This I will call a mathematical syllabus. Austwick lays out these topics and the relationships between them in a clear diagram, the style of which I have copied in Fig 2 so as to aid comparison.

The drawback of this approach within a service course is neatly summed up by Austwick. "The basic course ... covers the fundamental concepts ... but does not reach as far as the simplest usable techniques". Much "education" has this limit of giving an appreciation of the subject without going far enough to achieve any level of competence. An introductory course meant for those who will study further is rarely satisfactory as a self-contained elementary course.

Austwick then goes on to describe extensions oriented to particular Xologies. "The most valuable extension for sociology students would be through the Chi-square goodness of fit test and contingency tables". This then introduces a new danger, that the researcher with (perhaps) adequate knowledge of a limited range of tests will apply a test blindly, just because he knows it. The Chi-square test is a classic for overuse and misuse. It must be better to have a wider appreciation of statistical methods and objectives so as to know how to choose a test from a "cookbook" or to seek expert advice. Other frequently seen examples concern users still using "quick" but approximate calculations of low power published in the 40's and 50's when even a desk calculator was a luxury.

Teaching Statistics.

The alternative functional syllabus, Fig 2, was developed particularly for a course given to a diverse group of mature students undertaking one year research projects with supportive lectures on methods. In eight two-hour sessions it was possible to touch upon each of the topics and to relate them one to another from a number of starting points. The emphasis throughout is upon the reasons for choosing a statistical approach and the assumptions and implications of particular methods.

Within the time limit it was not possible to include instruction on the use of the computer to allow students to work through their own examples. The course was therefore linked to a general course run by the Computer Centre consisting of two sessions on computer access and the operating system and four on SPSS (each of one and a half hours). This course is discussed below. There were people who attended either the statistics or the SPSS course or both, so each course had to be self-contained although complementary.

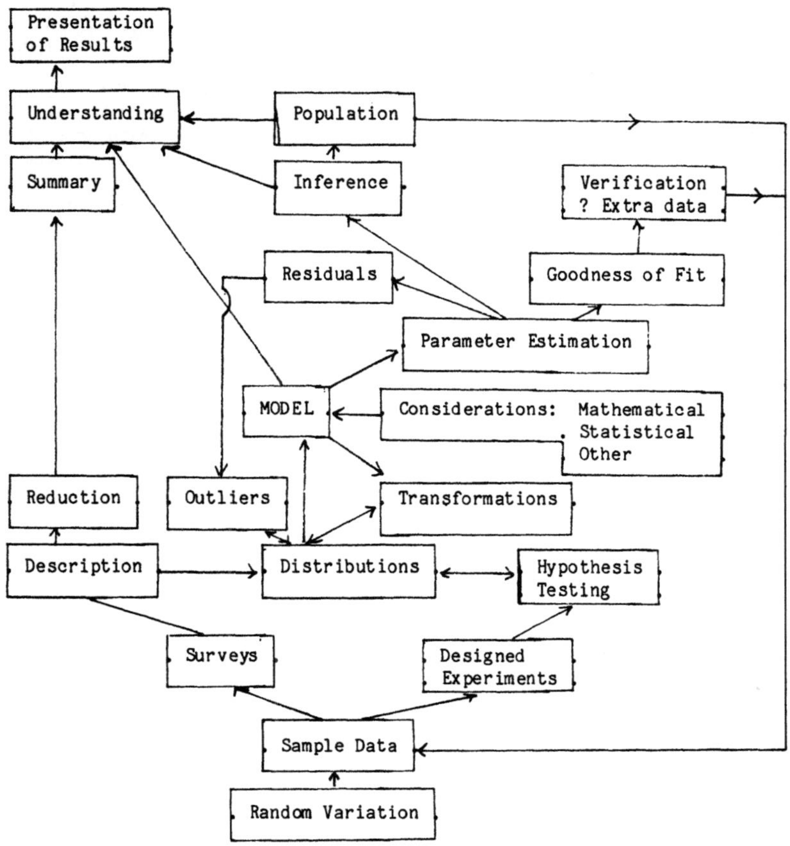

Fig 2. Analysis of a Function-Orientated First Course in Statistics.

Examples, within the statistics course, were given in the form of SPSS output for discussion. Points that were stressed were the distinction between descriptive and explanatory interpretations of possibly the same statistics and the distinction between methods for understanding data and for presenting conclusions to other people. Research was stressed to be an iterative process, where the results at any stage may suggest either a new approach or the need to modify a previous method.

Obvious examples would be choosing a data transformation or collapsing a contingency table.

There are few textbooks that usefully accompany this approach. My recommendation was Youngman (1979) but this book is in parts more detailed and sophisticated than is desirable. For example, comparison of grouped observations is discussed (pp 79-84) in an almost cryptic language that requires considerable statistical knowledge and experience to appreciate but the example that follows and the suggestions for presentation of results (pp 85-90) are far more accessible. On balance it is a "good buy" for the students, that will remain useful to them after the course.

If one were giving such a course to students with a background of mathematical statistics, so that the intention is to show them the pitfalls they will encounter when applying their knowledge, Cox and Snell (1981), Stoodley et al (1980) and Hunter (1981) are relevant and useful.

The reactions of the students were assessed by questionnaire after the course. All who replied found it interesting but only half thought it would be relevant in their current research. This was not too disappointing as the lectures took place after half their research period had passed. Most thought that it had changed their attitude to quantitative research and made it more likely they would use statistical methods in future research. Those who did not attend the SPSS course did not think they would be more likely to use a computer.

Teaching Computing.

The SPSS service course referred to has been given in most terms since 1978, i.e. roughly three times per year. Each time, it has attracted between five and fifteen students. The effect of this promotion can be seen in the graph of the ⟶

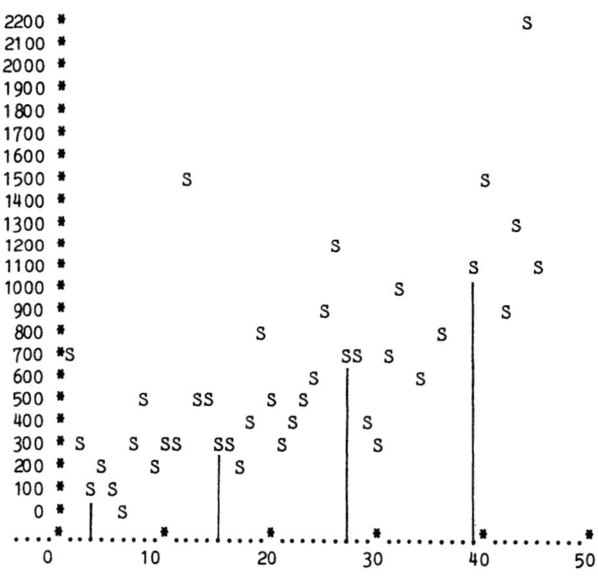

Fig 3. Monthly tally of calls of the SPSS macro on Hull University's ICL 1904S. This does not include SPSS used on other computers from Hull.
Month 1 is August 1978. Month 45 is April 1982. Minor variations can be related to starts and ends of terms. Lines show the beginning of each academic year.

monthly total of calls of SPSS, i.e. job-steps, Fig 3. In contrast, there has been no demand for Genstat courses and those who have had to resort to this package have not gained confidence or expertise enough to become enthusiastic. Fig 4 shows the monthly calls of Genstat. Note the difference of scale from Fig 3.

```
130 *         G
120 *
110 *   G
100 *
 90 *
 80 *
 70 *                              G
 60 *          G                  G        G
 50 *    GG
 40 *                           G    G
 30 *                          G       G
 20 *         G       GG      G  G            G G
 10 *     GGG    G G          |           G  G
  0 *GGG          G|GGG GGGGGG|            |   G
....*..|......*....|....*.....|...*........|*.........*
    0        10        20        30        40        50
```

Fig 4. Monthly calls of the GENSTAT macro over the same period as Fig 3.

The SPSS course has sections on the nature of data, data structures, coding, logical and physical organisation and the rationale of analysis in addition to the mere mechanics of data input, job control and SPSS commands. Indeed, the commands needed to calculate a variety of univariate statistics for a number of variates are so few so straightforward that after one session the student has a pattern job that he can adapt and use, even if he attends no further lectures.

To quote from the course notes, "the stages of data analysis by computer are:

Collection of data
Coding and inputting the data to the computer
CHECKS for completeness, typing errors and internal consistency
Descriptive and summary statistics
CHECKS for outliers and assumed distributions
LOOKING for relationships between variables
Transformations of the data to simplify relationships
CHECKS for underlying assumptions of models
Fitting mathematical models
Tests of fit. Confirmatory tests (on extra data?)
Inferences from acceptable models
Interpretation of results
Presentation of conclusions."

This list parallels the syllabus of Fig 2 but places an even greater stress on CHECKing and LOOKing at the data. The greatest benefit from the course is probably found by those people who are persuaded that statistical methods are not a "sausage machine" that one pours data into and receives infallible answers in return. The SPSS language provides very powerful tools for data checking and examination; just the procedures FREQUENCIES, CONDESCRIPTIVE, SCATTERGRAM, CROSSTABS and BREAKDOWN, in combination with data transformations, provide far more scope than most non-statisticians will need. Multivariate procedures, for example REGRESSION, ANOVA and FACTOR, are mentioned but with a caution that the user <u>must</u> undertake some background statistical appreciation.

Some of the examples used in this course illustrate analyses that SPSS does not explicitly provide but which, with a little thought and possibly some post-processing of output by hand or with a text editor, can be readily performed. Conversely, I also point out those times when SPSS is a dismal failure. There is no such thing as a perfect computer program; no one program or language can be said to be "the best" and recommended in all circumstances; all critical results should be CHECKED, which implies using at least two programs. (Reese, 1978)

The third course was given to forty final-year zoology students. In theory they had all completed a statistical methods course of one term's length and so the objective of this course was to introduce them to computers, both to give them a tool for use in research and to enhance their employment prospects. The same examples as they had studied in the statistics course were used; in some cases the methods required a high level of appreciation of programming, since neither the examples nor the package had been designed with the other in mind. However, the programs (command sequences) were all short enough to be written on the board for students to copy and execute line by line. As there were only twenty four terminals, students were expected to work in pairs but this was seen as beneficial.

GLIM (Baker and Nelder, 1978) was used as it allows all levels of work from "desk-calculator" to sophisticated modelling. GLIM has features that make it very attractive in teaching. Not only the data, but also executing code or stored code for subsequent execution on demand can be stored in a file. Examples can be set up so that part of a file is read (and executed), input is taken from the terminal and then more of the file is read. Thus a practical session can range from conversational hand-holding to the terse "data as described is in file X - analyse it and report your conclusions".

This course was not well received. Attendance was on the whole poor after the first few weeks. Part of this can be attributed to poor hardware reliability at that time; when they suffered two or three system breaks in each session, students failed to see any advantage in using the machine over hand calculation. A few had been to other computer appreciation courses where they had used BASIC and these students failed to see the advantage of the more powerful and specific language.

However, during the course it was apparent that the students' appreciation of statistics as a philosophy for data analysis was completely lacking. What they remembered of the statistics course was a random collection of names of tests and in future work I fear they would, in desperation to have their results published, resort to cookbook statistics.

I hope to revise this course to run in parallel with the introductory statistics, possibly at the end of the penultimate year for the next group.

References and sources.

Austwick et al. (1971), The First Course in Statistical Methods, Review of the I.S.I. Vol 39:3

Baker R.J. & Nelder J.A. (1978), GLIM Release 3 User Manual, NAG, Oxford.

Barnett V.D. (1973), Comparative Statistical Inference, Wiley.

Cox D. & Snell (1981), Applied Statistics: Principles and Examples, Chapman & Hall.

Hunter W.G. (1981), Six Cautionary Tales, Statistician Vol 30:2

JMB (1981), Joint Matriculation Board Examinations Council, GCE Syllabuses 1983, Manchester.

JRSS (1964), Symposium on the Teaching of Statistics, Series A. Vol 127 pp 199-233.

Pukkila T. & Puntanen (1980), Computer usage on first statistics courses, COMPSTAT 80.

Reese R.A. (1978), Hull University Computer Centre Newsletter, Vol 2:3 (Dec.)

Stoodley K.D.C. et al. (1980), Applied Statistical Techniques, Horwood, Chichester.

Youngman M.B. (1979), Analysing Social and Educational Research Data, McGraw-Hill.

Canonical Analysis with B-splines

J.v. Rijckevorsel, Erasmus University of Rotterdam, The Netherlands

SUMMARY

We show how principal components analysis and multiple correspondence analysis (De Leeuw, 1982) can be generalized respectively specialized by using a basis of B-splines (De Boor, 1978). In the first case we deal with a form of optimal scaling in non-metric principal components analysis (Young, 1981); in the latter case we refine the basis of stepfunctions by using first degree B-splines (De Leeuw et al., 1981). We also discuss the potential use and rationale of B-splines in this context. Both techniques are illustrated with an example. The algorithm and computer program are described in a companion paper by Coolen, De Leeuw, Van Rijckevorsel (1982).

KEYWORDS

B-splines, correspondence analysis, optimal scaling, principal components analysis.

1: Introduction

Principal Components Analysis (=PCA) and Multiple Correspondence Analysis (=MCA) are forms of canonical analysis (Saporta, 1980). A straightforward generalization of MCA is MCA with B-Splines (De Leeuw a.o. (1981), Gifi (1981) and straightforward generalizations of PCA are: PCA with Splines in general by Lafaye de Micheaux (1978) and by Besse (1980), PCA with B-Splines by De Leeuw et al.(1981) and PCA with integrated B-Splines by Winsberg et al.(1982). In this paper we follow the elegant and general integrated approach of De Leeuw (1982) for describing MCA and PCA. We will show how B-Splines fit in and we will give some possible reasons and according interpretations for the use of B-Splines in this context, illustrated with an example. An extensive description of the ALS algorithm we used is in the companion paper by Coolen et al. (1982).

2: Variable-subspaces and their bases

Say $L_1,\ldots,L_m$ are closed subspaces of a separable Hilbert space H with inner product $<.,.>$, norm $|.|$ and unit sphere S. We suppose that all L_j are finite dimensional with dimensionality q_j.

Such a subspace L_j we call the variable subspace. The dimension n of H can be finite of infinite. In this paper we assume that n in finite but this is not a necessary condition for the results. We can compute the matrix $R(y_1,\ldots,y_m)$, (mxm), with elements $r_{j\ell}(y_1,\ldots,y_m) = \langle y_j,y_\ell \rangle$ for every choice of elements $y_j \in L_jS$, the intersection of L_j and S. This matrix we call the correlation matrix; it is positive semi-definite and with diagonal elements equal to one. There exist several objective functions of such correlation matrices that we can optimize in a PCA context (Rao(1980), Gifi(1981)). Here we are only interested in the sum of the p dominant eigenvalues of the matrix $R(y_1,\ldots,y_m)$: $\sigma_p(y_1,\ldots,y_m)$. As a basis for L_j we choose a set of B(asis)-Splines.

3: B-Splines

A very good and complete introduction to splines with a wealth of applications, algorithms and derivations is given by De Boor (1978). The B-splines as we use them were introduced first by Curry and Schoenberg (1947, 1966). Splines stem from approximation theory, where they are used to approximate smooth functions. In our case the idea is to partition the domain of a continuous variable, or a variable we want to treat as continuous, into a number of adjoining intervals that merge in so called breakpoints $\ldots < t_{-2} < t_{-1} < t_0 < t_1 < t_2 \quad \ldots$
Within each interval the original variable is approximated by a polynomial of degree k - 1. Because different intervals can correspond with different polynomials the resulting approximating function is a piecewise polynomial.
If these piecewise polynomials have k - 2 derivatives in each of the breakpoints we call them splines of degree k - 1 (of order k). The space of all splines of degree k, with break-points t_i, is a linear space. If p_j is the number of interior knots, i.e. the number of breakpoints larger than the smallest and smaller than the largest observation, of variable j, then the dimension of the space L_j is $q_j = p_j + k - 1$. B-splines are a convenient basis for this space. They have local support, i.e. they are zero everywhere, except in k consecutive intervals, and they are linearly independent. We do not define them in general, because this is not needed for our purposes. For the kind of applications we have in mind we want to approximate L_j with low dimensional subspaces.
This means that we prefer to work with a small number of knots per variable and with B-splines of a low degree. For the situations where k=1 or k=2 the B-spline representations are easy to compute. For k=1 that B-splines are the stepfunctions:

$B^j_{s,1}(y_{ij}) = 1 \quad t_2 \leq y_{ij} < t_{s+1})$

$B^j_{s,1}(y_{ij}) = 0 \quad$ otherwise

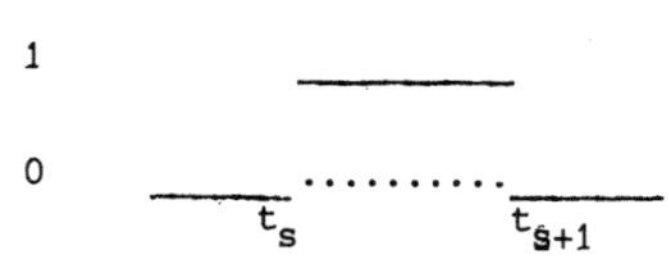

and for k=2 the B-splines are the piecewise linear functions:

$B^j_{s,2}(y_{ij}) = (t_{s+1}-t_s)^{-1}\ (y_{ij}-t_s) \quad t_s \leq y_{ij} < t_{s+1}$

$B^j_{s-1,2}(y_{ij}) = (t_{s+1}-t_s)^{-1}\ (t_{s+1}-y_{ij}) \quad t_{s+1} \leq y_{ij} < t_{s+2}$

$B^j_{s,2}(y_{ij}) = 0 \quad$ otherwise

4: PCA

Suppose we want to represent variable j with a spline function S_j that consists of a linear combination of q_j B-splines $B^j_{s,k}$ as defined in the last paragraph. These basis functions $B(y_j)$ form the columns of the pseudo indicator matrix S_j, (nxq_j). For k=1 S_j is a real (0,1) indicator matrix with one non-zero entry in every row. For k>1 S_j is a sparse pseudo indicator matrix with exactly k nonzero entries that add up to one for every row.
Define $G_j=D_j^{-1/2}S_j$ with $D_j=S_j^TS_j$. Than $y_j=G_ja_j$ is in L_jS iff the q_j-vector a_j satisfies $a_j^Ta_j=1$. Because $<B^j_{s,k},B^j_{s+1,k}> =0$ for $i \geq k$, D_j is a symmetric matrix and its diagonality depends on k. For k=1 D_j is unidiagonal, for k=2 D_j is tridiagonal, for k=3 D_j is pentadiagonal etc.. Now we can define the matrix $G_j^TG_\ell = C_{j\ell}$, (q_jxq_ℓ), the supermatrix $G=(G_1 \ |....| G_m)$ with $sq=\Sigma q_j$, and the supermatrix $C = \{C_{j\ell}\}$. (sqxsq). Due to De Leeuw (1982) we can compile the following results:
$R(y_1,...,y_m) = A^TCA$, were A contains the direct sum of the vector a_j and A(sqxm).
$\sigma_p(y_1,...,y_m) = \max \{T^TA^TCAT\}$ with $T^TT=I$ and T(mxp).
$\sigma p(y_1,...,y_m) = \max \{U^TCU\}$ with $U^TU=I$, U(sqxp) and $U_j=a_jt_j^T$;
t_j being a row of T and $U_j(q_jxp)$.
The according stationary equation to solve T for fixed A is $RT=T\Omega$ and to solve A for fixed T $\sum_{\ell=1}^{m} \gamma_{j\ell}\ C_{j\ell}\ a_\ell = \theta_ja_j$ with $\Gamma = \{\gamma_{j\ell}\}$ defined by $\Gamma=TT^T$.
$\theta=\text{diag}\ (T\Omega T^T)$ and $\text{tr}(\theta)=\text{tr}(\Omega)$.
If we use a standard linear basis with $a_j=\pm 1$, $RT=T\Omega$ is the familiar PCA equation. For nonmetric PCA (=NCA) a_j is the q_j-vector with scaled categories (or breakpoints). These two stationary equations are sufficient to construct convergent algorithms (Gifi(1981)).

5: Single PCA with B-Splines

If we try to interprete De Leeuw's results in the usual PCA terminology the matrix T (mxp) contains the correlations between the original variables and the p principal axes; a_j is the vector with optimally scaled breakpoints and U_j consists of the p-dimensional breakpoints quantifications. With Gifi(1981) we call these breakpoints quantifications single because of the blockwise rank-one representation $U_j=a_jt_j^T$ of every q_j dimensional subspace L_j. For p=1 this representation always has rank one. The induced scores for elements of H on the principal axes are equal to $X = \Sigma G_jU_j$ with $X = \{x_{iv}\}$, i=1,...,n and v=1,...,p.

An alternative approach where X is directly derived from the stationary equations is also possible (De Leeuw et al.(1981)); both methods lead to the same results in our situation (Gifi(1981)). To see how the optimal transformed variable looks like one should plot a_j and/or $G_j a_j$ against the original breakpoints resp. datapoints. For k=1 PCA with B-Splines is equal to the single nominal approach to ties for discrete data. See Young(1981) and Gifi(1981).

6: Multiple Correspondence Analysis with B-Splines

If we work with zero degree B-splines(k=1) we have the particular and well known situation of multiple correspondence analysis (De Leeuw et al.(1981)). The supermatrix $C=G^TG$ contains the scaled bivariate marginals of the breakpoints. In case of categorical data and the breakpoints coinciding with the category number the matrix C is also known as "le tableau de Burt". The objective function $\rho_p\ (y_1,\ldots,y_m)$ is equal to max $\{\text{tr } V^TCV\}$ and this is again the sum of the first p eigenvalues; the corresponding stationary equation is $Cv=\mu v$. V is the matrix containing the p-dimensional breakpoints quantifications and the induced scores of the n elements on the first p dimensions are collected in the matrix $X = \Sigma G_j V_j$ with $X = \{x_{iv}\}$ i=1,...,n and v=1,...,p. The geometric interpretations like "le principe barycentrique" and the chi square distances are not so useful for $k>1$. The p non-linear approximations of variable j, $G_j V_j$ give on the other hand much more information than their equivalent based on stepfunctions. There exists a certain relationship between the matrices V_j and the earlier mentioned matrices U_j. Both they contain p vectors of breakpoints quantifications. The difference is that the rank of V_j is maximally p ($p \leqslant q_j$) and the rank of U_j is maximally 1. This difference is trivial for p=1 and De Leeuw points out that V_j and U_j are then equal up to a normalization factor. This difference in rank also means that the dimensionality of the transformation subspace is always equal of greater in the multiple case than in the single case. For a multiple analysis we also cannot define the vector a_j anymore. In some restricted cases one can be more precise about the nature of the multiple quantifications. The a_j vector can then so to say be "replaced" by a $q_j x q_j$ matrix K_j, containing q_j different optimal transformations. It would be interesting to see whether one could enhance the robustness of De Leeuw's KPL-diagonalization by using first degree B-spline transformations in MCA.

7: Why Splines?

There exist three different kinds of reasons for us to use splines in PCA and MCA.

a) Computational reasons

b) Theoretical interest

c) Practical use.

ad a Computational aspects. B-splines are numerically stable; they can be computed with simple recursive formula's and because of their local support they generate banded linear systems that are relatively easy to solve (De Boor (1978)). Optimal transformations with B-splines fit elegantly in the framework of ALS algorithms for PCA and MCA (De Leeuw, Van Rijckevorsel, Van der Wouden (1981); Coolen, De Leeuw, Van Rijckevorsel (1982)). Splines in general are also easy to compute with additional contraints like monotony or convexity (Winsberg Ramsay(1982); Gifi(1981)).

ad b Theoretical interest. B-splines form an orthonormal basis (De Boor(1978)); They are more efficient in the approximation of (non-) linear PCA than polynomials or stepfunctions. The first are more rigid in the sense that their degree becomes prohibitively large for a good approximation; the latter are less efficient because the required number of steps becomes prohibitively large for a good approximation. The approximation with B-splines thus uses lower dimensional subspaces for transformation. (Gifi(1981); Besse(1980)). The same argument against stepfunctions also holds for approximation with a priori discretization (Lafaye de Micheaux(1978); Dauxois, Pousse(1976)). An M-spline, i.e. a non-normalized B-spline, has the characteristics of a probability density function and the integrated M-spline, called I-spline, has the characteristics of a probability distribution function (Winsberg et al.(1982)). For MCA with B-splines one may expect that De Leeuw's KPL diagonalization will work more readily for nearly linear data - structures with low-dimensional B-spline transformations than with low-dimensional stepfunctions.

ad c Data-analytical reasons. These are linked to theoretical reasons, especially in this case where we have not (yet) enough practical experience. B-splines can be attractive for the non-lineair analysis of continuous numerical variables; the obtained spline transformations will visualize the size and character of eventual non-linearities and for $k>1$ one can see how the approximated smooth function looks like. In the situation where $q_j \to n$, the single ordinal approaches, either primary of secondary (Young(1981)) and the stepfunction approach will give trivial results; PCA with B-splines will not. The reason that we use nonnegative B-splines in stead of, say, nonnegative monotone integrated B-splines is based on the fact that if there exists an optimal monotone transformation MCA, and thus also MCA with B-splines, will recover this; De Leeuw proved this for some situations and suggests it may be true for a wider class of data-structures. If there exists no such order than there is no need looking for it. Even in those situations where one is inclined to impose an a-priori order on a variable, like in preference analysis, B-spline transformations with a knot placement based on the marginal frequencies are likely to give a better fit in terms of eigenvalues for partially ordered, curvilinear or not totally linear preference functions, than the application of the single ordinal vector model as discussed a.o. by Heiser(1981). MCA with B-splines fits in a long French tradition of data analysis with advantages of terminology, interpretation and familiarity among users. The same reason, although to a lesser extent, holds for PCA with B-splines which is a further step in a more Anglo Saxon trend of PCA with optimal scaling. See Gifi (1981).

7: An Example

The permitted size of this paper does not allow us to show an example. Nevertheless we want to see how B-spline transformations look like and how they are interpreted. We analyzed a table from the consumers report (1980) with price,weight, engine-size and fuel consumption of 33 different cars. Winsberg et al.(1982) suggested that price is the most non-linear variable in this set. We plotted the single(=$G_j a_j$) and multiple(=$G_j U_{j1}$) B-spline transformations of price against the datapoints. The breakpoints are at \$4000, \$5000, \$6000, \$7000, \$8000.

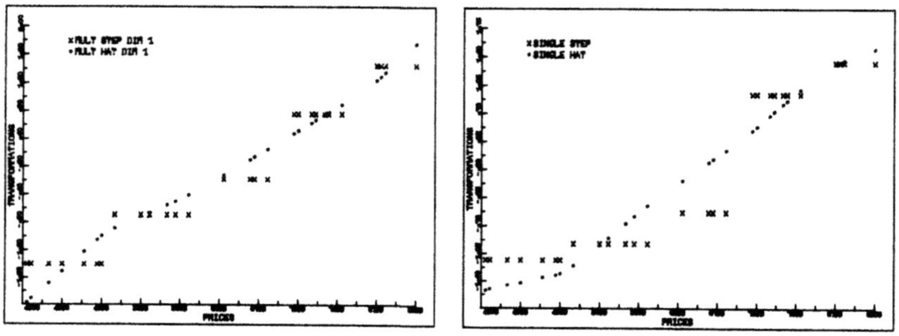

The non-linearity is caused by a greater variety in fuel consumption annex engine-size for middle class cars between \$5000 and \$8000, than for cheap cars (<\$5000) or expensive cars (>\$8000), independent of the main trend of increase of price, engine-size, weight and fuel consumption. For each class, thus defined, the hat (K=2) transformations are linear, even if they cover several knots, but between classes the transformations are not linear.

8: References

Coolen H.,De Leeuw, Van Rijckevorsel(1982)An algorithm for NLPCA , COMPSTAT 82,Wien.
Dauxois J.,Pousse A., Les analyses factorielles en calcul des probabilites et en statistique, Thèse, Université Paul Sabatier, Toulouse (1976)
De Leeuw J.,(1982), Non linear principal components analysis,COMPSTAT 82,Physica,Wien.
De Leeuw J.,Van Rijckevorsel,Van der Wouden (1981),Nonlinear principal components analysis with B-splines in:Methods of Operations Research,Volume 43,Hain.
De Boor C.(1978),A practical guide to splines, Springer New York.
Besse P.(1980), Deux exemples d'analyse en composantes filtrantes, Statistique et analyse des donnees3,5-15.
Gifi A.(1981), Nonlinear multivariate analysis,Dep. of datatheory, FSW, Leiden.
Heiser W. (1981) Unfolding of proximity data,Dep. of Datatheory, FSW, Leiden.
Winsberg S.,Ramsay J.(1982) Principal components with I-splines, forthcoming.
Young F.(1981),Quantitative analysis of qualitative data,Psychometrika,46,357-388.
Lafaye de Micheaux D.(1978),Approximation d'analyses non-canoniques de variables aléatoires, Thèse, Université de Nice.
Saporta G.(1980),About some remarkable properties of generalized canonical analysis, paper read at the spring meeting of the psychometric society in Groningen.

Detecting Outliers in Simultaneous Linear Models

A.E. Ronner, University of Groningen, The Netherlands

Abstract: The robustness literature provides many solutions to the problem of detecting and handling outliers. In this paper we study how to detect 'influential points' in a simultaneous linear model. The model describes the Dutch economy during the post-war period. Outliers in both explanatory and the dependent variables are located by using regression diagnostics. The (interactive) computer-program lists influential points by methods due to Belsley, Kuh and Welsch. The estimation method is modified two stage least squares. Finally simultaneous predictions are presented and compared to the 2SLS predictions and the realisations.

Keywords: Regression diagnostics, robustness, outliers and 2 SLS.

1 INTRODUCTION

In econometrics robust methods are of special interest, since the models are always an approximation of the phenomenon under study. Moreover, 'with the introduction of high spead computers and the frequent use of large-scale models, the researcher has become evenmore detached from intimite knowledge of his data', Belsley e.a. (1980). The authors develop several criteria in order to detect outliers in (large) single equation models. In this paper a simultaneous model is considered and both in the first and the second stage of 2SLS (two-stage least squares) outliers are studied. Our experience is that outliers occurring in the first stage did not seriously influence the estimates. The 'second stage outliers' are more interesting. Some of the outliers are clear candidates for the predicate error, those observations do not correspond to the fitted equation. On the other hand outliers in the X-direction together with near zero outliers in the Y-direction strongly support the estimated relation. We shall illustrate this in more detail in Section 3 where the several equations are presented. Section 2 is devoted to the introduction of the criteria.

2 THE MODEL AND THE CRITERIA

We start from the simple linear regression model

$$y_i = x_i\beta + e_i \qquad i = 1,\dots,n \qquad (2.1)$$

where y_i is the i-th observation on the dependent variable, x_i is a p-row vector of explanatory variables and $\beta \in \mathbb{R}^p$ is the vector of unknown regression coefficients. In a simultaneous model x_i also consists of endogenous variables. Let $\hat{\beta}$ be the LS estimator for β. In order to detect possible 'influential points' we have used four criteria, which are mostly based upon the principle of deleting one of the observations.

The first criterion measures the influence of one observation, say (x_i, y_i), upon the estimate $\hat{\beta}$. Let $\hat{\beta}[i]$ be the LS estimator for β when (x_i, y_i) is deleted. In order to standardize the difference $\hat{\beta}-\hat{\beta}[i]$ we divide by the estimated standard deviation of $\hat{\beta}_j$:

$$BET_j(i) := \{\hat{\beta}_j-\hat{\beta}_j[i]\} \;/\; \{s[i]\sqrt{(X^TX)^{-1}_{jj}}\} \tag{2.2}$$

where $s^2[i] := (n-p-1)^{-1} \sum_{k \neq i} (y_k - x_k\hat{\beta}[i])^2$.

Let X be the n × p matrix of explanatory variables. We consider the diagonal elements h_i of the projection matrix $H := X[X^TX]^{-1}X^T$. The elements h_i indicate 'the distance' between x_i and $\bar{x} := n^{-1}\sum_{i=1}^{n} x_i$. So we can calculate $h_i := x_i(X^TX)^{-1}x_i^T$. The second criterion is based upon the fit $\hat{y}_i := x_i\hat{\beta}$. Define

$$FIT(i) := \{x_i\hat{\beta}-x_i\hat{\beta}[i]\} \;/\; \{s[i]\sqrt{h_i}\}, \tag{2.3}$$

where $s^2[i]h_i$ is the estimated variance of $x_i\hat{\beta}$.

The influence of outliers in the residuals can be detected by the estimated residuals $\hat{e}_i := y_i - x_i\hat{\beta}$. Define

$$SR(i) := \hat{e}_i \;/\; \{s[i]\sqrt{1-h_i}\} \tag{2.4}$$

In case of non-stochastic regressors the standardized residual SR(i) has a t_{n-p-1} distribution.

The fourth criterion deals with 'multiple row effects'. Sometimes two outliers occur together and single row deletion methods are not decisive. The partial-leverage plots give the opportunity to detect more influential points at the same time. As well-known the k-th regression coefficient can be calculated in two steps. First calculate the regressions of y resp. the k-th column of X on X[k], X[k] being the n × (p-1) matrix obtained from X by deleting the k-th column. Let u_k and v_k be the columns of estimated residuals, respectively. Then $\hat{\beta}_k$ equals the simple regression coefficient obtained from the regression of u_k on v_k. The scatter diagram of u_k and v_k is called the 'partial leverage plot' for the estimate $\hat{\beta}_k$.

In the next section we apply the criteria to both the first and the second stage of 2SLS. It turned out that changing the first stage estimation procedure did not essentially improve the estimates and predictions, so we focus on the second stage model.

3 THE GRECON-MODEL 81-A

The model contains 9 fundamental endogenous variables giving 9 behaviour equations. Yearly data of the 1952 - 1979 period are available, cf. Voorhoeve e.a. (1981). The choice of the explanatory variables in each equation was made by trying to find the specification with the best fit by stepwise procedures. The model is completed by adding 14 definitional equations (cf. Ketellapper (1982), Voorhoeve e.a. (1981)). The first four equations are estimated by 2SLS since the diagnostic checks did not show irregular observations.

1. Consumption

$$c = \underset{(0.093)}{0.586}\, y^d_{g-\frac{1}{2}} + \underset{(0.020)}{0.058}\, c_c - \underset{(0.248)}{0.451}\, b'_{g-\frac{1}{2}} + \underset{(0.693)}{1.333} \qquad R^2 = 0.69$$

2. Government expenditure price

$$p_x = \underset{(0.118)}{0.857}\, W + \underset{(0.062)}{0.138}\, p_{m-\frac{1}{2}} + \underset{(0.109)}{0.352}\, \Delta p_c - \underset{(1.104)}{1.189} \qquad R^2 = 0.88$$

3. Export price

$$p_b = \underset{(0.080)}{0.757}\, p_m + \underset{(0.030)}{0.173}\, p_{c-\frac{1}{2}} - \underset{(0.390)}{0.482} \qquad R^2 = 0.98$$

4. Employment

$$a = \underset{(0.049)}{0.411}\, v'_{-\frac{1}{2}} - \underset{(0.042)}{0.199}\, W_{-\frac{1}{2}} - \underset{(0.177)}{0.371}\, iso'_{-\frac{1}{2}} + \underset{(0.520)}{0.649} \qquad R^2 = 0.81$$

More interesting are the following 5 equations. The diagnostic checks will show clear outliers. We have elaborated several estimation methods, such as down-weighing (Huber or p-norms with $1 \leq p < 2$) but strong rejection performed better. We have compared the estimators by means of ex-post predictions. The model is therefore estimated by using 25 observations (1952 - 1976) and then the mean square prediction error (MSPE) of the 1977 - 1979 observations is calculated, cf. Table 1.

5. Investment

The very low 2SLS R^2 of 0.60 shows that investments are rather difficult to explain and to predict. The partial leverage plots show many irregular observations, but particularly 1958 does not fit within the model. In this year the investments sharply dropped because of severe cuts in government expenditures.
We observe a remarkable drop in MSPE: MSPE(2SLS) = 23.4 and MSPE(2SLS excl. 7) = 2.93 cf. Table 1. Table 2 gives the values of the criteria FIT and BET$v'_{-\frac{1}{2}}$ and 1958 can clearly be detected. By deleting 1958 we obtain (2SLSC)

$$i_m = \underset{(0.550)}{0.978}\, v'_{-\frac{1}{2}} + \underset{(0.109)}{0.468}\, y^d_{b-\frac{1}{2}} - \underset{(3.307)}{2.067} \qquad 'R^2 = 0.62'$$

Table 1. Ex-post predictions of 2SLS and corrected 2SLS(2SLSC).

	eq. 5 i_m		eq. 6 m		eq. 7 p_i		eq. 8 p_c		eq. 9 W	
year	2SLS	2SLSC	2SLS	2SLSC	2SLS	2SLSC	2SLS	2SLSC	2SLS	2SLSC
1977	12.56	17.38	3.57	3.32	4.90	6.33	3.29	4.23	11.31	11.22
1978	0.95	3.77	6.85	6.61	3.03	3.27	2.61	3.04	8.22	7.58
1979	0.66	2.82	2.93	2.84	4.50	6.31	3.55	4.39	8.23	7.66
MSPE	23.94	2.93	3.46	3.28	6.00	3.25	3.34	1.24	4.80	3.64

6. Imports

The 2SLS estimates seem quite satisfying with $R^2 = 0.94$. Still diagnostic checks give useful additional information. The imports depend on the import prices p_m. As expected the oil crisis has had a large impact on the import prices. Influential points are 1974 and 1975 (cf. Table 2). Based on MSPE we use 2SLS (excl. 1975) and find (2SLSC)

$$m = \underset{(0.320)}{2.963}\, n' + \underset{(0.151)}{1.345}\, v' - \underset{(0.085)}{0.211}\, (p_m - p_v') - \underset{(0.988)}{0.330} \qquad 'R^2 = 0.94'$$

It is interesting to exclude 1974; than the coefficient of the price

Figure 1. Partial leverage plots.

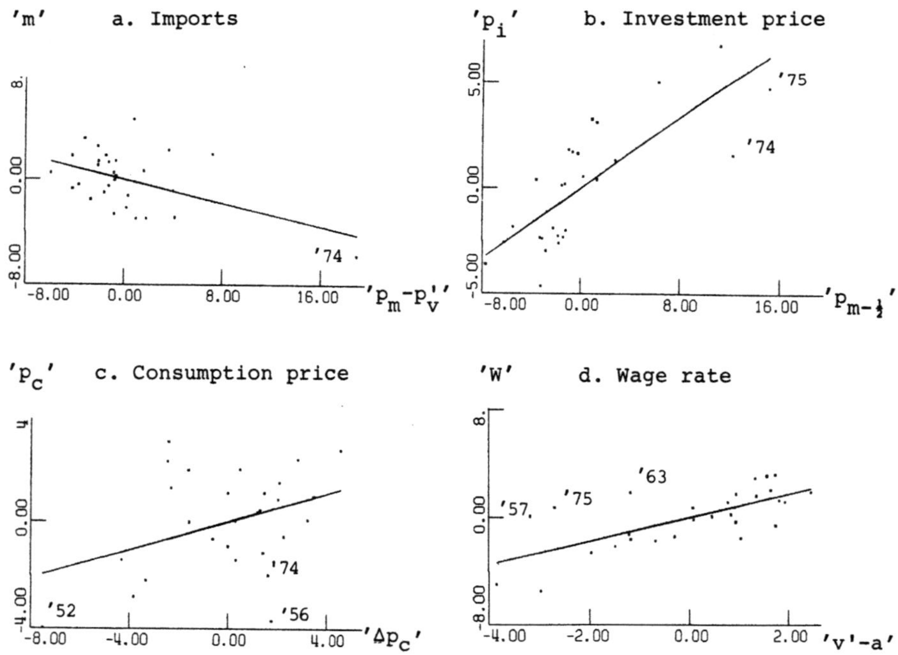

Table 2. Criteria for five behaviour equations.

	eq. 5 i_m		eq. 6 m		eq. 7 p_i	eq. 8 p_c	eq. 9 W
Year	FIT	BET$v'_{-\frac{1}{2}}$	FIT	BET(p_m-p_v')	BET$p_{m-\frac{1}{2}}$	SR	SR
1952	-0.08	0.04	-0.91	-0.01	0.63	-1.62*	-1.48*
1953	0.13	0.00	0.45	-0.10	0.25	-0.87	0.34
1954	-0.31	-0.04	0.29	-0.10	0.07	0.03	0.43
1955	0.33	0.19	-0.65	-0.22	-0.02	-0.07*	0.65
1956	0.10	0.08	0.51	0.08	-0.03	-2.56	-0.31*
1957	0.02*	-0.00*	-0.20	-0.01	0.00	-0.42	2.37
1958	-1.37	1.25	-0.29	0.13	-0.00	0.61	-1.12
1959	0.50	-0.38	0.17	-0.05	0.04	0.41	-0.14
1960	0.14	0.07	-0.58	-0.06	0.10	-0.85	-0.20
1961	0.17	0.12	0.10	-0.03	0.07	-0.50	-0.70
1962	0.20	0.03	0.01	-0.00	0.06	-0.64	-0.57
1963	-0.53	-0.22	0.11	0.63	-0.01	0.50	1.90
1964	0.29	0.20	-0.38	0.13	-0.00	-0.53	1.09
1965	-0.43	-0.30	-0.48	-0.10	0.04	-0.16	-1.33
1966	0.33	0.14	0.10	-0.04	-0.00	0.17	0.40
1967	0.02	-0.01	-0.13	0.04	0.05	0.22	-0.70
1968	-0.07	-0.02	0.09	-0.03	0.06	-0.26	-0.16
1969	-0.52	-0.37	0.04	-0.01	-0.05	0.05	1.08
1970	0.20	0.15	0.57	0.25	-0.02	-0.06	-0.19
1971	-0.23	-0.17	-0.02	0.00	-0.05	0.18	-0.31
1972	-0.33	-0.05	-0.22	0.14	-0.15	1.37*	-0.72
1973	0.16	-0.07	0.17*	-0.02*	0.07*	1.77*	1.14
1974	-0.15	0.07	-2.66	-2.43	-1.54*	-1.75	-0.15*
1975	0.92	-0.37	0.75	0.75	-0.71	1.10*	2.47
1976	-0.21	0.14	-0.38	-0.38	0.07	2.20	0.32
1977	0.38	-0.21	-0.10	-0.10	0.01	1.05	-1.60
1978	0.17	-0.12	-0.19	-0.19	0.05	1.02	-0.37
1979	0.11	-0.69	0.78	0.78	0.35	0.95	-0.51

variable p_m-p_v' drops to 0.144 (0.161)! The contribution of this variable is no longer significant and we see that p_m-p_v' is only because of the oil crisis accepted as an explanatory variable. The 2SLS (excl. 1974) equation, which shall not be used for the predictions, is

$$m = \underset{(0.333)}{2.971}\, n' + \underset{(0.157)}{1.265}\, v' - \underset{(0.161)}{0.144}\, (p_m-p_v') + \underset{(1.183)}{0.421} \qquad 'R^2 = 0.94'$$

7. Investment price

The 2SLS R^2 equals 0.75. The investment price is via the import price p_m strongly influenced by the oil crisis. We observe that BET$p_{m-\frac{1}{2}}$ is large for both 1974 and 1975. Using MSPE we decide to delete 1974 and 1975 and find (2SLSC)

$$p_i = \underset{(0.085)}{0.211}\, W + \underset{(0.079)}{0.724}\, p_{m-\frac{1}{2}} + \underset{(0.806)}{1.589} \qquad 'R^2 = 0.84'$$

8. Consumption price

The 2SLS R^2 equals 0.71. The partial leverage plots of Δp_c show outliers for 1952, 1956 and 1974, years of changes in the economic situation, Korean war, severe cuts in government expenditures and the oil crisis, cf. Figure 1.c. Deleting 1952, 1956 and 1974 yield 2SLSC:

$$p_c = \underset{(0.111)}{0.382}\, W + \underset{(0.071)}{0.327}\, p_{m-\frac{1}{2}} + \underset{(0.128)}{0.157}\, \Delta p_c + \underset{(1.045)}{0.519} \qquad 'R^2 = 0.79'$$

We observe that in 2SLSC the contribution of Δp_c is no longer significant.

9. Wage rate

The 2SLS R^2 equals 0.81. The partial leverage plots show a remarkable clustering of 1957, 1963 and 1975. We have already discussed the years 1957 and 1975, but also 1963 marks a change in the economic structure of the Netherlands. In 1964 wages started to increase rapidly and a period of economic growth began. By deleting 1957, 1963 and 1975 we find (2SLSC)

$$W = \underset{(0.100)}{0.940}\, p_{c-\frac{1}{2}} + \underset{(0.149)}{1.120}\, (v'-a)_{-\frac{1}{2}} + \underset{(0.286)}{0.837}\, b'_{g-\frac{1}{2}} - \underset{(0.347)}{1.035}\, \tilde{w}_{-1} - \underset{(0.883)}{1.142} \qquad 'R^2 = 0.91'$$

5. PREDICTIONS AND CONCLUSION

Using the same values for the predetermined variables of 1981 as Grecon 81-A the outlier corrected model yields the following ex-ante predictions for the variables which realisations are already available.

Table 3. Ex ante (1981) predictions of Grecon 81-A and the corrected model

Variable	c	p_b	a	i_m	m	p_c	W
Grecon	0.5	11.8	-0.1	-5.5	-1.1	4.5	3.7
Corrected model	-0.2	12.0	-0.1	-1.8	-0.9	6.7	3.3
Preliminary realisation	-3	15	-1.9	-11.5	-6	6.5	4.5

The regression diagnostics turned out to be a useful tool for a better understanding of the data. Moreover, if clear outliers occur, correction methods improve the prediction quality of a model.

REFERENCES

Belsley, D.A., Kuh, E and Welsch, R.E., (1980), 'Regression Diagnostics, Identifying Influential Data and Sources of Collinearity', John Wiley.

Ketellapper, R.H., (1982), 'The impact of observational errors on parameter estimation in econometrics, PhD. thesis, Groningen, The Netherlands.

Voorhoeve, W., Kooyman, M.A., Ketellapper, R.H.and Steerneman, A.G.M., (1981) 'The Grecon 81-A Model', Econometric Institute, Groningen, The Netherlands.

ACKNOWLEDGEMENT

I thank Gerard Kuper for writing the computerprogram.

List of Variables

A tilde (~) over a symbol indicates the absolute level of the variable. The symbol Δ in front of a variable indicates its first difference. Variables without these symbols represent relative first differences (expressed as a percentage). Moreover, $a_{-\frac{1}{2}}$ is defined by $a_{-\frac{1}{2}}(t) = \frac{1}{2}(a(t) + a(t-1))$.

$\tilde{a}$	labour volume in private enterprises (thousands of manyears)
b'_g	difference in direct tax receipts added to the balance of income transfers of private persons to and from the government expressed as a percentage of the disposable income of private persons.
$\tilde{c}$	private consumption (1975 prices)
$\tilde{c}^c$	consumption credits (1975 prices)
$\tilde{i}_m$	gross investments in fixed assets (excluding dwellings) by private enterprises (1975 prices)
iso'	difference in government deficit as a percentage of total output
$\tilde{m}$	imports of commodities (c.i.f.) (1975 prices)
n'	difference of inventories as a percentage of total output
$\tilde{p}_b$	price index of exports of commodities (1975=100)
$\tilde{p}_c$	price index of private consumption (1975=100)
p_i	price index of gross investments in fixed assets by private enterprises (1975=100)
$\tilde{p}_m$	price index of imports of commodities (1975=100)
p_x	price index of government expenditures (1975=100)
p_v'	price index of output of commodities (1975=100)
v'	total expenditure minus increase of inventories minus exports of services minus non-material government consumption (output of commodities) (1975 prices)
$\tilde{w}$	unemployment rate
W_d	average gross wages (thousands of Dutch guilders per manyear)
$\tilde{Y}^d_b$	disposible income of private enterprises (1975 prices)
$\tilde{Y}^d_g$	disposible income of private persons (1975 prices)

List of data (1952 - 1979)

i_m	-9.65	12.79	21.92	22.65	9.46	0.42	-16.35*	13.48	17.14
	8.16	5.42	-9.97	17.31	2.46	10.14	6.19	10.53	-1.20
	15.65	0.78	-7.89	10.69	-0.34	-7.54	-6.22	18.38	6.23
	3.82								
m	-10.56	18.28	24.02	8.69	13.93	2.88	-5.88	13.50	16.90
	7.54	5.74	9.57	15.94	5.64	6.01	6.49	12.72	15.45
	15.74	5.99	5.22	12.66	-0.10	-5.38	9.96	2.52	5.51
	5.67								
$p_m - p_v'$	-3.35	-5.50	-4.62	0.43	-0.88	-0.07	-6.19	-3.57	-0.07
	-3.08	-3.00	-0.31	-4.18	-1.53	-3.33	-3.01	-4.74	-2.19
	0.93	-2.98	-6.03	-0.99	18.74	-3.85	-1.99	-2.36	-4.44
	6.03								
Δp_c	-12.22*	-0.99	4.99	-3.04	-0.01	4.71	-3.13*	-1.53	0.98
	-0.04	-0.04	2.46	2.64	-3.26*	1.67	-1.99	-0.94	3.97
	-2.29	3.84	0.33	1.07	-0.03*	2.11	-2.49	-3.50	-1.16
	0.65								
W	5.68	4.08	8.97	8.95*	8.78	11.76	4.40*	2.37	8.15
	7.24	5.87	8.96	14.95*	11.14	10.97*	8.77	8.87	13.43
	13.22	13.13	12.59	15.59	15.67	12.81*	10.99	8.29	7.21
	6.16								

Least Squares Optimisation of General Log-likelihood Functions and Estimation of Separable Linear Parameters

G.J.S. Ross, Rothamsted Experimental Station, Harpenden, G.B.

Summary

Many log-likelihood functions may be expressed as sums of squares of generalised residuals. This allows problems in maximum likelihood estimation to be solved using least squares optimisation routines. When linear parameters are separately estimable they may be eliminated algebraically and the resulting residuals used directly in a least squares optimisation routine.

KEYWORDS Maximum likelihood estimation, least squares optimisation, separable linear parameters, non-linear estimation.

1. Introduction

Optimisation problems tend to be easier to solve when the objective function is a sum of squares. The special form of the objective function allows the computations to depend on the first derivatives of the component functions rather than on the second derivatives of the objective function. This means in practice that models are often fitted to data by least squares methods when statistical considerations indicate that errors are not Normally distributed and that maximum likelihood estimation should be used. This paper shows how this problem may be overcome, and how models for which linear parameters are separable can also be expressed as sums of squares of functions of the non-linear parameters only. These techniques make it possible to fit non-linear models in general statistical programs in a unified manner, so that the choice of error distribution does not affect the method of solution.

2. The log-likelihood function as a sum of squares

Given a set of n independent observations $\underset{\sim}{y}$ to be fitted to a model with expectations $\mu(\theta)$ which are functions of p unknown parameters $\underset{\sim}{\theta}$, and a probability distribution $\tilde{p}(\underset{\sim}{y}|\underset{\sim}{\mu})$ of known form, then p as a function of $\underset{\sim}{\theta}$ is the likelihood of $\underset{\sim}{\theta}$ given $\underset{\sim}{y}$. The method of maximum likelihood estimates the value of $\underset{\sim}{\theta}$ at which p is a maximum. In practice the related function $-\log(p)$ is minimised, and is known as the log-likelihood, $L(\underset{\sim}{\theta})$. When p is the Normal distribution with unknown variance σ^2 the log likelihood is

$$L(\underset{\sim}{\theta}) = \frac{1}{2\sigma^2} \sum_i (y_i-\mu_i(\underset{\sim}{\theta}))^2 + \frac{n}{2} \log(2\Pi\sigma^2) \qquad (1)$$

In this case minimisation of $L(\theta)$ is obtained by minimising the sum of squares of residuals, so that the least squares estimate of $\underset{\sim}{\theta}$ is also the maximum likelihood estimate.

Let $p(y|\mu)$ be any distribution with the following properties:

(1) The maximum value of $p(y|\mu)$ is attained when $\mu=y$.

(2) There are no local maxima, i.e. p decreases monotonely with $|\mu-y|$ in either direction.

Then the function $L_i(\theta)$-log $p(y_i|\mu_i)$ has an absolute minimum, L_{min}, when $\mu_i=y_i$ for each i and hence the function

$$e_i(\underset{\sim}{\theta}) = (L_i(\underset{\sim}{\theta}) - L_{min})^{1/2} .\mathrm{sign}(y_i-\mu_i) \qquad (2)$$

is a monotone increasing function of $y_i - \mu_i$ known as the generalised residual (Ross and Nelder (1977)).

The log likelihood is thus the sum of squares of the generalised residuals, apart from terms independent of $\underset{\sim}{\theta}$. Hence maximum likelihood estimation of $\underset{\sim}{\theta}$ is obtained by minimising $\Sigma\ e_i^2$.

Twice the minimised log likelihood is known as the deviance, so it is convenient to multiply the generalised residuals by $\sqrt{2}$ for comparison with weighted residuals based on the expected variance of y_i at $y_i = \mu_i$.

2.1. Examples of generalised residuals

Let y be a Poisson variable with mean μ. Then

$$e_i^2 = y_i \log(y_i/\mu_i) - y_i + \mu_i \qquad (3)$$

whereas the conventional weighted residual is $(y_i-\mu_i)/\sqrt{\mu_i} = r_i$, say. For example, if y = 2, the values of $e\sqrt{2}$ and r may be compared in the following table:

Table 1

μ_i	$e_i\sqrt{2}$	r_i
0.5	1.595	1.061
1.0	0.879	0.707
1.5	0.388	0.354
2.0	0.000	0.000
2.5	-0.328	-0.354
3.0	-0.615	-0.707
3.5	-0.873	-1.061

To differentiate e with respect to θ, note that $de_i/d\mu$ is continuous with respect to μ, and its value at $\mu = y$ is the negative square root of the weight function.

The following table lists the generalised residuals for some commonly used frequency distributions:

Table 2

Distribution	e_i^2	de_i/d_μ
Normal, variance σ^2	$(y-\mu)^2/\sigma^2$	$-1/\sigma$
Poisson	$y\log(y/\mu) - y + \mu$	$(1-y/\mu)/e_i$
Binomial, total n	$y\log(y/\mu)+(n-y)\log\dfrac{n-y}{n-\mu}$	$\dfrac{n(\mu-y)}{\mu(n-\mu)e_i}$
Negative Binomial, parameter k	$y\log(y/\mu) - (y+k)\log\dfrac{y-k}{\mu+k}$	$\dfrac{k(\mu-y)}{\mu(\mu+k)e_i}$
Gamma	$\log(\mu/y) + y/\mu-1$	$(\mu-y)/(\mu^2 e_i)$

2.2 Geometrical interpretation of generalised residuals

If $\underset{\sim}{y}$ is represented by a point Y in n-dimensional space, and the locus of all possible values of $\underset{\sim}{\mu}$ generated by the parameters $\underset{\sim}{\theta}$ is a p-dimensional subspace M, known as the solution locus, then the squared distance from Y to a point T on M is $\Sigma(y-\mu)^2$. Hence the least squares estimate of θ corresponds to the point U on M closest to Y, the perpendicular from Y onto M. (Fig.1.)

If maximum likelihood estimation is used the log likelihood associated with the point T is no longer the squared distance and the minimum measure is given by some other point V. To obtain an analogous representation it is only necessary to plot the generalised residuals, $\underset{\sim}{e}(\theta)$ which lie on a locus M' in a space in which the origin corresponds to the data point, Y'. The maximum likelihood estimate is then at V', the foot of the perpendicular from Y' to M'. (Fig.2.)

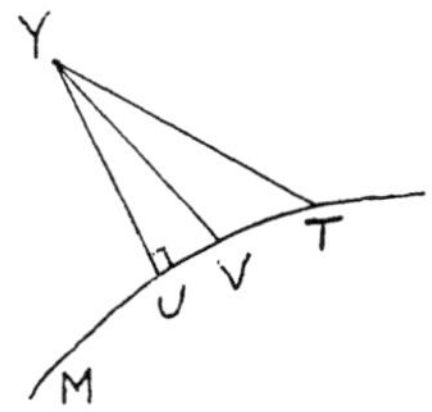

Fig.1

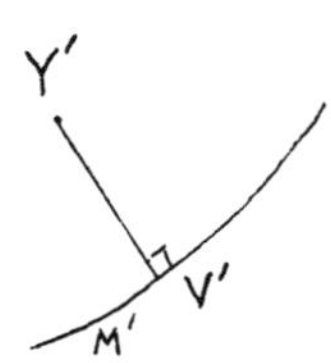

Fig.2

The Gauss-Newton method of minimising a sum of squares is the iterative procedure of dropping a perpendicular from Y onto the tangent to M at T_0, reading off parameter estimates on the tangent, and translating these into a new point, T_1 on M. (Fig.3.)

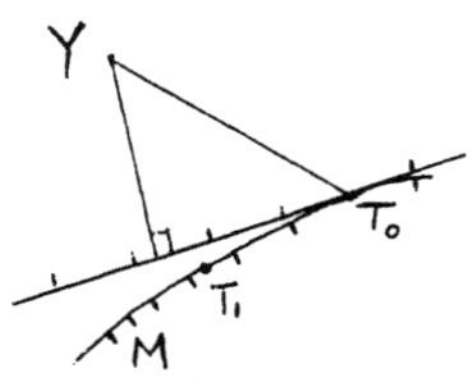

Fig.3

It is clear from Fig.3 that the success of the Gauss-Newton method depends on the linearity of the parameter system within M as well as the curvature of M and the proximity of Y to M. A stable parameter system is a great advantage for the success of the Gauss-Newton method (Ross (1970), Bates and Watts (1980)). Most practical least-squares algorithms start with the Gauss-Newton method but employ various safeguards to ensure convergence (Lill, (1976)).

2.3. Practical implementation of the method

Generalised residuals computed by equation (2) are stored in an array as required by the least squares algorithm being used. If derivatives are also required they may be computed using the formulae for $de/d\mu$ in section 2.1, multiplied by $d\mu/d\theta$. It is simpler to use methods that do not require derivatives, provided the step lengths used for differencing are appropriate. A suitable algorithm is the method of Peckham (1970) which is NAG algorithm E04FAF (NAG Library (1977)). The method is also being incorporated into MLP 3.07 and GENSTAT 3.04.

Comparison with other algorithms depends on many factors, in particular the linearity of the solution locus M' compared with M in Figs. 1 and 2. A study of the convergence rate of the Gauss-Newton Generalised Residuals (GNGR) and Iteratively Reweighted Least Squares (IRLS) for sigmoid response models with binomial variables (Probit analysis) showed that GNGR saved at least one iteration compared with IRLS and was very superior for poor initial estimates. But this is no guarantee that it is universally superior.

3. Models with separable linear parameters

Parameters are separable if given the first q parameters the remaining p-q may be estimated algebraically. The problem may then be reduced to one of optimising the first q non-linear parameters. The usual application is in least squares models of the form $\mu = \Sigma\beta_i f_i(\theta)$ in which the parameters β appear as linear regression coefficients for given $\underset{\sim}{\theta}$. However there are other examples, in particular:

1) Poisson distribution with $\mu = \beta f(\underset{\sim}{\theta})$, where β is estimated by $\Sigma y/\Sigma f$.

2) Multinomial distribution in which $\Sigma f(\underset{\sim}{\theta})=1$, as used in fitting frequency distributions to classified data, when the likelihood is the same as for the Poisson distribution using $\beta = \Sigma y$.

3) Gamma distribution with $\mu=\beta f(\underset{\sim}{\theta})$, where β is estimated by $\Sigma(y/f)/n$.

The advantages of separating parameters have long been known (Richards (1961), Ross (1970), Lawton and Sylvestre (1971)), but the method is not readily available in standard packages, apart from MLP and GENSTAT, in spite of the work of Golub and Pereyra (1973).

The vector of fitted values $\mu^*(\theta)$ obtained by substituting back the estimated linear parameters lies in a subspace M* of q dimensions within M in Fig.1. Since it includes the optimum point, the application of any least squares algorithm to the generalised residuals e* corresponding to μ^* will produce the correct solution. These residuals may be called partially fitted generalised residuals.

3.1 . Practical implementation

The main difference between the use of ordinary generalised residuals and partially fitted generalised residuals is that there are no conveniently computable formulae for the first derivatives of the latter and first differences must always be used. This restricts the choice of algorithm and requires working space for q+1 vectors of n residuals and first differences. However, this is allowed for in NAG routine E04FAF, and in MLP and GENSTAT, and so provided space is available the method is easy to implement.

For each call of the parameter values $\underset{\sim}{\theta}$ there are two passes through the data, one to compute the working variables $f_i(\underset{\sim}{\theta})$ and the linear parameters $\underset{\sim}{\beta}$, and a second pass to compute the generalised residuals from $\underset{\sim}{\beta}$ using equation (2).

3.2. Numerical examples

1) To fit the model $\mu = \beta_0 + \beta_1\theta^x$ with Normal errors to the data in Table 3, a single Gauss-Newton iteration is illustrated, starting at θ=.5

Table 3

x	y	$.5^x$	e*(.5)	$.51^x$	e*(.51)	de/dθ
0	3	1	.1478	1	.1258	-2.20
1	11	0.5	.0087	.51	.0645	5.58
2	14	0.25	-1.0609	.2601	-1.0468	1.41
3	18	0.125	.9043	.1327	.8565	-4.78
	$\beta_0(\theta)$	19.1304		19.3258		
	$\beta_1(\theta)$	- 16.2783		- 16.4516		

New estimate $= 0.5 - \Sigma(e^*(.5)\ de/d\theta) \,/\, \Sigma(de/d\theta)^2$
$= 0.6002$ (final estimate .618)

2) To fit model $\mu = \beta'(1+\theta x)$, with Poisson errors to data in Table 4, starting with $\theta=1$.

Table 4

x	y	f(1)	e*(1)	f(1.1)	e*(1.1)	de/dθ
0	25	1	-.5264	1	-.6183	-.919
1	16	0.5	.2825	.4762	.3483	.658
2	12	0.3333	.5184	.3125	.6087	.903
	$\beta(\theta)$	28.9091		29.6306		

New estimate = 0.456 (final estimate 0.547)

References

Bates,D.M. and Watts,D.G. (1980) Relative curvature measures of non-linearity. J.R.statist.Soc. B. 42 , 1-25

GENSTAT Manual (NAG, Oxford)

Golub,G.H. and Pereyra,V (1973). The differentiation of pseudo-inverses and non-linear least squares problems whose variables separate. SIAM J.Num.Anal. 10 , 413.

Lawton,W.H. and Sylvestre,E.A.(1971) Elimination of linear parameters in non-linear regression. Technometrics, 13 , 461-467.

Lill,S.A.(1976). A survey of methods for minimizing sums of squares of non-linear functions. In Optimization in Action , Ed.L.C.W.Dixon, Academic Press.

MLP Manual (1980). Rothamsted Experimental Station, Harpenden.

NAG Library Manual, Mark 6 (1977). Oxford.

Peckham,G (1970). A new method for minimizing a sum of squares without calculating gradients. Computer J. 13 , 418-420.

Richards,F.S.G. (1961) A method of maximum likelihood estimation. J.R.statist.Soc.B, 23 , 469-426.

Ross,G.J.S.and Nelder,J.A. (1977). In Annual Report of Rothamsted Experimental Station, Pt.1.p.304.

Ross,G.J.S. (1970) The efficient use of function minimization in non-linear maximum likelihood estimation. J.R.statist.Soc. C, 19 , 205-221.

Computational Procedures for Bounded-Influence Regression

A. Samarov and R.E. Welsch, Massachusetts Institute of Technology, U.S.A.

The ordinary least-squares (OLS) estimator of the coefficients in the classical linear regression model is efficient when the appropriate assumptions are valid. However, even small departures from the standard assumptions on the model may seriously damage OLS, causing inefficiency, bias, inconsistency, and incorrect tests of significance. Bounded-influence estimators attempt to insure that no small fraction of the data is overly influential in the determination of the estimated parameters or fitted values. In this paper we describe algorithms for computing several bounded-influence regression procedures. Although the kernel of the bounded-influence regression package (BIF) is written in portable FORTRAN, it is currently operational in fully interactive form in TROLL (a large statistical modeling system) providing the user with a variety of ways to get diagnostic information. Considerable output results from bounded-influence regression (data weights, robust distances in explanatory variable space, etc.) and we provide some graphical ways to present this information. The paper concludes with an example of the use of the above procedures.

Keywords: bounded-influence regression, iterative procedures for robust regression, graphical data analysis

1. Definition of estimators. The standard linear regression model is generally written as

$$Y = X\beta + \varepsilon,$$

where Y is an n×1 response variable, X is an n×p matrix containing p explanatory variables including an intercept as a first column, β is a p vector of unknown coefficients, and ε is a normal random n-vector with $E(\varepsilon) = 0$ and $E(\varepsilon\varepsilon^T) = \sigma^2 I$. These assumptions, particularly the normality assumption, should be regarded as defining the "central model", and since we fully expect small violations of these assumptions, our aim is to present an estimator which is not too sensitive to these small violations.

We consider a class of estimators $\hat{\theta} = (\hat{\beta},\hat{\sigma})$ defined as a value of $\theta = (\beta,\sigma)$ which minimizes (cf. Huber (1981)):

$$Q(\beta,\sigma) = d\cdot\sigma + \sum_{i=1}^{n} \rho\left(\frac{y_i - x_i^T\beta}{\sigma v(x_i)}\right) \sigma\, u(x_i) v^2(x_i) \qquad (1)$$

where x_i^T denotes the i-th row of the X matrix, $v(\cdot)$ and $u(\cdot)$ are positive functions, and $\rho(\cdot)$ is often called a robustifying criterion function.

The functions u(x) and v(x) are included in (1) to downweight the influence of outlying points in the X-space while preserving reasonably high efficiency of the estimates [see Krasker and Welsch (1982) and Maronna, Bustos, and Yohai (1979)].

The minimization of the function $Q(\beta,\sigma)$ may be viewed as a generalization of the least squares method for which $\rho(t) = t^2/2$, $d = (n-p)/2$, $u(\cdot) = v(\cdot) = 1$. It is usually assumed [Huber (1981)], that $\rho \geq 0$ is a convex function, and $0 < \lim_{|t|\to\infty} \rho(t)/|t| < \infty$.

With these assumptions, estimates minimizing (1) can equivalently be characterized by the p+1 equations:

$$\text{(a)} \sum_{i=1}^{n} u(x_i)v(x_i)\psi\left(\frac{r_i}{\hat{\sigma}v(x_i)}\right) x_i = 0, \quad \text{(b)} \sum_{i=1}^{n} u(x_i)v^2(x_i)\,\chi\left(\frac{r_i}{\hat{\sigma}v(x_i)}\right) = d, \qquad (2)$$

where $r_i = y_i - x_i^T\hat{\beta}$ is the residual, $\psi(\cdot) = \rho'(\cdot)$ and $\chi(t) = t\psi(t) - \rho(t)$. The parameter d in (2) is chosen to make the scale estimate $\hat{\sigma}$ asymptotically unbiased for normal errors.

The BIF program uses only Huber's psi-function [Huber (1981)]:

$$\psi_c(t) = \rho'(t) = \min(c, \max(t,-c)), \tag{3}$$

which has proved to be optimal in different contexts. The chi-function in the scale equation (2b) is chosen of the form $\chi_c(t) = \frac{1}{2}\psi_c^2(t)$ which corresponds to Huber's "proposal-2" estimates [Huber (1981)].

It is often convenient to rewrite the equations (2) in the weighted least-squares form:

$$\text{(a)} \sum_{i=1}^{n} u(x_i)\, w\left(\frac{r_i}{\hat{\sigma}v(x_i)}\right) r_i x_i = 0, \quad \text{(b)} \sum_{i=1}^{n} u(x_i)\, w^2\left(\frac{r_i}{\hat{\sigma}v(x_i)}\right) r_i^2 = 2d\hat{\sigma}^2 , \tag{4}$$

where $w(z) = \frac{\psi(z)}{z}$. The BIF program in TROLL provides the following four estimators obtained from (2) by different choices of the functions $u(\cdot)$ and $v(\cdot)$.

I. Huber M-estimators (Huber (1981)):

$$u(x) = v(x) = 1, \; d = d_H = \frac{n-p}{2}\, g_1(c),$$

where

$$g_1(z) = E\psi_z^2(\eta) = E\min(\eta^2, z^2) = z^2 + (1-z^2)(2\Phi(z)-1) - 2z\phi(z). \tag{5}$$

Here η is a standard normal variable, and $\phi(z)$ and $\Phi(z)$ denote its density and distribution function, respectively.

These estimates, even though they control for large residuals, are not qualitatively robust [see Hampel (1974)], in the sense that they may have an arbitrarily large asymptotic bias when the assumptions defining the "central model" are satisfied only approximately. The following classes of estimators have been proposed to correct this drawback of standard M-estimates for regression.

II. Mallows Estimates (Mallows (1973)):

$$v(x) \equiv 1, \; u(x) = \psi_b[(x^T\hat{A}^{-1}x)^{\frac{1}{2}}]/(x^T\hat{A}^{-1}x)^{\frac{1}{2}} , \; d = d_H \cdot \frac{1}{n} \sum_{i=1}^{n} u(x_i), \tag{6}$$

where $\psi_b(\cdot)$ is defined by (3) and the $p \times p$ matrix $\hat{A}$ satisfies the equation

$$\hat{A} = \frac{g_1(c)}{n} \cdot \sum_{i=1}^{n} u^2(x_i) x_i x_i^T. \tag{7}$$

Mallows' estimator guarantees qualitative robustness by downweighting outlying points in X-space. However, any downweighting of X-data that does not include some consideration of how the response fits the remainder of the data cannot produce efficient estimates, because the outlying points can contribute much to the precision of $\hat{\beta}$ and therefore increase the efficiency of the estimator.

III. Schweppe Bounded-Influence Estimator (Handschin, et al. (1975)):

$$u(x) \equiv 1, \; v^2(x_i) = 1-h_i, \; i=1,\ldots,n, \; d = \frac{(n-p)}{2n} \sum_{i=1}^{n} g_1(c\sqrt{1-h_i}), \tag{8}$$

where $h_i = x_i^T(X^TX)^{-1}x_i$ are the diagonal elements of the hat matrix.

IV. Krasker-Welsch Estimator (Krasker and Welsch (1982)):

$$u(x) \equiv 1, \; v^2(x) = 1/x^T\hat{A}^{-1}x, \; d = \frac{n-p}{2n} \cdot \sum_{i=1}^{n} g_1\left(c/(x_i^T\hat{A}^{-1}x_i)^{\frac{1}{2}}\right) , \tag{9}$$

where the p×p matrix $\hat{A}$ satisfies the equation

$$\hat{A} = \frac{1}{n} \sum_{i=1}^{n} g_1 \left(c/(x_i^T \hat{A}^{-1} x_i)^{\frac{1}{2}}\right) x_i x_i^T . \tag{10}$$

Krasker and Welsch (1982) show that so defined the estimator $\hat{\beta}$ satisfies the necessary conditions for achieving the highest efficiency among all weighted least squares estimators with sensitivity $\leq c$ (in the strong sense that its asymptotic covariance matrix differs from all others by a non-negative definite matrix). Because of this result, the Krasker-Welsch estimator has been adopted as the default estimator in BIF.

2. Computational Procedure. The general structure of the algorithm used in the BIF program consists of four major modules which we will consider in order.

I. Starting Values. The starting values $\hat{\beta}^{(0)}$ for the coefficients are always computed by OLS unless the user gives his own starting coefficients.

The starting scale is computed from starting residuals $r_i^{(0)} = y_i - x_i^T \beta^{(0)}$ robustly using MAD: $\hat{\sigma}^{(0)} = \text{med}\, |r_i^{(0)} - \text{med}(r_i^{(0)})|/.6745$. The diagonal elements h_i, i=1,...,n of the hat matrix $H = X(X^T X)^{-1} X^T$ are also computed at this stage. The program uses the QR-decomposition and related subroutines from LINPACK (see Dongarra, et al (1979)).

II. Robust x-Distances. Robust x-distances $d(x) = (x^T \hat{A}^{-1} x)^{\frac{1}{2}}$, needed for evaluating functions u(x) and v(x), are computed by iterative solution of a slightly modified form of the robust covariance M-estimator equation, see (7) or (10):

$$d^2(x_i) \cdot q(x_i) = n \cdot h_i(q), \quad i=1,\ldots n,$$

where q(•) is some weight function, e.g., $q(x) = g_1(c/d(x))$ for the Krasker-Welsch estimator and $q(x) = g_1(c) \cdot u^2(x) = g_1(c) \min(1, b^2/d^2(x))$ for the Mallows estimator ($g_1(c)$ is defined in (5)), and $h_i(q)$, i=1,...,n are the diagonal elements of the hat matrix corresponding to the weighted X-matrix $\text{diag}(q^{\frac{1}{2}}(x_i)) \cdot X$. At convergence of the iteration we have robust x-distances $d(x_i)$, and x-weights $q^{\frac{1}{2}}(x_i)$.

III. Iteration for Coefficients and Scale. This part constitutes the kernel of the algorithm and is essentially the procedure for minimizing the objective function $Q(\beta,\sigma)$ (1) or, equivalently, for solving the system (3).

Huber (1973), and Holland and Welsch (1977) proposed several iterative procedures for minimizing $Q(\beta,\sigma)$ with $u(x) = v(x) = 1$. The BIF program uses a modification of the simplest of those algorithms called algorithm H (see Huber 1981). Algorithm H is the standard iterative least squares algorithm but with so-called metrical Winsorization of the residuals. Unlike the iteratively reweighted least squares (IRWS) algorithm (see Holland and Welsch (1977)) it does not require performing the QR decomposition at each iteration step. Thus, algorithm H appears to be computationally preferable even though it requires more iterations to converge. The convergence of the algorithm under certain assumptions was proved by Huber (1981). Algorithm H can be defined as follows:

1) Compute starting values $\beta^{(o)}$, $\sigma^{(o)}$ and starting residuals $r_i^{(o)}$, i=1,...,n. Put k=0.
2) Compute Winsorized residuals $z_i^{(k)} = v(x_i)\, \psi_c\left(\frac{r_i^{(k)}}{\sigma^{(k)} v(x_i)}\right) \sigma^{(k)}$, i=1,...,n,

   where $\psi_c(\cdot)$ is defined in (3), and $v(x_i)$ is defined above for each estimator separately. Weights $w_i^{(k)} = z_i^{(k)} / r_i^{(k)}$ for all but Mallows, and $w_i^{(k)} = q^{\frac{1}{2}}(x_i) \cdot z_i^{(k)} / r_i^{(k)}$ for Mallows estimator are also computed at this stage.

3) Update the scale $(\sigma^{(k+1)})^2 = \frac{1}{d} \sum_{i=1}^{n} u(x_i)(z_i^{(k)})^2$ where $u(x_i)$ and d are defined above for each estimator separately.

4) Solve for $\Delta\beta^{(k)}$: $X^T X \Delta\beta^{(k)} = X^T Z^{(k)}$, where $Z^{(k)}$ is an n-vector with the components $z_i^{(k)}$, and update coefficient estimates: $\beta^{(k+1)} = \beta^{(k)} + \Delta\beta^{(k)}$.

5) Compute the convergence criterion for the coefficients $DEL = ((\Delta\beta^{(k)})^T V^{-1}(\Delta\beta^{(k)}))^{\frac{1}{2}}$, where V is an estimate of the covariance matrix of the coefficient estimates computed at stage II. Compute the convergence test for scale $DELSC = |\sigma^{(k+1)} - \sigma^{(k)}|/\max(1,\sigma^{(k)})$, and denote DELTA = max {DEL, DELSC}. Stop when DELTA $\leq$ SCNVCR where SCNVCR is a scale-free convergence tolerance. If DELTA >SCNVCR, put k = k+1 and go to 2).

At convergence we have final estimates of the coefficients β_F, scale σ_F, residuals R_F weights w_F.

IV. Covariance Matrices. Covariance matrices are computed by estimating the asymptotic covariance matrices from the sample using final residuals, weights, scale, and x-distances.

Marazzi (1980) developed the ROBETH package for robust and bounded-influence regression which implements similar estimation procedures. Differences between BIF and ROBETH include the computation of robust distances in factor space, computing the Mallows estimator, the choice of cut-off parameter, and the convergence criteria. BIF gives the user more flexibility in terms of the alternative choice of the cut-off parameter or efficiency, starting values, scale iteration, and the capability to interrupt the algorithm, examine intermediate results, and change some of the algorithm control parameters.

3. Computation of the Efficiency and Cut-off Parameters. To fully define the estimator the user has to specify the cut-off parameter which determines the sensitivity of the estimator, see Krasker and Welsch (1982). In practice we want to choose the cut-off parameter so that the asymptotic efficiency of the bounded-influence estimator is not too much lower than least-squares when we have ideal data for the use of least-squares. This usually means X is taken as given and the error structure is normal. The relative efficiency then would be obtained by comparing the covariance matrices of the OLS and bounded-influence estimates. There is no canonical way to compare two matrices. The trace, determinant, or extreme eigenvalue could be used. BIF uses the largest relative eigenvalue and the p-th root of the ratio of determinants to compute relative efficiency.

The cut-off parameter can be computed by the program from the user supplied efficiency. The computation involves the solution of a nonlinear equation which depends on the given X matrix. This computation may turn out to be very expensive in some cases, especially for the Mallows and Krasker-Welsch estimators. For the Mallows estimator two cut-off parameters b and c are needed such that $b \geq \sqrt{p/g_1(c)}$, and its efficiency is a product of two factors, one of which is the efficiency of the Huber estimator and depends on c only, and the second factor depends only on b. In order to solve the efficiency equation in this case we make these factors have equal efficiency.

For the Krasker-Welsch estimator the cut-off parameter c coincides with the bound on the sensitivity and should satisfy the inequality $c > \sqrt{p}$.

4. Computation of R^2 and F Statistics. The standard expressions for R^2, F, and other statistics theoretically justified in ordinary least squares are not valid for bounded-influence regression. The theory of these statistics for robust and bounded-influence regression is the subject of ongoing research (see, for example, Ronchetti

(1982)). Here we present certain preliminary formulas for R^2 and F statistics which are implemented in the BIF algorithm.

F statistics for testing the zero slope regression are computed from the final coefficients β and covariance matrix V as follows: $F = \frac{1}{p-1}\hat{\beta}_{p-1} V_{p-1}^{-1} \hat{\beta}_{p-1}$, where $\hat{\beta}_{p-1}$ is (p-1) vector of the last (p-1) components of $\hat{\beta}$, and V_{p-1} is (p-1)×(p-1) matrix obtained from V by deleting the first row and column.

R^2 and $\bar{R}^2$ statistics are computed as follows: $\bar{R}^2 = 1 - \sigma_p^2/\sigma_1^2$ where σ_p^2 is the final scale of the bounded-influence p-variable regression while σ_1^2 is the similar scale when only the intercept is robustly fitted with the cut-off parameter being chosen to achieve the p-variate regression efficiency at the central model. The R^2 statistic is computed from $\bar{R}^2$ as usual: $R^2 = 1 - (1-\bar{R}^2)\cdot(n-p)/(n-1)$.

5. Example. Bounded-influence regression generates considerably more output than OLS regression. Since the weights in such procedures depend upon residuals and (in some cases) distances, it is natural to try to summarize all of this information in one display or plot. This can be accomplished by plotting the absolute value of the scaled residuals $|r_i|/\hat{\sigma}$ on the vertical axis and the distances $d(x_i)$ on the horizontal axis. Contours of constant weight can be superimposed, giving a rough idea of the final weight assigned to each observation in order to bound the influence while at the same time providing information about residuals and distances.

In order to illustrate the use of these algorithms and plots we give some results based on an example described in Krasker and Welsch (1982). The data is a subset of that contained in McDonald and Ayers (1978) and consists of age-adjusted mortality in 60 U.S.M.A. (the response) and five explanatory variables: percent nonwhite (NONW), average years of education (EDUC), population per square mile (POPDEN), precipitation (RAIN), and the logarithm of the sulfur dioxide concentration (LOGSO 2)

In order to compare the effects of the various estimators, we decided to downweight the same fraction of the data for each. Huber regression was run with c=1.345, the approximate cut-off for 95 percent asymptotic efficiency. The resulting average weight was .948. We then found cut-offs for Schweppe (1.485) and Krasker-Welsch (4.2) that yielded the same average weight. For the Mallows estimator the two cut-off parameters were chosen to minimize the mean square error of the worst linear combination while yielding the same average weight (b=4.14, c=1.51).

The coefficient estimates and standard errors for all five regressions are given in Table 1. The most notable changes occur in NONW, POPDEN, and RAIN when least-squares is compared to the Krasker-Welsch estimator. In Table 2 we show the final weights for the four methods for cities where one or more methods gave $w_i<1$. The contour plots for Krasker-Welsch and Mallows, Figure 1a and 1b, show quite clearly four outlying cities: Lancaster (28), Miami (32), New Orleans (37), and York (59).

Acknowledgements. This work was supported in part by National Science Foundation Grants SOC79-08865 and MCS77-26902; Department of Energy Grant EX-76-A-01-2295; and a Grant from I.B.M. Corporation. Part of the programming for BIF was done by Stephen Peters and Leonard Roseman. We thank Alan Minkoff for helping us to produce contour plots, and Karen Glennon for typing this paper.

References:

Dongarra, J.J., Moler, C.B., Bunch, J.R., Stewart, G.W. (1979), Linpack User's Guide, *SIAM*, Philadelphia.

Hampel, F.R. (1974), "The Influence Curve and Its Role in Robust Estimation," *J. Amer. Statist. Assoc.*, 69, 383-394.

Handschin, E., Kohlas, J., Fiechter, A., Schweppe, F. (1975), "Bad Data Analysis for Power System State Estimation," *IEEE Transactions on Power Apparatus and Systems*, PAS-94, 2, 329-337.

Holland, P.W. and Welsch, R.E. (1977), "Robust Regression Using Iteratively Reweighted Least-Squares," *Communications in Statistics*, A6, 813-827.

Huber, P.J. (1981), *Robust Statistics*, Wiley, New York.

Krasker, W.S. and R.E. Welsch (1982), "Efficient Bounded-Influence Regression Estimation," to appear in *J. Amer. Statist. Assoc.*, September 1982.
Mallows, C.L. (1973), "Influence Functions," Talk at NBER Conference on Robust Regression, Cambridge, MA.
Marazzi, A. (1980), "ROBETH: A Subroutine Library for Robust Statistical Procedures; Compstat-80: Proceedings in Computational Statistics, 577-583, Physica-Verlag, Wien.
Maronna, R.A. and Bustos, O., Yohai, V.J. (1979), "Bias- and Efficiency-Robustness of General M-estimators for Regression with Random Carriers," in *Smoothing Techniques for Curve Estimation*, eds. T. Gasser and M. Rosenblatt, Lecture Notes in Mathematics No. 757, Berlin-Heidelberg-New York: Springer Verlag, 91-116.
McDonald, G.C. and Ayers, J.A. (1978), "Some Applications of the Chernoff Faces': A Technique for Graphically Representing Multivariate Data." In *Graphical Representation of Multivariate Data*, P.C.C. Wand (ed.), New York: Academic Press.
Ronchetti, E. (1982), "Robust Testing in Linear Models: The infinitesimal approach," unpublished abstract of Ph.D. thesis, ETH, Zurich.

Table 1. Coefficient Estimates

	NONW	EDUC	POPDEN	RAIN	LOGSO2	CONST	$\hat{\sigma}$
L.S.	3.350 (0.59)	-13.296 (6.969)	.00283 (0.00376)	1.637 (0.62)	13.778 (3.82)	930.28 (96.15)	36.39
Huber	2.839 (0.48)	-12.933 (5.26)	0.00395 (0.00283)	1.868 (0.46)	14.902 (2.88)	915.66 (72.50)	31.18
Schweppe	2.788 (0.52)	-13.338 (7.24)	0.00497 (0.00614)	1.888 (0.51)	14.376 (5.15)	918.15 (93.07)	30.27
Mallows	2.794 (0.46)	-13.996 (5.42)	0.00573 (0.00295)	1.870 (0.48)	13.844 (2.97)	925.13 (74.70)	30.25
Krasker-Welsch	2.600 (0.68)	-13.679 (6.12)	0.00714 (0.00469)	2.010 (0.44)	13.601 (4.10)	915.34 (80.99)	29.84

Table 2. Final Weights

	2 Albany	6 Birmingham	9 Buffalo	28 Lancaster	31 Memphis	32 Miami	35 Nashville	37 New Orleans	53 Toledo	57 Washington	59 York
Huber	0.54	1.00	0.80	0.41	1.00	0.56	0.97	0.35	0.85	0.88	0.55
Schweppe	0.58	1.00	0.88	0.41	1.00	0.50	1.00	0.33	0.89	0.91	0.42
Mallows	0.60	1.00	0.99	0.44	1.00	0.51	1.00	0.37	0.93	0.94	0.33
Krasker-Welsch	1.00	0.89	0.99	0.38	0.95	0.29	1.00	0.22	1.00	1.00	0.19

a)

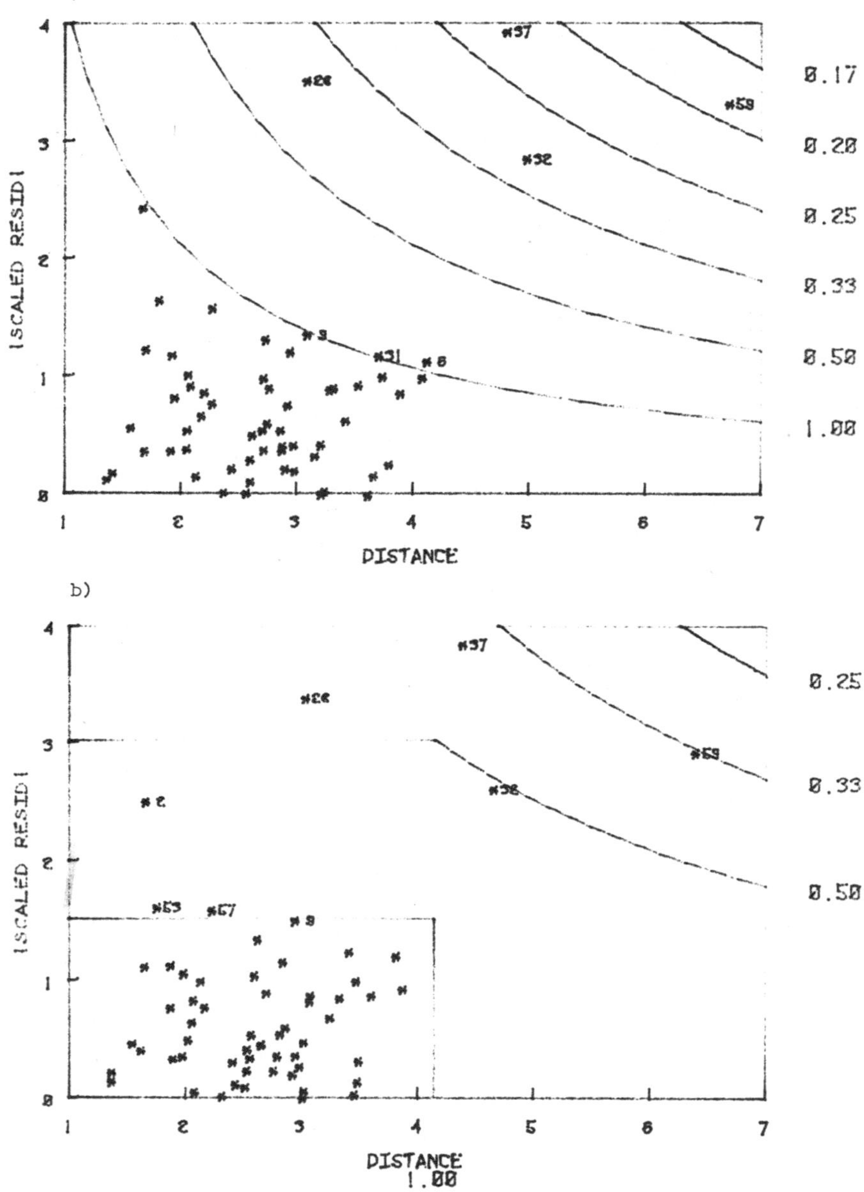

b)

Figure 1. Contour plots for a) Krasker-Welsch and b) Mallows estimators. A contour is a curve of constant weight whose value is denoted on the right end of the contour.

MUDFOLD: Multiple Stochastic Unidimensional Unfolding

W.H.v. Schuur and **I.W. Molenaar**, University of Groningen, The Netherlands

Summary

A unidimensional probabilistic unfolding scale consisting of a maximal subset of stimuli, given a 'pick k out of n' data collection or transformation, is found by an agglomerative search procedure. A coefficient of scalability is defined analoguous to Loevinger's H, on the basis of observed and expected frequency of the error pattern in all triples of the scale. From the triples not represented in the scale, a second etc. scale may be found on the same principles.

Keywords: multiple scaling, unidimensional unfolding, preference representation, unfolding, nonparametric scaling model.

Introduction

A unidimensional unfolding scale is defined as the numerical representation of a number of elements of two sets (often subjects and stimuli; from now on we will use these terms), in such a way that (preference) orderings by each subject of the stimuli are monotonely related to the distance between the subject point and the stimulus point on the dimension (e.g. Coombs, 1964).

In many empirical situations the deterministic unidimensional model does not hold for all elements in both sets. Generally, three strategies have been applied to overcome this problem (not necessarily mutually exclusive):

a) use a stochastic model (e.g. Bechtel, 1968, or Zinnes and Griggs, 1974);

b) use a multidimensional model (e.g. Carroll, 1972, or Heiser, 1981);

c) represent a maximal subset of stimuli in one dimension (e.g. Mokken, 1971 for an analoguous approach in cumulative scaling).

However, some assumptions underlying multidimensional models (e.g. all stimuli have a numerical value on each dimension, all dimensions are used independently, simultaneously and compensatorily) do not necessarily describe the cognitive process that underlies the formation of preference judgments.

In our opinion the first and third strategy can be well defended. Many idiosyncratic factors may play a role in the preference judgments of respondents, so that a joint representation of subjects and stimuli demands a stochastic approach. The third strategy has the advantage that the dimensional interpretation is not dependent upon rotation and that - especially in exploratory situations - a best fitting set of indicators can be found that contains the most prototypical stimuli for the underlying dimension.

Our approach in this paper is thus a stochastic unidimensional model for a maximal subset of stimuli and for all subjects.

A possible consecutive step was described by Van Schuur (1979) as the 'floater model'. Here each stimulus is representable in one of a number of disjoint sets (e.g. unidimensional unfolding scales), and the preference ordering follows a two-step procedure: a) for each set (scale) determine the most preferred stimulus; b) for all most preferred stimuli (one for each set) determine the overall most preferred stimulus.

Data

Each subject has to pick k out of n stimuli. At this stage k has to be the same for all subjects, missing data are not allowed and the values $2 \leq k \leq 5$ are implemented. According to a single peaked preference model, the stimuli picked should be adjacent on the underlying dimension. Hence a dichotomous data matrix, consisting of "1"'s (indicating that the subject mentions the stimulus among the k most preferred) and "0"'s (otherwise), can be permuted in such a way that the "1"'s of the response patterns form a parallellogram pattern.

Strategy of analysis

A best fitting maximal subset of stimuli will be found by adding the best fitting (p+1)st stimulus to a p-stimulus scale ($p \geq 3$), unless the fit is worse than some user defined lower boundary. The p-stimulus scale may be derived either from a user defined start set of from a best fitting triple of stimuli.

A fitting triple of stimuli will be called an 'elementary scale' as the response pattern to at least three stimuli is needed to falsify the model for the data. In a stochastic model, falsification does not merely imply 'counting errors', but 'comparing the number of errors observed to the number of errors expected under some model of statistical independence'. We therefore need to specify how to count errors, as well as how to calculate the number of errors expected under statistical independence, and finally compare these two in a measure of goodness of fit.

Definition of error

In a triple, in the correct order, only one of the eight possible response patterns (the pattern 101) falsifies the model. This pattern will be called an 'error pattern'. Only patterns in which non-adjacent stimuli are picked are in error. Hence, for instance neither the pattern 000, nor the pattern 010 is in error with respect to this triple. In the last case the (k-1) remaining picked stimuli need not be part of the same unfolding scale. But if they are, the middle stimulus of the 010 pattern will form one or more error patterns with the other picked stimuli, in other triples.

In a fourtuple, all possible response patterns with error are: 1010, 0101, 1001, 1101 and 1011. We consider the first two of these patterns as being less in error than the remaining three, as in the last three patterns both the first and last stimuli are picked, but one or two of the middle stimuli are not picked. If the four stimuli are denoted as A, B, C and D, the pattern 1010 implies that the triple ABC is in error. Similarly: 0101 (BCD); 1001 (ABD and ACD); 1101 (ACD and BCD); 1011 (ABC and ABD). Thus, the frequency of errors in an ordered fourtuple can be obtained by counting the errors in all four triples of the fourtuple in the order of the fourtuple. This definition can be generalized for a p-tuple ($4 < p \leq n$). Hence, for all $\binom{n}{3}$ possible triples in all three possible orderings (each of the three stimuli may be the middle one) the frequency of the error pattern is counted.

Expected number of errors for independence

The number of errors observed should be less than expected under a null model in which the subjects show no preference for picking adjacent stimuli. Such a model will be based on the observed stimulus popularities $p(i) = N_i/N$, where N_i of the N subjects have included stimulus i in their picked subset of k stimuli.

The simplest model, in which the expected error frequency of a triple (say ABC) would equal $p(A)(1-p(B))p(C)$ violates the condition that each subject picks exactly k stimuli. This is remedied by the following two-step procedure. First, a quasi-independence model known from the loglinear model theory (Bishop, Fienberg and Holland, 1975, sec. 3.5) is applied to find the expected frequency of all $\binom{n}{k}$ k-tuples: by iterative proportional fitting parameter values f(j) for j=1,2,...,n can be found such that the expected frequency $f(i_1).f(i_2)...f(i_k)$ of a k-tuple $(i_1,i_2,...,i_k)$ satisfies the marginal conditions.

Next, the expected frequency of the pattern 101 for a triple ABC, say, is simply obtained by adding expected frequencies for all k-tuples including A and C but not B. If e.g. k=4 and n=6, a 101 pattern for ABC implies that ACDE, ACDF or ACEF were picked.

Goodness of fit

As a measure of goodness of fit Loevinger's H (see Mokken, 1971) is used

$$H = 1 - \frac{E(obs)}{E(exp)} .$$

E(obs) and E(exp) are the observed and expected number of errors, as defined in the previous sections. In order to evaluate the representability of individual items in the scale, a goodness of fit measure for an individual item will be defined as:

$$H_i = 1 - \frac{E_i(\text{obs})}{E_i(\text{exp})} \quad ,$$

where E_i(obs) and E_i(exp) are the sum of all errors in patterns involving item i.

Finding the best triple

As the statistic H may be positive for an elementary scale ABC in all its three essentially different orders (e.g. because the items are too far away from each other, so that the patterns 100, 010 and 001 are overrepresented and the patterns 011, 101 and 110 are all underrepresented), we define the best fitting triple as a triple for which the ordering of the stimuli on the scale is unambiguous. That is, not only should H_{ABC} be larger than some user specified lower boundary, but also should H_{BAC} as well as H_{ACB} be negative.

The best fitting triple is that triple that satisfies the above condition, and that in addition allows the representation of most people on the basis of two or three adjacent most preferred stimuli on the scale (i.e. with response patterns 111, 110 and 011).

Adding the best (p+1)st stimulus to a scalable subset of p stimuli

It should be noted that there are no theoretical objections for a (p+1)st stimulus to be placed between the extreme stimuli of the p-tuple. So, for adding a stimulus X to the best triple ABC, not only XABC and ABCX are investigated, but also AXBC and ABCX.

For each of the n-p remaining stimuli, for each of the p+1 possible positions of the (p+1)st stimulus, we proceed as follows:

a) all new $\binom{p}{2}$ triples, consisting of the (p+1)st stimulus and 2 stimuli from the p-tuple should have a positive statistic H;
b) the value for the statistic H for the scale as a whole can be found by adding the sum of errors observed and the sum of errors expected from the new triples to the given appropriate values of the p-stimulus scale;
c) for each item we keep it at the best fitting position in the p-tuple;
d) we keep the stimulus with highest statistic H as the (p+1)st stimulus to extend the p-tuple, and try to add the next stimulus, unless H drops below a user defined lower boudary.

Generalization: rank k out of n

The definition of error in an ordered triple is the same as in the pick-situation, with the addition that patterns ACB and CAB also contradict the J-scale ABC.

In a larger scale some additional inconsistencies may arise because of metric information. For instance: in the J-scale ABCD either the triple CBA or the triple BCD is inadmissible, depending upon whether d_{AB} is larger or smaller than d_{CD}.

Observed and expected error frequencies can in principle be computed in a similar way as for the pick k out of n case. Even when the subjects were instructed to rank their k most preferred stimuli, however, there are reasons to apply the preceding algorithm for searching a maximal scalable subset by treating the data as pick data under neglection of the ranking. The fact that each k-tuple contains k! ordered k-tuples not only drastically increases the computer time involved in the search, but it also leads to small observed and expected error frequencies for many ordered k-tuples and thus to unreliable estimates of the H-coefficients, unless extremely large samples of subjects have been interrogated. The generalization to the rank case has therefore been given a low priority.

Using the scale as a measurement instrument

Once the stimuli are ordered on an underlying continuum, it is possible to assign (ordinal) scale values to respondents as the mean values of the rank scores of the stimuli preferred. Note that the respondents who do not prefer any of the stimuli on the scale cannot be represented. They do not get a scale value.

Once a scale is found, it is in principle possible to investigate whether it fullfills the stronger assumptions of a metric probabilistic model, for which interval scale values can be assigned (e.g. Sixtl, 1973).

Computer program

Under guidance of the authors, PASCAL programs have been written by A. Bruinsma and H. Guldenmond. A row- and column wise ordered N by k data matrix (for N respondents the rank numbers of the k most preferred stimuli) is taken as input to calculate the p(i)'s and to generate a dichotomous N by n data matrix. A run of 1159 respondents picking 3 out of 15 stimuli took 0.7 sec. on a CDC Cyber 170/760 computer. The p(i)'s are used as input to calculate the expected error frequencies. For 15 stimuli this took 48 iterations in 7.7 sec.

The dichotomous data matrix and the file containing the expected error frequencies are used to count the observed number of errors to calculate the H-value for all $\binom{n}{3}$ triples in all three orders, to determine the best triple and to add the best (p+1)st stimulus to the best p-tuple. For 1159 respondents and 15 stimuli a run took 14.4 sec. on a CDC Cyber (most of this time goes into counting the observed errors for all triples).

Applications

Interpretable results have been found for the preferences expressed by Dutch Members of Parliament for the political parties in Parliament. The method has also been used to find unfolding scales for 'importance of political problems'. Results of this will be discussed in the first author's forthcoming thesis (Van Schuur, forthcoming).

Literature

Bechtel, G.G. (1968), Folded and unfolded scaling from preferential paired comparisons, Journal of Mathematical Psychology, 5, 333-357.

Bishop, Y.M.M., S.E. Fienberg & P.W. Holland (1975), Discrete multivariate analysis, Theory and Practice, MIT, Cambridge Mass.

Carroll, J.D. (1972), Individual differences and multidimensional scaling, in: R.N. Shepard, A.K. Romney & S.B. Nerlove (eds.), Multidimensional scaling, Theory and applications in the behavioral sciences, Vol. I, Theory Seminar Press, New York, 114-155.

Coombs, C.H. (1964), A theory of data, Wiley: New York.

Heiser, W.J. (1981), Unfolding analysis of proximity data, dissertation, Leyden University.

Mokken, R.J. (1971), A theory and procedure of scale analysis, Mouton: Den Haag.

Van Schuur, W.H. (1979), Perception and evaluation of political problems, with an application of multiple scaling, paper prepared for presentation at the XIth World Congress of I.P.S.A. in Moscow, 1979.

Van Schuur, W.H. (forthcoming) Perception and evaluation of political problems, with an application of multiple scaling (thesis).

Sixtl, F. (1973), Probabilistic unfolding, Psychometrika, 38, 235-248.

Zinnes, J.L. & Griggs, R.A. (1974), Probabilistic multidimensional unfolding analysis, Psychometrika, 39, 327-350.

The Estimation of Missing Values in Highly Correlated Data

R.C. Tabony, Meteorological Office, Bracknell, G.B.

Summary

Various methods of estimating missing values in highly correlated climatological data are examined. Any generalised method is likely to be based on a correlation matrix, but the incompleteness of the data introduces problems with this approach. These are illustrated by program BMDPAM of the BMDP suite, which produces estimates worse than those using traditional methods based on single station comparisons. Principal component analysis is considered likely to be the best statistical tool for estimating missing values among highly correlated data. The high quality correlation matrix required as input can be obtained by using a simple estimating procedure to produce a preliminary set of complete data. A simple technique was devised for climatological data in the UK, and was found to give estimates similar to those obtained from an eigenvector scheme used for quality control purposes.

Key Words

Missing data, high correlations, eigenvectors, climatology.

1. Introduction

The problem of missing data is very common and a simple and effective procedure for eliminating it would be of considerable practical value.

The data analysed in this paper are monthly means and extremes of temperature and sunshine for climatological stations in the UK during the period 1959 to 1979. These data are highly correlated - correlation coefficients between neighbouring stations average 0.94 and range between about 0.80 and 0.995. In common with many other forms of data, climatological observations display a pronounced seasonal variation, and so all the data cannot be considered to belong to just one population. The proportion of missing data is of the order of 30%, and a number of station records do not overlap.

The BMDP suite of statistical software described by Dixon and Brown (1979) contains a program BMDPAM whose functions include the elimination of missing data. This contains a number of options which were thoroughly tested, but all produced estimates which were worse than those obtained from traditional methods based on single station comparisons. A simple technique based on linear regression was therefore devised and tested.

2. Review of techniques

2.1 Problems associated with generalised methods based on the correlation matrix

A general approach to the estimation of missing data will commonly start from the correlation matrix, but this introduces the following difficulties.

(i) If two station records do not overlap, the correlation between them has to be estimated. In the subsequent smoothing designed to obtain a positive semi-definite correlation matrix, any errors in the estimated correlations are shared amongst the others.

(ii) Selection of the 'best' predictor variables should depend not only on the inter-correlations between the variables, but also on the length of the overlap between them. The highest sample correlations are likely to be associated with variables with only a short period of overlap.

(iii) In some statistical packages, the correlation matrix is constructed from standardised anomalies based on means and standard deviations derived from all available data for each variable (climatological station). The mean and standard deviation for each variable will therefore be based on disparate cases (years) and, for climatological data at least, the computed correlations will tend to be too low.

2.2 Total v Partial Correlation

One of the simplest ways of estimating missing observations is to express them as a linear combination of observations from neighbouring stations. A basic problem is to decide whether the weights associated with the neighbours should be determined on the basis of partial or total correlation (Both stepwise and multiple linear regression are based on partial correlations).

An observation at a given station may be expressed as

linear relation x observation at a neighbour + systematic (geographic) difference + random difference.

The linear relation with the first neighbour will be taken into account in both the partial and total correlation approaches. Stepwise regression, by representing each significant systematic difference by only one station, fails to reduce random errors. By taking into account observations from a large number of neighbours, the total correlation approach is able to reduce random errors, but is unable to cope with systematic differences. The behaviour of multiple linear regression depends on the inter-correlations between the variables supplied; if these are low, the results will be similar to those obtained from stepwise regression, while if they are high, the expected values of the estimates will be similar to those derived from total correlations.

When highly correlated data are analysed using multiple linear regression, the equations used to determine the regression coefficients (weights) associated with each station are ill-conditioned, and the coefficients produced have large standard errors. Ridge regression is a technique designed to overcome this problem, and is well described by Meisner (1979). It is equivalent to dividing the off-diagonal elements of a correlation matrix by 1+K, where K is known as the ridge parameter. The stability of the regression coefficients, however, is obtained at the expense of a loss of variance in the final estimates.

When the inter-correlations between the variables are as high as for climatological data in the UK, neighbours selected on the basis of their total correlation with a test station are likely to yield better estimates than those based on partial correlation; stepwise regression is likely to produce estimated based on only one neighbour, while a combination of multiple linear and ridge regression leads to a loss of variance.

2.3 Optimum and linear interpolation

Optimum and linear interpolation are well described by Gandin (1970). In linear interpolation, estimates at a station are obtained by fitting a plane to observations from the surrounding stations. Hopkins (1977) applied this technique to temperature and sunshine data in East Anglia, where the uniform terrain enabled the required conditions to be most nearly satisfied. These results are compared with those obtained by other methods in section 5.

2.4 Principal component analysis

Principal component or eigenvector analysis is fully described by Kendall (1975). The leading eigenvectors will represent systematic differences between the stations while the random differences are consigned to higher order components. By reconstituting the data from only the leading eigenvectors, therefore, the genuine differences between stations are retained while the noise is ignored. As thus described, the technique seems an ideal means of estimating missing values. The disadvantage is that a complete and positive semi-definite correlation matrix is required so, if the data are incomplete, the problems outlined in section 2.1 are encountered.

These problems can probably be overcome by using a maximum likelihood routine to estimate any missing correlations. Alternatively, a simple estimating procedure (such as linear regression against the most highly correlated variable) may be used to obtain a preliminary set of complete data. The correlation matrix derived from this can then be submitted to principal component analysis which is used to obtain a more refined set of estimates.

This procedure is here recommended as the best general approach to the estimation of missing values from highly correlated data. A statistically simple technique based on total correlations, and described in section 4, was designed to act as a first step in this procedure. The estimates obtained, however, were found to be similar to those obtained from an eigenvector scheme used for the quality control of daily climatological data, so their further refinement seemed unnecessary.

3. Estimating using BMDPAM

3.1 The options

The main options are

(i) SIMPLE. Linear regression against the mostly highly correlated variable.

(ii) REGR. Multiple linear regression.

(iii) STEP. Stepwise regression.

(iv) TWOSTEP. Stepwise regression limited to a maximum of 2 variables.

All the options may be combined with ridge regression, while the level of significance used to retain or reject variables in stepwise regression may be varied by use of an 'F to enter' criterion.

3.2 Testing of 1979 version

The 1979 version of BMDPAM was tested on minimum temperatures in January at Corwen in North Wales from 1959 to 1979. The program was supplied with data from 50 neighbouring stations in which observations from the test station were restricted to different periods of years. The results were very poor with no single option emerging as significantly better than the remainder. RMS errors were of the order of 3.3°C, and were accompanied by a considerable loss of variance.

The estimates obtained from BMDPAM were poor mainly because they were derived from a correlation matrix which suffered from the problems described in section 2.1. Option SIMPLE has two major faults:-

(i) In selecting the best neighbour, it takes no account of the length of overlap between stations.

(ii) If two stations do not overlap, the correlation between them is assigned a default value of zero. This introduces inconsistencies into the correlation matrix and when these are smoothed out, a considerable lowering of the correlations takes place. When observations are standardised, the gradient of the linear regression between two variables is equal to the correlation between them. BMDPAM makes use of this relation, and the reduced value of the correlation is reflected in a decreased slope of the regression line. It is this procedure which leads to the loss of variance which is such a feature of the estimates produced by BMDPAM.

Next consider STEP and TWOSTEP. For the UK climatological network, the first neighbour would commonly account for such a large proportion of the variance that none of the other variables would have partial correlations which were significantly different from zero. In BMDPAM, however, the fit of the first neighbour is not good, and there is more residual variance then there ought to be. In addition, the F-to-enter criterion is not adjusted to take account of the amount of missing data. Thus some neighbours with high partial correlations (arising by chance from the brevity of the overlaps) are erroneously accepted as predictor variables.

3.3 Maximum likelihood estimation of missing correlations (1981 version)

In the 1981 release, (Dixon, 1981), a great improvement is effected by using a maximum likelihood routine to estimate missing correlations. The correlations, however, are calculated using means and standard deviations based on all available data for each station and, for climatological data, this results in correlations which are too low. Thus although the maximum likelihood routine is capable of producing estimates of missing correlations which are of the same general level as those present, in BMDPAM its effectiveness in producing unbiased values is compromised by the procedure used to calculate the correlations.

4. Traditional and proposed techniques

The traditional (UK) method of estimating missing climatological values involves assuming a constant difference between stations for temperature and a constant ratio for sunshine. The main points of the proposed technique are as follows:

(i) The 12 'best' neighbours are selected on the basis of their (total) correlation with the test station after adjustments have been made according to the length of overlap.

(ii) For every month of the year, data from each neighbour are regressed against that for the test station. The slope of the regression is set equal to the ratio of the standard deviations (ie. the moderating effect of correlation is ignored).

(iii) To ensure a regular seasonal variation in the relations between stations, the regression coefficients are smoothed over a number of months.

(iv) Estimates for the test station are made separately from each neighbour. These are combined to form a final estimate by attaching a weight $1/i$ to the ith ranking neighbour.

5. Comparison of estimates obtained from various techniques

For the stations and climatic elements examined, RMS errors of estimates obtained from option SIMPLE of BMDPAM were around twice as large as those obtained from the traditional method, which in turn were 40% larger than those associated with the proposed technique. Estimates derived from the proposed technique were similar to those obtained from principal component analysis (Spackman, 1980), and, in East Anglia, from linear interpolation (Hopkins, 1977).

6. Conclusions

It is very difficult to write a generalised program to estimate missing data. An obvious starting point is the correlation matrix, but difficulties are created by variations in the length of overlap between variables, leading in extreme cases to missing correlations. In climatology, the seasonal variation also makes it difficult for all the data to be taken into account. The program BMDPAM suffered from all these difficulties. The statistical techniques made available were not very suited to highly correlated data, and the program was not designed to cope with large amounts of missing data from a small number of cases.

It is suggested that the best general technique for estimating missing values amongst a set of highly correlated variables will be based on principal component analysis. The high quality correlation matrix required can probably be obtained from the sample correlations combined with a maximum likelihood routine to estimate the missing correlations. Alternatively, a simple estimating procedure (such as linear regression against the most highly correlated variable) may be used to produce a preliminarly set of complete data, and the correlation matrix may be computed from that.

A statistically simple technique, **relying** on the availability of a large number of highly correlated variables, was devised to produce estimates of climatological data in the UK. It was found to give similar results to an eigenvector scheme used for quality control purposes, so further refinement of the estimates seemed unnecessary.

Acknowledgment

The BMDP suite of programs were developed at the Health Sciences Computing Facility, UCLA, which was sponsored by NIH Special Research Resources Grant RR3.

References

Dixon, W.J. (1981), BMDP Statistical software 1981. Univ. California Press, Berkeley.

Dixon, W.J. and Brown, M.B. (1979), BMDP-79. Biomedical Computer Programs. P series. Univ. California Press, Berkeley.

Gandin, L. S. (1970), The planning of meteorological station networks. WMO Tech Note 111, TP149, No 265.

Hopkins, J.S. (1977), The spatial variability of temperature and sunshine over uniform terrain. Met Mag., 106, 278-292.

Kendall, M.G. (1975), Multivariate Analysis. Charles Griffin & Co., London.

Meisner, B.N. (1979), Ridge regression - time extrapolation applied to Hawaiian rainfall normals. J. App. Met., 18, 904-912.

Spackman, E.A. (1980), Areal quality control of daily climatological data using station factor scores. Met O 3 Tech Note No 6, Meteorological Office, Bracknell.

Relational Design of Statistical Databases

A.M. Tjoa, University of Vienna, Austria
R.R. Wagner, University of Linz, Austria

Summary:

A design concept for statistical databases based on the relational model is introduced. The external view of the end-user is the universal relation. Therefore it is possible for the end-user to formulate queries solely by specifications of attributes stored in the database. The conceptional model for the database administrator is the extended relational model RM/T which is developed by Codd in 1979.

Keywords: statistical databases, query language, relational model

INTRODUCTION

Evaluating the data from different fields of research e.g. from planned experiments in biology, agriculture or medical sciences -it always appears to be necessary to develop software-support which is user-oriented and also easy to handle for the user. The main task of this software-support is the manipulation of stored data. Therefore a user-oriented data manipulation language which is dialogue-supported (attribute-oriented language) is developed in Tjoa-Wagner(1979a). This data manipulation language is based on the relational model which was introduced by Codd(1970) with some extensions which make it possible for the user to formulate queries solely by specifications of attributes stored in the database. In this language it is not necessary to know in which normalized relation an attribute is contained. For this purpose an algorithm for the automatical construction of relational joins is used for the preparation of queries (see also Tjoa-Wagner (1979b)).

A main task of this paper is the introduction of a design method for statistical databases which allows the formulation of queries in attribute-oriented languages. Another aim of the paper is the introduction of the statistical security management facilities in the environment of a data base system with the properties described above. Although it is proposed that only aggregate information be revealed to users, this does not imply that there is only aggregate information

in the database. On the contrary, the conceptual model of the statistical data base should be similar to the conceptual model of any other general-purpose database. This requirement is necessary since for some users (e.g. the database administrator) the statistical database is a normal database and these users should have access not only to aggregate informations but also to all information stored in the database. On the other hand to enforce integrity and validity in the database it is also necessary to have total information about the environment. A first step such a statistical database is given in Chin-Ozsoyoglu (1981).
In the conceptual model in an own part statistical constraints are defined. In these constraints e.g. the properties for which statistical information is to revealed are defined; the allowable types of statistical information (COUNT,MEAN,MEDIAN,MAX,MIN etc.) can be specified.

THE DATABASE ARCHITECTURE

To obtain physical and logical data-independence a modern statistical database must be based on a three level database architecture. For the special purpose of attribute oriented languages the adequate model for the external level is a relational model based on the universal relation. The user view is a universal relation resp. a projection of the universal relation. The conceptual model of the database design proposed in this paper is the RM/T-model introduced by Codd (1979) with some extensions concerning statistical constraints.

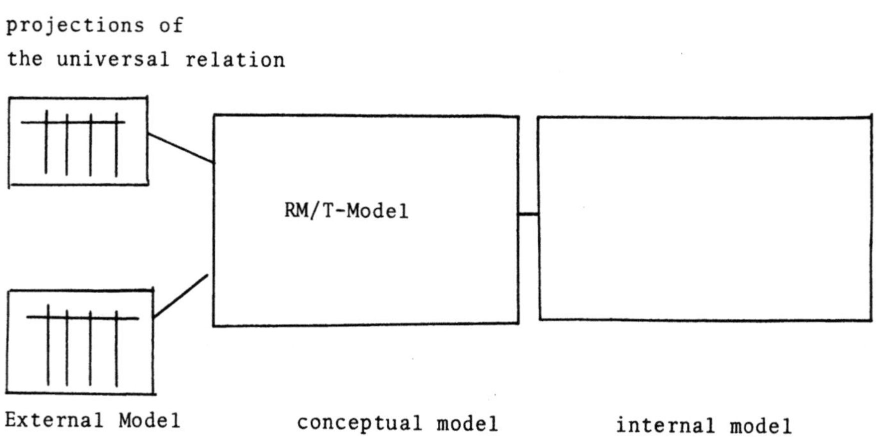

Fig. 1: Database Architecture

THE EXTERNAL MODEL

Given a set of relational schemas $RS_1(X_1;E_1),\dots,RS_n(X_n;E_n)$ where $X_1,\dots,X_n$ are attribute sets and $E_1,\dots,E_n$ are the set of constraints (e.g. functional dependencies, multivalued dependencies, first order hierarchical dependencies). The universal relational schema:

$$RS_U(ATR;E_U)$$

is defined by:

$$ATR = \bigcup_{i=1}^{n} X_i$$

$$E_u = \bigcup_{i=1}^{n} E_n$$

Given a set of instances $R_1 \in RS_1(X_1;E_1),\dots,R_n \in RS_n(X_n;E_n)$; there exists a universal relation instance $R_U \in RS_U(ATR;E_U)$ with the projections $R_U[X_i]=R_i(X_i)$ for $i=1,\dots,n$. (see also Beeri et al (1978))

The universal relation makes it possible for the end user to specify his queries by means of only one relation which contains all relevant attributes.

CONCEPTUAL MODEL

The conceptual model proposed in this paper is based on RM/T. The great advantage of this concept is given by the possibility to use ISA-hierarchies which are very important for statistical data bases.

By means of the property relation in RM/T one can formulate statistical constraints for subtypes.

Therefore the PG-relation in RM/T must be extended by an authorization attribute.

Example:

ISA-Hierarchy:

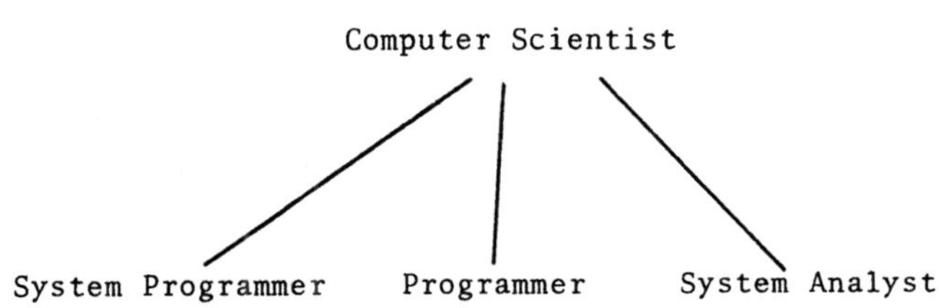

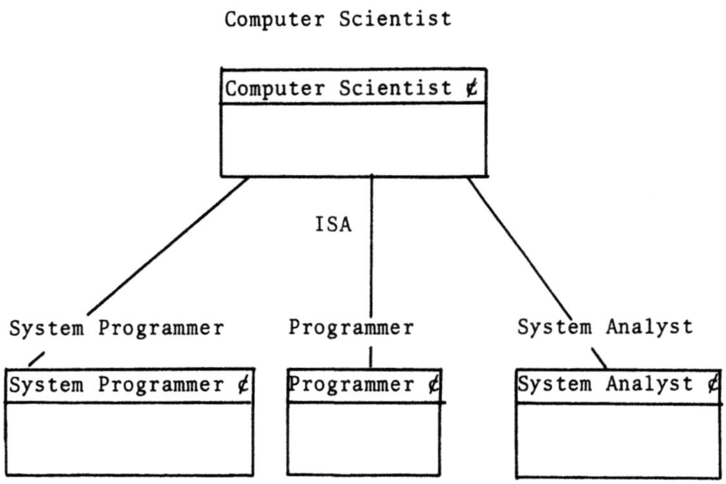

Property Relations:

P1	Programmer ¢	Salary

P2	Programmer ¢	Number of Programms writtem in 1982

Extended PG-relations:

SUP	SUB	User Group	Authorization
Programmer	P1	1	Max
Programmer	P2	2	Count Mean
Programmer	P2	1	No
Programmer	P2	2	Mean

It should be emphasized that although the RM/T model is used it is possible to held the surrogate-attributes of the E-domain to be hidden for the user.

Furthermore it should be mentioned, that the use of P-relations and the PG-relations in the conceptual model is important since as it is treated in Tjoa-Wagner(1980b) P-relations resp. PG-relations can easily be transformed into so called transposed files.

CONCLUDING REMARKS

The practical relevance of the described statistical database was studied in a pilot-project. In the pilot-project the system RELORD Tjoa et al 1979b was implemented for medical data (carcinoma of the oral cavity). In this test data-database-system the internal model is realized by using so called catalogue-relations i.e. the relations stored in the database are described by relations. For the physical realization B-trees are used. Since the RELORD-system was implemented in PL/1, the file management system was supported by REGIONAL organizations.

REFERENCES

Beeri C.,Bernstein P.A.,Goodman N(1978),A sophisticate's Introduction to database normalization theory,
Proc. Very Large Databases,Berlin.

Chin F.Y., Oszsoyoglu G. (1981),Statistical database design,
ACM Transactions on Database Systems, 6,1.

Codd E.F. (1970), A relational model for large shared data banks,
CACM 20

Codd E.F. (1979), Extending the database relational model to capture more meaning, ACM-Transactions on Database Systems, 4,4.

Tjoa A M., Wagner R.R. (1979a), Some more dependencies on the relational model and their relevance to queries on relational databases, Proc. EURO IFIP Conference, North-Holland, Amsterdam.

Tjoa A M., Wagner R.R., Ferr H.,Perndl C., Schmittner G., RELORD-Database management system, Technical Report,Univ.Linz.

Tjoa A M., Wagner R.R. (1980a), Some considerations on the relevance of relational data-manipulation-languages for medical research, Proc. Medinfo 80, IFIP, North-Holland, Amsterdam.

Tjoa A M., Wagner R.R. (1980b), A general concept for the simulation of interactions on statistical data bases, Proc. Computational Statistics COMPSTAT 80, Physika Verlag.

A Forward Multivariate Regression Procedure Based on the Maximization of the RV Coefficient

M. Vallée, Dominion Engineering, Montréal, Canada
P. Robert, Université de Montréal, Canada

SUMMARY.

The RV Coefficient is shown to be an acceptable criterion for variable selection in multivariate linear regression. An efficient computational procedure for forward selection based on this criterion is detailed. Finally, an example is given.

Keywords: RV coefficient, multivariate regression, forward regression.

1. The RV coefficient

Let X be a n×p data matrix of values of p explicative variables measured on n individuals and Y a n×q matrix of values of q variables to be predicted for the same n individuals. We assume X and Y to be centered, i.e. $1'X = 0'$ and $1'Y = 0'$ where $1' = (1\ 1 \ldots 1)$.

The matrix X induces a configuration C(X) of n points in $\mathbb{R}^p$ whose relative distances are characterized by the matrix $XX'/\{tr(XX')^2\}^{\frac{1}{2}}$, in which the denominator is introduced to make this characteristic matrix independent of a global change of scale. Using the matrix scalar product tr(A'B) and its induced norm ($\{trA'A\}^{\frac{1}{2}}$), a distance between the two such configurations C(X) and C(Y) can be defined as the distance between their characteristic matrices:

$$dist[C(X),C(Y)] = \{2-2RV(X,Y)\}^{\frac{1}{2}}$$

where

$$RV(X,Y) = \frac{tr\ XX'YY'}{\{tr(XX')^2 \cdot tr(YY')^2\}^{\frac{1}{2}}} .$$

A detailed account of this approach is to be found in Robert & Escoufier (1976). The sample RV coefficient is shown to be a measure of similarity between two multidimensional configurations of points which has the usual properties of a squared correlation coefficient (dist[C(X),C(Y)] is a decreasing function of RV(X,Y); RV(X,Y) = RV(Y,X); $0 \le R(X,Y) \le 1$; RV(XA,Y) = RV(X,Y) when A is orthogonal; RV(X,Y) = 0 iff X'Y = 0).

2. Multivariate regression using the RV coefficient

From the point of view of the RV coefficient, fitting the general linear model $Y = XB + E$ by multivariate regression is seen as the minimization over linear transformations B of the distance between the configurations $C(XB)$ and $C(Y)$ or equivalently, the maximization of

$$RV(XB,Y) = \frac{\text{tr } B'X'YY'XB}{\{\text{tr}(B'X'XB)^2 \cdot \text{tr}(Y'Y)^2\}^{\frac{1}{2}}}$$

Since we will like to partition $B'X'XB$ into uncorrelated blocks, we will require uncorrolated new variables. The problem is then to maximize $RV(XB,Y)$ under the constraint $B'X'XB = D$, where D is an unspecified diagonal matrix. Note that the maximal value of $RV(XB,Y)$ is not altered by this constraint since for a given B_0, there always exists an orthogonal matrix A such that $RV(XB_0A,Y) = RV(XB_0,Y)$ and $A'B_0'X'XB_0A$ is diagonal.

Using a diagonal matrix M of Lagrange multipliers it can be shown that the maximum is attained for any $\hat{B}$ and M which satisfy

$$X'YY'X\hat{B} = X'X\hat{B}M \quad \text{and} \quad \hat{B}'X'X\hat{B} = D \ .$$

Then $RV(XB,Y) = \{\text{tr}DM/\text{tr}D^2 \cdot \text{tr}(Y'Y)^2\}^{\frac{1}{2}}$, which is maximum when $D = M$ and $\hat{B}$ is any solution of the generalized eigenvalue problem $X'YY'XB = X'XBM$.

$$\max_B RV(XB,Y) = RV(X\hat{B},Y) = \{\text{tr}M^2/\text{tr}(Y'Y)^2\}^{\frac{1}{2}} = \{\text{tr}\underline{SSR}^2/\text{tr}\underline{SST}^2\}^{\frac{1}{2}}$$

where the diagonal element of $\underline{SSR} = B'X'XB = M$ and $\underline{SST} = Y'Y$ are sums of squares due to the regression and total sums of squares, respectively. Since the eigenvalues of the above problem are the same as those of the problem $Y'X(X'X)^-X'YW = WM$, we have

$$RV(X\hat{B},Y) = \{\text{tr}(Y'X(X'X)^-X'Y)^2/\text{tr}(Y'Y)^2\}^{\frac{1}{2}} ,$$

of form independent of $\hat{B}$.

It can be seen that the usual projection of Y on the column space of X, $\hat{B}_0 = (X'X)^-X'Y$ is a particular solution. It can also be seen that when Y is unidimensional, $RV(X\hat{B},Y)$ is equal to the multiple correlation coefficient.

3. The forward regression algorithm

The aim of any regression algorithm is to determine an "optimal" subset of (say r) variables from X to be retained as regressors. Let Z be the $n\times r$ matrix of values of the retained regressors and $\bar{Z}$ be the $n\times(p-r)$ matrix of values of the left out variables; then for some permutation matrix P, $X = (Z\ \bar{Z})P$.

Using the RV coefficient as the optimality criterion, one would seek the submatrix Z of X such that the solution $\{\hat{B},M\}$ to the induced generalized eigenvalue problem $Z'YY'ZB = Z'ZBM$ maximizes the RV coefficient $RV(Z\hat{B},Y)$ subject to $\hat{B}'Z'Z\hat{B} = M$. From the above, it is verified that Z maximizes $\text{tr}(Y'Z(Z'Z)^-Z'Y)^2$.

Because of the impractibility of direct determination of the optimal submatrix Z, forward, backward or stepwise selection procedures have been developped. We only describe here what was found to be an efficient algorithm for the forward method.

Assume r variables have been selected. Without loss of generality, assume that the corresponding matrix Z is formed by the first r columns of X and that $Z'Z$ is non-singular (any forward method will stop if addition of a new regressor would make $Z'Z$ singular). Assume that the first r steps of the Choleski decomposition (Noble, 1969) to have been performed on $S = (X\ Y)'X$ to express S as

$$S = \begin{pmatrix} Z'Z & Z'\bar{Z} \\ \bar{Z}'Z & \bar{Z}'\bar{Z} \\ Y'Z & Y'\bar{Z} \end{pmatrix} = \begin{pmatrix} R_r \\ Q_r \\ L_r \end{pmatrix} (R_r' Q_r') + \begin{pmatrix} 0 & 0 \\ 0 & C_r \\ 0 & E_r \end{pmatrix}$$

where R_r is a $(r\times r)$ full rank lower triangular matrix. It can be shown that the RV criterion $RV(ZB,Y)$ is equal to $\{tr(L_r'L_r)^2/tr(Y'Y)^2\}^{\frac{1}{2}}$. Thus the best next variable to be included in the regression is the one that will maximize the increment to $tr(L_r'L_r)^2$. As shown below, we shall also make use of the matrix $H_r = L_r'E_r$.

Now, let $\bar{Z} = (z\ Z^*)$, where z is the vector of the n observations for the next variable to be included. Performing another step of the Choleski decomposition will write S as above but in terms of R_{r+1}, Q_{r+1}, E_{r+1} (the latter being $q\times(p-r-1)$). When C_r, E_r and H_r are partitioned as

$$C_r = \begin{pmatrix} k^2 & kv' \\ kv & c \end{pmatrix}, \quad E_r = (ku\ E), \quad H_r = (h\ H),$$

where k is a scalar, v, u, h are respectively vectors of dimension $(p-r-1)$, q and r, the updating of L, C, E and H are found to be

$$L_{r+1} = (L_r\ u) \qquad C_{r+1} = C - vv'$$

$$E_{r+1} = E - uv' \qquad H_{r+1}' = (H_r' - k^{-1}vh'\ \ E_{r+1}'\ u)\ .$$

Denote by A^j the j-th column of a matrix A. Then the index of the best variable to be added turns out to be the one that maximizes, over $j = 1, \dots (p-r)$,

$$(2\|H_r^j\|^2 + \|E_r^j\|^2/c_j)\ ,$$

where c_j is the j-th diagonal element of C_r and $\|.\|$ denotes the Euclidion norm. To initialize the procedure, from $r = 0$, use $C = X'X$, $E = Y'X$ and enter in the regression the j-th variable for which $\|E^j\|^2/c_j$ is maximum over $j = 1, \dots, p$.

In terms of the given partitions, the least square estimator of B

is $\hat{B}_0 = (LR^{-1})'$; the multiple correlation coefficient between the i-th variable defining Y and the chosen regressors is $\|(L')^i\|^2/\|Y^i\|^2$; a test of the significance level of the last variable chosen will use the matrix of squares due to this variable:

$$\underline{SSH} = \underline{SSE}_r - \underline{SSE}_{r+1} = L_{r+1} L_{r+1} - L_r L_r' = uu'.$$

The algorithm requires $O(p^2q) + O(p^3)$ elementary arithmetic operations to enter all p variables. With $q \le p$, this is less than the cost of inverting a $p{\times}p$ matrix.

This forward, as well as the backward and stepwise procedures have been implemented on a CDC-173, using standard FORTRAN, with efficient management of row and column indices and zero-treshold values. Details can be obtained from the authors.

4. An example

Purposely, the example presented here involves a small number of variables and can be treated rather easily with classical methods of data exploration. The reader should keep in mind the increased complexity that would be met in dealing with a larger number of variables.

We use data from W.G. Waltz et alt. (1948) on nicotine in cured bright tobacco as related to mineral element composition. The sample consists of 25 units on which nine measurements were taken: three regressand variables (Y_1 = consumability index of a cigarette, Y_2 = % sugar and Y_3 = % nicotine in a tobacco leaf) and six regressors which are the percentage contents of six minerals in a tobacco leaf (X_1 = % nitrogen, X_2 = % chloride, X_3 = % potassium, X_4 = % phosphorus, X_5 = % calcium, X_6 = % magnesium).

A look at the corrolation matrix (Table 1) shows that variable X_4, is, globally, the least correlated with Y and is a candidate for elimination. The fact that the variables in the set $\{X_1, X_3, X_5, X_6\}$ exhibit non-negligible pairwise correlations suggests possible elimination within this subset.

	Y_1	Y_2	Y_3	X_1	X_2	X_3	X_4	X_5
Y_2	-.32							
Y_3	.22	-.70						
X_1	.23	-.71	-.77					
X_2	-.62	.43	-.27	.09				
X_3	.49	.19	-.29	-.01	-.09			
X_4	-.32	.24	-.05	-.11	.07	-.21		
X_5	-.12	-.48	.69	.63	.14	-.59	.01	
X_6	-.31	-.53	.73	.60	.12	-.61	.15	.76

Table 1. Correlation matrix

A principal component analysis of X shows that the two smallest eigenvalues are, in magnitude, less than .005 and account together for less than 1% of the trace of $X'X$. Thus, dropping two variables could remove this apparent and undesirable colinearity in X.

Stepwise multiple regressions for each component of Y will give the results shown in Table 2. If the level of significance of an F-to-enter is set at .10, it is found that all three regressions will stop with only three variables entered:

- for Y_1 : X_2 X_3 X_5 (X_6) in that order,
- for Y_2 : X_1 X_2 X_3 (X_4)
- for Y_3 : X_1 X_6 X_2 (X_5)

where the fourth variables to enter are indicated in parentheses. This suggest keeping the subsets

$$\{X_1, X_2, X_3\} \text{ or } \{X_1, X_2, X_6\} .$$

Regressand	Regressors entered with value of R^2					
	1st	2nd	3rd	4th	5th	6th
Y_1	X_2	X_3	X_5	X_6	X_4	X_1
	.39	.57	.64	.68	.69	.70
Y_2	X_1	X_2	X_3	X_4	X_6	X_5
	.50	.63	.68	.72	.73	.73
Y_3	X_1	X_6	X_2	X_5	X_3	X_4
	.59	.71	.79	.80	.80	.80

Table 2. Results of stepwise multiple regressions with values of squared multiple correlations R^2.

A study of multiple correlations (not shown here) among X_1, X_3, X_5 and X_6 suggests that X_1 and X_3 best represent this subset of four variables, thus pleading for $\{X_1, X_2, X_3\}$. On the other hand, Table 1 shows that X_1 and X_6 have higher linear relationships with Y. In the absence of other considerations, such as cost or technical difficulties of certain measurements, one would likely retain the subset $\{X_1, X_2, X_6\}$, with X_4 as a possible fourth regressor.

The proposed forward algorithm was applied to this data. As shown in Table 3, the variables were selected in the order $X_1, X_2, X_6, X_4, X_3, X_5$. The successive values of the RV coefficient indicate that three regressors are sufficient; they are precisely $\{X_1, X_2, X_6\}$.

Table 4 shows, for the four subsets $\{X_1, X_2, X_6\}$, $\{X_1, X_2, X_3\}$, $\{X_1, X_2, X_6, X_4\}$ and $\{X_1, X_2, X_3, X_4\}$ the values of the resulting RV coefficients and the values of R^2, the squared multiple correlations, with Y_1, Y_2 and Y_3. These values are to be compared with the optimal values of R^2 for subsets with the same number of regressors, as given in Table 2.

This example and others worked out by the authors show that the Rv-coefficient as used in the proposed forward algorithm is efficient in selecting good subsets of explanatory variables in multivariate regressions.

Regressors (in order of entry)	RV value	% of Max. RV value
X_1	.52	68%
$X_1\ X_2$	.65	85%
$X_1\ X_2\ X_6$	.72	94%
$X_1\ X_2\ X_6\ X_4$	.763	99.3%
$X_1\ X_2\ X_6\ X_4\ X_3$	.765	99.6%
$X_1\ X_2\ X_6\ X_4\ X_3\ X_5$	.768	100%

Table 3. Summary of foward selection using the RV coefficient (as per the proposed algorithm)

Selected variables	RV	Values of R^2 with Y_1	Y_2	Y_3
$X_1\ X_2\ X_6$	.72	.61	.68	.79
$X_1\ X_2\ X_3$	.70	.61	.68	.73
$X_1\ X_2\ X_6\ X_4$	.76	.63	.72	.79
$X_1\ X_2\ X_3\ X_4$	.74	.64	.72	.73

Table 4. RV and R^2 values for selected subsets of regressors.

References

Noble B. (1969), Applied Linear Algebra, Prentice-Hall, New York.

Robert P., Escoufier Y. (1976), A unifying tool for linear multivariate analysis: the RV-coefficient, J.R.S.S., ser.C., 25, 257-265.

Waltz W.G., Reid W.A., Colwell W.E. (1948), Sugar and nicotine in cured bright tobacco as related to mineral element composition, Proc. Soil Sci. Soc. Am., 13, 385-387.

Evaluation of Statistical Software Based on Empirical User Research

H. Wilke, Free University of Berlin, F.R.G.

Summary: Almost 10 years of research activities went into the evaluation of statistical software. The outcome so far shows that a reorientation is necessary. Future research should broadly address problems that work as constraints on user productivity. This includes local support infrastructure, training, computer oriented research methodology as well as the quality of the software itself. Empirical user research is the proposed method to learn more about current computing practices.

Keywords: statistical software package, software evaluation, statistical software research, empirical user research

Introduction

First organized attempts to a systematic evaluation of statistical software can be traced back to the early 70's, when the American Statistical Association formed the "Committee on the Evaluation of Statistical Program Packages". Thus forming an impressive history of 10 years of thinking, researching and writing on this subject. The most recent highlights of this 'evaluation movement' are the volumes edited by Ivor Francis (1979 and 1981) that contain formal and informal descriptions and evaluations of many systems.

The amount of work that went into these and earlier products is certainly impressive. Unfortunately there seems to be an unhealthy balance between human investments and relevance of the outcome. Software evaluation has remained an esoteric field of study with almost no recognition among the users of the software. It has contributed very little to the understanding of the nature of data analysis packages, their rôle in various contexts of applied research and the people using the systems. The advent of statistical analysis packages has changed the course of empirical research, computing, education, job structure and many peoples everyday life. The evaluation movement so far was unable to recognize the changes and to contribute to their rationale understanding and meaningful development.

This critique might be considered unfair as the goal of the movement never was to analyse the packages as the complex phenomenon in research life as they are. The view on the packages chosen by the key persons in the field was by far more limited. Thus encyclopaedic lists of software products have been generated as well as hundreds of articles comparing selected features of selected programs by selected criteria. Their inherent irrelevance to the current community of statistical package users should be overcome by a new approach that considers evaluation of statistical packages as part of a general statistical software research.

Statistical Software Evaluation so far

A systematic approach to the evaluation of statistical packages was initiated by statisticians who were concerned about the quality of the systems that became heavily used in the early 70's. Their findings were intended to help in two directions:

- give users hints as to which package to prefer
- give producers feedback with regard to weak parts of their systems.

The concept behind this was to act as a 'consumer organization'. Given the uncertain quality of software at that time and a low level of information about alternative products among the users this was certainly a valid approach. It proved to be very productive in the following years and an impressive stack of publications was produced. Thus some of the goals were achieved:

- it has been contributed to improvements of the systems
- a better market transparency for buyers was achieved.

Nevertheless the research remained of limited use to the package user community. Many of the urgent questions of software usage were never really addressed. The overall relevance of the research findings gives rise to the question on how to proceed in the future. Statistical software evaluation in the tradition of the ASA Committee seems to be at a dead end today.

The approach that has been followed so far had to suffer from two limitations which have their roots in the constellation of problems ten years ago:

- evaluators personal views and background
- concept as a consumer organization.

The evaluation researcher had mostly a background as a statistician and probably experience with the subroutine libraries of the 60's. Consequently in the evaluation literature a user was mostly conceptualized as a person experienced in methods and statistics looking for a special feature to ease his calculations (e.g. a test). This is untypical for an average user of a package and led to a biased view in several respects.

The modern integrated packages want to provide their user with a complete data analysis environment; he is guided by the data analysis philosophy and the range of features of the system. A move from one package to another to perform a special test is quite unlikely under this approach. Users do apply these systems not just for the purpose of statistical computation but for datamanagement, datadocumentation and report generation. A modern package can support the entire phase of research data processing.

This totality of a package as opposed to a subroutine library was never really incorporated in the approach of the evaluation movement. The single features comparison of the comparative evaluation literature failed to realize this very nature of packages. To assess the quality of a feature it has to be seen as an integrated part of the data analysis philosophy materialized as a certain package.

The consumer orientation reduced the dimensions of package use to just one: vendor - buyer. But productive use of a system on a qualitatively high level depends on many more factors than just the product itself. The local working environment is at least as important: support infrastructure, education and more. Rather than considering all factors that influence research productivity with statistical software the relatively narrow possibilities of academic influence on vendors were examined.

These historic shortcomings of the evaluation approach should be overcome. The topic on hand is by far too relevant to leave statistical software evaluation at the dead end where it is now.

Statistical Software Research

One should leave the view of the early seventies and address the needs of the 80's: Statistical software research as an empirical field of study that addresses the full range of problems that is connected with the current use of packages. The goal should be to contribute to the productivity of software usage by deepening the understanding of its potentials and its prerequisites. Among those the quality of the software itself is just one factor among others that affect software productivity.

Research methodology: Computing has to be an integrated part of the research process. From the design of a questionaire to the final report project management could rely on the support of research software. What is needed is a revised research methodology that makes use of the facilities in a systematic way. But the use of computers not just to calculate regression coefficients but rather as an information processing machine requires a revised concept of the role of statistical software in the research process. The capability of this software needs to be exploited within a larger framework of computerized information processing. Research on this topic needs to expand beyond the mere look at a single feature of a single software products. Statistical theory and practice should develop from a batch-philosophy to a support of a creative dialogue with the data with the help of intelligent software. Tukeys EDA is certainly a step in this direction. But more can be done.

Presentation of results is more and more understood to be an integrated part of a data analysis package. But very little is known about the perception of graphics and tables by human eyes. So far they are something 'extra', neat but not really necessary. They could become far more relevant if one would know how: A theory of graphics and table perception and their role in data analysis is still outstanding.

Support infrastructure: A productive use of statistical software in research requires an effective, local support infrastructure with trained consultants, documentation and utility routines. All this has to be organized to meet users needs and several alternative models exist: central service at the computer center, statistical laboratory, library, medical department or just nothing. Little empirical evidence on the virtues of these models exists so far.

The availability of software and modes of their usage mostly depends on computer center policy. Here typically the influence of statistical software users is small and causes often dissatisfaction with this central service. Better organizational procedures must be developed to help computer center personal to understand us and observe our requirements with regard to software, manpower and hardware (disc space, terminals, high quality printer).

Education: Software can not be used productively without trained personal. Currently the use of statistical software is unsufficiently incorporated in the teaching of statistics and methods of empirical research. Purely technical seminars at computer centers can hardly replace a thorough integration into existing curricula. To make this possible, adjustments on the side of the software as well as the course contents are required. Research in this field should try some models and provide empirical clues beyond mere personal opinions.

Carreer patterns of statistical software experts are still very unclear. Recruitment of qualified persons is often a difficult task; but after employment they often receive little academic recognition. In face of the crucial role that computing meanwhile plays for many empirical projects it is certainly inadequate to put programmers in a row with a secretary as it still happens in forewords. But what is the proper role of computing people in research? Definitions of roles, self-perception and carreer patterns are still open.

Software: Software requirements to accomplish the goals have been discussed in the literature. But the general problem of user friendlyness has been unsufficiently covered. The philosophy of a package is to provide an integrated toolkit to users that are not computing experts. Thus software should be powerful AND easy to learn and to use. The language should be systematic, well documented with good error diagnostics. Convenient interfaces among systems should support the use of several systems in parallel.
Research on these topics seems to be in full swing but more in the context of microcomputers, text processing systems and data base query languages.
An improved user friendlyness of statistical software, operating systems and editors will certainly make life easier and thus improve productivity. A prerequisite to this process is a better understanding on how people and systems interact. Again, statistical software research should accept the challenge and provide more answers than it has done in the past.

Statistical software research should address these questions and approach computing not in terms of isolated products and features any longer but rather observe it as the complex phenomenon in research practice as it is. A point of departure might be empirical research on real life performance: current use and misuse of the systems can systematically be analyzed. Empirical data on user behavior are available and can point in two directions:
- work as an indicator of users needs
- " " " " " systematic limitations in software or its environment.

This could help to identify constraints on productivity and remove them. Thus statistical software research could start to play the relevant role in the development of research methology it deserves.

References: Ivor Francis (ed.) A Comparison of Statistical Software, Voorburg (IASC) 1979

Ivor Francis (ed.) Statistical Software - A Comparative Review, New York (North Holland) 1981

Sound and Exploratory Data Analysis

S.R. Wilson, The Australian National University, Canberra, Australia

Summary: Most techniques used in the exploratory analysis of data rely on visual representations. It is well known that such representations have certain limitations and also can be misleading. A new and novel technique which overcomes some of the limitations of visual methods is proposed for the exploratory analysis of data. This technique gives an auditory representation of the data. An example of the use of the technique for multivariate data analysis will be given.

Keywords : Aural, sound, exploratory data analysis

1. Introduction

Most techniques used in the exploratory analysis of data rely on visual representations. There have been many recent advances in the use of graphical methods for statistical analyses of multidimensional data. These include glyphs, profiles, stars/polygons, function plots, faces, boxes, constellation graphs and trees and castles. References to these techniques can be found in Fienberg (1979) and Kleiner and Hartigan (1981). Dependent on the particular context, each of these graphical tools has its uses. Each also has its limitations and can be misleading. All statistical data analysts gain, over time, knowledge of the ways in which our visual sense can be mislead. Here another tool is proposed whose use will supplement, although not supplant other tools. Its inherent advantage is its use of another sense, namely our aural sense.

Our aural sense highlights *different* aspects of the data to those highlighted by our visual sense. This tenet can be used to advantage in exploratory data analysis. Independently of the system to be described, another method for the audio display of multidimensional data has been under development by Bly (1982) and Yeung (1980). Their technique encodes the multivariate observation into a single note using music characteristics such as pitch, loudness, direction, duration and so forth. To many, such sounds are most dissident. Nevertheless these researchers have shown that after training people can discriminate data based on such sounds.

As Thisted (1981) says: "Personal computers offer new and fundamentally different computing tools around which new methods for exploratory data analysis can be developed". Such methods include sound representations.

2. Brief Outline of the Aural Representation

The essence of this representation is to first round the data values appropriately, (which is a practice advocated by many statistical data analysts; see

Ehrenberg, 1981). Then the numbers are placed in 1-1 correspondence with the notes of the C-major arpeggio; (see fig. 1).

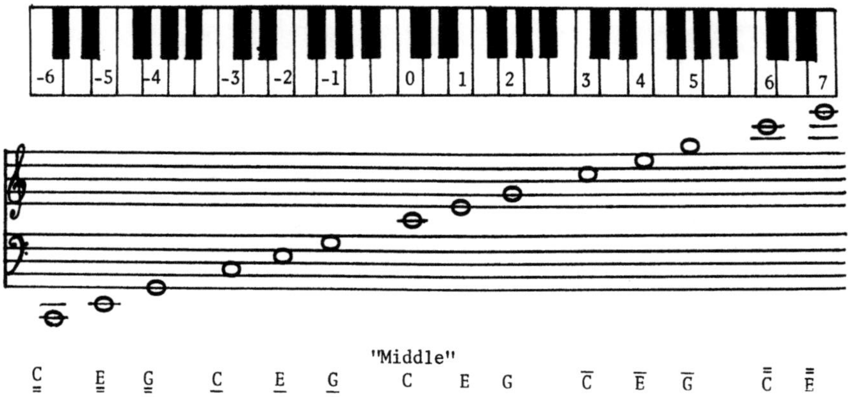

Figure 1. Sound Encoding System

It is quite straightforward using an appropriate sound board attached to a micro (say), to output these encoded data values as a "melody". These melodies sound quite pleasant to the (Western-trained, not tone deaf) ear. Finally, it is remarked that it is no longer necessary to be facing in one direction when analysing data in this (exploratory) way.

This encoding system has been found to be particularly useful for exploratory contingency table analysis (Wilson, 1982). Due to space limitations, instead just a simple multivariate data set will be considered.

3. A Simple Example : Fisher's Iris Data.

Measurements of the sepal length and width and the petal length and width of 50 plants for each of 3 types of iris were first analysed by Fisher (1936). Anderson (1935) originally collected these data with the following question in mind:

> "On the simple assumption that if current theories are true, one should be able to find evidence of continuing evolution in any group of plants, I have been going around the world looking as sharply as possible at variation in irises. On any theory of evolution the difference between individuals get somehow built up, in time, into the differences between species".

To examine these data, first the four measurements for each plant are rounded and these values are given in Table 1. The mean values for each of the 3 types of iris can be heard by listening to the representation given in fig. 2. They are quite strikingly different, (as has been shown using many other techniques).

TABLE 1

Iris setosa								*Iris versicolor*								*Iris virginica*							
S	S	P	P	S	S	P	P	S	S	P	P	S	S	P	P	S	S	P	P	S	S	P	P
L	W	L	W	L	W	L	W	L	W	L	W	L	W	L	W	L	W	L	W	L	W	L	W
-3	1	-4	-5	-3	-1	-4	-5	2	0	1	0	1	-1	1	0	1	0	3	5	3	0	3	2
-3	-1	-4	-5	-3	1	-4	-4	1	0	1	1	2	-2	1	0	-1	-2	2	2	0	-2	1	2
-4	0	-4	-5	-2	1	-4	-5	2	0	2	1	2	-1	2	2	3	-1	3	3	0	-1	2	2
-4	0	-4	-5	-2	1	-4	-5	-2	-4	0	0	0	-1	1	1	1	-1	3	3	1	-2	3	3
-3	2	-4	-5	-4	0	-4	-5	1	-2	1	1	-1	-2	-1	-1	1	-1	3	4	3	-1	3	1
-2	3	-4	-4	-3	0	-4	-5	-1	-2	1	0	-2	-3	0	-1	4	-1	4	3	3	-2	4	2
-4	1	-4	-4	-2	1	-4	-4	1	0	1	1	-2	-3	0	-1	-3	-3	1	2	5	2	4	3
-3	1	-4	-5	-2	4	-4	-5	-3	-3	-1	-1	-1	-2	0	0	+3	-1	4	2	1	-2	3	4
-5	-1	-4	-5	-2	4	-4	-5	1	-1	1	0	0	-2	2	1	2	-3	3	2	1	-2	2	1
-3	0	-4	-5	-3	0	-4	-5	-2	-2	0	0	-2	-1	1	1	3	2	4	5	0	-2	3	0
-2	3	-4	-5	-3	0	-5	-5	-3	-5	-1	-1	0	1	1	1	1	0	2	3	4	-1	4	4
-3	1	-4	-5	-2	1	-4	-5	-1	-1	0	1	2	0	1	1	1	-2	2	2	1	1	3	5
-3	-1	-4	-5	-3	2	-4	-5	0	-4	0	-1	1	-4	1	0	2	-1	3	3	1	0	3	2
-5	-1	-5	-5	-5	-1	-4	-5	0	-1	1	0	-1	-1	0	0	-1	-2	2	3	0	-1	1	2
-1	3	-5	-5	-3	1	-4	-5	-1	-1	-1	0	-2	-3	0	0	-1	-2	2	5	2	0	2	3
-1	5	-4	-4	-3	1	-4	-4	2	0	1	0	-2	-2	1	0	1	0	2	4	2	0	3	5
-2	3	-4	-4	-4	-4	-4	-6*	-1	-1	1	1	0	-1	1	0	1	-1	3	2	2	0	2	4
-3	1	-4	-4	-5	0	-4	-5	-1	-2	0	-1	-1	-2	0	0	4	2	5	4	-1	-2	2	3
-1	3	-4	-4	-3	1	-4	-3	0	-4	1	1	-3	-4	-1	-1	4	-2	5	4	2	0	3	4
-3	3	-4	-4	-3	2	-3	-4	-1	-2	0	1	-1	-2	0	0	0	-4	2	1	2	0	3	5
-2	1	-4	-5	-3	-1	-4	-4	-1	0	1	2	-1	-1	0	0	2	0	3	4	2	-1	2	4
-3	3	-4	-4	-3	2	-4	-5	0	-2	0	0	-1	-1	0	0	-1	-2	2	3	1	-3	2	2
-4	3	-5	-5	-4	0	-4	-5	1	-3	2	1	-0	-1	1	0	4	-2	5	3	1	-1	2	3
-3	0	-4	-3	-2	2	-4	-5	0	-2	1	0	-3	-3	-2	-1	1	-2	2	2	0	1	2	4
-3	1	-3	-5	-3	0	-4	-5	1	-1	1	0	-1	-2	0	0	2	0	3	3	-1	-1	2	2
Means				-3	1	4	-4					0	-2	1	0					1	-1	3	3

Listening to the *Iris setosa* type data (by encoding from Table 1 using fig. 1), the melodies are heard to be fairly similar, indicating that the data are relatively homogeneous. The only exception is observation 42(*) and careful inspection of the original data indicates a possible misprint/misrecording of its sepal width value. (The other observation noted by Gnanadesikan (1977) is merely a more "extreme" example of others of this type.)

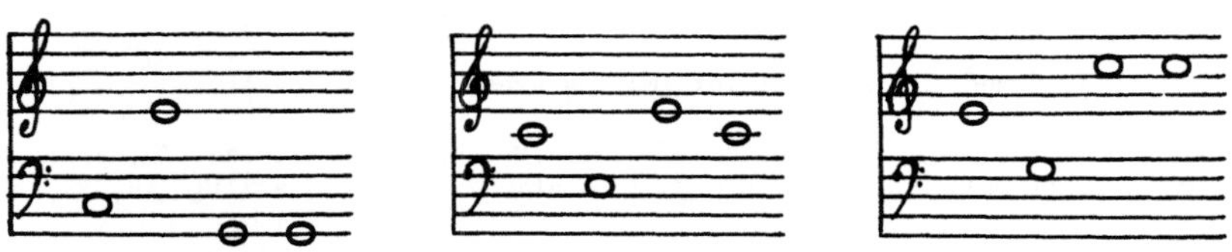

Iris setosa *Iris versicolor* *Iris virginica*

Figure 2. Sound representation of the means

Listening to the individual *Iris versicolor* type melodies, it is heard that there is relatively more variation from one to another. With careful listening one becomes aware of two subgroupings, one with melodies sounding like that of Fig. 3a, the other with melodies like Fig. 3b. (Subgroup 1 comprises observations 8, 10, 12, 15, 17, 21, 35, 36, 39, 46, 47, 49, and subgroup 2, the remainder.) Doing a discriminant analysis on these two groupings, a chi-square value of 41.7 on 4 degrees of freedom is obtained. However, confirmation of the subgroup hypothesis for these data would need careful evaluation in light of new data.

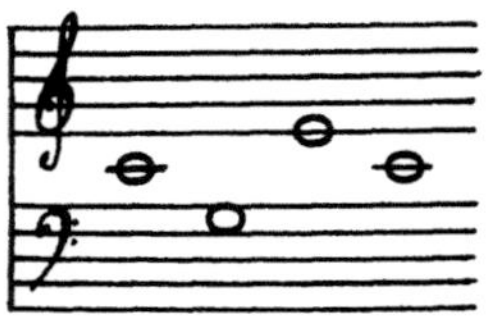

Iris versicolor

Figure 3a. Sound representation of Subgroup 1 mean

Figure 3b. Sound representation of Subgroup 2 mean

Listening to the type *Iris virginica* data, two possible subgroupings are heard. (Subgroup 1 comprises observations 6, 8, 18, 19, 23, 26, 30, 31, 32 and subgroup 2, the remainder. A chi-square value of 50.0 was obtained for testing the discrimination between these two possible subgroups.) Further details are given elsewhere (Wilson, 1982).

It is interesting to note that both the types *I.versiolor* and *I.virginica* are more varied in colour than *I.setosa* and that both these types show, in an exploratory way, evidence of being composed of two subgroups, while type *I.setosa* appears to be quite homogeneous (with the possible exception of one outlier). This supports the evidence for continuing evolution in this group of plants, for which Anderson was originally seeking. It also should be remarked that each subgroup is more like its own type than like any other subgroup or group.

References

Anderson E. (1935), The irises of the Gaspé Peninsula, Bull.Amer.Iris Soc., 59, 2-5.

Bly S. (1982), Sound and Computer Information Presentation, Thesis in preparation.

Ehrenberg A.S.C. (1981), The problem of numeracy, Amer.Statistician, 35, 67-71.

Fienberg S.E. (1979), Graphical methods in statistics, Amer.Statistician, 33, 165-178.

Fisher R.A. (1936), The use of multiple measurements in taxonomic problems, Ann.Eugenics 7, 179-188.

Gnanadesikan R. (1977), Methods for Statistical Data Analysis of Multivariate Observations, John Wiley: New York.

Kleiner B.and Hartigan J.A. (1981), Representing points in many dimensions by trees and castles, J.Am.Statist.Ass., 76, 260-276.

Thisted R.A. (1981), The effect of personal computers on statistical practice, Thirteenth Symposium on the Interface between Computer Science and Statistics, 12-13 March 1981, Pittsburgh, Pennsylvania.

Wilson, S.R. (1982), Sound and Exploratory Data Analysis, Manuscript in preparation.

Yeung E.S. (1980), Pattern recognition by audio representation of multivariate analytical data, Anal.Chem., 52, 1120-1123.

Monotone Splines: A Family of Transformations Useful for Data Analysis

S. Winsberg, Université de Montréal, Canada
J.O. Ramsay, McGill University, Montréal, Canada

A class of transformations useful for data analysis is discussed. Monotone splines are a powerful family of smooth transformations amenable to statistical treatment. In this paper we will describe this class of transformation and then present applications in multiple regression, principal components analysis, and regression analysis of categorical data.

Key words: monotone transformations; splines; regression; principal components; categorical data.

Introduction

It is often useful to seek empirical transformations which optimize a given objective function. The choice of this function depends on the aims of the data analysis. Benefits of transforming are better fit, simpler models, and improved ease of interpretation. The study of the transformations may also be an important goal.

In the past two approaches have been taken to the problem of seeking empirical transformations. In the first (Box and Cox, 1964; Andrews, 1971), the use of power transformations, is often not sufficiently flexible. In the other approach isotonic regression is applied resulting in a step function (Kruskal, 1965; Kruskal and Shepard, 1974) which is neither smooth, nor expressible in terms of a limited number of parameters. Monotone splines as an alternative to these approaches are a powerful family of smooth transformations amenable to statistical treatment which are extremely flexible and represented by a limited number of parameters. The generation and application of monotone splines are discussed by Winsberg and Ramsay (1980, 1981). Here we shall briefly describe this class of transformation and then present applications in multiple regression, principal components analysis and in the analysis of categorical data.

Monotone Spline Transformations

Let the monotone spline be defined on the interval (a, b). It will consist of monotone polynomials of degree k joined at points interior to the interval at which we assume for simplicity there are k-1 derivatives. Using the notation of de Boor (1978) this determines a knot sequence $\underline{t}$ in which the first k knots are equal to a, the last k knots are equal to b, and the rest are set equal to the points of junction for the polynomial segments. The total number of knots is equal to n+k where n must be at least k and specifies the number of degrees of freedom of the monotone spline.

We generate a family of monotone splines by first forming a basis for all splines of degree k-1 on (a, b). This M-spline basis for a given knot sequence and order k consists of n basis splines. From the family of M-splines we generate the family of I-splines (integrated splines): $I_{qk}(x) = \int_a^x M_{qk}(u)\,du$. From the nonnegativity of the M-splines we have the monotonicity of the I-splines. Moreover the M-splines are normalized such that $I_{qk}(b) = 1$. So, one may represent a monotone spline function by $F(x) = \Sigma c_q I_{qk}(x) \quad c_q \geq 0,\ q=1,\ldots,n$

Our experience leads us to favor the use of k=2 along with a few interior knots. Interior knots should be placed where one expects to find the greatest nonlinearity and so as to include some data points between consecutive knots. We have found that this choice is both parsimonious and retains sufficient flexibility.

A Multiple Regression Problem

The data consist of N observations x_{ij} (i=1,.,N; j=0,.,p) on a dependent variable and on each of p independent variables. The problem is to find a set of transformations F_j and possibly a constant term such that an objective function is optimized with respect to $F_j(x_{ij})$ subject to monotonicity constraints on F_j. Only monotonically increasing functions need be considered. If necessary a prior order-reversing transformation may be performed.

The transformed dependent variable observations F_0 are assumed independently distributed as $N(\mu_i, \sigma)$ where $\mu_i = \sum_j F_j(x_{ij}) + c_0$. Maximum likelihood estimation is then a nonlinear programming problem with $n_0+\ldots+n_p-1 = L$ parameters, subject to one linear equality constraint to fix scale and L-1 linear inequality constraints of the form $c_{jq} \geqslant 0$. We used the penalty function approach described by Fiacco and McCormick (1968) to solve the resultant nonlinear programming problem with constraints.

A Principal Components Problem

Consider a set of data consisting of the measurements of N objects with respect to p variables. Bilinear methods have been used to represent the objects and the variables in a space of dimensionality $r<p$. We assume that a monotone transformation of the values in each column of the N x p data matrix is determined by a linear combination of a new set of values characterizing each row object. We represent the monotone transformation as a nonnegative combination of I-splines.

Let X^t denote the N x p matrix of original observations and denote the matrix of transformed observations as $Y^t(X^t)$. The elements of $Y^t(X^t)$ are $y_j(x_{ij})$, $i=1,..,N$; $j=1,..,p$. The $y_j(x_{ij})$ are a monotone transformation of the x_{ij}. Denote the column centered matrix as Y^t_{cc}. We propose that $\underline{y}_i(\underline{x}_i)$ is distributed as $MN(\underline{\mu}_i, \sigma I)$, where $\underline{\mu}_i = A\underline{f}_i$, $\underline{f}_i$ being an r dimensional vector of factor scores and A a p x r factor pattern matrix. Thus $\hat{Y}^t_{cc} = F^t A^t$. This factorization is determined only to an affine transformation. Since this model is a generalization of the Eckart-Young model (Eckart and Young, 1936), we call this the EY-T model, the Eckart-Young model with transformation.

The parameters are divided into two sets: the transformation parameters and the structural parameters. Starting with the original data we iterate, alternating between a conditional solution for one set of parameters given the other set until the solution stabilizes. The objective function used to estimate the transformation parameters is the log likelihood. As in the regression problem a penalty function approach is used to keep the transformation parameters nonnegative.

Analysis of Categorical Data

Sometimes data are measurements consisting of a single response taking values in a limited set of ordered categories. We shall develop models and methods of analysis appropriate to this type of categorical data. If we denote m_{rg} as the frequency with which object r is placed in category g and $\pi_{rg}=\frac{1}{N_r}E(m_{rg})$ where N_r is the total number of responses for object r, then the general model for this type of data may be written as:

$$\pi_{rg} = H\,[\tau_g - s(\underline{x}_r, \underline{\theta})] - H\,[\tau_{g-1} - s(\underline{x}_r, \underline{\theta})] \qquad (1)$$

where $\pi_{rg}>0$ and $\sum_{g=1}^{G} \pi_{rg}=1$ $r=1,.,R$ $g=1,.,G$; $\underline{\theta}$ is a vector of parameters to be estimated; $\underline{x}_r$ is a vector of known values for object r; τ_g is the upper boundary for category g. Particular functions such as the inverse logistic have been used to represent H. We represent the function H as $H=F.\phi$ where F is a linear combination of I-splines and ϕ is a known function $\phi:R\to(0,1)$. We focus on linear models for s but the application of monotone splines may be easily extended to other types of models as well.

The problem is to estimate the object parameters $\underline{\theta}$, the category boundary parameters $\underline{\tau}$, and the transformation parameters $\underline{c}$ such that some objective function is optimized. We use the statistic suggested by Neyman (1949)

$$\chi^2_N = \Sigma_r\ \Sigma_g\ \frac{[m_{rg} - N_r\,\pi_{rg}\,(\xi_1,\ldots.\xi_q)]^2}{m_{rg}} \qquad (2)$$

for the objective function to be minimized. This modified chi square statistic has an asymptotic χ^2 distribution if the postulated model is true. The constraints that $\pi_{rg}>0$ $r=1,.,R$; $g=1,.,G$ and that $\sum_{g=1}^{G} \pi_{rg}=1$ $r=1,.,R$ are met provided that $c_i>0$ $i=1,\ldots,n$ and $\sum_{i=1}^{n} c_i=1$. We have a nonlinear programming problem subject to one linear equality and n inequality constraints. Again a penalty function was used.

Example: Principal Components Analysis of Automobiles

Data on thirty three automobiles originally published in Consumer Reports (1980) were analyzed. We retained five of the reported characteristics: a) price,

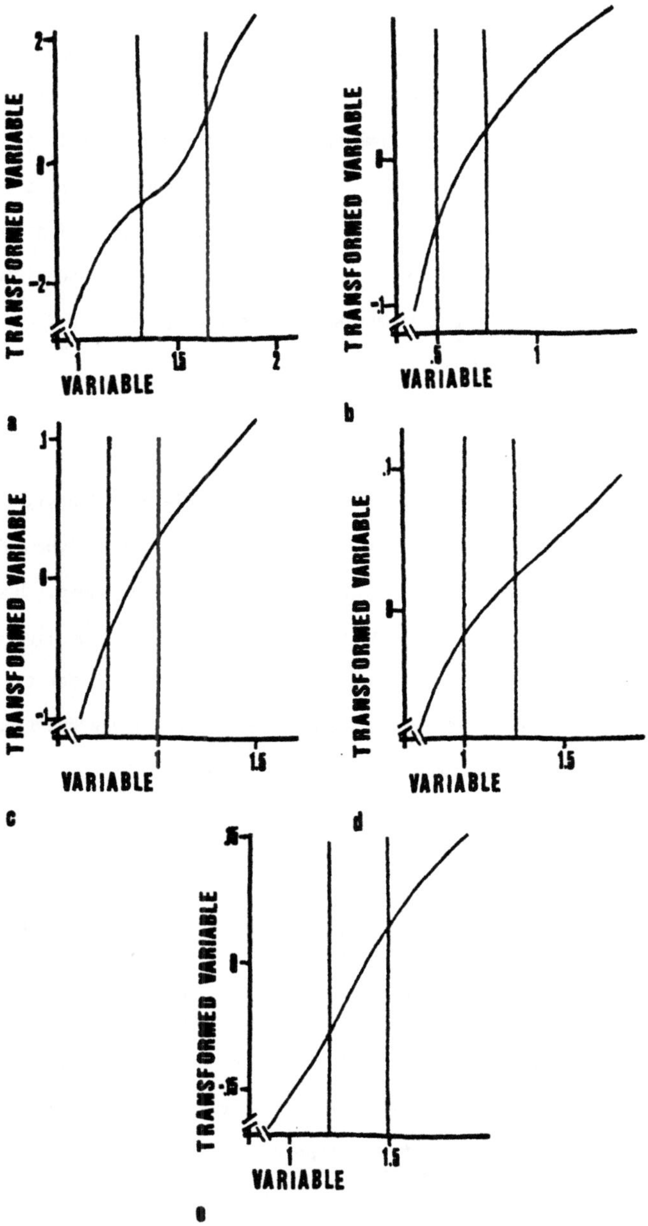

Figure 1. Recovered transformations for automobiles. Vertical lines are knot positions. a) price b) displacement c) mpg city d) mpg highway e) weight.

b) engine displacement, miles per gallon for c) city and d) highway driving and e) weight, rescaling each variable. Order two splines with two interior knots per variable were used. The first two components contain .997 of the total variance as compared to .940 for the untransformed data. The recovered transformations are shown in Figure 1. Of note is the need to transform miles per gallon to something like the reciprocal which makes more physical sense and the nonlinearities in the transformation for price. The plateau in the transformation for price occurs in the neighborhood of 6000 dollars.

Discussion

Integrated basic splines appear to be an extremely useful basis for representing monotone transformations. Since splines are piecewise polynomials their behavior in one region does not determine their behavior everywhere. However, since they are everywhere simple polynomials, they are computationally convenient, and the resulting curve is smooth. Finally integrated splines offer the advantage for some applications that they can be constrained to have the properties of a cumulative distribution function.

References

Andrews, D.F., Gnandesikan, R. & Warner, J.L. (1971), Transformations of multivariate data, Biometrika, 27, 825-840.

Box, G.E.P. & Cox, D.R. (1964), An analysis with transformations (with discussion), J.R. Statist. Soc. B, 26, 211-252.

Consumer Reports, April 1980, 45, no 4.

de Boor, C. (1978), A practical guide to splines, Springer-Verlag, New York.

Eckart, C. & Young, G. (1936), The approximation of one matrix by another of lower rank, Psychometrika, 1, 211-218.

Fiacco, A.V. & McCormick, G.P. (1968), Nonlinear programming: sequential unconstrained minimization techniques, Wiley, New York.

Kruskal, J.B. & Shepard, R.N. (1974), A nonmetric variety of linear factor analysis, Psychometrika, 39, 123-157.

Neyman, J. (1949), Contributions to the theory of the X^2 test. Proceedings of the First Berkeley Symposium on Mathematical Statistics and Probability. (ed. J. Neyman). 230-273. University of California Press, Berkeley.

Winsberg, S. & Ramsay, J.O. (1980), Monotonic transformations to additivity using splines, Biometrika, 67, 669-674.

Winsberg, S. & Ramsay, J.O. (1981), Analysis of pairwise preference data using integrated B-splines, Psychometrika, 46, 171-186.

Results of the Application of Multimodal Optimization Algorithms Based on Statistical Models

A. Žilinskas, Lithuanian SSR Academy of Sciences, USSR

Summary

Difficulties in applying well-known optimization techniques to some problems of optimal design are discussed. It is shown that multimodal optimization algorithms based on statistical models of objective functions are efficient for such difficult practical problems.

Key words: Multimodal optimization, Statistical models, Computer aided design, Stochastic optimization

Introduction

Algorithms for global optimization are usually constructed based on some statistical approach. Recent results of Archetti (1980), Mockus (1980), Schagen (1980), Žilinskas (1980a) show that algorithms based on statistical models of complicated multimodal functions are most efficient. A comparative analysis by Dixon and Szegö (1978) of optimization results of some functions widely used for global algorithm testing has also demonstrated efficiency of the algorithms of such class.

Justification of the usage of statistical models in multimodal optimization theory is given by Žilinskas (1978): information on an objective function and the uncertainty of its behaviour have been axiomatically formalized and a statistical model of an objective function representing the suggested axioms has been constructed. Using a model of an objective function the results obtained by minimization procedure may be interpreted, as well as the results of the current steps forecasted; nevertheless, the problem of the construction of a minimization algorithm remains not trivial. Since the optimal algorithms (Bayesian or minimax) are too complicated for computer realization, various simplifications are suggested (e.g. Mockus (1980)); however, it is difficult to evaluate their accuracy. Therefore, it was suggested by Žilinskas (1980b) to construct an algorithm axiomatically, formalizing simple and intuitively obvious requirements to the algorithm at a current minimization step. An axiomatic approach to multimodal optimization is summed up by Žilinskas (1982). Efficiency of these algorithms was investigated by Žilinskas (1980,1981) using known test functions and it was shown there that they were comparatively efficient regarding the number of evaluations of the objective function. Therefore the most rational field of applications of constructed algorithms is optimization of complicated multimodal functions, one evaluation of which consumes a great amount of computer time, and dimensionality of which equals approximately 10. Optimization problems of such type occur in some problems of optimal design. The difficulties in applying known local algorithms to these problems are discussed below. It is shown that these problems may be solved by global optimization algorithms and that the algorithms constructed using a statistical model of an objective function are most efficient.

Optimal design of magnetic deflection systems (MDS)

Minimization of the aberration of an electron beam by choosing currents in sections of MDS is an important part of the computer aided design of MDS. An algorithm for computing an objective function was developed by Grigas and Gudonavičiute (1980). The objective function $f(x)$, $x=(x_1,\ldots,x_{13})$, $0 \leq x_i \leq 200$, $i=\overline{1,13}$ is defined by a complicated algorithm including numerical integration of the system of differential equations of the electron movement in the magnetic field of MDS. An analytical investigation of the characteristics of $f(\cdot)$, including the number of local minima and regions of their attraction, is difficult. Additional difficulties in the choice of an algorithm arise in connection with the great amount of computer time needed for one evaluation of $f(\cdot)$; it is equal to 21 seconds for the computer BESM-6. Besides, the errors of numerical solution of the system of differential equations may be essential for the numerical differentiation of $f(\cdot)$.

Application of local optimization methods to this problem was not satisfactory, e.g. for the initial point x^0, $f(x^0)=0.545$ which corresponds to the uniform distribution of the currents ($x_i^0=18.4, i=\overline{1,13}$) the modification of a variable metric algorithm after two hours' calculation reached value 0.544 (the number of evaluations of $f(\cdot)$ was equal to 410). The failure of the variable metric algorithm may be explained by inaccurate and expensive numerical differentiation of $f(\cdot)$. The simplex algorithm, FORTRAN codes of which are given by

Himmelblau (1972), does not require the calculation of derivatives and it is widely known as an efficient algorithm for solving engineearing optimization problems. After one hour's calculation (with the initial point x^0) this algorithm reached value 0.371 and after two hours value 0.336. Minimization with two other initial points generated randomly has led to worse results.

It is interesting to apply multimodal optimization algorithms to this problem. In an experiment the following algorithms were used: 1) the search on LP_τ grid which is approximately uniform, 2) the algorithm by Torn (1978), 3) a simple Bayesian algorithm by Mockus (1980), 4) the algorithm axiomatically constructed by Žilinskas (1980a) basing on a stochastic model of an objective function, 5) the coordinate algorithm constructed on the base of the one-dimensional multimodal algorithm by Žilinskas (1981) with the initial point x^0.

The results of the experiment are given in Table 1. Algorithms 1-4 minimized $f(\cdot)$ during one hour; the best found value is denoted by f_G. From the best found point local minimization was carried out using the codes of the simplex algorithm FLEXIPLEX given by Himmelblau (1972); the reached value is denoted by f_{opt}. The total number of the evaluations of $f(\cdot)$ is denoted by N. Algorithm 5 is terminated after one coordinate cycle and local minimization was not carried out; the time of minimization equals ~65 minutes.

Table 1

Algorithm	1	2	3	4	5
f_G	0.505	0.413	0.435	0.461	0.205
f_{opt}	0.282	0.287	0.312	0.261	
N	403	399	380	308	233

The results of Table 1 show that for the solution of this problem algorithm 5 is most efficient. Such conclusion was unexpected. Because of the failure of local algorithms the structure of $f(\cdot)$ must have been very complicated. Therefore the structure of $f(\cdot)$ was investigated using the algorithm of Šaltenis and Radvilavičiute (1977). The 150 results of $f(\cdot)$ evaluated on the LP_τ grid were analysed. The main structural characteristics of $f(\cdot)$ are given in Table 2. These results show that the interaction of variables is not essential; the other five variables and all the interactions form only 28 % of the full dispersion of the values of f(·). These results show that f(·) is approximately separable and therefore they explain the efficiency of algorithm 5 for the minimization of the investigated problem.

Table 2

No of variable	13	12	11	1	2	3	4	5
Structural characteristics	20.8	15.8	9.1	6.9	6.7	5.1	4.5	2.9

The failure of local algorithms may be explained by the results of investigation of f(·) optimum sensitivity while comparing them with the results of Table 2. The deviation from the optimum is less than 5 %, when $140 \leq x_{13} \leq 200$, $80 \leq x_{12} \leq 180$, $0 \leq x_{11} \leq 15$ and when the other variables differ from the optimal value less than by 10 %. These results show that the average character of the relief of f(·) essentially differs from its character at the optimum point. This feature of f(·) as well as the presence of local minima may cause the failure of local algorithms.

Optimal design of pigmental mixtures

The design of pigmental mixtures with given spectral or colour characteristics is an urgent problem of colour technology. This problem may be formulated as a minimization one, where variables x_i, $i=\overline{1,n}$, are mass proportions of the pigments and f(·) is the difference between the spectral or colour characteristics of the considered mixture and the given standard. The algorithms of f(·) calculation and the examples of necessary data are given by Barauskas et al. (1980). It is also shown there that local algorithms are not efficient for this problem. Since the minimization results essentially depend on the initial point, it is necessary to repeat local minimization several times using randomly generated initial points. Therefore the time needed for solving such problem is about 2 hours.

It is interesting to apply multimodal optimization algorithms to this problem. The number of nodes of LP grid was equal to 3000. Parameters of algorithm 2 were P=20, ND=1, RD=10. The number of evaluations by global algorithm 3 was equal to 500. Parameters of algorithm 4 were chosen by the algorithm automatically. Local descent after global search was carried out by modification of the variable metric algorithm by Tiešis (1975) because f(•) is rather smooth in this case; parameters of the algorithm were the same for all cases. The meaning of f(•) was the colour distance between the mixture and the given standard. The versions corresponding to standards No 2 and 3 were minimized; n=9 and the necessary data are given by Barauskas et al.(1980). The minimization results are given in Table 3, where T denotes minimization time in seconds (computer BESM-6 was used).

Table 3

No of version	2			3		
Algorithm	f_{opt}	N	T	f_{opt}	N	T
1	11.77	3328	1161	7.54	3508	1134
2	11.77	8007	401	2.78	9826	479
3	11.80	910	892	4.21	1324	913
4	11.77	1966	276	$4 \cdot 10^{-5}$	3839	343

Conclusions

Some optimization problems which are difficult to solve by usual local optimization techniques , may be efficiently solved by statistical multimodal algorithms. Algorithms constructed axiomatically on the base of a statistical model of an objective function are preferable for solving minimization problems dimensionality of which equals approximately 10 and the evaluation of the objective function of which consumes a great amount of computer time.

References

Archetti F.(1980). A probabilistic algorithm for global optimization problems with a dimensionality reduction technique, Lecture Notes in Control and Information Sciences, 23, 36-42.

Barauskas et al.(1980). Investigation of optimization problems of the synthesis of pigmental compositions, Optimal Decision Theory, Vilnius, 6, 57-74, (in Russian)

Dixon L.C.W. and Szegö G.P.(1978). The global optimization problem: an introduction, In: Towards Global Optimization 2, eds.Dixon L.C.W. and Szegö G.P., North Holland, 1-15.

Grigas F.J. and Gudonavičiute V.V.(1980). The calculation of the distribution law of ampere-turns of the magnetic deflection systems, Proceedings of Highschools of Lith.SSR, Radioelectronics, 16, No 2, 117-121 (in Russian).

Himmelblau D.(1972). Applied nonlinear programming, McGraw-Hill Book Company.

Mockus I.B.(1980). The simple Bayesian algorithm for the multidimensional global optimization, In: Numerical Techniques for Stochastic Systems, eds. Archetti F. and Cugiani M., North Holland, 369-377.

Shagen I.P.(1980). Stochastic interpolation applied to the optimization of expensive objective functions, COMPSTAT 1980, 300-307.

Šaltenis V. and Radvilavičiute J.(1977). The separation of the main variables in optimization problems, Optimal Decision Theory, Vilnius, 3, 57-71 (in Russian).

Tiešis V.(1975). Variable metric method for local minimization of multidimensional functions with simple constrains, Computers, Kaunas, 111-114 (in Russian).

Torn A.(1978). A program for global optimization, multistart with clustering (M.S.C.), Rapporter fran datacentralen vid Abo Akademi, Turku, No 7.

Žilinskas A.(1978). On statistical models for multimodal optimization, Math.Operat.Stat., ser.Statistics, 9, 255-266.

Žilinskas A.(1980a). On the use of statistical models of multimodal functions for the construction of the optimization algorithms, Lecture Notes in Control and Information Sciences, 23, 138-147.

Žilinskas A.(1980b). The use of statistical models for construction of multimodal optimization algorithms, Third Czechoslovak-Soviet-Hungarian Seminar on Information Theory, Akademia, Prague, 219-224.

Žilinskas A.(1981). Two algorithms for one-dimensional multimodal minimization, Math.Operat.Stat., ser.Optimization, 12, 53-63.

Žilinskas A.(1982). Axiomatic approach to statistical models and their use in multimodal optimization theory, Mathematical Programming, 22, 104-116.

Address-List of Authors

Anderson, A.J.B., University of Aberdeen, Aberdeen AB9 2UB, G.B.
Baker, R.J., Rothamsted Experimental Station, Harpenden, Herts. AL5 2JQ, G.B.
Blåfield, E., University of Jyväskylä, Seminaarindatu 15, 40100 Jyväskylä 10, Finland
Bliss, J., Chelsea College, University of London, G.B.
Blumenthal, S., I.U.T. de Paris, Avenue de Versaille, 75016 Paris, France
Bremner, J.M., University of Canterbury, Kent CT2 7NF, G.B.
Caliński, T., Academy of Agriculture, ul. Wojska Polskiego 28, 60-637 Poznań, Poland
Catot, J.M., see *Marti, M.*
Celeux, G., I.N.R.I.A., Domaine de Voluceau, Rocquencourt B.P. 105 – 78153 Le Chesnay Cedex, France
Chambers, J.M., Bell Laboratories, Murray Hill, New Jersey 07974, U.S.A.
Chan, T.F., Yale University, New Haven, CT 06520, U.S.A.
Clarke, M.R.B., see *Baker, R.J.*
Clark, S., Timberlake Clark Ltd., 22 Westerdale, Chelmsford CM1 5 UN, G.B.
Collomb, G., Paul Sabatier University, 118 route de Narbonne, 31062 Toulouse Cedex, France
Dekker, A.L., Statistical Institute for Asia and the Pacific, Akasaka P.O. Box 13, Tokyo 107, Japan
Diday, E., I.N.R.I.A., Domaine de Voluceau, Rocquencourt B.P. 105, 78153 Le Chesnay Cedex, France
Digby, P.G.N., Rothamsted Experimental Station, Harpenden, Herts. AL5 2JQ, G.B.
Dixie, J.M., Office of Population Censuses and Surveys, Titchfield Fareham, Hants PO15 5RR, G.B.
Djindjian, F., Musée antiquites nationales, 32, rue Boussingault, 75013 Paris, France
Dutter, R., Technical University Graz, Hamerlinggasse 6, 8010 Graz, Austria
Eddy, W.F., Carnegie-Mellon University, Pittsburg, PA 15213, U.S.A.
Fahrmeier, L., University of Regensburg, Universitätsstr. 31, 8400 Regensburg, F.R.G.
Ford, B., see *Hague, S.J.*
Francis, I., Cornell University, U.S.A.
Friedman, J.H., Stanford University, Stanford, California 94305, U.S.A.
Futagami, K., see *Kobayashi, Y.*
Gautier, J.-M., C.O.R.E.F., 204, Rond-Point du Pont de Sèvres, 92516 Boulogne Billancourt Cedex, France
Gentle, J.E., IMSL, Inc., 7500 Bellaire, Houston, TX 77036, U.S.A.
Gerard, J., Institut National de la Statistique et des Etudes Economiques, 18, boulevard Adolphe-Pinard, 75675 Paris Cedex 14, France
Gilchrist, R., Polytechnic of North London, Holloway Road, London N7 8DB, G.B.
Golub, G.H., Stanford University, Stanford, CA 94305, U.S.A.
Gordesch, J., Free University of Berlin, Specktstraße 15, 1000 Berlin 33, F.R.G.

Green, M., see *Baker, R.J.*
Grize, F., University of Neuchâtel, 7 Pierre-A-Mazel, 2000 Neuchâtel, Switzerland
Grosbras, J.M., see *Gerard, J.*
Grossmann, W., University of Vienna, Rathausstraße 19, 1010 Vienna, Austria
Hague, S.J., NAG Central Office, Mayfield House, 256 Banbury Road, Oxford OX2 7DE, G.B.
Hájek, P., Math. Inst. ČSAV, 115 67 Prague, Czechoslovakia
Haux, R., University of Heidelberg, Im Neuenheimer Feld 325, 6900 Heidelberg 1, F.R.G.
Hext, G.R., Civil Service College, Sunningdale Park, Ascot, Berks SL5 OQE, G.B.
Ibrahim, A., see *Schektman, Y.*
Ikeda, H., see *Kobayashi, Y.*
Ivánek, J., School of economics, 130 67 Prague, Czechoslovakia
Jockin, J., see *Schektman, Y.*
Joiner, B.L., University of Wisconsin, 1210 W. Dayton St., Madison, Wisconsin 53706, U.S.A.
Kredler, Ch., Technical University of Munich, Arcisstr. 21, 8000 München 2, F.R.G.
Krzyśko, M., Adam Mickiewicz Universitiy, Matejki 48/49, 60-769 Poznań, Poland
Kobayashi, Y., Hiroshima University, 1-1-89 Higashi-senda, Naka, Hiroshima 730, Japan
Korhonen, P., Helsinki School of Economics, Runeberginkatu 14–16, 00100 Helsinki 10 Finland
Läuter, H., Akademie der Wissenschaften, Institut für Mathematik, Mohrenstr. 39, 108 Berlin, G.D.R.
Lambert, T.W., see *Hague, S.J.*
Lauro, N., University of Napoli, Via Partenope 36, 80210 Napoli, Italy
Lebart, L., C.R.E.D.O.C., 142, Rue du Chevaleret, 75634 Paris Cedex 13, France
Lechevallier, Y., see *Celeux, G.*
Leeuw, J. de, Rijksuniversiteit te Leiden, Breestraat 70, 2311 CS Leiden, The Netherlands
Leredde, H., University of Paris-Nord, Avenue Jean-Baptiste Clément, 93430 Villetaneuse, France
LeRoux, S., C.N.R.Z., 78350 Jouy en Josas, France
LeVeque, R.J., see *Golub, G.H.*
Linde, A.v.d., Free University of Berlin, Arnimallee 2–6, 1000 Berlin 33, F.R.G.
Mallet, J.L., Ecole Nationale Superieure de Geologie, BP 452, 54000 Nancy, France
Marti, M., Polytechnic University of Barcelona, Diagonal, 647 Barcelona 28, Spain
McDonald, J.A., see *Friedman, J.H.*
McNicol, J.W., Scottish Crop Research Inst., Pentlandfield, Roslin Midlothian EH25 9RF G.B.
Mélard, G., Université Libre de Bruxelles, Boulevard du Triomphe, 1050 Bruxelles, Belgium
Messean, A., see *LeRoux, S.*

Molenaar, I.W., see *Schuur, W.H.v.*
Momirović, K., University of Zagreb, Engelsova bb, 41000 Zagreb, Yugoslavia
Morineau, A., Cepremap, 142, rue du Chevaleret, 75013 Paris, France
Murphy, B.P., University of Western Australia, Nedlands, W.A. 6000, Australia
Mustonen, S., University of Helsinki, Aleksanterindatu 7, 00100 Helsinki, Finland
Ng, S.C.M., see *McNicol, J.W.*
Nin, G., C.R.I.G., 2 Boulevard Andre Aune, 13006 Marseille, France
Novák, M., Czechoslovak Academy of Sciences, P.O.B. 5, 18200 Prague 8, Czechoslovakia
Obgorn, J., see *Bliss, J.*
Paass, G., Gesellschaft für Mathematik und Datenverarbeitung, St. Augustin, F.R.G.
Pastor, J., see *Schektman, Y.*
Payne, R.W., see *Digby, P.G.N.*
Pflug, G.Ch., see *Grossmann, W.*
Pokorný, D., Czechoslovak Academy of Sciences, CS, 142 20 Prague, Vídeňská 1083, Czechoslovakia
Polasek, W., University of Vienna, Rooseveltplatz 6/2, 1090 Vienna, Austria
Prat, A., see *Marti, M.*
Robert, P., Université de Montréal, CP. 6128, Succ A., Montréal H3C 3J7, Canada
Ramsay, J.O., McGill University, Montréal, P.Q., Canada
Raphalen, M., I.R.I.S.A., Campus de Beaulien, 35042 Rennes Cedex, France
Reese, R.A., University of Hull, Hull HU6 7RX, G.B.
Rijckevorsel, J.v., Erasmus University of Rotterdam, Rotterdam, The Netherlands
Ronner, A.E., University of Groningen, P.O. Box 800, 9700 AV Groningen. The Netherlands
Ross, G.J.S., Rothamsted Experimental Station, Harpenden, Herts, AL5.2JQ, G.B.
Samorov, A., Massachusetts Institute of Technology, 1 Amherst Street, Building E 40–148, Cambridge, Mass 02139, U.S.A.
Saporta, G., Université Rene Descartes, 143, Avenue de Versailles, 75016 Paris, France
Schektman, Y., Paul Sabatier University, 118, route de Narbonne, 31062 Toulouse Cedex, France
Schuur, W.H.v., University of Groningen, Oude Boteringestraat 23, 9712 GC Groningen, The Netherlands
Slater, M., see *Baker, R.J.*
Sprenger, C.J.A., University of Amsterdam, The Netherlands
Ŝtalec, J., see *Momirović, K.*
Stitt, F.W., clinical data research services inc., 340 Lombard Street, San Francisco, CA 94133, U.S.A.
Stokman, F.N., University of Groningen, Oude Boteringestraat 23, 9712 GC Groningen, The Netherlands
Stützle, W., see *Friedman, J.H.*
Sylwestrowicz, J.D., University of London, Mile End Road, London E1 4NS, G.B.

Tabony, R.C., Meteorological Office, London Road, Bracknell, Berkshire RG 12 2SZ, G.B.

Tjoa, A.M., University of Vienna, Rathausstraße 19, 1010 Vienna, Austria

Vallée, M., Dominion Engineering, Montréal, Canada

Vielle, D., see *Schektman, Y.*

Vila, J.-P., see *LeRoux, S.*

Wagner, R.R., University of Linz, 4040 Linz, Austria

Welsch, R.E., see *Samorov, A.*

White, R.P., see *Baker, R.J.*

Wilke, H., Free University of Berlin, Babelsberger Str. 14–16, 1000 Berlin 31, F.R.G.

Wilson, S.R., Australian National University, PO. Box 4, Canberra, ACT 2600, Australia

Winsberg, S., Université de Montréal, CP 6128, Succ. A., Montréal H3C 3J7, Canada

Zakkajšek, E., University of Ljubjana, Yugoslavia

Žilinskas, A., Lithuanian SSR Academy of Sciences, 232000 Vilnius, MTP-1, Lenino pr. 3, USSR

GPSR Compliance

The European Union's (EU) General Product Safety Regulation (GPSR) is a set of rules that requires consumer products to be safe and our obligations to ensure this.

If you have any concerns about our products, you can contact us on ProductSafety@springernature.com

In case Publisher is established outside the EU, the EU authorized representative is:

Springer Nature Customer Service Center GmbH
Europaplatz 3
69115 Heidelberg, Germany

Batch number: 10584556

Printed by Printforce, the Netherlands